Springer Water

The book series Springer Water comprises a broad portfolio of multi- and interdisciplinary scientific books, aiming at researchers, students, and everyone interested in water-related science. The series includes peer-reviewed monographs, edited volumes, textbooks, and conference proceedings. Its volumes combine all kinds of water-related research areas, such as: the movement, distribution and quality of freshwater; water resources; the quality and pollution of water and its influence on health; the water industry including drinking water, wastewater, and desalination services and technologies; water history; as well as water management and the governmental, political, developmental, and ethical aspects of water.

More information about this series at http://www.springer.com/series/13419

Osman Abdalla · Anvar Kacimov
Mingjie Chen · Ali Al-Maktoumi
Talal Al-Hosni · Ian Clark
Editors

Water Resources in Arid Areas: The Way Forward

Springer

Editors
Osman Abdalla
Earth Sciences Department,
College of Science
Sultan Qaboos University
AlKhod
Oman

Anvar Kacimov
Sultan Qaboos University
AlKhod
Oman

Mingjie Chen
Water Research Center
Sultan Qaboos University
AlKhod
Oman

Ali Al-Maktoumi
Sultan Qaboos University
AlKhod
Oman

Talal Al-Hosni
Sultan Quaboos University
AlKhod
Oman

Ian Clark
Department of Earth and Environmental
Sciences
University of Ottawa
Ottawa, ON
Canada

ISSN 2364-6934 ISSN 2364-8198 (electronic)
Springer Water
ISBN 978-3-319-51855-8 ISBN 978-3-319-51856-5 (eBook)
DOI 10.1007/978-3-319-51856-5

Library of Congress Control Number: 2017936464

Printed on acid-free paper

This Springer imprint is published by Springer Nature
The registered company is Springer International Publishing AG
The registered company address is: Gewerbestrasse 11, 6330 Cham, Switzerland

To communities in arid regions striving for sustainable and healthy water supply.

Preface

The exponentially growing demand for water due to population growth, water-intensive diets, and rising of living standards has considerably stressed water resources worldwide. The challenge is greater in arid areas where evaporation significantly exceeds precipitation, and thus natural water resources are depleting. The water budget deficit in arid areas, the high cost of water supply, and the essential need for food and associated energy value among other challenges all need to be scientifically addressed to propose solutions to world current and future water problems. Multidisciplinary and interdisciplinary fundamental and applied scientific research is essential to contribute to solve water problems. Thus, this book "Water Resources in Arid Areas: The Way Forward" addresses diverse water issues in arid regions through gathering selected outstanding contributions presented at the International Water Conference "Water Resources in Arid Areas (IWC 2016)," which was held in Muscat, Oman, in March 2016. This book presents to the reader different examples of applied and fundamental evolving water science that will hopefully enlighten decision-makers, planners, and communities in making sound judgments for better management of water resources.

The book contains 6 main parts with a total of 28 chapters representing the contributions to different water issues in arid regions. The following themes represent the 6 parts and appear in the book in the following sequential order:

Climate and Water Resources
Groundwater Resources
Water Resources Management
Salinity and Desalination
Wastewater Treatment and Reuse
Agriculture and Irrigation Management

A few decades ago, the arid regions (mostly located in developing areas) were entirely relying on external expertise to assess/understand local and regional water resources challenges and to suggest and implement solutions. The region has now developed fair scientific and technological independence evident from the contributions published in this book. The various applications of isotopes, satellites, exploration techniques, advancements in water treatment and desalination, hydrological modeling among others presented in this book demonstrate this development. The arid regions are playing an important part in the equation of global climatic changes, and the effects are also seen in the region similar to polar areas; therefore, arid regions can be an important player in the international global climatic changes policy and mitigation developments. It is our hope that through presenting these examples and case studies that reader will gain a broader understanding of the current research and developments in water resources of arid areas.

Thanks

Osman Abdalla
Anvar Kacimov
Mingjie Chen
Ali Al-Maktoumi
Talal Al-Hosni
Ian Clark

Acknowledgements

This book assembles selected papers presented at the conference "Water Resources in Arid Areas: The Way Forward" organized by the Water Research Center (WRC) at the Sultan Qaboos University (SQU), Oman. The highly appreciated generous support provided by SQU for the conference has paved the ground for this book. The review of papers conducted by a pool of experts in various fields of water science (names of reviewers are listed at page XIX) has significantly improved the quality and provided constructive criticism important for quality assurance. We would like to also express our deepest thanks and gratitude to the authors of the selected papers for their valuable contribution. We are also indebted to the office work and support of Ms. Maria Diana Austria, the WRC office coordinator.

Contents

About the Editors

Dr. Osman Abdalla holds B.Sc. (1988) and M.Sc. (1993) in geology from University of Khartoum and Ph.D. (2000) in hydrogeology from University of Technology, Berlin. He is the director of Water Research Center at Sultan Qaboos University since 2012 and associate professor of hydrogeology at the Department of Earth Sciences, College of Science. His research bridges several areas of physical and chemical hydrogeology in arid areas and intends to develop a solid platform for hydrogeological and environmental research and training of an international standard with emphasis on groundwater recharge and discharge. Dr. Abdalla has established many international and regional collaborations, been awarded over 10 major grants including His Majesty Trust Fund, and published several articles in reputable international journals.

Dr. Anvar Kacimov He completed his B.Sc., M.Sc., Ph.D. (fluid mechanics) from Kazan University, USSR, 1982, 1987. He worked at Kazan University, Departments of Seepage and Mathematical Analysis for the period 1982–1998. Since 1998, he worked at SQU, Oman, as assistant–associate full professor, Department of Soils, Water and Agricultural Eng; administratively, HoD (2007–2012); director of Water Research Centre (2011–2012); and dean (2012–2015). His areas of interest include fluid, heat and mass transfer through porous media (soils, aquifers, oil formations) and of applications in hydroecology, soil physics, irrigation and drainage, hydraulic and hydrologic engineering, groundwater hydrology, fluid mechanics, and reservoir engineering. He has published 128 papers in refereed journals (in English) and has co-authored two books (in Russian). He has been awarded Omani Green Research Award with a Special Commendation of Mitsubishi Corporation (2010) and Best Reviewer of Vadose Zone Journal (2005) and J. of Irrigation and Drainage Eng., ASCE (2013).

Dr. Mingjie Chen holds a bachelor degree in environmental engineering from Tsinghua University (1997) and a master degree in environmental sciences from Peking University in China (2000). Dr. Chen got his Ph.D. degree in Environmental Sciences/Hydrogeology at the University of California, Santa Barbara in USA in 2005. During his postdoc research from 2005 to 2008 at Los Alamos National Laboratory in USA, Dr. Chen developed an innovative intrusive method to quantify uncertainty of multiphase flow and reactive transport in heterogeneous subsurface area. In late 2008, Dr. Chen accepted a research assistant professor position at Tufts University in USA, leading a numerical modeling team in collaboration with a laboratory experimental team to study bio-enhanced PCE-DNAPL dissolution by anaerobic mixed cultures. At the end of 2010, Dr. Chen joined Lawrence Livermore National Laboratory (LLNL) in USA as an earth scientist. He is a PI or co-investigator of numerous projects funded by Department of Energy (DOE of USA) on underground fossil energy and radioactive waste. Owing to his excellent performance, Dr. Chen was rewarded Directorate Award by Physical and Life Science Directorate of LLNL in 2013. Since 2014, Dr. Chen has been a senior hydrogeologist (associate professor rank) in Water Research Center at Sultan Qaboos University, Oman, while he still remained an associate role of hydrogeology in LLNL of USA. Dr. Chen is mainly responsible for developing research programs in groundwater resources and geothermal energy using high-performance computing. Dr. Chen commits duties of organizing committees of water-related international workshops and conferences, proposal-review panel, and international journal editors.

Dr. Ali Al-Maktoumi holds a B.Sc. in soils and water (1998) and an M.Sc. in soils and water Management from Sultan Qaboos University in Oman (2001). He received a scholarship from Oman government in 2003 to continue with his Ph.D. studies in environmental engineering (Water Resources) at the University of Queensland in Australia. After his Ph.D. in 2007, Al-Maktoumi joined the Department of Soils, Water and Agricultural Engineering at Sultan Qaboos University as an assistant professor in the area of hydrology. He contributed to teaching a wide range of courses in the area of arid zone hydrology and water resources management at both B.Sc. and M.Sc. levels. Al-Maktoumi worked as a consultant in a project "Groundwater contamination in a golden mine site in North Queensland, Australia for the period 2007–2008". Through a number of awarded grants, he established scientific collaboration with Utrecht University, Delft University of Technology, UNESCO-IHE, California Institute of Technology, Jet Prolusion Laboratory-NASA, University of Nebraska-Lincoln, University of Putra Malaysia, and University of Jordan. Al-Maktoumi organized a number of training courses in the area of numerical modeling and co-organized a number of international conferences. In research, Al-Maktoumi focuses in feasibility of managed aquifer recharge using treated wastewater in MENA region along with enhancement of recharge dams efficiency. Al-Maktoumi's developed experience in those fields is reflected in his publications.

Dr. Talal Al-Hosni holds a B.Sc. in earth sciences (1999) from Sultan Qaboos University (Oman) and an M.Sc. in hydrogeology (2001) from Birmingham University (UK). He received a scholarship from Oman government in 2003 to continue with his Ph.D. study in chemical Hydrogeology at Melbourne University (Australia). After his Ph.D. in 2007, Al-Hosni joined the Department of Earth Sciences at Sultan Qaboos University as an assistant professor in the area of hydrogeology and environmental Geology. He worked as a theme supervisor (Omani Land) in the Omani Encyclopedia for the period 2008–2009. Since 2012, he is a member of the scientific committee for the Oman Mountains Atlas Project. His areas of interest include mainly hot springs, groundwater recharge, intra-aquifers mixing, and usage of bottled water and its impact. Al-Hosni contributed to teaching a number of courses including environmental geology and hydrogeology and developed a number of training courses in the area of EIA of mining and groundwater management.

Dr. Ian Clark is a professor in the Department of Earth and Environmental Sciences at the University of Ottawa. Professor Clark completed a bachelor's of science degree in earth sciences and a master's of science degree in hydrogeology at the University of Waterloo, followed by his doctoral degree at the Université de Paris-Sud (Orsay) in isotope hydrogeology and paleoclimatology. Since his earliest work on geothermal systems in western Canada, Ian's research has focused on the integration of geochemistry and isotopes to address the questions on the origin, age, paleoclimatic context, and geochemical history of groundwater and solutes in natural and contaminated settings. He continues to work with his graduate students in diverse hydrogeological environments, ranging from groundwater dynamics in permafrost in the Arctic or beneath the deserts of Oman, to contamination of water resources, dispersion of radionuclides in the environment, and the burial of nuclear waste. Dr. Clark and his colleagues established, just this year, new facilities in the Advanced Research Complex (ARC) at the University of Ottawa for training and analysis in the geosciences. The ARC includes laboratories for geochemistry stable isotopes, tritium, noble gases, and accelerator mass spectrometry (AMS) for radiocarbon and other radioisotopes. Professor Clark teaches geochemistry and environmental isotopes in hydrology and recently published a new undergraduate textbook, Groundwater Geochemistry and Isotopes, which complements his graduate-level textbook (co-authored with Professor Peter Fritz): "Environmental Isotopes in Hydrogeology."

Reviewers List

Abdelaziz Hirich, Irrigation and Greenhouse Production Specialist, International Center for Biosaline Agriculture, U.A.E.
Abdolmajid Lababpour, Assistant Professor, Department of Industrial and Environmental Biotechnology, National Institute of Genetic Engineering and Biotechnology (NIGEB), Iran
Abdulrahim Al-Ismaili, Assistant Professor, Department of Soils, Water and Agricultural Engineering, College of Agricultural and Marine Sciences, SQU, Oman
Ahmed Al-Busaidi, Researcher, Department of Soils, Water and Agricultural Engineering, College of Agricultural and Marine Sciences, SQU, Oman
Ahmed Douaik, Principal Chief Engineer, (Geo)Statistics, Geomatics, and Agroclimatology, Research Unit on Environment and Conservation of Natural Resources, Regional Center of Rabat, National Institute of Agricultural Research (INRA), Morocco
Ahmed Ibrahim Ramzi Ibrahim, Researcher, Head, Department of Digital Mapping, National Authority for Remote Sensing and Space Sciences (NARSS), Egypt
Ala'a H. Al-Muhtaseb, Associate Professor, Petroleum and Chemical Engineering Department, College of Engineering, SQU, Oman
Alaa El-Sadek, Associate Professor, Arabian Gulf University, Kingdom of Bahrain, Bahrain
Arash Malekian, Associate Professor, University of Tehran, Iran
Ashok K. Alva, Supervisory Research Soil Scientist, USDA ARS, USA
Azizzllah A. Izady, Research Scientist, Water Research Center, Sultan Qaboos University, Oman
Balai Chandra Das, Assistant Professor in Geography, Krishnagar Government College, India
Brahim Askri, Assistant Professor, Civil Engineering, Department of Built and Natural Environment, Caledonian College of Engineering, Oman
Daniel Moraitis, Assistant Professor, Department of Earth Science, College of Science, SQU, Oman

Deia Eldin Osman Hagahmed Mohamed, Environmental Consultant/Environmental Management Section Head, Petrodar Operating Company, Sudan
Elsayed Abu El Ella, Professor, Geology Department, Faculty of Science, Assiut University, Egypt
Eman Hasan, Associate Professor, National Water Research Center, Egypt
Furat Ahmed Mahmood Al-Faraj, Research Fellow, The University of Salford, Greater Manchester, School of Computing, Science and Engineering, Civil Engineering Research Group, UK
Galip Yuce, Professor, Hacettepe University, Faculty of Engineering, Department of Geological Engineering, Hydrogeology Division, Turkey
George N. Zaimes, Lecturer, Department of Forestry and Natural Environment Management, Greece
Hemesiri Kotagama, Assistant Professor, Natural Resource Economics, College of Agricultural and Marine Sciences, SQU, Oman
Himan Shahabi, Assistant Professor in Remote Sensing, Department of Geomorphology, Faculty of Natural Resources, University of Kurdistan, Iran
Insaf Mekki, Assistant Professor, Institut National de Recherche en Génie Rural, Eaux et Forêts (INRGREF), Tunisia
Iswar Chandra Das, Earth Sciences and Geoinformation Scientist, National Remote Sensing Centre (NRSC), Indian Space Research Organization (ISRO), India
K.S. Kasiviswanathan, Postdoctoral Research Scholar, Department of Civil Engineering, University of Calgary, Canada
Khadijah Semhi, Consultant, Department of Earth Sciences, Sultan Qaboos University, Oman
Khalid Salim Al-Mashikh, Director, Ministry of Regional Municipalities and Water Resources, Oman
M.E. Banihabib, Associate Professor and Head of Department, University of Tehran, Iran
M. Farhad Howladar, Associate Professor, Department of Petroleum and Mining Engineering, Faculty of Applied Science and Engineering, Bangladesh
Mahmoud Abd El-Gelil, Assistant Professor, Department of Civil Engineering and Architecture, College of Engineering, SQU, Oman
Mahmoud Hozien, Professor, National Research Centre, Egypt
Mansour Al-Haddabi, Associate Researcher, Department of Soils, Water and Agricultural Engineering, College of Agricultural and Marine Sciences, SQU, Oman
Matthew Herod, Environmental Risk Assessment Officer, Canadian Nuclear Safety Commission, Canada
Micòl Mastrocicco, Contract Researcher, Physics and Earth Sciences Department, University of Ferrara, Italy
Mohamed Chiban, Assistant Professor, Ibn Zohr University, Morocco
Mohamed Meddi, Professor, Ministry of Higher Education and Scientific Research —Higher National School of Hydraulics, Algeria
Mostafa Ghasemi, Senior Lecturer/Research Fellow, Fuel Cell Institute of Universiti, Malaysia

Mushtaque Ahmed, Associate Professor, Department of Soils, Water and Agricultural Engineering, College of Agricultural and Marine Sciences, SQU, Oman
Nadhir Al-Ansari, Professor, Department of Civil, Environmental and Natural Resources Engineering, Lulea University of Technology, Sweden
Norhan Abd. Rahman, Associate Professor, Taibah University, Saudi Arabia; Universiti Teknologi Malaysia, Malaysia
Pradeep Naik, Superintending Hydrogeologist, Ministry of Water Resources, India, India
R. Arthur James, Assistant Professor and Head, Department of Marine Science, Bharathidasan University, India
Raafat Abd El-Ghany Abo-Mandour Abd El-Daiem, Toxicology Fellow, Emergency Hospital—Mansoura University, Egypt
Ramadan Abdelaziz, Researcher and Lecturer, TU Bergakademie Freiberg, Germany
Rashid Umar, Associate Professor, Aligarh Muslim University, India
Said Al-Ismaily, Assistant Professor, Department of Soils, Water and Agricultural Engineering, College of Agricultural and Marine Sciences, SQU, Oman
Sid Ali Ouadfeul, Associate Professor, Department of Geophysics, Geology and Reservoir Engineering, Algerian Petroleum Institute-IAP Corporate University, Algeria
Slim Zekri, Associate Professor, Natural Resource Economics, College of Agricultural and Marine Sciences, SQU, Oman
Sundarajan Narasimman, Associate Professor, Department of Earth Science, College of Science, SQU, Oman
Tariq Gibreel, Assistant Professor, Natural Resource Economics, College of Agricultural and Marine Sciences, SQU, Oman
Yaseen Al-Mulla, Associate Professor, Department of Soils, Water and Agricultural Engineering, College of Agricultural and Marine Sciences, SQU, Oman
Yassine Abdel Rahman Charabi, Associate Professor, Department of Geography, College of Arts and Social Sciences, Oman
Zakaria Atia Mohamed, Professor, Botany Department, Faculty of Science, Sohag University, Egypt

Part I
Climate and Water Resources

Paleoclimatic Registers from Semi-arid Costal Sediments of Southeastern India: A Multi Proxy Approach

Anburaj Vidyasakar, Helena Sant'Ovaia, Linto Alappat, P. Morthekai, Seshachalam Srinivasalu, A.K. Singhvi, Ferreira Jorge and Celeste Gomes

Abstract The red sand dunes appear along the south east, -west coast of Tamil Nadu, India between the latitudes and longitudes of 8°07′56″N to 8°22′11″N; 77° 19′84″E to 77°53′40″E. The dune sands from this region were studied through magnetic methods such as magnetic susceptibility measurements and acquisition of isothermal remanent magnetization, geochemistry and X-ray diffraction methods. Optically stimulated luminescence (OSL) dating method was used to constrain the chronology of deposits. Three sections were excavated up to 5–9.5 m with one inland deposit (TPV) and two near coastal sections (THOP and MUT).

Celeste Gomes—deceased

A. Vidyasakar (✉) · S. Srinivasalu
Department of Geology, Faculty of Science and Humanities,
Anna University, Chennai, India
e-mail: a.vidyasakar@gmail.com

A. Vidyasakar · H. Sant'Ovaia
Pole of the Faculty of Sciences, Earth Sciences Institute,
Rua do Campo Alegre, Porto 4169-007, Portugal

L. Alappat
Department of Geology and Environmental Science,
Christ College, Irinjalakuda 680125, Kerala, India

P. Morthekai
Luminescence Dating Laboratory, Birbal Sahni Institute of Palaeobotany,
53 University Road, Lucknow 226007, India

A.K. Singhvi
Planetary and Geoscience Division, Physical Research Laboratory,
Ahmedabad 380009, India

F. Jorge
LaboratórioNacional de Energia e Geologia, I.P/Rua da Amieira, Apartado 1089,
S. Mamede de Infesta, Porto 4466-901, Portugal

C. Gomes
CGUC, Department of Earth Sciences, Faculty of Sciences and Technology,
University of Coimbra, Largo Marquês de Pombal, Coimbra 3000-272, Portugal

O. Abdalla et al. (eds.), *Water Resources in Arid Areas: The Way Forward*,
Springer Water, DOI 10.1007/978-3-319-51856-5_1

The magnetic parameters show both significant contribution of hematite structures and indicate the presence of multi-domain magnetite or mixed mineral contents of magnetite and anti-ferromagnetic minerals in the sample. The occurrence of magnetite in THOP and TPV sections is possibly due to the marine sediments transported by sturdy onshore winds. In XRD data, correlation analysis indicated TPV and MUT sections have a similar type of deposition and THOP did not show any positive correlation with TPV and MUT and even with its own deposition. In comparison with geochemistry data, χ variation and OSL dates, it was shown that the sample MUT21 (200 cm) with an OSL age of 14 ± 2 ka indicated deposition during the humid interval and at $\sim 17 \pm 2$ to $\sim 19 \pm 2$ ka MUT61 (600 cm) depicts the dry period of deposition.

Keywords Magnetic parameters · X-ray diffraction · Optically stimulated luminescence · Geochemistry · Teri sands · Hematite · Magnetite

1 Introduction

The environmental changes in a region may happen due to climatic variability or as a result of human intervention, which usually lasts for an extended period. Moreover, research on various aspects preceding the climate change suggested that climatic variability has occurred sporadically in the past, notably through a quaternary period in response to different forcing's, e.g., orbital variations and solar activity changes (insolation) (Wolff et al. 2010; Sanchez Goñi and Harrison 2010). It is to be noted that a change in the climate disturbs the coastal areas around the globe causing changes in sea level. Therefore, in the context of Quaternary climate change, it is ideal to study sediment archives along the coast. The study may hold substantial evidence of coastline advance and retreat in adherence to past variations regarding sea level and its landscape response (e.g., Islam and Tooley 1999; Alappat et al. 2015; Brückner (1988, 1989)). Also, reconstruction of the paleo-sea level is a long and intricate process. Hence, it involves reminiscing the fragments of evidence left in nature, recording of those fragments through dating of the event in question and final interpretation of available data (Morner 2010). The red dunes on the southeast coast of India is a potential terrestrial archive of climate change and provide vital information about the landscape response to such variations. Red dunes are referred to as 'teri sands or teris' in the literature and are categorized as inland, coastal and teri (near-coastal) dunes by earlier researchers (Joseph et al. 1997, 1999). According to the standard Tamil dictionary, teri means 'a heap of sand.' Some researchers (Jayangondaperumal et al. 2012) used the term 'wasteland' because the local people deemed it as useless for agriculture.

The foremost intent of the present study is to draw a distinct line between previous dry and wet climatic intervals using various proxies like magnetic studies, Optically Stimulated Luminescence (OSL) dating, X-ray Diffraction (XRD) and geochemical analysis. The soil magnetic parameters are relatively quick and easy to obtain information compared to geochemical, sedimentological and paleontological data. It is helpful in obtaining information about a broad range of climatic and environmental changes (Maher and Thompson 1999). OSL dating has the number of advantages as a chronological tool in coastal environments over other methods, considering its wide dating range, direct dating of events and provides ages in calendar years (Jacobs et al. 2008). It also includes the use of most abundant minerals like quartz and feldspar as dosimeters for age determination, which makes its wide application in the majority of terrestrial sedimentary archives. The qualitative analysis of sediments has been carried out to define major and minor constituent minerals present in the samples collected at different depths in the coastal area by XRD technique. The geochemical analysis of samples can be of greater help in understanding the sediment dynamics underlying behind a particular geological system (Albarède 2003).

2 Study Area

The study area is located in the southwest and southeast coast of Tamil Nadu, India between the latitudes and longitudes of 8°07′56″N to 8°22′11″N; 77°19′84″E to 77° 53′40″E. Three representative sections were excavated in the area as indicated in Fig. 1, viz-a-viz; Muttom (MUT; 8°07′56″N, 77°19′84″E), Thopuvilla (THOP; 8° 19′57″N, 77°57′27″E) and Edayanvilai (TPV; 8°22′11″N, 77°53′40″E). The various morphological settings from east and west coast will be further discussed in the following section. At Muttom, Thopuvilla and Edayanvilai, a 9.5, 7.4 and 4 m sections respectively were excavated to reach up to the palaeo-surface.

2.1 Regional Climate

The study area falls under the semi- arid and subtropical climatic conditions. Muttom, Thoppuvillai and Edayanvillai receive most of its rainfall during the summer (during SW monsoon). The climate has been classified according to Koppen and Geiger climate classification as Aw (Peetal et al. 2007). In Muttom, the average annual precipitation is 1093 mm and the variation in precipitation between the wettest and driest month is 178 mm. The average temperature in Muttom is recorded as 29 °C. The lowest average annual temperature in the year occurs in December when it is around 26 °C. In Thopuvillai and Edayanvillai average annual rainfall is 729 mm, and among the wettest and driest months, the change in precipitation is 199 mm. With an average temperature of ~31 °C, May is the hottest month of the year with a mean maximum of 38 °C. The coldest month is the January, with temperatures averaging 26.2 °C (www.climate-data.org).

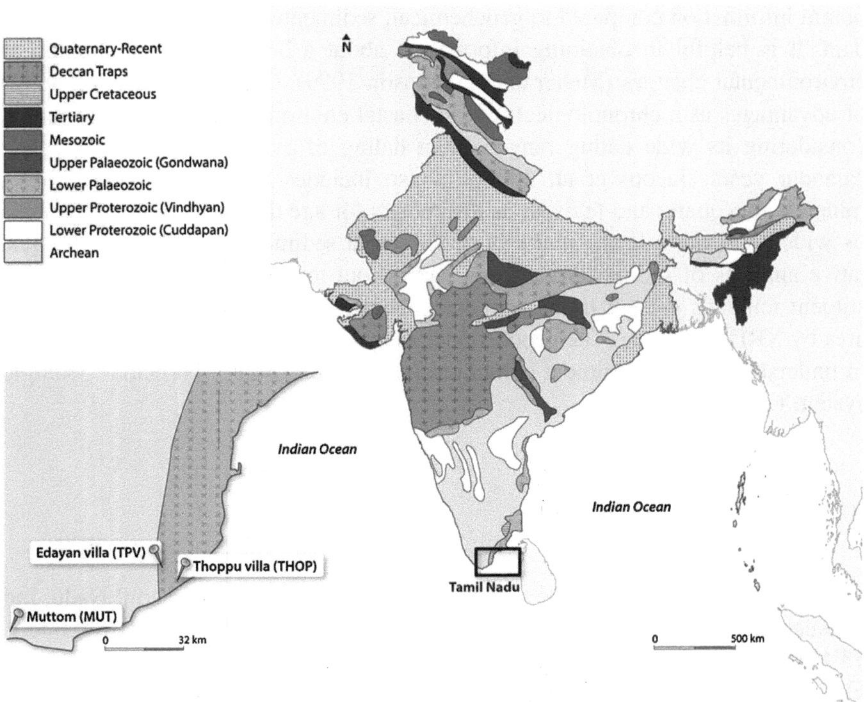

Fig. 1 Map of the study area with sample location in southeast-southwest coast of India

2.2 Regional Stratigraphy and Geomorphology

The teri sands in the area cover an area of up to ~500 km^2 and thickness at places reach up to ~12 m (Gardner 1986). The igneous and metamorphic rocks that found below ovari series are from Archean period. The Pleistocene coastal sediments (Ovari Series) sits on Archaean granulitic rocks, most of which are peninsular gneiss, garnet-sillimanite-graphite gneiss and charnockite gneiss (Jayangondaperumal et al. 2012). Idindakarai Series (Fig. 2) overlies the Ovari Series and comprises terrigenous grains of gravel and sand size and shell fragments of shallow marine origin (Jayangondaperumal et al. 2012). The Kanyakumari Series that superimposes on Idindakarai Series consist of fossil coastal dunes concreted with calcium carbonate sediments and also with marine sediments. (Jayangondaperumal et al. 2012). In addition to that, Poochikkadu Series is covered with calcrete and some marine sediments (Jayangondaperumal et al. 2012). The Inland teri sands found in big patches are dark in colour and discontinuous, while coastal teri sands are continuous and light coloured. Marine deposits of Mandapam Series overlie the teri deposits (Jayangondaperumal et al. 2012). Along the shoreward direction, teri sands rest on either of the crystalline basement, the Ovari marine sandstone, aeolianite of

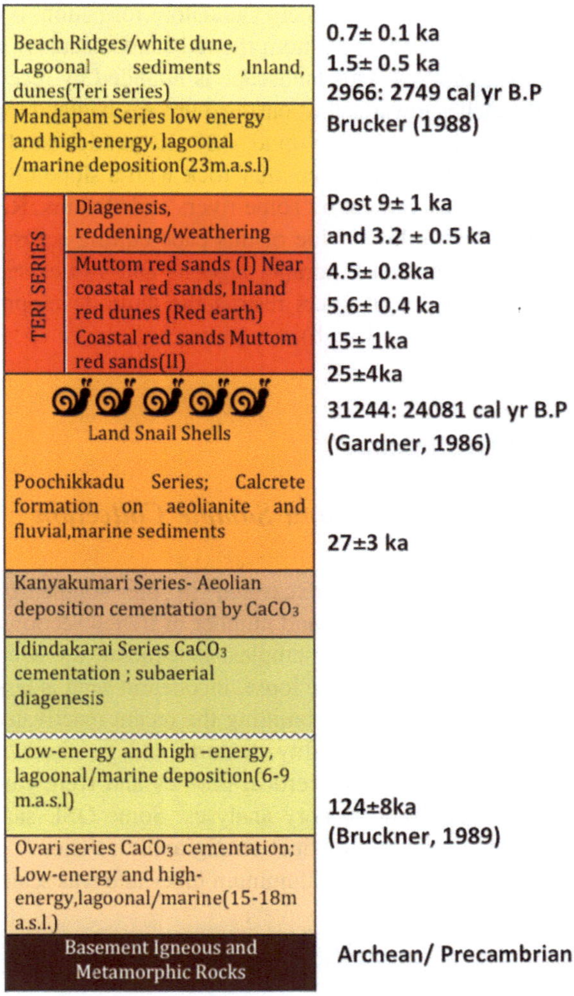

Fig. 2 Regional stratigraphy from Jayangondaperumal et al. (2012) and some modified ages from Allapat et al. (2013a, b)

Kanyakumari or Poochikkadu Series (Gardner 1986; Joseph et al. 1997). Perceiving sediment texture Joseph et al. (1997), inferred that teri sediments were originated from the wide-open continental shelf during a phase of low relative sea level, and these deposits were transported by high landward winds. According to chandrasekharan and Murugan (2001) sands are rich in heavy minerals like ilmenite, rutile, zircon, garnet, monazite and sillimanite, suggest their source from Precambrian Khondalite, charnockites and granite gneisses rocks.

The geology of the area comprises Archaean granulitic rocks, predominantly peninsular gneiss (garnet-sillimanite-graphite gneiss) and charnockite gneiss (Alappatet al. 2013a, b). The littoral area has Pleistocene and recent fluvial, fluviomarine and marine deposits (Alappat et al. 2013a, b). The rocks of Mio-Pliocene age (early Neogene) namely Cuddalore formation, is exposed towards the northeast at Muttom. Also, at Muttom the red sands occur as a dune against a headland into the Arabian Sea. This landscape is bounded by recent dunes, beach ridges, swales and beaches towards the coast and unveils badland topography with weathering and gully erosion subjecting up to ~10–12 m thick sand unit at its central part. The red sands overlie laterized country rock with a slope towards SE. Moreover, dunes on the east coast are cut by some rivers and streams. River Valliyar joins the Arabian Sea at the northwest side of this cape cutting across the dunes at the coast. Some discontinuous coast, parallel lagoons and swales separate the near coastal dunes. The floodplains of rivers restrict the dune development as discontinuous patches (Alappat et al. 2013a, b).

3 Methodology

3.1 Field Work and Sample Collection

Three profiles were excavated in the southeast and southwest coast using mechanical excavators and number of trench sections were made at different levels to collect the required samples. In general, the sections were excavated in areas were the thickness of the loose, incoherent sand is less than 1 m thick to reach up to the older sand units by omitting the recent reactivated sand at places. The samples for magnetic susceptibility, textural and geochemical analyses were collected at 10 cm intervals along vertical profiles and were carefully packed in sample bags and labeled for laboratory analyses. Some OSL samples was collected from different units having distinct stratigraphic relations. Samples for OSL dating was collected in 10 × 2 in. aluminum tubes and was sealed light tight immediately after sample collection.

3.2 Laboratory Analyses

3.2.1 Geochemical Analysis

According to Saravanan (2012, p. 43) the samples were dried in a hot air furnace at 60 °C to eliminate the moisture and 100 g of samples were taken by repeated coning and quartering to obtain a homogeneous representative sample. The following procedures were performed:

(i) *Clay fraction removal*: The samples were soaked in water with 2 g of sodium hexametaphosphate ($(NaPO_3)_6$) overnight and then washed in water (325 sieve mesh) to eliminate the clay portions. The samples were weighed and weight loss was taken as the weight of clay.

(ii) *Organic matter removal*: The samples were treated with hydrogen peroxide (H_2O_2) overnight, and rinsed with water (325 sieve mesh) until a clear column of water without any turbidity was achieved. The samples were then dried and weighed, and weight loss was taken as the weight of the organic material.

(iii) *Carbonate removal*: The samples were treated with 10% hydrochloric acid (HCl) to the carbonates in the sediments. After washing and drying, the samples were weighed, and the weight loss was taken as the weight of carbonates.

(iv) *Iron removal*: The samples were treated with 10% nitric acid (HNO_3) over night and were then washed (325 sieve mesh) with water until a clear column of water was achieved. The samples were then dried and weighed and the weight loss was taken as the weight of Fe.

(v) *Grain size analyses*: The grain size may have an impact in accordance with heavy mineral composition, usually fine-to-medium grained sand yield the optimum heavy mineral assemblages. Sieving was carried out in ASTM at the 1/2Φ interval. The sieve sets are stacked in descending order with respect to their sizes and were shaken using a mechanical sieve shaker continuously for about 20 min. During sieving, proper care was taken to minimize the sand loss from sieve sets. The samples collected on each sieve were taken separately and weights of individual fractions were tabulated. The sieved sands of 80, 100 and 120 meshes were kept separately for heavy mineral studies. It is a basic premise of sedimentology that every sedimentary unit is formed as a result of its response to a certain set of environmental conditions (Blatt et al. 1980). The geochemical analysis has been obtained from Department of Geology, Anna University, Chennai, Tamil Nadu, India.

3.2.2 Magnetic Measurement

To ascertain the strength and type of environmental magnetic carriers in samples, magnetic parameters were measured and scrutinized. Before measuring the low-field magnetic susceptibility, samples were dried at 40 °C and packed in the Department of Geology, Anna University, Chennai, Tamil Nadu, India. Maher (1986) cited the importance of drying samples at low temperature before investigation to evade any possible changes in mineralogy. The specific or mass susceptibility χ (measured in m^3/kg units) is expressed as the ratio of material magnetization and 'J' per mass unit to the weak external magnetic field, H: $J = \chi H$. Magnetic susceptibility has been measured using KLY-4S Kappabridge equipment in Earth Sciences Institute at the University of Porto, Portugal. In order to know whether the instrument is working correctly, a minimum of three susceptibility measurements of each sample was taken, and the average value was used for the study.

The Remanence assimilated by a sample exposed to a direct magnetic field at ambient temperature is termed as Isothermal Remanent Magnetization (IRM). Moreover, IRM has been measured in a minispin fluxgate magnetometer (Molspin Ltd.) after magnetization in a pulse magnetizer (Molspin Ltd.) in the Department of Earth Sciences, University of Coimbra, Portugal. The IRM acquired in the magnetic field of one tesla (T) was defined as Saturation Isothermal Remanent Magnetization (SIRM). Thus, IRM was imparted at fields of 0/12.5/25/50/75/100/150/200/250/300/500/700/900 mT and up to 1000 mT and backfield was imparted up to 1000 mT. To find the relative influence of antiferromagnetic minerals, Hard IRM using the formula HIRM = 0.5 × (SIRM + $IRM_{-300\ mT}$) (Alargarsamy 2009), wherein SIRM = IRM_{1T} (Lourenco et al. 2012) was calculated and Hard% (= HIRM/SIRM × 100) has been determined (Lourenco et al. 2014). The S-ratio parameters were used to determine the relative contribution of hematite versus magnetite by dividing each IRM value with corresponding value from SIRM ($S_{-100} = |IRM_{-100\ mT}/SIRM|$ and $S_{-300} = |IRM_{-300\ mT}/SIRM|$) (Thompson and Oldfield 1986; Maher and Thompson 1999; Evans and Heller 2003). The SIRM/χ is the ratio between Saturation Isothermal Remanent Magnetization and magnetic susceptibility.

3.2.3 X-ray Diffraction Analysis

The mineral composition was determined in non-oriented powder mounts for bulk sample analysis. In addition to that, XRD measurements with Pro Alfa 1 equipment were made at Laboratório Nacional de Energia e Geologia, I.P. The samples used for magnetic measurements were used for the XRD measurements. The selected samples have undergone coning and quartering to ensure they are homogeneous to evade errors during XRD analysis. The estimates of mineral abundances were based on subsequent peak intensities. Hence, for semi-quantification of identified principal clay minerals, peak areas of specific reflections were calculated and weighed (Lapa and Reis 1977). Finally, High Score Plus software was used to correct the intensities and to correlate between each sample in the present study through adopting Pan Analytical X'Pert Pro Alfa1 model, which is 200 times precision-centric and faster than old models. This X-ray Diffraction model requires minimum 4 grams sample with the grain size of 400–450 mesh is kept as the ideal of a sediment sample. Samples are nicely grained and powder mounts are prepared with back loading firm pressure. In this X-ray Diffraction model, Copper tube is used with the wavelength of 1.541874 Å and goniometer radius of 240 mm. The samples analysis time take is 15 s per step.

3.2.4 Optically Stimulated Luminescence (OSL)

The OSL dating was used to reveal the characteristics of coastal deposits from various geographical and geological settings and to study the relationship between periods of sediment accretion and environmental change. During the collection of OSL dating samples, the samples were protected from the sunlight exposure using thick black blankets (Fig. 3). OSL dating was carried out at Physical Research Laboratory, Ahmedabad, Gujarat, India. The selected cores were sampled and processed under controlled red light laboratory conditions for luminescence studies. The superficial 1.5 cm were not sampled to prevent possible contamination of bleached and/or disturbed outer parts while retrieving the core. The quartz grain fractions of 100–150 μm or 150–200 μm used from every sample were separated using the standard laboratory preparation procedure and dry sieving (Aitken 1998). Thus, after extracting the grain size of 100–200 μm, samples were treated with 0.1 N hydrochloric acid (HCl) to remove the presence of carbonates, 0.01 N sodium oxalate ($C_2Na_2O_4$) and 30% hydrogen peroxide (H_2O_2) were used to remove clay coatings as well as to segregate the grains and to remove the organic matter from samples. The quartz-rich fraction (<2.70–2.62 g cm^{-3}) was separated by density separation using an aqueous solution of sodium polytungstate ($3Na_2WO_4$ $9WO_3$ H_2O). Also, the quartz fraction was etched using 40% hydrofluoric acid (HF) for 60 min to remove feldspar contamination. The HF etching also removed the outermost layer of quartz grains to evade the impact of alpha particles in coarse grain sand. The samples were then washed and neutralized with distilled water and sieved again with respective mesh to get rid off grains that had become smaller. The sieved grains were mounted on stainless steel aliquots in a uniform thin layer and fixed with silicon spray shortly before measurement. The methodological details and SAR protocol optimization of red dune sand are detailed in Alappat et al. (2013a, b).

The single aliquot regenerative dose (SAR) protocol for quartz (Murray and Wintle 2000; Wintle and Murray 2006) was used to determine the equivalent dose of the samples. An automated Riso TL/OSL DA-20 reader attached to a 90Sr/90Y beta source was used for the OSL measurements. The quartz grains were stimulated by a blue light emitting diodes (LED) (470 ± 30 nm). A Hoya U-340 (7.5 mm) filter was placed between the photomultiplier and sample to detect luminescence signals from quartz grains. The dose recovery test at different preheat temperature shows that the given dose was recovered effectively at a preheat temperature of 200 °C with a cut heat of 160 °C. The De values were calculated using signals integrated using the first 0.64 s and an early background was subtracted corresponding to 1.12–2.4 s.

The dose rates were calculated from the activity concentrations of decay chains 238U, 232Th and 40K measured using high-resolution gamma spectrometry (High Purity Germanium (HPGe)) at Physical Research Laboratory, Ahmedabad. The dose-rate conversion factors of Guérin et al. (2011) and beta attenuation factors of Mejdahl (1979) were applied for calculation. The cosmic dose rate was calculated based on the method proposed by Prescott and Hutton (1994).

Fig. 3 General view of Teri sands, trench sampling and description how OSL samples are were collected

4 Results

4.1 Magnetic Parameters

The magnetic susceptibility, χ, reflects total mineralogical composition of sediments with an existing contribution from ferromagnetic minerals, which have much higher susceptibility values than paramagnetic and diamagnetic minerals such as clay or quartz (e.g., Verosub and Roberts 1995; Dekkers 1997; Maher and Thompson 1999; Evans and Heller 2003). The χ, SIRM, S-ratios, SIRM/χ and HIRM% variations in the different selection sites for some representative samples of each section were presented in Table 1. In THOP area, magnetic susceptibility values range between 11.67 and 55.16 $\times$ 10^{-8} m^3 kg^{-1} (mean = 36.46) and SIRM between 0.58 and 3.11 $\times$ 10^{-3} A m^2 kg^{-1} (mean = 2.10) (Vidyasakar et al. 2015). In TPV area, magnetic susceptibility values range between 11.31 and 67.40 $\times$ 10^{-8} m^3 kg^{-1} (mean = 54.25) and SIRM between 1.04 and 7.74 $\times$ 10^{-3} A m^2 kg^{-1} (mean = 6.20) (Vidyasakar et al. 2015). In MUT area, magnetic susceptibility values were between 5.66 and 83.76 $\times$ 10^{-8} m^3 kg^{-1} (mean = 34.23) and SIRM between 0.12 and 6.50 $\times$ 10^{-3} A m^2 kg^{-1} (mean = 2.95). According Vidyasakar et al. (2015) in THOP, the S_{-300} ($S_{-300} = |IRM_{-300\ mT}/SIRM|$) close to 1 indicates that the saturation of the magnetization is obtained to a field close to 300 mT, which is typical of a magnetic structure with the magnetic behavior of magnetite. On the other hand, the HIRM% parameter is proportional to the content of hematite. In the case of THOP, the mean value of HIRM is around 10% (lower than in TPV and MUT sections), which indicates the low fraction of hematite-structure. Even so, in the THOP section the sample THOP15 shows a singular magnetic behavior with a low S_{-300} ratio and a high HIRM content, which point out a hematite-structure contribution. At the depth THOP 21 to THOP 31 where we have calcrete formation binded along with dead marine fossil, which reveals the sudden sea level change and the oxidation played important role of hematite contribution.

In TPV and MUT area samples, the absence of or a small amount of magnetite-like structure is substantiated by values of HIRM%, which are always higher than 10%, endorsing the significant contribution of canted antiferromagnetic (hematite) structures. The SIRM/χ ratio depends on the composition and grain size of magnetic particles. Moreover, when the magnetic mineralogy is homogeneous, this ratio revealed changes in the grain size assemblage of ferrimagnetic minerals (Thompson and Oldfield 1986; Moreno et al. 2003). The closer the value of S-300 ratio is to one, the bigger is the contribution of magnetite to magnetic mineralogy and the smaller is the contribution of other iron silicates or iron oxides. This means that the closer the value of S-300 ratio is to one, the higher the homogeneity of magnetic mineralogy is. When comparing THOP, TPV and MUT, the mean ratio of S-300 ratio is higher for THOP section, indicating the higher homogeneity of magnetic mineralogy.

As the magnetic mineralogy indicated by the S_{-300} ratio is not homogeneous in samples from TPV and MUT areas, SIRM/χ ratio only can be considered in THOP

Table 1 Results of magnetic parameters in different profiles

Depth (cm)	Samples	χ (× 10^{-8} m^3/kg)	SIRM(× 10^{-3} A m^2/kg)	S-300	S-100	HIRM %	SIRM/χ (kA/m)
10	THOP2-L1	52.49	3.08	0.83	0.48	8.34	5.86
50	THOP6-L1	49.38	2.45	0.84	0.48	7.80	4.97
80	THOP9-L1	55.16	2.90	0.77	0.33	11.27	5.26
150	THOP16-L1	41.49	3.11	0.82	0.40	8.81	7.49
170	THOP18-L1	49.51	3.00	0.81	0.40	9.74	6.07
210	THOP22-L1	48.57	2.76	0.80	0.40	9.84	5.69
220	THOP1-L2	11.67	0.58	0.78	0.31	11.09	4.94
260	THOP5-L2	30.81	2.52	0.82	0.44	8.87	8.17
300	THOP9-L2	36.41	1.98	0.83	0.49	8.62	5.43
330	THOP12-L2	30.91	2.30	0.81	0.36	9.69	7.45
360	THOP3-L3	32.34	2.37	0.78	0.25	11.00	7.32
400	THOP7-L3	34.47	2.88	0.73	0.07	13.36	8.36
440	THOP11-L3	27.89	1.68	0.73	0.22	13.36	6.02
480	THOP15-L3	24.53	1.27	0.57	0.23	21.68	5.19
540	THOP21-L3	21.95	1.04	0.80	0.34	9.87	4.75
570	THOP3-L4	39.75	2.06	0.81	0.49	9.68	5.19
590	THOP5-L4	44.74	1.88	0.84	0.52	8.15	4.21
640	TH0P10-L4	35.69	1.54	0.81	0.50	9.40	4.33
670	THOP13-L4	27.90	1.51	0.82	0.47	8.84	5.41
710	THOP17-L4	33.65	1.16	0.82	0.50	9.04	3.44
	Mean	36.47	2.10	0.79	0.38	10.42	5.78
0	TPV-1	57.41	6.55	0.74	0.21	12.99	114.13
40	TPV5	56.72	7.67	0.75	0.15	12.58	13.52
90	TPV10	59.67	6.91	0.77	0.21	11.47	11.59
140	TPV15	55.85	6.56	0.74	0.21	13.14	11.74
230	TPV24	61.24	6.36	0.77	0.23	11.35	10.39
290	TPV30	64.43	6.73	0.76	0.24	11.78	10.45
350	TPV36	67.40	7.74	0.74	0.27	12.76	11.49
400	TPV41	11.31	1.04	0.76	0.25	12.15	9.21
	Mean	54.25	6.20	0.75	0.22	12.28	24.06
0	MUT1	47.60	5.53	0.80	0.32	9.76	11.62
40	MUT5	35.17	4.54	0.76	0.22	12.12	12.91
90	MUT10	29.00	3.41	0.73	0.12	13.60	11.74
150	MUT16	52.77	4.75	0.80	0.36	10.10	9.01
170	MUT18	74.05	6.50	0.84	0.48	7.96	8.78
200	MUT21	28.27	3.57	0.71	0.06	14.57	12.64
230	MUT24	47.15	0.12	0.77	0.30	11.27	10.71
280	MUT29	34.56	3.17	0.72	0.15	13.81	9.16
310	MUT32	22.62	2.65	0.66	0.32	17.12	11.73

(continued)

Table 1 (continued)

Depth (cm)	Samples	χ ($\times$ 10^{-8} m^3/kg)	SIRM($\times$ 10^{-3} A m^2/kg)	S-300	S-100	HIRM %	SIRM/χ (kA/m)
350	MUT36	78.80	5.85	0.86	0.52	7.19	7.43
370	MUT38	19.85	2.16	0.62	0.07	12.15	10.89
390	MUT40	83.76	5.79	0.87	0.51	6.58	6.91
440	MUT45	17.06	1.52	0.57	0.12	21.49	8.89
550	MUT56	12.79	1.05	0.53	0.18	23.46	8.19
680	MUT69	9.73	0.69	0.52	0.17	24.00	7.11
750	MUT76	8.74	0.61	0.51	0.14	24.56	6.99
840	MUT85	8.56	0.73	0.53	0.16	23.65	8.56
940	MUT95	5.66	0.46	0.48	0.18	25.87	8.05
	Mean	34.23	2.95	0.68	0.24	15.51	9.52

THOP and TPV data source from Vidyasakar et al. (2015)

section. According to Thompson and Oldfield (1986) the mean SIRM/χ value of 10 kA m^{-1} indicated a magnetite grain size of 5 μm. However, Sandgren and Thompson (1990) indicated that a value of 6.40 kA m^{-1} corresponded to a magnetite grain size of 8 μm. As stated by these authors, mean SIRM/χ value of 5.78 kA m^{-1} obtained in THOP area samples indicated an average grain size of ferrimagnetic particles of ca. 8 μm in THOP. IRM acquisition curves are necessary to estimate the characteristic coercivity of ferromagnetic structures. The trend of IRM possession curves depends on the relative concentration of low-coercivity in magnetite type minerals and high-coercivity in hematite type minerals respectively (Thompson and Oldfield 1986). In THOP area samples, IRM acquisition curve showed saturation in fields between 500 and 600 mT followed by a small increase in intensity by increased fields, suggesting that these samples may contain both multidomain magnetite structures and antiferromagnetic minerals such as hematite. In MUT and TPV area samples, IRM acquisition curve showed saturation is not obtained at applied fields, which pointed towards the presence of a canted antiferromagnetic phase such as hematite (Fig. 4). The remanence coercivity parameter, B_{OCR} was obtained from IRM curves (Fig. 4). Also, THOP area samples showed B_{OCR} that varies from 50 mT to 75 mT depicting multidomain magnetite or magnetite and antiferromagnetic minerals; other THOP samples present B_{OCR} between 25 mT and 50 mT endorsed that these samples are magnetite bearing (Vidyasakar et al. 2015). In TPV samples, B_{OCR} changing from 50 mT to 75 mT indicated the existence of a multidomain magnetite or mixed mineral contents of magnetite and antiferromagnetic minerals (Vidyasakar et al. 2015). Also, MUT area samples showed (Fig. 4) the B_{OCR} variation from 100 mT to 150 mT and in which some varies from 150 mT to 200 mT indicating the presence of antiferromagnetic

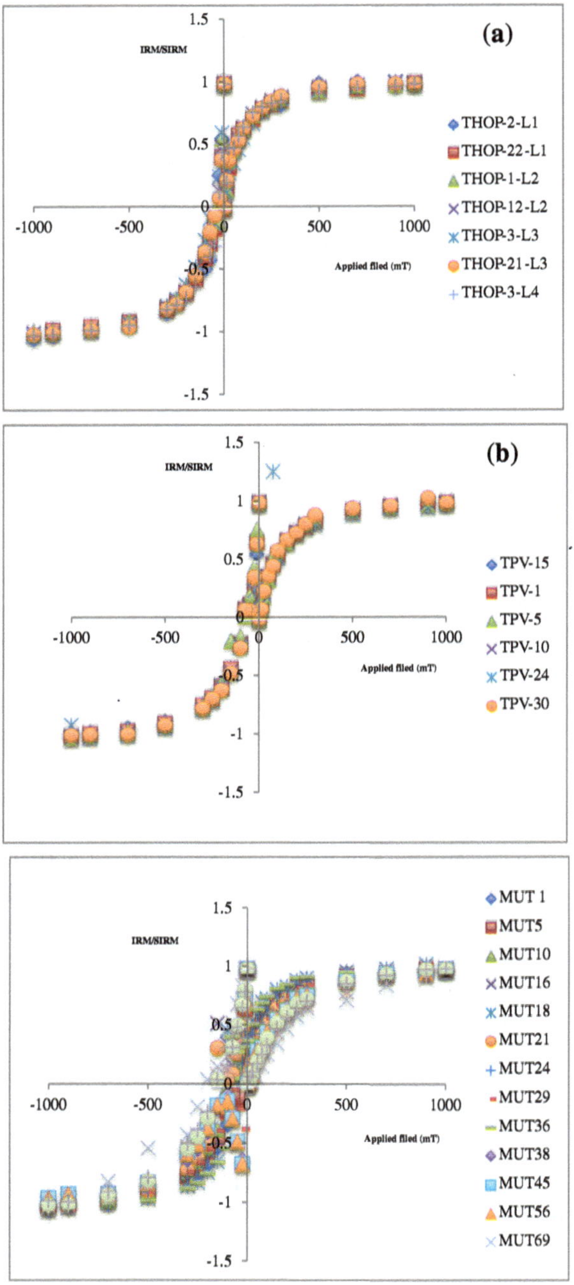

Fig. 4 IRM normalized (IRM/SIRM) acquisition curves for the representative samples from **a** THOP, **b** TPV and **c** MUT. THOP and TPV data source from Vidyasakar et al. (2015)

minerals. Moreover, MUT area samples show B_{OCR} of <100 mT stating that they bear multidomain magnetite or mixed mineral contents of magnetite and antiferromagnetic minerals.

4.2 Variation of Magnetic Parameters

The depth changes and magnetic parameters of three profiles are presented in (Fig. 5). In THOP profile, at 220 cm one can see a decrease of χ, SIRM and S_{-300} and a slight increase of HIRM%. However, in TPV profile, χ and SIRM did not change with depth and seemed to be homogeneous till paleo-surface. Finally, in Muttom profile, there was a sudden increase of χ, SIRM, S_{-300} and S_{-100} at depths of 170, 350 and 390 cm and a steep decrease at depths of 200, 310, 370 and 410 cm respectively. It is obvious that below the depth of 400 cm at Muttom profile, there is no significant change in the profile.

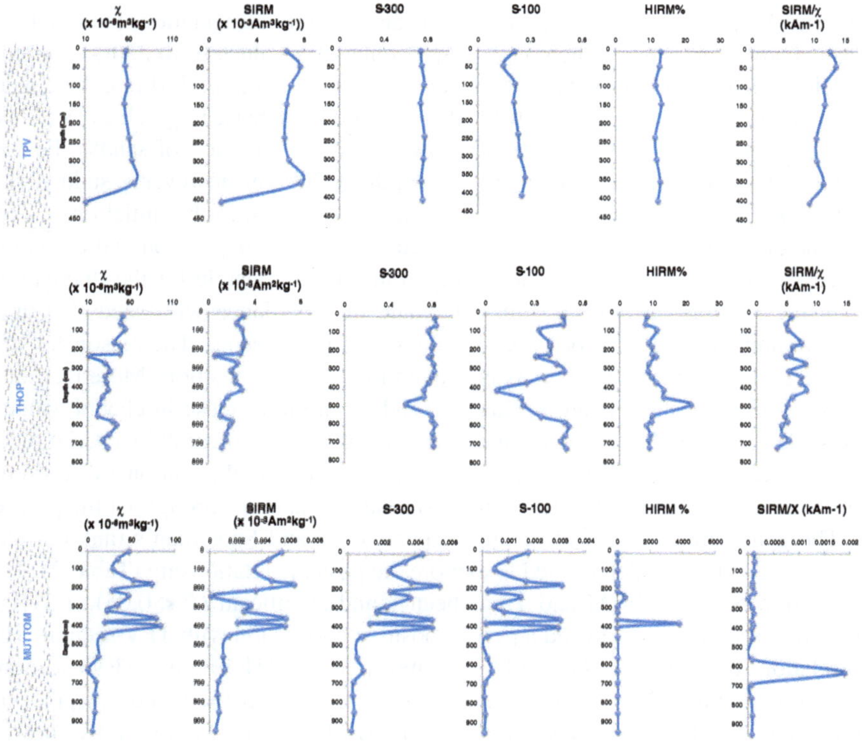

Fig. 5 Variation of magnetic parameters along each profile. TPV data source from Vidyasakar et al. (2015)

4.3 Relationship Between Magnetic Parameters and Geochemical Data

In the analogy with geochemistry data, discrepancy of χ frequency hangs upon the deposition within the profile. The marine sediments are very subtle indicators of temporal variations in concentration and grain size of deposited terrigenous/lithogenous material. These sediments would have been deposited both in dry and wet periods and dependent on environment, concentration and grain size of deposition, which may have led to their deposition. By relating both the parameters through depth wise, we could determine between dry and wet of the depositional depth.

The TPV section (Fig. 6) indicated χ variations for the whole section seem to be nearly uniform with little variations towards end depth of 370 cm because of $CaCO_3$ cementation. It is to be noted that other than sand and clay (%) persists more or less uniform with minimal fluctuations in between, which are predominantly at the depth of 50 cm. The Fe (%) appears to be uniform till the depth and high-value variations in the last sample (350 cm) may be because of $CaCO_3$ cementation. Thus, THOP is the one section that showed a heterogeneous deposition from 220 to 330 cm (Fig. 6) because of shell fragments and calcrete cementation. Invariably as it is evident from (Fig. 6) there is a sharp decline in χ values from 220 to 330 cm, which is because of high values of $CaCO_3$ (%). At the depth of 550 cm, there is an increase in sand%, decrease in $CaCO_3$ (%) values and increase in χ values. Notably, clay has a steady rise when there is a downfall in the variation of sand%. Fe (%) contents have no big fluctuation till the depth of 200 cm however a sudden rise occurs till the depth of 600 cm, and then the values return to the initial ones. This may be due the sudden sea level changes during the time deposition and oxidation process combined with high wind energy could be responsible for the fluctuation.

In the Muttom section, χ values fluctuate in accordance with sand (%) until 400 cm and from 410 cm onwards χ values remain the same. The reason for this can be attributed to homogeneous concentration in the deposition. Moreover, at a depth of 430 cm, there is an increase in sand % and a decrease in clay % values. Thus, clay seems to have a steady increase when there is a downfall in a variation of sand %. $CaCO_3$ (%), OM (%) and Fe (%) have related fluctuation till 180 cm; particularly Fe (%) and $CaCO_3$ (%) have similar fluctuations throughout the profile.

The geochemical data with magnetic parameters and Spearman's rank correlation coefficients were determined to explore the possible relationship (Table 2). The correlation between SIRM and χ has been found significant ($P < 0.05$) in THOP and MUT sections and showed a perfect positive correlation with TPV section. The correlation between $CaCO_3$ and χ is negative in THOP and TPV sections (Vidyasakar et al. 2015). As expected, in MUT section, HIRM% and S_{-300} present significant ($P < 0.05$) negative correlation. In MUT section, S ratios and χ showed significant ($P < 0.05$) positive correlation. The $CaCO_3$ and S ratios showed

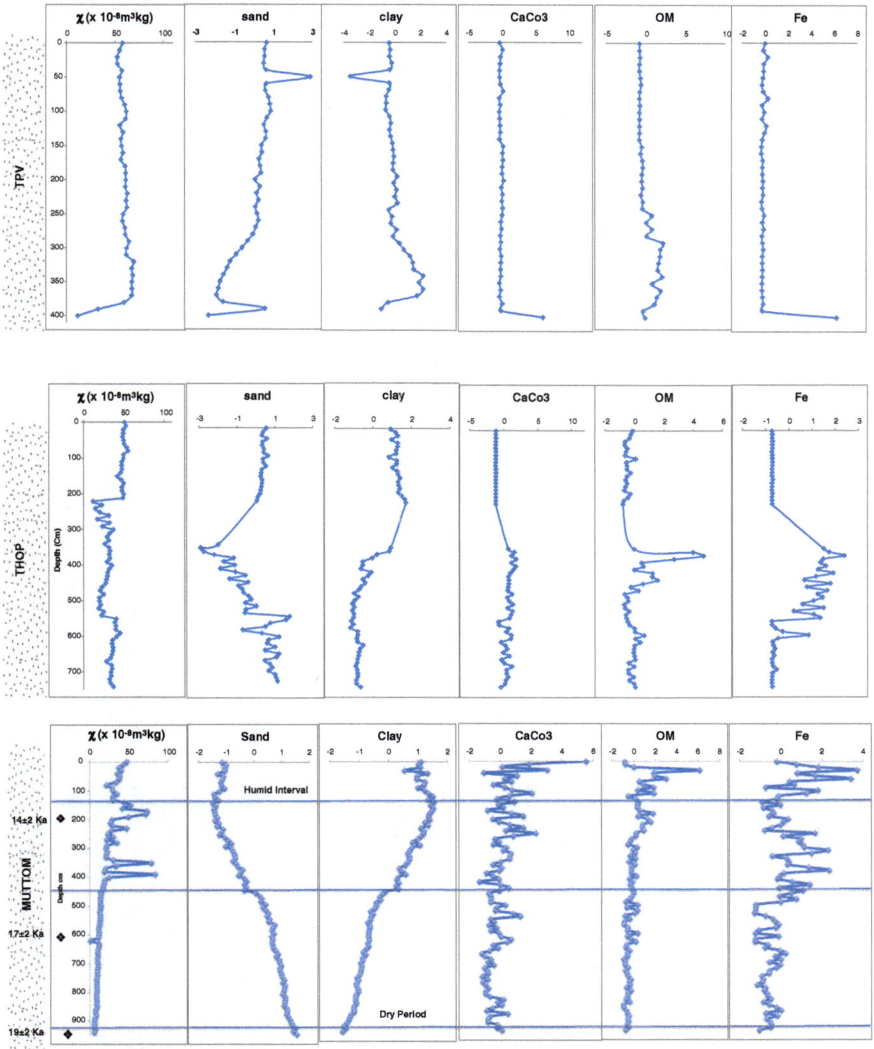

Fig. 6 Variation of magnetic susceptibility, sand%, clay%, CaCO3%, Fe% and OM (normalized values) along each profile. OSL dates in MUT section along with humid and dry period line. THOP and TPV data source from Vidyasakar et al. (2015)

significant ($P < 0.05$) negative correlation in THOP and negative correlation in TPV section. In these three sections, no significant correlation was found between OM and χ. In all the sections, the correlation between χ and clay is significant ($P < 0.05$). In MUT section, sand and S ratios showed significant negative and positive correlations with silt and χ respectively. In MUT section, OM showed positive correlation with clay and negative correlation with sand.

Table 2 Spearman's rank correlation coefficients for geochemical data and magnetic parameters THOP, N = 20; MUT, N = 19 and TPV, N = 8)

	χ	SIRM	S-300	S-100	HIRM%	SIRM/X	Clay	Sand	Silt	OM	Fe	$CaCO_3$
MUT												
χ	1.0000											
SIRM	0.840*	1.0000										
S-300	0.956*	0.872*	1.0000									
S-100	0.953*	0.801*	0.923*	1.0000								
HIRM%	−0.1083	−0.1238	−0.1212	−0.1757	1.0000							
SIRM/χ	−0.3125	−0.1663	−0.1933	−0.1465	−0.0588	1.0000						
Clay	0.677*	0.723*	0.791*	0.546*	0.0221	−0.2248	1.0000					
Sand	−0.668*	−0.715*	−0.783*	−0.534*	−0.0349	0.2246	−0.999*	1.0000				
Silt	0.519*	0.4359	0.4104	0.3946	0.2190	−0.2840	0.4232	−0.4362	1.0000			
OM	0.1832	0.2552	0.3231	0.0613	−0.0562	−0.0848	0.620*	−0.622*	0.0091	1.0000		
Fe	0.2165	0.3117	0.2067	0.1095	0.3588	−0.3034	0.4806	−0.4853	0.3373	0.1131	1.0000	
$CaCO_3$	0.0788	0.1258	0.2304	0.1232	0.0171	0.0959	0.2152	−0.2301	−0.0607	−0.2185	−0.0603	1.0000
THOP												
χ	1.0000											
SIRM	0.750*	1.0000										
S-300	0.4260	0.1830	1.0000									
S-100	0.4080	−0.0600	0.836*	1.0000								
HIRM%	−0.4260	−0.1830	−1.000*	−0.836*	1.0000							
SIRM/χ	0.0560	0.672*	−0.1800	−0.504*	0.1800	1.0000						
Clay	0.625*	0.729*	0.0570	−0.2540	−0.0570	0.571*	1.0000					
Sand	0.2210	−0.0430	0.4390	0.564*	−0.4390	−0.3250	−0.1890	1.0000				
Silt	−0.2040	−0.4290	0.4390	0.7060	−0.4390	−0.552*	−0.4900	0.3880	1.0000			
OM	−0.0320	−0.1050	−0.0450	0.1340	0.0450	−0.2180	−0.4430	−0.0430	0.3630	1.0000		

(continued)

Table 2 (continued)

	χ	SIRM	S-300	S-100	HIRM%	SIRM/X	Clay	Sand	Silt	OM	Fe	$CaCO_3$
MUT												
Fe	−0.543*	−0.486*	−0.475*	−0.3210	0.475*	−0.2360	−0.486*	−0.5360	0.0910	0.490*	1.0000	
$CaCO_3$	−0.657*	−0.479*	−0.2860	−0.0790	0.2860	−0.1500	−0.725*	−0.4000	0.2910	0.527*	0.654*	1.0000
TPV												
χ	1.0000											
SIRM	0.5000	1.0000										
S-300	0.3330	0.1830	1.0000									
S-100	0.4760	−0.0600	0.3100	1.0000								
HIRM%	−0.3330	−0.183 0	−1.000*	−0.3100	1.0000							
SIRM/χ	−0.143 0	0.672*	−0.6190	−0.762*	0.6190	1.0000						
Clay	0.738*	0.729*	0.0000	0.2620	0.0000	−0.1190	1.0000					
Sand	−0.0950	−0.0430	−0.0480	−0.762*	0.0480	0.762*	−0.3100	1.0000				
Silt	0.0480	−0.4290	−0.3100	0.5710	0.3100	−0.3330	0.1190	−0.741*	1.0000			
OM	0.5480	−0.1050	0.5000	0.833*	−0.5000	−0.786*	0.3810	−0.738*	0.5950	1.0000		
Fe	−0.667 0	−0.486*	−0.3570	−0.1900	0.3570	0.1900	−0.741*	−0.0480	0.1900	−0.2380	1.0000	
$CaCO_3$	−0.262 0	−0.479 *	0.738*	0.2140	−0.738*	−0.738*	−0.3330	−0.3100	−0.1190	0.3810	0.1430	1.0000

Values are significant at $*p < 0.05$

THOP and TPV data source from Vidyasakar et al. (2015)

4.4 Principal Component Analysis (PCA) 33

In PCA correlation, X-ray diffraction data was used to find out the similar type of mineral deposition in sampling section. This would help us to get an idea about the deposition between Southeast coast and southwest cost of a sampling area. At 940 cm depth, Muttom sample (Table 3) had 81% correlation between Edayanvillai (TPV) at the depth of 230–330 cm. Moreover, TPV had showed 84% positive correlation between 10 and 230 cm. The PCA correlation confirmed that TPV and MUT almost have a similar type of mineral deposition. Interestingly, THOP did not showed positive correlation with MUT and TPV. Muttom had 88% positive correlation between 180 and 350 cm respectively. Surprisingly, THOP have not showed any positive correlation with PCA. This correlation analysis indicated that THOP may have gone through ample wash over during the rise in sea level or a seasonal Arroyo/Gulch (Fig. 3) could have taken place over deposition. It is to be noted that THOP is the only sample that is pale yellow compared to TPV and MUT samples, which are darker red.

4.5 Mineral Inference with XRD Analysis

The present research study used X-ray diffraction as a tool to find out the minerals and its percentage of available quantity at different depths. In doing so, one can roughly determine the type of deposition that has taken place and its occurrence. In TPV section (Table 4), quartz, rutile, zircon, kaolinite, hematite, magnetite and sillimanite are present. In this section, presence of quartz is dominant than other minerals. The minerals that are present in TPV section are usually found in semi-arid/dry condition. The Aeolian process must have also made a significant role during the TPV deposition. In THOP section, quartz, calcite, rutile, zircon, pyrope and sylvite are present. At 30 cm, the minerals are supported by Aeolian process, which showed arid deposition with heavy wind flow. Furthermore, at 280 cm, calcite mineral depicted the dominance due to shells of dead marine organisms. This supported the presence of paleo minerals (pyrope and sylvite) in abundance in the sea level at a depth of 510 cm. These minerals were formed during dry condition. Also, in MUT section (Table 4) quartz, almandine, kaolinite, hematite, and halloysite are present. In this section, hematite and kaolinite are dominant among other minerals. These minerals are formed during wet-to-dry period. This may be because of ancient paleo sea level or due to primordial lagoon formation. Thus, during Last Glacial Maximum (LGM), continental shelves were wide-open with a vast reservoir of sediments with strong landward winds during (NE) winter monsoon and sediments coasted inland until 11.4 ka to form coastal aeolian teri deposits. The Coastal teri reddening happened by in situ weathering that started at <11.4–5.6 ka during a humid climate (Jayangondaperumal et al. 2012).

Table 3 PCA correlation between section depositions

Samples	TPV41	TPV24	TPV1	MUT95	MUT36	MUT19	THOP20_ L4	THOP18_L3	THOP7_L2	THOP4_L1
TPV41	100	34	34	35	29	27	48	44	54	37
TPV24	34	100	84	81*	66	69	58	34	34	55
TPV1	34	84	100	79	67	69	56	35	33	53
MUT95	35	81*	79*	100	65	67	65	37	37	57
MUT36	29	66	67	65	100	88*	49	24	26	38
MUT19	27	69	69	67	88*	100	46	26	28	41
THOP20_L4	48	58	56	65	49	46	100	47	46	60
THOP18_L3	44	34	35	37	24	26	47	100	54	59
THOP7_L2	54	34	33	37	26	28	46	54	100	43
THOP4_L1	37	55	53	57	38	41	60	59	43	100

* Positive correlation

Table 4 List of minerals found by XRD analysis in various sections at different depth

Ref.no	Depth/min	Quartz	Calcite	Rutile	Zircon	Pyrope	Almandine	Sylvite	Kaolanite	Hematite	Halloysite	Xenotime	Ilmenite	Magnesite	Sillmanite	Magnetite
TPV1	0 CM	✓	X	✓	✓	X	X	X	✓	✓	X	X	X	✓	X	X
TPV24	230 cm	✓	X	X	✓	X	X	X	✓	✓	X	X	X	X	✓	X
TPV57	560 cm	✓	✓	✓	X	X	X	X	X	✓	X	X	X	X	X	X
THOP/L1/4	30 cm	✓	X	✓	✓	X	✓	X	X	X	X	X	X	X	X	X
THOP/L2/7	280 cm	✓	✓	✓	X	X	X	X	X	X	X	X	X	X	X	X
THOP/L3/18	510 cm	✓	✓	✓	X	✓	X	✓	X	X	X	X	X	X	X	X
THOP/L4/20	740 cm	✓	✓	✓	✓	X	X	X	X	X	X	X	X	X	X	X
MUT'19	180 cm	✓	X	X	X	X	✓	X	✓	✓	✓	✓	X	X	X	X
MUT36	350 cm	✓	X	X	X	X	X	X	✓	✓	X	X	X	X	X	X
MUT95	940 cm	✓	X	X	X	X	✓	X	X	✓	✓	X	X	X	X	X

THOP and TPV data source from Vidyasakar et al. (2015)

4.6 Depositional Period—OSL Chronology

Coastal depositions are very sensitive to the environment and have a particular capacity to store the untold story of the period/inference in deposition. Therefore, with limited resources (Fig. 6) the researchers have compared χ, Sand%, Clay%, $CaCO_3$%, OM% and Fe% with each other to draw a distinct line between dry and wet periods. Muttom section (Fig. 6) has three OSL dates, which would let us know the approximate age of deposition in accordance with depth. The previous terrestrial record (Table 5) from this region supported the line drawn between the humid and dry period in this section.

According to Folk (1976) red beds are developed from minerals having source of abundant iron, either heavy minerals or ferruginous clays. They are mostly originated from heavy minerals when deposited above the water table for easier access to Oxygen.

4.7 Comparison Between Magnetic Parameters and XRD Analysis

In XRD analysis, we have determined various minerals that are present at different depths. In Table 4 one can see that magnetite is not present in any of the following samples. However, magnetic parameters such as remanence coercivity (Fig. 3) were able to pick minor quantities of magnetite present in the samples. The presence of magnetite could not be assessed by XRD due to its less availability in the sample, which implied magnetic measurements are more precise and faster than X-ray diffraction as per the magnetic mineralogy studies.

Table 5 Previous terrestrial climate records in India

Epoch	Time interval (Ka)	Climatic proxy/method	Location	Inference	Reference
Late Pleistocene–Early Holocene	14.5–7	Pollen	Berijam lake, Palani Hills, S. India	Warm and humid climate	Bera et al. (1995)
	16	$S1^3$ C of peat	Nilgiri Hills, S. India	Arid condition	Rajagopalan et al. (1997)
	20–17	Pollen	Berijam lake, Palani Hills, S. India	Cold and dry climatic condition	Bera et al. (1995)
	20–11.0	OSL dating	S.margin of Thar Dessert	Dry climatic condition	Juyal et al. (2006)

5 Discussion

- Magnetic parameters showed the significant contribution of canted antiferromagnetic (hematite) structures to sands. In adherence to the findings of Gardner (1981) on Teri sands, hematite originated from chemical alteration of ilmenite and garnet. Thus indicated wet, humid climate and high temperature due to oxidation of magnetite.
- The remanence coercivity parameter B_{OCR} indicated in TPV samples indicated the presence of a multidomain magnetite or mixed mineral contents of magnetite and antiferromagnetic minerals. Finally, some THOP samples showed B_{OCR} indicating multidomain magnetite or magnetite and antiferromagnetic minerals. In other THOP samples presence of B_{OCR} confirmed that these minerals are magnetite bearing. The presence of a magnetite-structure in THOP and TPV samples supported values of S_{-300} close to unity, 0.79 and 0.75, respectively. The presence of magnetite was however not detected by X-ray diffraction, even so magnetic measurements point out the ferromagnetic behavior of THOP and TPV samples.
- The MUT samples showed the highest quantity of hematite and smallest quantity of magnetite as indicated by highest values of HIRM% and lowest of S_{-300} and SIRM respectively.
- The SIRM and χ have showed positive correlations with MUT, THOP and TPV sections due to the magnetic signal of ferromagnetic minerals. The negative correlations between $CaCO_3$ and χ in THOP and TPV sections are justified by diamagnetism of carbonaceous materials. The HIRM% and S_{-300} are found negative and has a significant correlation in all sections, which has been described using the fact that first parameter is associated with the existence of antiferromagnetic minerals and second with ferromagnetic minerals. There were no correlations found between Fe (%) and χ in MUT and TPV sections, which has been quite surprising. However, this may be related to the fact that whole content of Fe was determined instead of FeO and Fe_2O_3 contents, which may correlate with magnetite and hematite content. Also, OM and χ have no significant correlation, which is a bit strange, but this has been justified in accordance with Lourenço et al. (2014). The positive correlation between χ with clay in TPV, THOP and MUT sections, based in Spearman correlation, lead to magnetic susceptibility, which has been dominantly meticulous by differences in the concentration of ferromagnetic minerals and paramagnetic minerals carried by clay matrix.
- The mean SIRM/χ value of 5.78 kA m^{-1} recorded in the present study in THOP section indicated the average grain size of ferromagnetic particles of ca is 8 μm.
- The presence of magnetite in THOP (near coastal) and TPV sections are probably due to oceanic origin (biogenic magnetite particles), which has been carried away by strong offshore winds (Vidyasakar et al. 2015). Also, the

presence of biogenic magnetite in sediments was mentioned by several authors (Kirschvink and Chang 1984).
- XRD investigation showed the minerals that exist in TPV section are commonly found to be in semi-arid/slow dry conditions.
- XRD proves that Aeolian process must have played a significant role in TPV deposition.
- In XRD analysis, THOP section at 280 cm depicted calcite mineral is dominant due to shells of dead marine organisms. It also supported the presence of paleo minerals (pyrope and sylvite) in abundance in the sea level at a depth of 510 cm. However, these minerals were formed during dry conditions.
- PCA correlation showed TPV and MUT sections have a similar type of depositions. Also, THOP have not showed any positive correlation with PCA. The correlation analysis indicated that THOP area would have gone through several wash over during rise in sea level or storm surge being nearer to coastal belt, hence, showed negative correlations.

6 Conclusion

The investigation in connection with features of magnetic parameters and its association with OSL, X-ray diffraction and geochemistry data made the subsequent conclusions possible:

- The studied sections of Teri sands show significant contribution of hematite presence due to oxidation of magnetite.
- In MUT, $\sim 14 \pm 2$ ka depositions showed the humid interval at $\sim 17 \pm 2$ to $\sim 19 \pm 2$ ka, which displays depositions during the dry period.
- In TPV, aeolian process played a significant role and the minerals that are found in that section are usually formed in Semi-arid/Dry condition.
- The three sections of XRD data were correlated with each other and found that MUT and TPV have a similar type of deposition. Moreover, THOP may have gone through a wash during some natural calamities or Arroyo/Gulch would have occurred.
- It has been a fact that during the Last Glacial Maximum (LGM) continental shelves were exposed to a vast reservoir of sediments with strong landward winds (NE/E) to form aeolian teri. The NE winter monsoon forms the fluvial teri sediments.
- The SW summer monsoon also plays a vital role in carrying sediments from one place to another, but it depends on the availability of loose sediments (minerals) at the time of transport.
- The colour red in the sediments is a distinct indicator of climate that prevailed at the time of diagenetic/pedogenetic transformation.

Acknowledgements The researchers are thankful to Dr. Ana Lourenço for their valuable advice and comments during the work. We are very thankful to Indian Journal of Geo-Marine science (IJMS) by allowing us to use the few images and tables from their journal. Special thanks to Ms. Sujita Ganaraj for her support in documentation assistance and Mr. Arul Britto for their timely during sample collection in this distinct field. Lastly, acknowledgements to Erasmus Mundus (INDIA4EU II) scholarship officials for their continuous financial support to the research study of Mr. Vidyasakar Anburaj (Indi1200057) during 2013–2015.

References

Aitken MJ (1998) Introduction to optical dating. Oxford University Press, Oxford 262

Alagarsamy R (2009) Environmental magnetism and application in the continental shelf sediments of India. Mar Environ Res 68:49–58

Alappat L, Palaniandy S, Shukla AD, Trivikramji KP, Singhvi AK (2013a) Chronology of red dune aggradations of South India and its paleo-environmental significance. Geochronometria 40(4):274–282

Alappat L, Morthekai P, Vidyasakar Srinivasalu S, Sandode SJ, Reddy DV, Singhvi AK (2013b) Chronology of deposition of coastal Red dunes (Teri sands) in south India and its paleoenvironmental implictions. PAGES-YSM, Goa (O)

Alappat L, Frechen M, Kumar SS, Babu DSS, Ravurb R, Tsukamotoa S (2015) Evidence of late Holocene shoreline progradation in the coast of Kerala, South India obtained from OSL dating of palaeo-beach ridges. Geomorphol 245:73–86

Albarède F (2003) Geochemistry: an introduction. Cambridge University Press, p 356

Blatt H, Middleton G, Murray R (1980) Origin of sedimentary rocks. Prentice-Hall, Englewood Cliffs, p 782

Bera SK, Gupta HP, Farooqui A (1995) Berijam Lake. 20,000 yrs. Sequence of paleofloristics and paleoenvironment in Palni Hills, South India. Geophytology 26:99–104

Brückner H (1988) Indicators for formerly high sea levels along the east coast of India and on the Andaman Islands. Hamburger Geographische Studien 44:47–72

Brückner H (1989) Late quaternary shorelines of India. In: Scott DB (ed) Late Quaternary sea level correlation and applications. Kluwer Academic, New York, pp 169–194

Chandrasekharan S, Murugan C (2001) Heavy minerals in the beach and the coastal red sands (Teris) of Tamilnadu. Special issue on Beach and Inland Heavy Mineral Sand Deposits of India, Exploration and Research for Atomic Minerals 13:87–109

Dekkers MJ (1997) Environmental magnetism: an introduction. Geol Mijnbouw 76:163–182

Evans ME, Heller F (2003) Environmental magnetism. Academic Press, p 213

Folk RL (1976) Reddening of desert sands; Simpson Desert, N.T., Australia. J Sedim Res 46 (3):604–615

Gardner RAM (1981) Reddening of dune sands—evidence from southeast India. Earth Surf Proc Land 6:459–468

Gardner RAM (1986) Quaternary coastal sediments and stratigraphy Southeast India. Man Environ X, 51–72

Guérin G, Mercier N, Adamiec G (2011) Dose-rate conversion factors: update. Ancient TL 29:5–8

Islam MS, Tooley MJ (1999) Coastal and sea-level changes during the Holocene in Bangladesh. Quat Int 55:61–75

Jacobs Z, Wintle AG, Roberts RG, Duller GAT (2008) Equivalent dose distributions from single grains of quartz at Sibudu, South Africa: context, causes and consequences for optical dating of archaeological deposits. J Archaeol Sci 35:1808–1820

Jayangondaperumal R, Murari MK, Sivasubramanian P, Chandrasekar N, Singhvi AK (2012) Luminescence dating of fluvial and coastal red sediments in the SE coast, India, and implications for paleoenvironmental changes and dune reddening. Quat Res 77(3):468–481

Joseph S, Thrivikramaji KP, Anirudhan S (1997) Textural parameters, discriminant analysis and depositional environments of the Teri sands, southern Tamil Nadu. J Geol Soc India 50 (3):323–329

Joseph S, Thrivikramji KP, Anirudhan S (1999) Mud content, clay minerals and oxidation states of iron in Teris of southern Tamil Nadu. Implication to the origin of redness. J Indian Assoc Sedimentol 18:83–94

Juyal N, Chamyal LS, Bhandari S, Bhushan R, Singhvi AK (2006) Continental record of the southwest monsoon during the last 130 ka: evidence from the southern margin of the Thar Desert, India. Quat Sci Rev 25(19–20):2632–2650

Kirschvink JL, Chang SBR (1984) Ultrafine–grained magnetite in deep-sea sediments: Possible bacterial magnetofossils. Geology 12:559–562

Lapa ML, Reis RP (1977) Contribuicâo para o estudo dos minerais argilosos em formações sedimentares da Orla Meso-Cenozóica Ocidental. Memórias e Notícias, Pub Mus Lab Min Geol Univ Coimbra 83:3–25

Lourenço AM, Rocha F, Gomes CR (2012) Relationships between magnetic parameters, chemical composition and clay minerals of topsoils near Coimbra, central Portugal. Nat Hazards Earth Syst Sci 12:2545–2555

Lourenço AM, Sequeira E, Sant'Ovaia H, Gomes CR (2014) Magnetic, geochemical and pedological characterisation of soil profiles from different environments and geological backgrounds near Coimbra, Portugal. Geoderma 213:408–418

Maher BA (1986) Characterisation of soils by mineral magnetic measurements. Phys Earth Plan Int 42:76–92

Maher BA, Thompson R (1999) Quaternary climates, environments and magnetism. Cambridge University Press, 390 p

Mejdahl V (1979) Thermoluminescence dating. Beta dose attenuation in quartz grains. Archaeometry 21:61–63

Moreno E, Sagnotti L, Dinare's-Turell J, Winkler A, Cascella A (2003) Biomonitoring of traffic air pollution in Rome using magnetic properties of tree leaves, Atmos Environ 37:2967–2977

Morner Nils-Axel (2010) Some problem in the reconstruction of mean sea level and its changes with time. Quat Int 221:3–8

Murray AS, Wintle AG (2000) Luminescence dating of quartz using an improved single-aliquot regenerative-dose protocol. Radiat Meas 32:57–73

Peel MC, Finlayson BL, McMahon TA (2007) Updated world map of the KÖppen-Geiger climate classification. Hydrol Earth Syst Sci Discuss 4(2):439–473

Prescott JR, Hutton JT (1994) Cosmic ray contributions to dose rates for luminescence and ESR dating. Large depths and long-term time variations. Radiat Meas 23:497–500

Rajagopalan G, sukumar R, Ramesh R, Pant RK, Rajagopalan G (1997) Late quaternary vegetational and climate changes from tropical peats in southern India—an extended record up to 40000 years BP. Curr Sci 73(1):60–63

Sanchez Goñi MF, Harrison SP (2010) Vegetation response to millennial-scale variability during the last glacial. Quat Sci Rev Spec 29:2823–2980

Sandgren P, Thompson R (1990) Mineral magnetic characteristics of podzolic soils developed on sand dunes in the Lake Gosciaz catchment, central Poland. Phys Earth Planet Inter 60:297–313

Saravanan P (2012) Depositional characteristics of recent and past washover sand sheets from cauvery delta coast, Tamil Nadu, India. PhD thesis. Anna University

Thompson R, Oldfield F (1986) Environmental magnetism. Allen & Unwin Press, London, p 227

Verosub KL, Roberts AP (1995) Environmental magnetism: Past, present, and future. J Geophys Res 100(B2):2175–2192

Vidyasakar A, Sant'Ovaia H, Gomes CR, Lourenço AM, Srinivasalu S (2015) A Magnetic and Geochemical Characterization of Red Dune Sands (teri sands) of Tamil Nadu coast. Indian J Mar Sci 44(9):1382–1392

Wintle AG, Murray AS (2006) A review of quartz optically stimulated luminescence characteristics and their relevance in single-aliquot regeneration dating protocols. Radiat Meas 41(4):369–391

Wolff EW, Chappellaz J, Blunier T, Rasmussen SO, Svensson A (2010) Millennial-scale variability during the last glacial: the ice core record. Quat Sci Rev 29:2828–2838.

http://en.climate-data.org

Collective Impact of Upstream Anthropogenic Interventions and Prolonged Droughts on Downstream Basin's Development in Arid and Semi-arid Areas: The Diyala Transboundary Basin

Furat A.M. Al-Faraj

Abstract The integrated management of transboundary waters has increasingly becoming a heavy burden facing decision-makers and water managers, especially in downstream countries. Human-intervention activities upstream combined with prolonged droughts have intensified the challenges encounter the water governance in a sustainable manner. This study quantifies the combined effects of upstream man-made modifications and basin-wide extended droughts on temporal river flow paradigms of the downstream riparian country. The Diyala watershed of about 32,000 km^2 shared between Iraq and Iran was chosen as an example case study. The Indicators of the Hydrologic Alteration (IHA) and the Range of Variability Approach (RVA) were adopted to characterise the streamflow alteration. Findings reveal that the joint impact has destructively influenced the development of the middle and lower portions of the basin in the downstream country, including the security of the irrigated agriculture, domestic and industrial water demands, and prompted people to leave their homes and lands coupled with growing conflicts between tribes. Prolonged severe droughts were marked between 1999 and 2015. The size and magnitude of the joint impact are anticipated to increase in the foreseeable future when under construction and future planned water withdrawal facilities upstream will be commissioned. The two successive acute droughts of (1999–2001) and (2008–2009) twinned with upstream regulation practices have hindered the socio-economic activities and deteriorated the environmental system of the middle and lower portions of the Diyala basin in the lower country.

Keywords Transboundary basin · Climate alteration · Hydrologic indicator · Sustainable management · Human-induced disturbance

F.A.M. Al-Faraj (✉)
School of the Built Environment, The University of Salford, Salford, Greater Manchester M5 4WT, UK
e-mail: f.a.m.al-faraj1@salford.ac.uk

O. Abdalla et al. (eds.), *Water Resources in Arid Areas: The Way Forward*, Springer Water, DOI 10.1007/978-3-319-51856-5_2

1 Introduction

Being a downstream riparian country across all of its transboundary rivers, with over 75% of its surface fresh water originating from outside its territory, Iraq has increasingly becoming vulnerable to the unilateral water abstraction and water policies of the upstream riparian countries, particularly Turkey and Iran. The climate change is making the level of vulnerability even worse. The IPCC (2014) reported that many arid and semi-arid areas (e.g., Mediterranean region, Western United States and South Africa) would encounter a decline in water resources due to potential impacts of climate change and climate variability. The growing water demands for agriculture, household and industrial purposes associated with climate change impacts have underlined the importance of a deep understanding of the adverse collective influence of upstream human pressures and climate alteration and variability at a regional scale on the runoff volume and quality of water of the downstream country (Al-Faraj and Scholz 2014; Al-Faraj and Al-Dabbagh 2015).

Increasing water demands in the territory of each riparian country and the differences in interests and water policies among riparian countries are expected to bring about increasing challenges to cooperation, possibly even increasing conflicts and tensions. For basins that entirely lie within the countries' territories, mitigation of the climate change impacts will be hard enough; however, when basins cross borders, bringing in multiple political entities and actors, integrated management of shared water resources in a changing climate and variability will be largely challenging. Iraq is a typical example of this second case. As more than 75% of its water resources originate in the upper riparian countries, Iraq is highly sensitive to the water withdrawal and regulation practices and unilateral management policies upstream.

This chapter examines the temporal streamflow alteration of the Diyala flow paradigm of the downstream country Iraq due to the mutual impact of a spectrum of upstream anthropogenic regulations and prolonged droughts at transboundary scale.

2 Materials and Methodology

2.1 Case Study Area

The Diyala basin is a transboundary river basin shared between Iraq and Iran (Fig. 1). The basin is intensively equipped with dams, extensive irrigation practices, fish farms, hydraulic water diversions and inter-basin water transfer tunnels. Moreover, Daryan dam, which is currently under construction primarily aims to divert a considerable amount of the river flow (about 1.4 billion m^3/year) to the South-western part of Iran for irrigation, through the 48 km long Nosoud Water Conveyance Tunnel. The dam will also have a 210 MW hydroelectric power station. Both Daryan dam and Nosoud inter-basin water transfer tunnel are expected to be commissioned in 2018 (Al-Faraj and Scholz 2014, 2015).

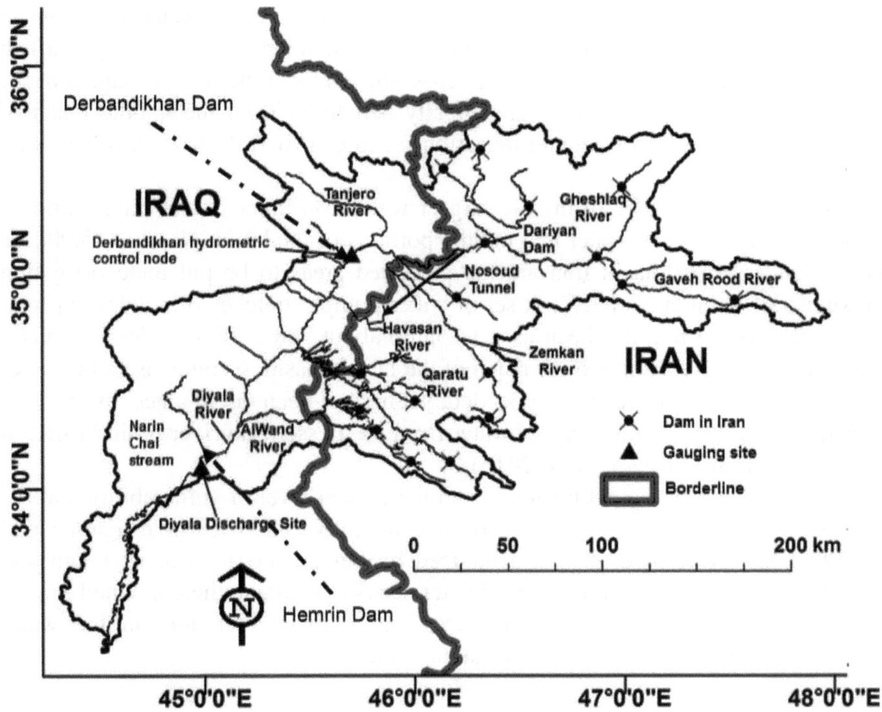

Fig. 1 The transboundary Diyala river basin (After Al-Faraj et al. 2015)

Precipitation usually falls between October and May, mainly spans November to April. The dry season usually extends from June to September. The basin lies between the latitudes 33.216°N and 35.833°N, and the longitudes 44.500°E and 46.833°E. It drains about 32,000 km^2, of which 43% (about 14,000 km^2) lies in Iraq and 57% (approximately 18,000 km^2) in Iran.

The upper portion of the basin drains about 17,900 km^2 of which approximately 25% stretches in Iraq. The upper portion is largely devoted to irrigated agriculture, fisheries, domestic and industrial water supply and water conveyance networks, including inter-basin water transfer tunnels to poor-water basins (Al-Faraj and Scholz 2014, 2015).

Concerning the downstream country Iraq, the main river corridor equipped with two key dams: the Derbandikhan dam, which controls the upper reach and Hemrin dam that monitors the most downstream site of the middle Diyala portion. Derbandikhan dam is located approximately 65 km southeast of Sulaimaniya and 230 km northeast of Baghdad. The middle Diyala stretches between Derbandikhan and Hemrin dams (Fig. 1), draining about 11,900 km^2 of the transboundary basin. It receives runoffs from three main water courses that emerge in Iran (Havasan, Qaratu and AlWand). Some seasonal streams on the right bank of the river (e.g., Narin Chai stream flowing into the Hemrin reservoir) in Iraq are slightly contributing to the total runoff of the river.

The three water courses are equipped with a web of dams and water diversions in Iran, which dedicated for irrigation, domestic and industrial water supply and fish farms. Irrigation projects are expanded across the banks of the river's main corridor in Iraq. The irrigation networks are directly benefiting from the middle reach of the river by gravity and pumping irrigation facilities (MoWR 2010; Al-Faraj et al. 2015).

The agriculture sector forms the largest water consumer of the transboundary waters, particularly, the upper and middle portions of the basin. Nearly 445,000 ha in Iraq and 131,000 ha in Iran are the projected areas to be put under irrigation facilities (Al-Faraj et al. 2015). A set of water pumps stretches on the left bank of the Tigris river in Iraq has been used to pump about 288 million m^3/year of water from the Tigris river to the lower reach of the Diyala basin to compensate the water shortage. Moreover, it was recommended, yet has not been put in place, to annually transfer about 945.4 million m^3 of water from the Lesser Zab river basin in Iraq to the Diyala basin (Al-Faraj et al. 2015).

The basin's development upstream and the critical level of vulnerability that the downstream country Iraq has reached have urgently called for investigating and assessing the hydrologic alteration induced by human interference upstream and climate change and variability. This chapter seeks to examine the combined impact of upstream human-induced interferences and climate alteration on the water resources availability of the lower riparian country.

2.2 *Data Availability*

Considering the availability of flow data, the Derbandikhan hydrometric station in Iraq was chosen as a promising control gauging site to examine the combined effect of the human pressures upstream and climate change on the flow regime of the downstream country. Daily runoffs observed between 1931 and 2013 were examined. The dataset covers both the unaltered flow condition (1931–1982) and the altered flow regime (1983–2013). The impaired period of 1999–2013 was more broadly analysed taking into account two acute drought periods (1999–2001) and (2008–2009). The data were gathered from various sources (Harza and Binnie 1958, 1959, 1963; USGS 2010; MoAg and WR-KRG 2013). The effective regulation of the river in the upper country began in 1983 (Al-Faraj and Scholz 2014). Thus, this year was used to split the time series (1931–2013) into natural (undisturbed) and disturbed flow regimes.

2.3 *Characterization of Hydrologic Alteration*

In this chapter, the Indicators of Hydrologic Alteration (IHA) associated with the Range of Variability Approach (RVA) was adopted to assess the hydrologic shift.

The IHA and the RAV are well documented in the literature (Richter et al. 1996, 1997, 1998; Shiau and Wu 2004a, b; Zuo and Liang 2015). The Indicators IHA are broadly used, which consider a full range of undisturbed natural flow variability, including magnitude, frequency, timing, duration and change rate; the 33 parameters of IHA were classified into five domains of hydrologic features. In order to obtain the flow regime target with IHA, the RVA method was introduced to assess the hydrologic anomaly attributed to hydraulic control works. The method for assessing hydrologic alteration is based on the differences in the flow paradigm between two defined periods at a certain hydrometric control node.

The long-term yearly average runoffs for the undisturbed natural condition (1931–1982), and the regulated periods (1983–2013) and (1999–2013) were 172.8 m^3/s (5.45 billion m^3), 130.1 m^3/s (4.10 billion m^3) and 77.8 m^3/s (2.45 billion m^3), respectively. The corresponding drops in the long-term yearly average runoffs relative to the long-term yearly average runoff of the unimpaired flow regime were about −25 and −55%, respectively. The minimum, mean, maximum daily runoffs and the standard deviation of the natural flow regime over the period (1931–1982) were 1, 174.4, 5816 and 221.5 m^3/s, respectively. The prominent annual variations in the natural runoffs were primarily due to the nature of the water years (i.e. dry, normal and wet). The rate of alteration in the natural annual flow mainly depends on the degree of annual precipitation shift with respect to the long-term yearly average precipitation.

Concerning the impaired period (1983–2013), the corresponding values were 1, 130.1, 2446 and 163.8 m^3/s, respectively. The last fifteen hydrologic years (1999–2013) indicate significant drop where the minimum, mean, maximum daily runoffs and the standard deviation were 1, 77.8, 2446 and 106.8 m^3/s, respectively.

Figure 2 shows a comparison between the unimpaired and the impaired long-term mean monthly flows. The deviations of the long-term mean monthly flows for the altered period (1983–2013) are ranging from −7.2% observed in November to −36.1% perceived in September (Table 1). Remarkable anomalies were associated with the long-term mean monthly runoffs of the impaired period (1999–2013). The alterations were between −45.1% in November and −61.4% in September (Table 1). The considerable shifts associated with this period are

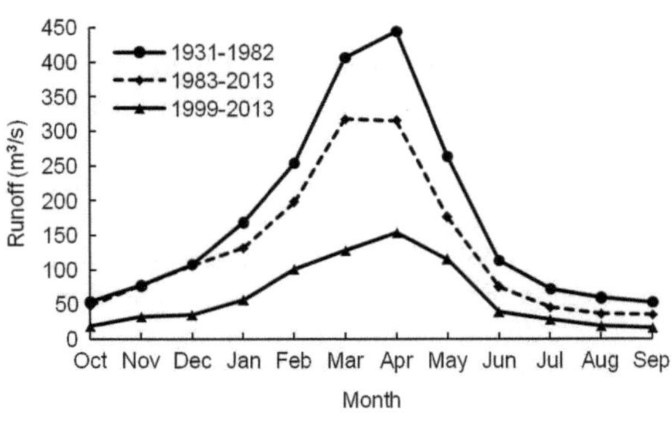

Fig. 2 A comparison between the long-term mean monthly runoffs of the undisturbed period and disturbed periods

attributed to the combined effects of droughts (1999–2001) and (2008–2009), and upstream river regulation arrangements and storage facilities.

2.4 Comparison of Monthly Percentiles Between Pre- and Post-alteration Periods

Table 2 shows the alterations in the flow percentiles over the periods of 1983–2013 and 1999–2013. Figure 3 compares the monthly percentiles (10, 25, 50, 75, and 90%) between the undisturbed and disturbed time series. The monthly 10th percentiles (Fig. 3a) presented a decline ranged from −33.4% in February to −70.6% in August for the period of 1983–2013, and drops between −37.2% perceived in February and −80.4% observed in August and September for the 1999–2013. For the monthly 25th percentiles (Fig. 3b), the first period of 1983–2013 was associated with departures between −23.1% in November to −45.4% in June. The corresponding shifts for 1999–2013 were between −45.6 and −76.9%. Concerning the monthly median flows (Fig. 3c), the anomalies associated with the 1983–2013 were ranged from −7.2% witnessed in November to −36.1% obtained in September. The corresponding alterations for 1999–2013 were between −45.1 and −61.4%. The alterations associated with the 75% flow percentiles for 1983–2013 ranged from −2.2% observed in October to −36.4% perceived in August. Unlike, November and December were accompanied by a positive shift of +1.8 and +15.7%, respectively. The corresponding deviations associated with the period of 1999–2013 were between −27.2% observed in February and −74.7% noticed in September (Fig. 3d). As far as the 90% percentiles for the 1983–2013 are concerned, the shifts ranged from −12.6% perceived in March to −40.7% in August. November, December and January were associated with positive alteration between +4.2 and 58.5%. The deviations linked to the period 1999–2013 were between −0.5% obtained in October and −74.1% witnessed in September (Fig. 3e).

2.5 Magnitude of Annual Extreme Conditions

The results of the Range of Variability Approach (RVA) analysis are given in Table 1. Low to moderate deviation was linked to the post-impact period of 1983–2013. The moderate deviation was perceived in January, July and September. The corresponding hydrologic alterations were −34.1, −35.9, and −36.1%, respectively. The moderate degree of hydrologic alteration was also associated with the regulated period of (1999–2013). The hydrologic alterations were between (−45.1%) observed in November and (−61.4%) noticed in September.

The extreme minima flows of the period (1983–2013) experienced low to moderate deviations. High shift of −75% was linked to 1-day minima flow over the

Table 1 Statistics of the hydrologic alteration for the middle RVA category of the Diyala River

Parameter group #1	HA (%) (1983–2013)	Degree of alteration	HA (%) (1999–2013)	Degree of alteration	Parameter group #2	HA (%) (1983–2013)	Degree of alteration	HA (%) (1999–2013)	Degree of alteration
October	−25.0	Low	−56.8	Moderate	1−day minimum	−37.5	Moderate	−75.0	High
November	−7.2	Low	−45.1	Moderate	3-day minimum	−34.1	Moderate	−56.5	Moderate
December	−25.7	Low	−58.1	Moderate	7-day minimum	−34.4	Moderate	−55.4	Moderate
January	−34.1	Moderate	−54.2	Moderate	30-day minimum	−26.0	Low	−56.9	Moderate
February	−18.8	Low	−49.0	Moderate	90-day minimum	−28.7	Low	−59.3	Moderate
March	−31.9	Low	−60.4	Moderate	1-day maximum	−14.3	Low	−54.5	Moderate
April	−17.0	Low	−55.9	Moderate	3-day maximum	−25.4	Low	−51.0	Moderate
May	−27.9	Low	−46.5	Moderate	7-day maximum	−23.5	Low	−50.5	Moderate
June	−32.5	Low	−61.3	Moderate	30-day maximum	−11.5	Low	−47.6	Moderate
July	−35.9	Moderate	−57.3	Moderate	90-day maximum	−20.5	Low	−48.1	Moderate
August	−21.3	Low	−57.3	Moderate					
September	−36.1	Moderate	−61.4	Moderate					
Parameter group #3									
Julian date of annual minima	Pre-impact (275)	Post-impact 1983–2013 (260)	Post-impact 1999–2013 (251)						
Julian date of annual maxima	Pre-impact (76)	Post-impact 1983–2013 (78)	Post-impact 1999–2013 (86)						
Parameter group #4									
Low pulse count	Pre-impact (2)	Post-impact 1983–2013 (8)	Post-impact 1999–2013 (6)						
Low pulse duration	Pre-impact (9.5)	Post-impact 1983–2013 (3)	Post-impact 1999–2013 (4)						
High pulse count	Pre-impact (6)	Post-impact 1983–2013 (6)	Post-impact 1999–2013 (5)						
High pulse duration	Pre-impact (3.5)	Post-impact 1983–2013 (3)	Post-impact 1999–2013 (2.5)						
Parameter group #5									
Rise rate	Pre-impact (2)	Post-impact 1983–2013 (8)	Post-impact 1999–2013 (4)						
Fall rate	Pre-impact (2)	Post-impact 1983–2013 (8)	Post-impact 1999–2013 (−5)						
Number of reversals	Pre-impact (2)	Post-impact 1983–2013 (8)	Post-impact 1999–2013 (177)						

Table 2 Alteration (%) in flow percentiles relative to the flow percentiles of the natural unimpaired flow regime

WY	Alteration (%)									
	1983–2013	1999–2013	1983–2013	1999–2013	1983–2013	1999–2013	1983–2013	1999–2013	1983–2013	1999–2013
Month	Percentiles									
	10%		25%		50%		75%		90%	
October	−60.7	−68.3	−27.6	−69.5	−25.0	−56.8	−2.2	−66.9	−22.7	−0.5
November	−35.3	−54.7	−23.1	−45.6	−7.2	−45.1	+1.8	−53.5	+18.4	−51.1
December	−34.4	−39.0	−41.7	−48.3	−25.7	−58.1	+15.7	−49.0	+58.5	−32.1
January	−43.3	−51.7	−36.7	−53.4	−34.1	−54.2	−6.9	−53.1	+4.2	−36.6
February	−33.4	−37.2	−40.7	−49.7	−18.8	−49.0	−13.0	−27.7	−15.7	−20.5
March	−40.6	−49.1	−43.5	−51.9	−31.9	−60.4	−11.3	−48.7	−12.6	−32.0
April	−59.5	−73.8	−33.8	−67.4	−17.0	−55.9	−21.1	−47.8	−27.8	−57.9
May	−64.7	−72.0	−24.7	−72.6	−27.9	−46.5	−30.1	−53.9	−40.4	−64.5
June	−62.6	−70.9	−45.4	−76.9	−32.5	−61.3	−31.6	−60.7	−32.3	−64.4
July	−65.0	−79.4	−28.7	−72.0	−35.9	−57.3	−26.8	−61.7	−35.1	−66.0
August	−70.6	−80.4	−29.0	−70.1	−21.3	−57.3	−36.4	−60.9	−40.7	−70.1
September	−69.9	−80.4	−31.9	−70.2	−36.1	−61.4	−28.6	−74.7	−28.6	−74.1

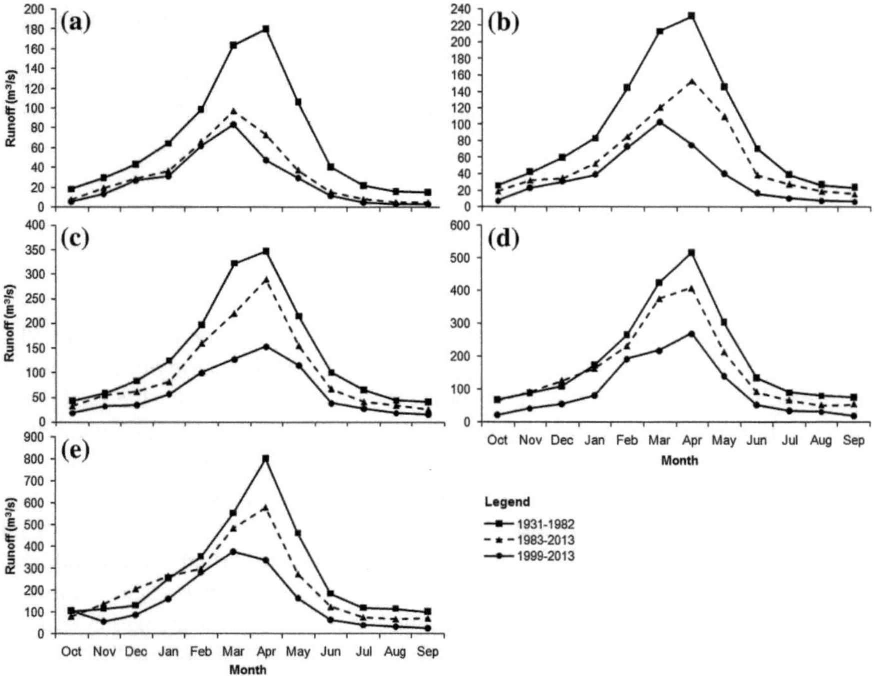

Fig. 3 Alterations in monthly flow percentiles relative to the reference natural monthly flow percentiles: **a** 10th percentiles; **b** 25th percentile; **c** median; **d** 75th percentile; and **e** 90th percentile

altered period of (1999–2013), while 3- to 90-day minima flows were incorporated with moderate deviation between −47.6 and −59.3%. Low alteration was associated with the parameters of 1- to 90-day maximum flows for the impaired period of (1983–2013), while moderate deviation was linked to the corresponding parameters for the altered period of (1999–2013).

The mean Julian date of annual 1-day minimum moved forward from the 275th day in the undisturbed time span to the 260th day of the disturbed period of (1983–2013) and to 251st day in the impaired period of 1999–2013. The mean Julian date of annual 1-day maximum impaired was later than that unimpaired, moving from 76th day in the undisturbed period to 78th day in the impaired period of 1983–2013 and to 86th day in the disturbed period of 1999–2013.

The number of the low pulse counts in the regulated periods was higher than in the unregulated period while the high pulse counts of the impaired period of 1983–2013 were equal to those in the pre-impact period while the impacted period of 1999–2013 was associated with lower counts. The duration of high and low pulses both decreased; e.g. the duration of low pulses dropped from 9.5 days in the pre-impact period to 3 and 4 days in the impaired period of 1983–2013 and 1999–2013, respectively. The duration of high pulses landed from 3.5 days in the unimpaired period to 3, and 2.5 days for the 1st and 2nd impaired periods, respectively.

The number of reversals was increased from 2 in the undisturbed period to 8 in the impaired period of 1983–2013 and to 177 for the regulated period of 1999–2013.

3 Conclusions

The flow regime of the Diyala watershed has experienced immense alteration. The construction of a web of dams, water abstraction for irrigated agriculture and public water supply, and water conveyance networks within and outside the basin have notably shifted the natural flow regime of the river. Exploring the extent to which human interventions would support water managers to manage the water resources in an integrated mode. The rapid growth of public water supply and irrigation demands, particularly in transboundary watersheds coupled with the effect of climate change and variability have increasingly called for a further understanding of the joint impact of the river regulation arrangements upstream and multi-year droughts at the basin scale, on the hydrologic regime of the lower riparian country. River regulation activities associated with successive droughts substantially shifted the flow characteristics of the Diyala River. The rates of water abstraction and diverted flow in the upper riparian country are expected to escalate and the degree of vulnerability of the downstream riparian country is expected to develop. The drought spells of 1999–2001 and 2008–2009 coupled with watershed development upstream, have resulted in severe modifications in the hydrologic regime of the river and produced a remarkable adverse impact on water availability of the downstream riparian country. The descent in the long-term yearly average altered runoffs relative to the long-term mean annual undisturbed flow was about −25% for the disturbed period (1983–2013) and −55% for the impaired period (1999–2013). The deviations of the long-term mean monthly flows for the post-alteration time window (1983–2013) ranged between −7.2% observed in November to −36.1% perceived in September. Concerning the second altered period (1999–2013), the alterations ranged between −45.1% in November and −61.4% in September. Results reveal that the man-made changes upstream associated with consecutive severe droughts have resulted in a notable negative impact on flow characteristics of the downstream country Iraq. This has influenced the development of the middle and lower portions of the basin in Iraq.

Acknowledgements The author would like to express his great appreciation to the Ministry of Agriculture and Water Resources in Iraqi Kurdistan, General Directorate of Dams and Reservoirs and Derbandikhan Dam Commission for their full support in data provision.

References

Al-Faraj FAM, Al-Dabbagh BNS (2015) Assessment of collective impact of upstream watershed development and basin-wide successive droughts on downstream flow regime: the Lesser Zab transboundary basin. J Hydrol 530(2015):419–430. doi:10.1016/j.jhydrol.2015.09.074

Al-Faraj FAM, Scholz M (2014) Assessment of temporal hydrologic anomalies coupled with drought impact for a transboundary river flow regime: the Diyala watershed case study. J Hydrol 517:64–73

Al-Faraj FAM, Scholz M (2015) Impact of upstream anthropogenic river regulation on downstream water availability in transboundary river watersheds. Int J Water Res Dev 31 (1):28–49. doi:10.1080/07900627.2014.924395

Al-Faraj FAM, Tigkas D, Scholz M (2015) Sensitivity of irrigation requirements to improvement in irrigation system performance efficiency. Example of transboundary watershed. In: The 9th World Congress of the European Water Resources Association (EWRA), Istanbul, Turkey between 10 and 13 June 2015

Intergovernmental Panel on Climate Change. Climate Change (2014) Impacts, adaptation, and vulnerability. Available online: http://www.ipcc.ch/report/ar5/wg2. Accessed 25 Oct 2015

Harza Engineering Company and Binnie & Partners (1959) Hydrological survey of Iraq; discharges for selected gauging stations in Iraq, Report. Harza Engineering Company, Chicago, IL; Binnie & Partners, London

Harza Engineering Company and Binnie & Partners (1963) Hydrological survey of Iraq; main report. Harza Engineering Company, Chicago, IL; Binnie & Partners, London

Harza Engineering Company and Binnie & Partners (2014) Hydrological survey of Iraq; discharges for selected gauging stations in Iraq, report. Harza Engineering Company, Chicago, IL; Binnie & Partners, London, Water 2014, 6 3048

Ministry of Agriculture and Water Resources-Kurdistan Regional Government (MA and WR-KRG) (2013) Various official data sets, general directorate of dams and reservoirs and directorate for operation of Derbandikhan; MA and WR-KRG: Erbil and Sulaymaniyah, Iraq

Richter BD, Baumgartner JV, Powel J, Braun D (1996) A method for assessing hydrologic alteration within ecosystems. Conserv Biol 10:1163–1174

Richter BD, Baumgartner JV, Wigington R, Braun DP (1997) How much water does a river need? Freshw Biol 37(1):231–249

Richter BD, Baumgartner JV, Powel J, Braun D (1998) A spatial assessment of hydrologic alteration within a river network. Regul Rivers Res Manage 14:329–340

Shiau JT, Wu FC (2004a) Assessment of hydrologic alterations caused by chi-chi diversion Weir in Chou-Shui Creek, Taiwan: opportunities for restoring natural flow conditions. River Res Appl 20(4):401–412

Shiau JT, Wu FC (2004b) Feasible diversion and instream flow release using range of variability approach. J Water Resour Plan Manage 130(5):395–404

USGS (2010) Stream gage descriptions and streamflow statistics for sites in the Tigris River and Euphrates River Basins, Iraq. A technical report prepared in cooperation with MoWR and MoAg&WR-KRG under the auspices of the U.S. Department of Defence, Task Force for Business and Stability Operations

Zuo Q, Liang S (2015) Effects of dams on river flow regime based on IHA/RVA. Remote sensing and GIS for hydrology and water resources (IAHS Publ. 368, 2015). In: Proceedings RSHS14 and ICGRHWE14, Guangzhou, China, August 2014. doi:10.5194/piahs-368-275-2015. Available online http://www.proc-iahs.net/368/275/2015/piahs-368-275-2015.pdf. Accessed 30 Nov 2015

Scenarios Based Climate Projection for Oman Water Resources

Sultan Al-Yahyai, Yassine Charabi, Said Al-Sarmi and Juma Al-Maskari

Abstract Oman is characterized by its arid climate. This is natural, in view of its location along the Tropic of Cancer, in a zone dominated by the subsident limbs of the Northern Hemisphere Hadley Cell circulation. However, the coastal position of Oman at the south-eastern end of the Arabian Peninsula, plus its high mountain range (Hajar Mountains) in the north both help to induce quite significant annual mean rainfall in a few favored areas. There are several factors that determine the overall impact of the climate change. Precipitation will be one of the most critical factors. Compared to air temperature, precipitation is difficult parameter to predict. For every degree centigrade, global scale researches showed that water vapor will increases by 7%. Based on these researches, it is less clear how this will that will impact the total change on precipitation. On the other hand, an increase of about 1–2% on the total volume of precipitation is likely to happen for each degree of warming. The main objective of this paper is to investigate the effect of climate change on the water resources and rainfall in Oman. It presents climate projection based on different Representative Concentration Pathways (RCPs) namely RCP2.6, RCP4.5, RCP6.0 and RCP8.5. The climate projection was based on Hadley Centre Global Environment Model version 2 (HadGEM2) downscaled to 1 km resolution. It was projected that slight dryness will take place over Hajar Mountains based on RCP 2.6 during 2041–2060. The effect of the dryness is projected to cover wider

S. Al-Yahyai (✉)
Information Technology Department, Mazoon Electricity Company, 1229, Muscat 131, Muscat, Oman
e-mail: s.alyahyai@gmail.com

Y. Charabi
Department of Geography, Sultan Qaboos University, Muscat, Oman
e-mail: yassine@squ.edu.om

S. Al-Sarmi · J. Al-Maskari
Directorate General of Meteorology, Public Authority for Civil Aviation, Muscat, Oman
e-mail: s.alsarmi@met.gov.om

J. Al-Maskari
e-mail: j.almaskari@met.gov.om

O. Abdalla et al. (eds.), *Water Resources in Arid Areas: The Way Forward*, Springer Water, DOI 10.1007/978-3-319-51856-5_3

areas based on the other RCPs scenarios except Dhofar region. RCP 8.5 projected wetter climate over Dhofar region during the period of 2061–2080.

Keywords Oman · Climate projection · Rainfall · RCPs · HadGEM2

1 Introduction

The Sultanate is characterized by arid climate (Charabi 2009). This is natural, in view of its location along the Tropic of Cancer, in a zone dominated by the subsident limbs of the Northern Hemisphere Hadley Cell circulation. However, the coastal position of Oman at the south-eastern end of the Arabian Peninsula, plus its high mountain range (Hajar Mountains) in the north both help to induce quite significant annual mean rainfall in a few favored areas (Al-Maskari 2006). Refer to Fig. 1 for the elevation and location of the main regions.

Figure 2 shows the annual average precipitation during the period of (1950–2000). It can be seen that the highest rainfall amount is seen over the Hajar mountains of about 300 mm. the figure shows also the annual average amount of rainfall over the southern Oman due to the summer monsoon.

Most of Oman's water (almost 75%) is supplied by major desalination plants connected to transmission systems to supply water to the main areas. About 20% of the water comes from wells—some in major wellfields connected to the main systems and others, typically single wells or small wellfields, supplying isolated systems mainly in rural areas. The remaining water supplies are from small desalination plants (about 5%) generally supplying small local distribution networks and tanker filling stations. Figure 3 shows the water production by source during 2014.

Water distributed across the national water systems increased by almost 14 million m^3 in 2014 compared to the previous years representing an increase of about 8% as shown in Fig. 4.

Climate change in Oman has been recently investigated on several studies. Choudri et al. (2013) investigated the effect of the Climate change on the farming activates over Al-Suwayq Wilayat. They illustrated the vulnerability and adaptation experiences of farmers to the climate change. Local people reported a perception of a change on the sowing period from July to September. This change is believed to be caused the increase on the frequency of the drought and shortening of the rainfall period. Vulnerability assessment of climate change impacts on the water resources in Al Jabal Al Akhdar was presented on Al-Kalbani et al. (2014). The guidelines prepared by United Nations Environment Programme (UNEP) was used to conduct this assessment. It was concluded that the is a clear signal of high level of stress is experienced on the local water resource system.

The main objective of this paper is to investigate the effect of climate change on the water resources and rainfall in Oman. It presents climate projections based on

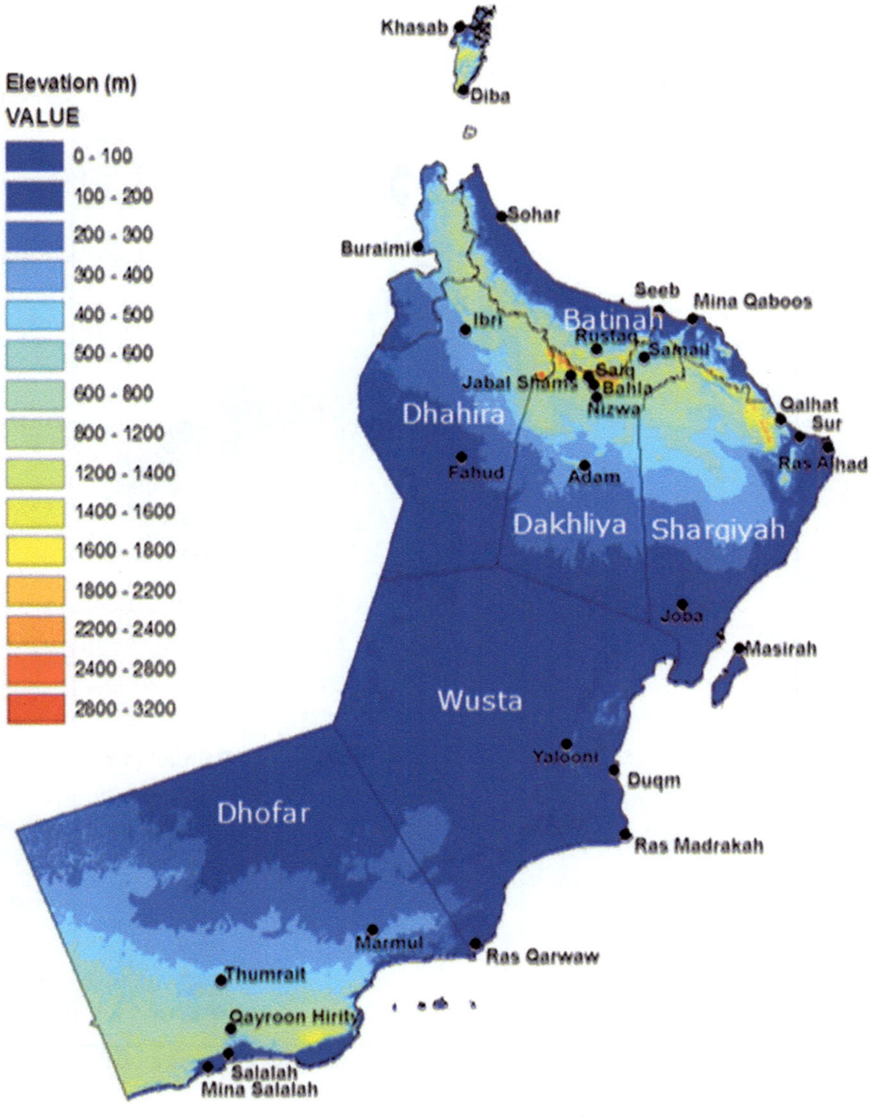

Fig. 1 Elevation on (m) and the names of the main regions

different Representative Concentration Pathways (RCPs) namely RCP2.6, RCP4.5, RCP6.0 and RCP8.5. The climate projection was based on Hadley Centre Global Environment Model version 2 (HadGEM2) downscaled to 1 km resolution. The results of this study can be used as an input for researches on the effect of climate change on fields like energy and agriculture. In addition, it can be used by decision makers to adopt strategies and develop mitigation plans based on different

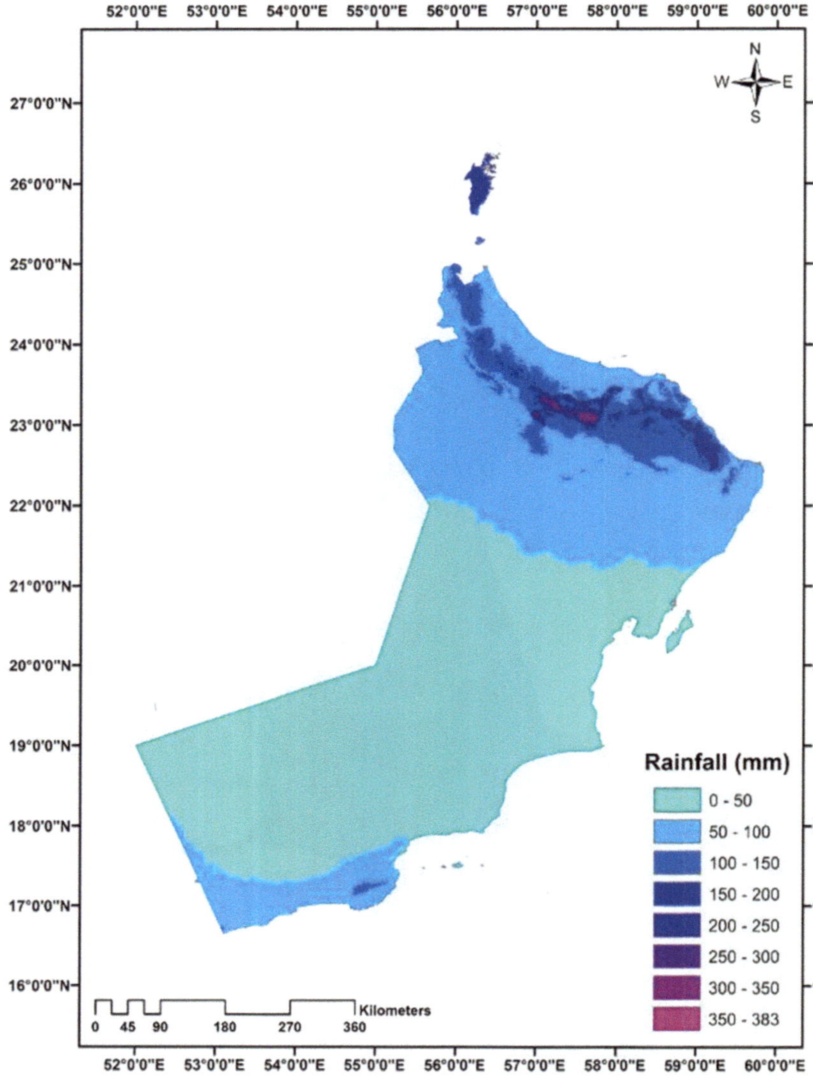

Fig. 2 Baseline annual average precipitation (1950–2000)

scenarios. The reset of the paper is organized as follow. Section 2 presents the main rainfall regimes and mechanisms in Oman. Section 3 presents the data and methodology used on this study. The main findings and results of the climate change production are presented on Sect. 4. Section 5 summarizes and concludes the paper.

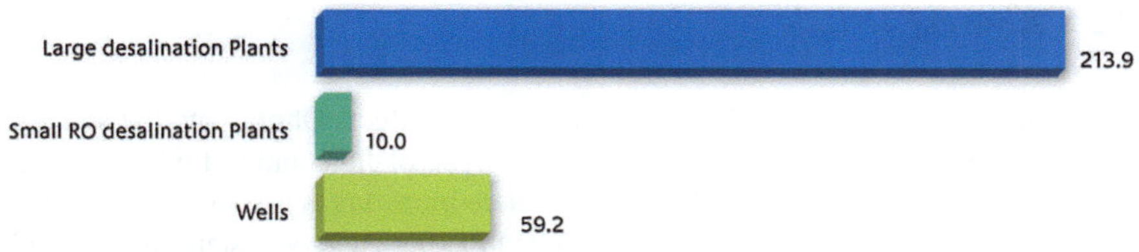

Fig. 3 Water production by source during 2014 (PAEW 2014)

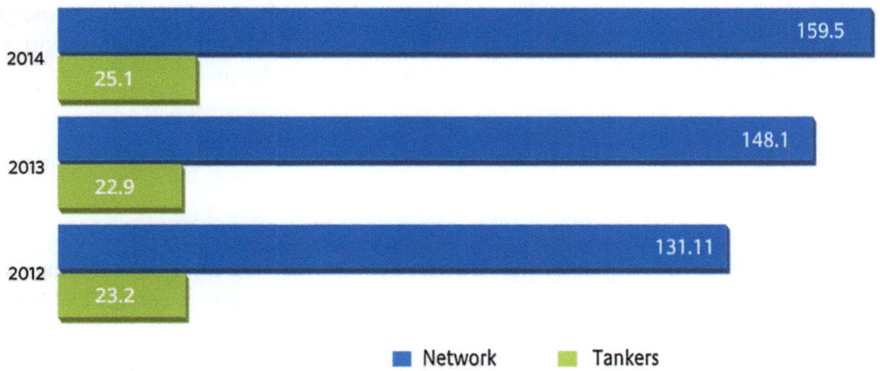

Fig. 4 Water distribution (Million m^3) during 2012–2014 (PAEW 2014)

2 Oman's Rainfall Regimes and Mechanisms

There are four mean mechanisms that bring rain to Oman.

2.1 Mobile Upper Troughs

Due to the strong zonal flow (every winter) in the Northern Hemisphere, mobile upper troughs or westerly depressions as they are sometimes called, starts over North Africa and the Mediterranean. Troughs or low-pressure systems are more active during December to February but occasionally extend to March (Al-Maskari 2006). On average there are four to five troughs a month, these troughs can bring widespread rain and flash floods especially over the Hajar Mountains. However, it has to be stressed that the these conditions are the exception. On many cases the upper trough is identified only by patchy cloud and then a wind shift associated by increasing surface pressure. Strong dust-rising wind (Shamal) might also developed in the northern part of Oman. Shamal wind may extend to reach Dhofar region.

2.2 *Khareef Drizzle*

India summer monsoon, locally known as Khareef, affects Dhofar and the southern coast of Oman during the period from the last week of June and to mid-September. Unlike India and other parts of the tropics, where monsoon is associated with heavy rain and thunderstorms, southern part of Oman experience persistent drizzle, light rain and rarely thunderstorm activity. Such activity turns the area green for over three month and attracts local and regional tourism.

Dhofar region and southern cost of Oman received less than 5 mm in a 24 h period during monsoon. According to climate data from the Oman Meteorology Department, the average rainfall during the monsoon period at Salalah airport is about 60 mm. Most of that is received during July and August.

2.3 *Tropical Cyclones*

Tropical cyclones develop over the Arabian Sea but they are rarely reach Oman at tropical cyclone intensity. It is very rarely that tropical cyclones extend to Oman Sea Sea (Al-Maskari 2010).

After the 1890 cyclone, tropical cyclone Gonu was the first destructive tropical cyclone to affect Muscat. There are two tropical cyclone season in the northern India ocean namely the pre-monsoonal period (May-June) and post-monsoonal period (October–November) (Membery 1985), Fig. 5 shows the number of tropical storms and cyclones affected Arabian Sea during 1801–2007 (Al-Maskari 2010).

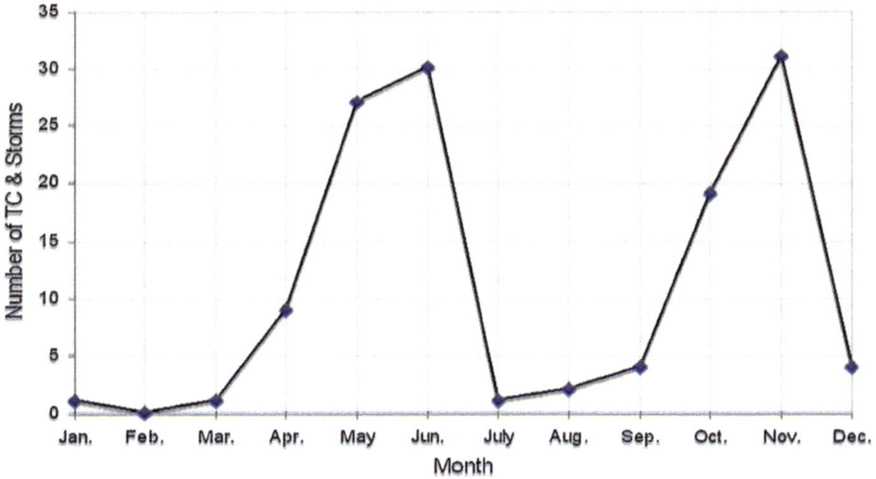

Fig. 5 Number of tropical storms and cyclones affecting the Arabian Sea, 1801–2007 (Al-Maskari 2010)

2.4 Orographic Convective Rain

During the summer months, a mixing zone, known as the Inter-Tropical Convergence Zone (ITCZ) is to be found along the Omani coast from Dhofar to Masirah Island. Usually the thermodynamics are such that the overriding influence of the tropical continental air from the Arabian interior is to inhibit the formation of any summer-time convection. An exception however has to be made for the Hajar Mountains where local sea and land breezes converge (Pedgley 1970), and strong isolation on elevated surfaces allow convection to break through the inversion.

According to Pedgley, the ITCZ may extend its position over Dakhliya region and the Hajar Mountains during summer months July and August. Therefore, a low-level, moist south-easterly wind from the Arabian Sea converges a relatively dry northerly/north-westerly wind from the Arabian Gulf (see for example, Fig. 2 in Pedgley (1970)). However, Al-Maskari (2010) showed that, the collision between the moist flow from the Arabian Sea and sea-breeze from Gulf of Oman trigger most convection over the Hajar Mountains. In fact, according to Al-Maskari (2010), dry northerly/northwseterlies act to suppress convection even if they flow comes approaches the mountains from the Arabian Gulf. Because of intense heating over the Hajar (rocky) Mountain surfaces, convective clouds form over the mountain peaks (the highest peak is about 10,000 ft). This results on thunderstorms over a relatively localized limited area of northern Oman. These thunderstones are classified by short duration, high intensity and are occasionally associated hail storms. Normally this activity occurs during the afternoon and early evening, with the cloud dissipating overnight. The Hajar Mountains plays the most important role in enhancing convection and precipitation over northern Oman. Al-Maskari (2010), used model simulations to show that without the Hajar Mountains summer precipitation would be significantly suppressed.

A climate projection describes the likelihood that a change will happen several decades to centuries in the future based on some conditions. The Intergovernmental Panel on Climate Change (IPCC) publishes assessment reports that summarized their findings based on the IPCC emission scenarios. The latest report is the 5th Assessment Report (AR5) published in 2014. AR5 provides up to date information about the latest research and knowledge related to climate change.

The future concentration of greenhouse gases and other pollutants is a key information to study the impact of climate change in future. Natural and man-made sources of emissions plays grate role on the expected concentration of those gases.

The Representative Concentration Pathways (**RCP**) are generated from different scenarios (Eyring et al. 2009). RCPs consists of projected scenarios on the change of the radiative forcing caused by the Atmospheric composition. This change on between the incoming radiation and outgoing one is used as an input to the climate modelling. Four RCPs were identified and named according to their total radiative forcing in 2100 namely RCP2.6, RCP4.5, RCP6.0 and RCP8.5 (IPCC 2007). Climate modelers at Global Circulation Models (GCM) used the four RCPs to conduct to develop the scenarios for the IPCC's Fifth Assessment Report.

The output of the GCM will then be used as boundary conditions for higher resolution Regional Climate Models (RCM).

3 Data Set and Methodology

Climate projection simulation requires an expensive computational power which is not available on many meteorological services. Global climate centers develop and maintain a global model simulation data.

Global Climate Models (GCMs) are used to generate the future and past climate. Different GCMs might give different simulation results. Based on the concentration of the GHGs, the simulated climate change will differ from one GCM to another. The projected GHG concentration future is normally described by one emission scenario. Therefore, the projected climate depend on the GSM and its associated emission scenario. GCM consists of complex numerical model equations that requires super computers to be solved. Due to the nature of the model equations, the world is divided into different grid cells with causer resolution (2°–3°). Because of the GCM's model resolution, then cannot resolve important small scale phenomena. To address this problem and based on the availability of computational power, global model simulations are used to derive regional scale climate model. Regional Climate Models are able to simulate high resolution phenomena over small spatial and temporal scales. This method requires a huge computational power which is not available at most of the regional centers (Hessami 2004).

To overcome the issue of computational power, downscaling is used. Model downscaling is implemented in a number of different ways. Some approaches use the weather truth observation to relate the larger-scale climate parameters such as atmospheric pressure at 800 hpa to the local surface parameter such as rainfall. The results of this relationship is then used to tune GCM output. This tuning method is based on the assumption that that GCMs perform better on predicting larger-scale parameters. In addition, it is also assumed that the relationships between the large scale and surface parameters remain valid in a changed climate prediction.

The data developed for this work was generated through calibration and downscaling of the second version of Hadley Centre Global Environment Model (HadGEM2). Different model configuration with the same model physics are used to develop the HadGEM2 data sets. HadGEM2 is atmosphere-ocean coupled model (Haywood et al. 2016). The model extends its vertical layers (38 layers) to resolve the stratosphere (up to 40 km). The model is also using dynamic vegetation scheme, ocean biology model and atmospheric chemistry configuration (Bellouin et al. 2011).

In the horizontal direction, the model is running at 1.25° × 1.875° resolution (Collins et al. 2011). Surface and aircraft emissions of ozone are used to simulated tropospheric ozone precursors and reactive gases (Haywood et al. 2016). Time varying change in total solar irradiance is used to represent climate forcing (Jones et al. 2011).

HadGEM2 data sets were used in the IPCC FA5 report (Collins et al. 2014).

HadGEM2 is used to generate the current climate and the projected climate based on different RCPs scenarios. The quality of GCMs prediction vary from one place to another. Therefore, the relation between the current climate and future prediction can't be applied equally in all places. Therefore, observations from all over the world are used to calculate the differences and then use these differences to calibrate the current climate and future projection. In the next step, the calibrated current and future climate are downscaled and interpolated to a grid at high resolution of about 1 km. This step is based on the assumption that the change in climate is relatively stable over space. Figure 6 describe the methodology in graphical representation. The following sections will presents the climate change projection based on different RCP's scenarios. Climate change results are presented for the period 2041–2060 and 2061–2080 for rainfall.

In this work, HadGEM2 was used as GCM to provide the required RCPs scenarios. In addition, it has been used to derive the current global model that will be used to calculate the climate change for each scenario. As HadGEM2 is a global model, weather observation from the Global Telecommunication Network GTS is used to calculate the anomaly and then apply bias correction and calibration techniques to improve the quality of the model. Then the corrected model simulation (current and RCPs scenarios) are downscaled and interpolated to 1 km resolution. Finally, Oman domain is extracted and manipulated to calculate the climate change values for different weather parameters.

Andy et al. (2009) presented an analysis of 27-year rainfall data in Oman. Table 1 shows comparison between the 27-years rainfall analysis and HadGEM2 model reanalysis (Fig. 1).

The table shows a spatial agreement between the model reanalysis and the observed rainfall analysis.

4 Results and Discussion

4.1 Rainfall Change (2041–2060)

Figure 7 shows the changes on rainfall during the period 2041–2060 based on RCP 2.6 (top left), RCP 4.5 (top right), RCP6.0 (bottom left) and RCP 8.5 (bottom right). Positive values imply that the projected period is wetter than the baseline period (1950–2000). On the other hand, negative values imply that the projected period dryer than the baseline period. RCP 2.6 projected a slight dryness (10 mm) over Hajar Mountains and nearby region and no change is projected for the other regions on the country. Slight (10–20 mm) wetter climate is projected over north of Musandam.

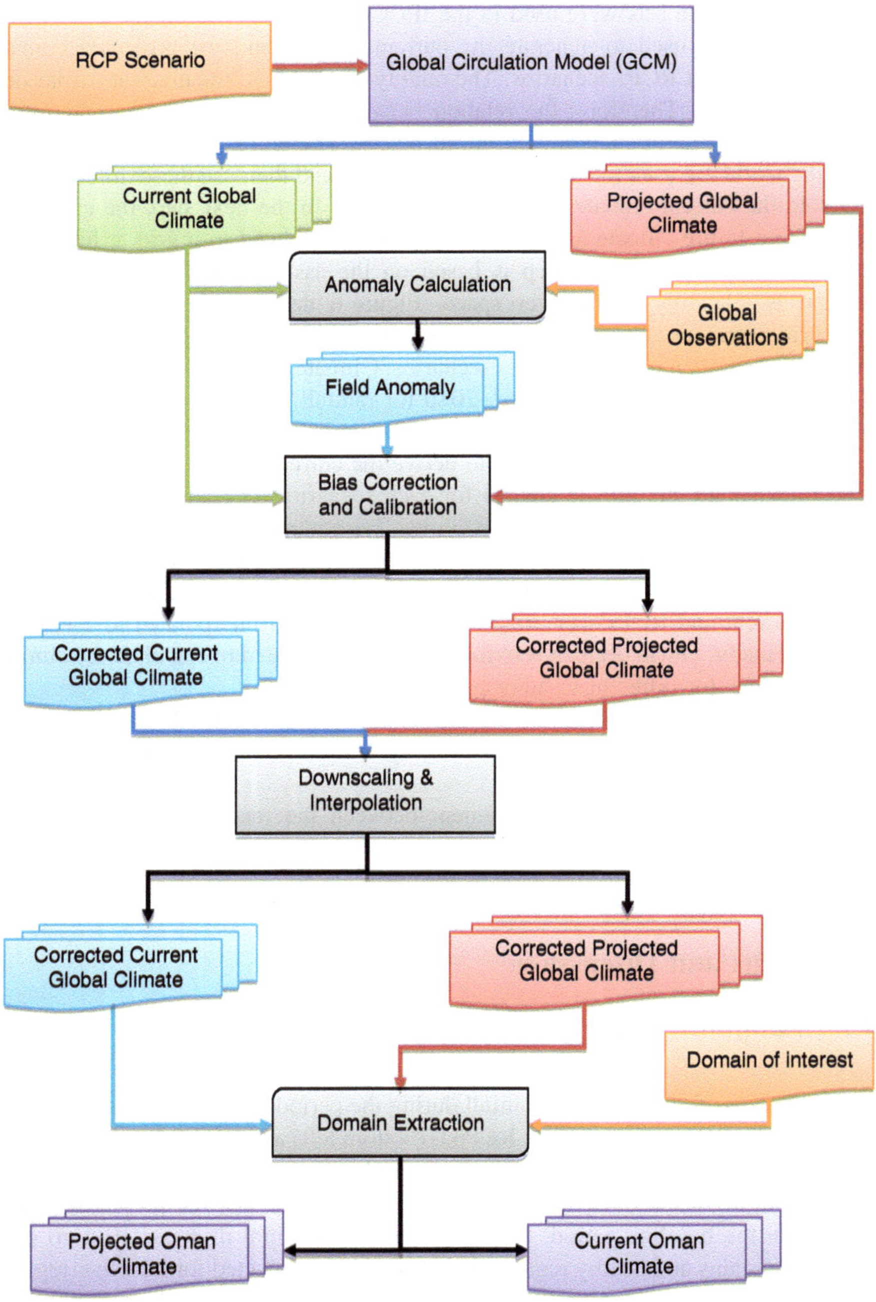

Fig. 6 Downscaling methodology

Table 1 Spatial agreement between the model reanalysis and the observed rainfall analysis

City	27-years rainfall analysis (Andy et al. 2009) (mm)	HadGEM2 model reanalysis (This study), range (mm)
Ibri	90	50–100
Salalah	112	100–150
Masirah	70	0–50
Buraimi	65	50–11
Saiq	330	300–350
Samail	84	100–150
Seeb	80	50–100
Ibra	122	100–150
Sur	79	50–100

Based on RCP 4.5, wider dryness (10–20 mm) is projected over most of the country. Based on this scenario, wetter climate is projected over Musandam and no change from the baseline is projected over Dhofar region. It is worth to notice that intensified (more than 40 mm) increase of rainfall is projected over Musandam based on RCP 6.0 scenario. North coastal areas and Dhofar regions are projected to have no change on the rainfall amount. Unlike all other RCP's scenarios, RCP 8.5 is the only scenario which projected dryness over Musandam for the period 2041–2060. RCP 6.0 and RCP 8.5 are comparable for the other regions over the country.

4.2 *Rainfall Change (2061–2080)*

Figure 8 shows the change on the rainfall amount during 2061–2080 based on RCP 2.6 (top left), RCP 4.5 (top right), RCP6.0 (bottom left) and RCP 8.5 (bottom right). During the period 2061–2080, RCP 2.6 and RCP 8.5 projected wetter climate over Musandam while RCP 4.5 and RCP 6.0 projected dryer climate compared to 2041–2060. Over Hajar Mountains, it was only RCP 2.6 which projected wetter period while other scenarios projected no change to take place during 2061–2080 compared to 2041–2060. Finally, all scenarios projected wetter climate over Dhofar region during the period 2061–2080 compared to 2041–2060. Figure 9 shows the differences on rainfall amount between the period 2061–2080 and 2041–2060 based on RCP 2.6 (top left), RCP 4.5 (top right), RCP6.0 (bottom left) and RCP 8.5 (bottom right).

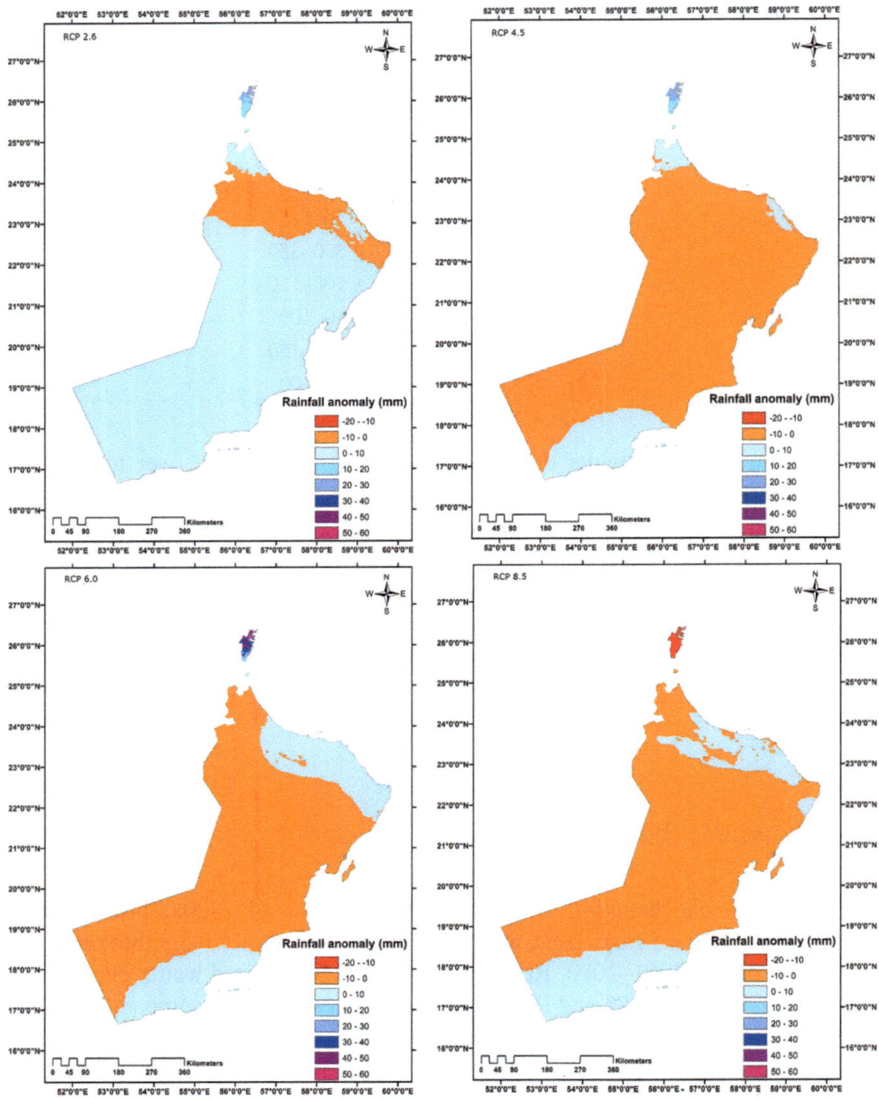

Fig. 7 Rainfall change (2041–2060) based on RCP 2.6 (*top left*), RCP 4.5 (*top right*), RCP 6.0 (*bottom left*) and RCP 8.5 (*bottom right*)

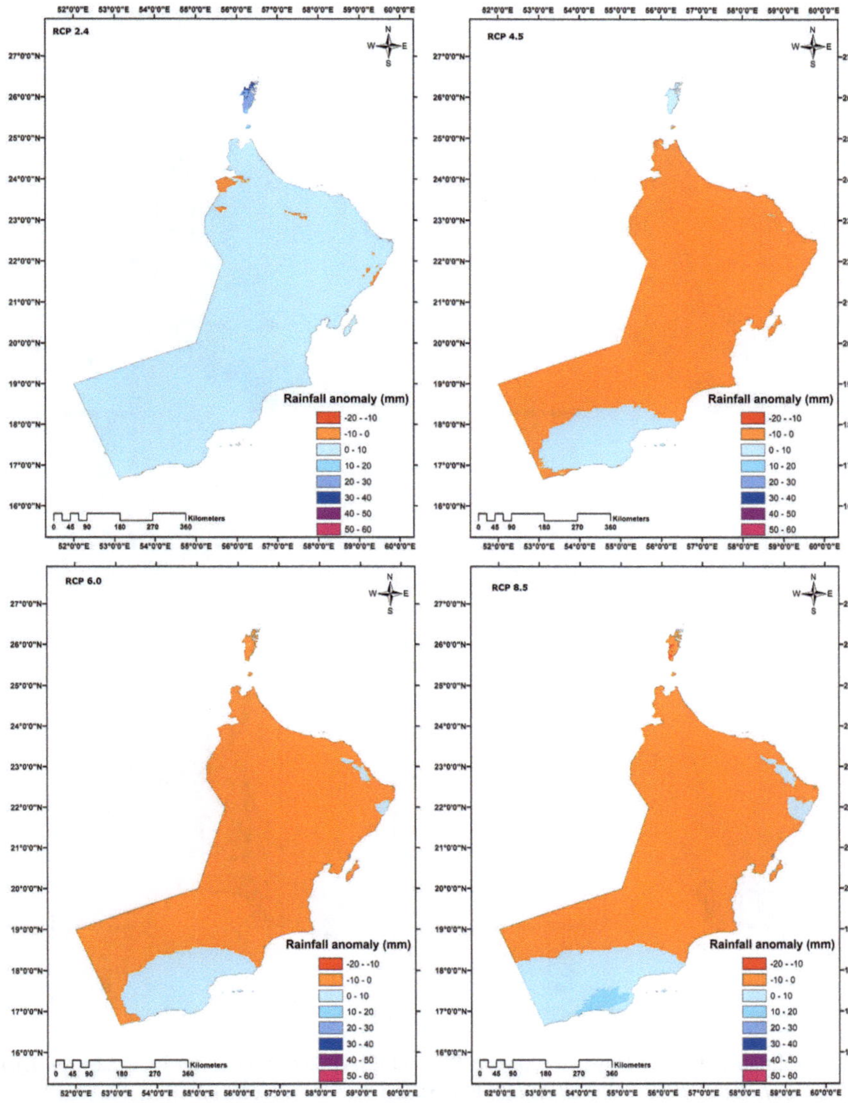

Fig. 8 Rainfall change (2061–2080) based on RCP 2.6 (*top left*), RCP 4.5 (*top right*), RCP 6.0 (*bottom left*) and RCP 8.5 (*bottom right*)

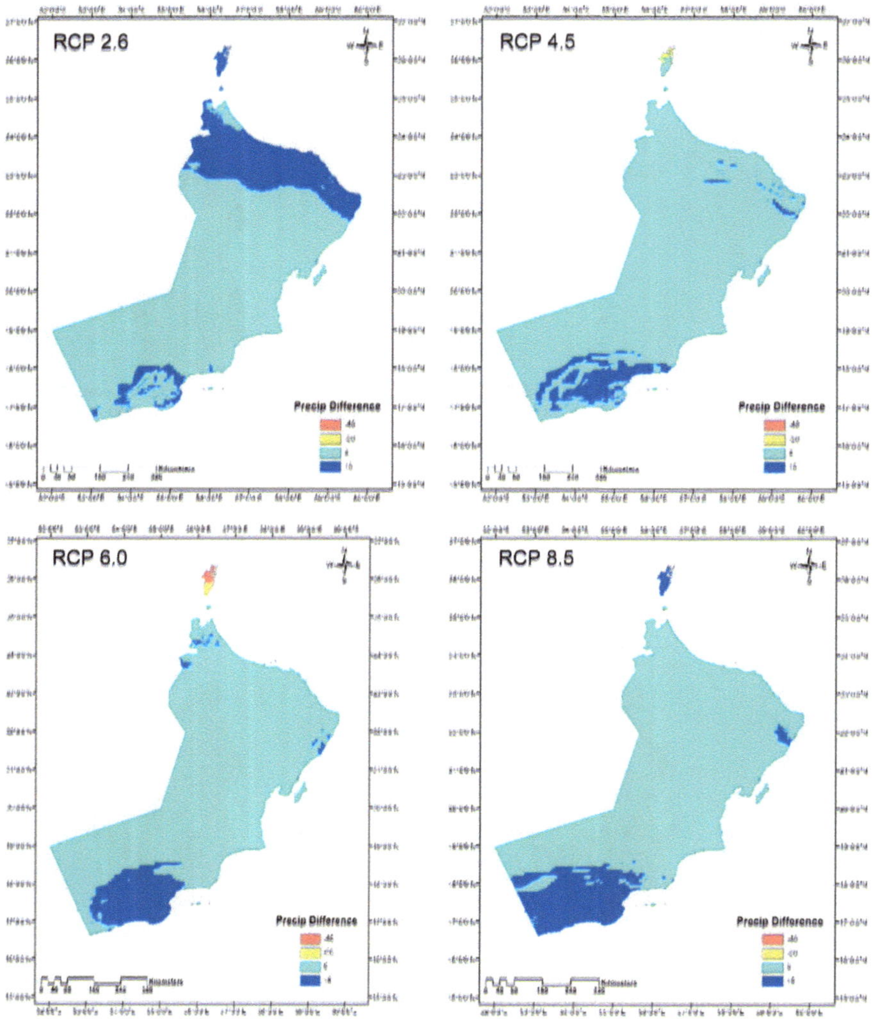

Fig. 9 Rainfall difference between (2061–2080) and (2041–2060) based on RCP 2.6 (*top left*), RCP 4.5 (*top right*), RCP 6.0 (*bottom left*) and RCP 8.5 (*bottom right*)

5 Conclusion

The main objective of this study was to investigate the effect of climate change on the water resources and rainfall in Oman. It presents climate projection based on different Representative Concentration Pathways (RCPs) namely RCP2.6, RCP4.5, RCP6.0 and RCP8.5. The climate projection was based on Hadley Centre Global Environment Model version 2 (HadGEM2) downscaled to 1 km resolution. It was concluded that Oman will be affected by climate change in its rainfall and water resources. It was projected that slight dryness will take place over Hajr Mountains based on RCP 2.6. The effect of the dryness is projected to cover wider areas based on the other RCPs scenarios except Dhofar region. RCP 8.5 projected wetter climate over Dhofar region during the period of 2061–2080.

The result of this study can be used as an input for researches on the effect of climate change on fields like energy and agriculture. In addition, it can be used by decision makers to adopt strategies and develop mitigation plans based on different scenarios.

References

Al-Kalbani M, Price M, Abahussain A, Ahmed M, O'Higgins T (2014) Vulnerability assessment of environmental and climate change impacts on water resources in Al Jabal Al Akhdar, Sultanate of Oman. Water 6(10):3118–3135

Al-Maskari J (2006) Ph.D. thesis: processes of convection and airflow over the Hajar Mountains. Leeds University, Leeds, UK

Al-Maskari J (2010) How the National Forecasting Centre in Oman dealt with tropical cyclone Gonu. Indian Ocean tropical cyclones and climate change, Netherlands

Andy K, Atsu D, Ganiga T (2009) Analysis of a 27-year rainfall data (1977–2003) in the Sultanate of Oman. Int J Climatol 29(1):605–617

Bellouin N, Rae J, Jones A, Johnson C, Haywood JM, Boucher O (2011) Aerosol forcing in the CMIP5 simulations by HadGEM2-ES and the role of ammonium nitrate. J Geophys Res 116: D20206. doi:10.1029/2011JD016074

Charabi Y (2009) Summer monsoon variability: teleconexion to ENSO and IOD. Atmos Res 91(2): 105–117

Choudri B, Al-busaidi A, Ahmed M (2013) Climate change, vulnerability and adaptation experiences of farmers in Al-Suwayq Wilayat, Sultanate of Oman. Int J Clim Change Strat Manage 5(4):445–454

Collins W, Bellouin N, Doutriaux-Boucher M, Gedney N, Halloranl P, Hinton T, Jones C, Hughes J, Jones C, Joshi M, Liddicoat S, Martin G, Connor F, Rae J, Senior C, Sitch S, Totterdell I, Wiltshire A, Woodward S (2011) Development and evaluation of an earth-system model—HadGEM2. Geosci Model Dev 4(1):1051–1075

Collins W, Bellouin N, Doutriaux-Boucher M, Gedney N, Hinton T, Jones C, Liddicoat S, Martin G, Connor F, Rae J, Senior C, Totterdell I, Woodward S, Reichler T, Kim J (2014) Evaluation of the HadGEM2 model. Met Office Hadley Centre Technical Note no. HCTN 74, available from Met Office, FitzRoy Road, Exeter EX1 3 PB http://www.metoffice.gov.uk/publications/HCTN/index.html

Eyring V, Isaksen I, Berntsen T, Collins, Corbett J, Endresen O, Grainger R, Moldanova J, Schlager H, Stevenson D (2009) Transport impacts on atmosphere and climate Shipping. Atmos Environ doi:10.1016/j.atmosenv.2009.04.059

Haywood J, Andy J, Nick D, Sean M, Michael V, Alejandro B, Matt H, Ben K, Jason C, Shingo W, Graeme S (2016) The impact of equilibrating hemispheric albedos on tropical performance in the HadGEM2-ES coupled climate model. Geophys Res Lett 43(1):395–403

Hessami M, Quarda T., Gachon P, StHailaire A, Selva F, Bobee B (2004) Evaluation of statistical downscaling method over several regions of eastern Canada. In: 57th Canadian water resources association annual congress, pp 150–155

IPCC (2007) Towards new scenarios for analysis of emissions, climate change, impacts, and response strategies. IPCC expert meeting report

Jones C, Hughes J, Bellouin N, Hardiman S, Jones G, Knight J, Liddicoat S, O'Connor F, Andres R, Bell C, Boo K, Bozzo A, Butchart N, Cadule P, Corbin K, Doutriaux-Boucher M, Friedlingstein P, Gornall J, Gray L, Halloran P, Hurtt G, Ingram W, Lamarque J, Law R, Meinshausen M, Osprey S, Palin E, Parsons L, Raddatz T, Sanderson M, Sellar A, Schurer A, Valdes P, Wood N, Woodward S, Yoshioka M, Zerroukat M (2011) The HadGEM2-ES implementation of CMIP5 centennial simulations. Geosci Model Dev 4(1):543–570

Membery D (1985) A unique August cyclonic storm crosses Arabia. Weather 40(1):108–115

Pedgley D (1970) The climate of interior Oman. Meteorol Mag 99(1171):29–37

Public authority for Electricity and Water PAEW (2014) Annual report 2014, pp 27–35

Study of Rainfall Variations in Tessa Subwatershed of Medjerda River in Tunisia

Sahar Abidi, Olfa Hajji and Hamadi Habaieb

Abstract The spatial and temporal study of rainfall variation is very important for efficient management of water resource. This subject is discussed in Tessa watershed, a subcatchement of Medjerda the only perennial River in Tunisia. 13 pluviometric gauge stations have been chosen to collect data from 1984 to 2014. Three scales were considered; monthly, seasonal and annual. Gap in data were reconstructed by the mean ponderation method. We tested the homogeneity of the series by the Pettitt, Buiscand and SNHT test via XlStat software. The ruptures are detected in the months of June, July, August and October, in the season of autumn and spring of the years 1994–1995 and 2000–2001. In order to create a spatial distribution of rainfall patterns, two interpolations methods were used; isohyet and Thiessen polygons via the software ArcGIS. In fact, the analysis results illustrate an alternation of wet and dry episodes with very marked excess rainfall, a gradual increase in rainfall. These rainfall fluctuations have contributed to weakening the natural environment. The results show the rainfall pattern of the watershed is a temperate continental regime.

Keywords Rainfall variability · Tessa watershed · Homogenous tests · Spatial interpolation · GIS

1 Introduction

The importance of the River basins is hilighted in water uses which is collected from runoff. In fact water is used for fresh drinking water, irrigation, hydropower generation, etc. Face to climate change in the world, Tunisia is at risk of drought in

S. Abidi (✉)
Sylvo-Pastoral Resources Laboratory, Sylvo-Pastoral Institute of Tabarka, Jendouba, Tunisia
e-mail: sahar.abidi@yahoo.fr

O. Hajji · H. Habaieb
Agronomy National Institute of Tunisia, Tunis, Tunisia

O. Abdalla et al. (eds.), *Water Resources in Arid Areas: The Way Forward*,
Springer Water, DOI 10.1007/978-3-319-51856-5_4

this several years. Given that Tunisia resources is 83% from Medjerda River, the conserving of water is becoming an important subject that the government is concentrated on. For many expert in the world, conserving water is made by managing river basins sustainably so that building a strategy of available resources use, is needed. Medjerda basin contain several tributaries which some of theme are controlled by hydraulic managements.

Based on the recent reports of climate change and its extreme variability, an understanding of temporal and spatial characteristics of rainfall is crucial to plane a management of water resources. As In Tunisia, the economy is based essentially on the agriculture. Such the problem of drought, Tunisia experienced also the flooding in many past years. So that such study is important, as Michaelides et al. (2009) express, in agricultural planning, flood frequency analysis, flood hazard mapping, hydrological modeling, water resource assessments and other environmental assessments. For Tunisia the rainfall concern in major the precipitation. So that in this paper we focus on studying spatial and temporal variation of precipitation.

Many researchers used the Geographical Information System (GIS) to analyze and mapping the spatial distribution and the trend of rainfall. DE (1987) define the GIS as a system for treating the data. It produces useful data and displays tabular and spatial information to assist decision-makers (Kaijuka 2007). Fiedler (2003) demonstrated how Thiessen polygons and isohyets maps can be utilized to determine station weights for hydrologic modeling.

This study utilized a case study approach on Tessa catchment covering 2374.64 km^2. The overall goal of this paper concerned approving the suitability of GIS as a tool for rainfall analysis to plane a strategy for water management. The data was collected at 13 stations across the study area. The study period elongated from September 1984 to August 2014.

2 Study Area

The Wadi Tessa crossed the northwestern part of Tunisia and is located in the south of Upper Tell region at the right bank of the Medjerda (Fig. 1). It is the main Tunisian river, originates in the semi-arid Atlas Mountains of eastern Algeria. Tessa watershed begins in the high plains of the Atlas Tunisian, Makthar delegations (Siliana governorate), El Ksour and Dehmani (Kef governorate) and then flows to the North from the delegation of Sidi Bou Rouis up at its confluence with Oued Medjerda to 124 m altitude. Tessa is localized between latitudes 36° 16′–36° 36′ North and meridians 8° 52′ West and 9° 04′ East.

The climate of the study area is Mediterranean throughout the year, and the rainfall is irregular; the average annual rainfall of the basin varies from 85 to 771 mm. The rainy season are winter and spring. The rainy month is March.

In Tunisia, the flood problem arise the only perennial river, Medjerda, in particularly the plain of Bou Salem where meets three tributaries which Tessa is among its. An hydrometric study of Tessa Wadi (ISL 2012), reveals that three largest

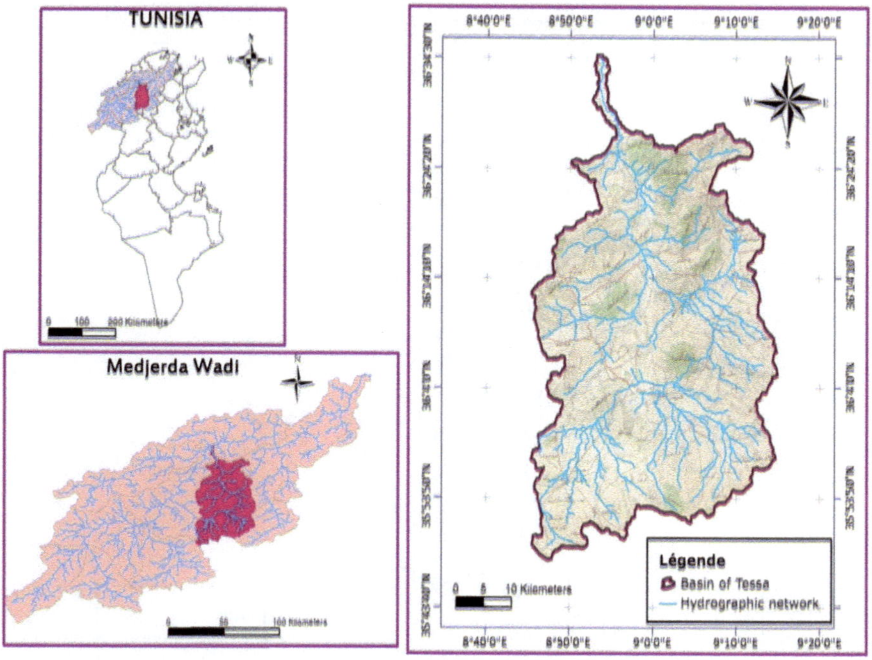

Fig. 1 The location of the study area

floods occurred in September and the registered flow rates are high for such a catchment area.

3 Methodology

The methodology has been divided into three major parts as data collection, data processing, and results. The analysis of data is redressed in three scales: monthly, seasonal and annual.

3.1 Data Collection

13 stations of rainfall data has been collected between 1984 and 2014 from Water resources Direction. These stations are distributed irregularly throughout the study area and represent the longest period available for the highest quality data covering the area. Figure 2 shows the spatial distribution of rainfall stations. Seasonal and annual rainfall data series of Tessa basin were computed using monthly rainfall data of rain gauge stations.

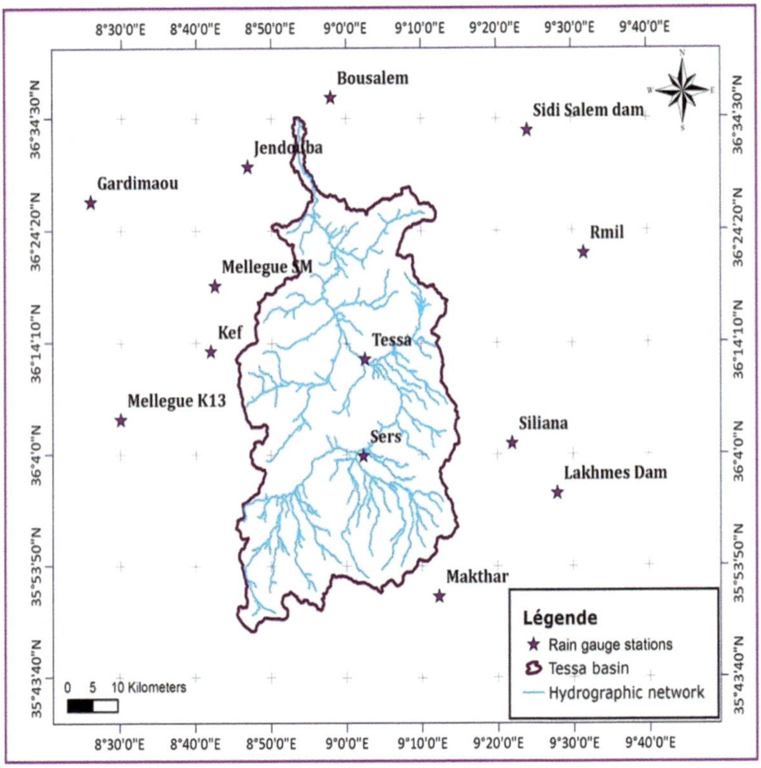

Fig. 2 Rain gauge stations location in Tessa watershed

3.2 Detection of Ruptures Within the Rainfall Series

To detect breaks (ruptures) in rainfall patterns, the Pettitt (1979), Buishand (1982) and SNHT (1986) tests were applied to the rainfall in the basin. The Pettitt-test is a non-parametric test based on the Mann–Whitney test. The existence of abrupt changes of some statistical parameters of hydrological series, especially their average is a possible cause of discontinuity in the series. This test allows an optimal partition of a time series into as many subsets as possible. The Buishand's test is a parametric test. It is applied for chronic with at least 10 values and whose distribution is normal. This test assumes a non-change of the series variance. The SNHT test (Standard Normal Homogeneity Test) was developed by Alexandersson (1986) to detect a change in a series of rainfall data. The test is applied to a series of ratios that compare the observations of a measuring station with the average of several stations.

3.3 Index of Nicholson

Nicholson et al. (1988) defined an index which is calculated for each year and expresses it self as follows:

$$\mathrm{Ip} = \frac{Xi - \bar{X}}{\sigma}$$

Where Xi: Height of rain per year i (mm); $\overline{X}$: average of the period of study (mm); and σ: standard deviation of the rain over the period of study.

3.4 Computation of Average Rainfall

A Thiessen map was created for the 13 rain gauge stations using ArcMap 10.0, generating a network of 13 polygons associated to each station and considered as a sub-area (Schwab et al. 1995). The area for each polygon was calculated in order to determine the mean rainfall (monthly, seasonal and annual) in Tessa basin.

The second method to determine the areas of influence of the punctual rainfall is the isohyet. An isohyetal is a line joining places where the rainfall amounts are equal on a rainfall map of a basin (Punmia et al. 1995). An isohyetal map showing contours of equal rainfall is more accurate picture of the rainfall over the basin. The isoheytal map was created through interpolation techniques named Spline in Arc GIS 10.0 platform. Two Splines methods were used and compared: the Regularized and the Tension one. The Regularized Spline creates a smooth, gradually changing surface. The regularizing parameter is in fact employed to achieve a smoother solution: e.g. a small value results in a close approximation of the data, while a large one results in a smoother solution (Gousie and Franklin 2005). The Tension Spline creates a less smooth surface with values more constrained by the sample data range: changing the value of the tension parameter tunes the surface from a stiff plate into an elastic sheet (Mitas et al. 1997). Each method had a weight coefficients variables.

To choose the suitable method of Spline interpolation, we dressed the isohyet map of 1984 Rainfall by the two methods and with different weight coefficients. The Root mean square error (RMSE) was calculated to determine the optimum Spline method with optimum weight coefficients.

4 Results

The outcome of the study revealed various facts about the spatial-temporal variation of rainfall in the basin. Table 1 presents the characteristics of the 13 rainfall stations chosen for this research.

4.1 Detection of Ruptures

The results of the ruptures detection in the rainfall series are gathered in Table 2. The application of Pettit, Buishand and SNHT tests, detect rupture in 7 stations.

We note that the tests, at the monthly and annual scale, detect ruptures of the same series in the same years. Pettitt's and Buishand's tests give breaks for rainfall series of Kef, Jendouba and Bou Salem in all scales. The SNHT test, detect ruptures at the seasonal scale for Sers and at the annual scale for Bou Salem station.

We present an example of ruptures detection with the Pettit test at the annual scale of Kef station (Fig. 3).

Table 1 Characteristics of the rainfall station

No.	Name of station	Code[a]	Latitude (N)	Longitude (E)
1	B. Lakmes	1485076724	9.46	36.00
2	B. Sidi Salem	1485077221	9.40	36.56
3	Mellegue SM	1485167223	8.71	36.32
4	Gardimaou	1485286422	8.43	36.45
5	Kef	1485361223	8.70	36.23
6	Maktar	1485410224	9.20	35.85
7	Mellegue K13	1485499023	8.50	36.12
8	Sers	1485588723	9.03	36.06
9	Siliana	1485676424	9.36	36.08
10	Jendouba	1485699022	8.78	36.50
11	Tessa	1485764323	9.62	36.31
12	Bou Salem	1485701822	8.96	36.61
13	Oued Rmil	1485505324	9.53	36.37

[a]Inside each basin, the rainfall stations are numbered with compounds codifications of ten digits including:

- The first three digits indicate the continent (1 for Africa) and country (48 for Tunisia) belonging to the resort
- The next five digits indicate the mechanographical number of the station data (code included in this table) with the first digit is the number of basin to which belongs the station (5 to the basin of the Medjerda)

Table 2 Results of ruptures detection

Station	Monthly			Seasonal			Annual		
	Pettit	Buishand	SNHT	Pettitt	Buishand	SNHT	Pettitt	Buishand	SNHT
Kef	July 95	October 03	–	Spring 99	Spring 99	–	1995	2002	–
Gardimaou	–	August 95	–	–	–	–	1995	–	–
Jendouba	August 95	August 95	–	Autumn 95	Autumn 95	–	1995	1995	–
Bou Salem	June 02	June 02	–	Autumn 97	Autumn 97	–	2002	2002	2002
B. Lakmes	–	June 02	–	–	–	–	–	2002	–
Siliana	–	October 03	–	–	–	–	–	–	–
Sers	–	–	–	–	Autumn 97	Autumn 90	–	–	–

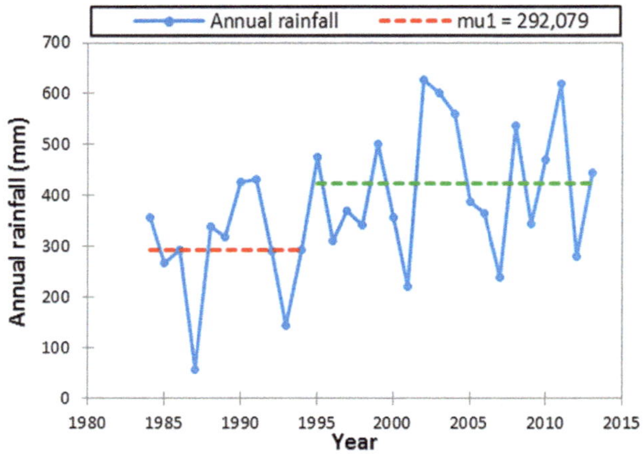

Fig. 3 Results of Pettit application to the annual rainfall of Kef

4.2 Index of Nicholson

Analysis of rainfall indexes allowed us to identify two distinct periods in all rainfall series: period of wet years and a second period of dry years. These results are summarized in Table 3.

The persistence of dry years is flushed in the period:

- From 1984 to 1998 for 54% of the stations,
- From 1984 to 1999 for 15% of the stations,
- From 1984 to 1996 or 1997 for the rest.

The majority had 14 years deficits and 16 wets years. The years 1998/1999 and mark what is commonly called climate disruption.

Table 3 Results of rainfall index

Station	Dry periods	Wet periods
Gardimaou	1985–2000	2001–2014
B. Sidi Salem	1985–1995	1996–2014
Kef	1985–1998	1999–2014
K 13	1985–1997	1998–2014
Oued Rmil	1985–1999	2000–2014
Siliana	1985–1999	2000–2014
Jendouba	1985–1999	2000–2014
Bou Salem	1985–1999	2000–2014
Tessa	1985–1999	2000–2014
B. Lakhmaș	1985–1999	2000–2014
Mellegue SM	1985–1999	2000–2014
Sers	1985–2000	2001–2014

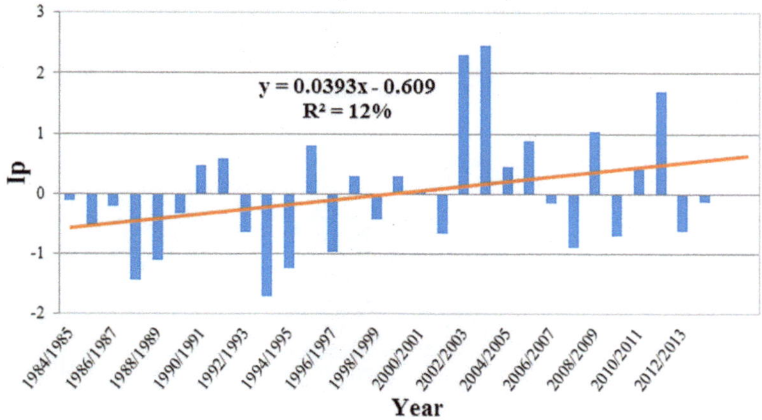

Fig. 4 Precipitation variation in Tessa station during 30 years

We present below the annual change in precipitation expressed in terms of the rainfall index in Tessa station (Fig. 4).

The visual analysis of the figure below shows an alternation of dry and wet periods of variable durations. We can detect in global two periods: a dry period from 1985 to 1999 and a wet period from 2000 to 2014.

In the first period, we note some wet years such in 1991, 1992 and 1996 that the index is low than 1. Since 2000, the increase in the rainfall is perceptible, but it does not arrive yet at the level of the surpluses and especially the wet interlude is of rather short duration. Year 2003 is made conspicuous by an important pluviometer and exceptional falls of snow.

4.3 Variation in the Average Rainfall

The variation of mean rainfall was studied for the basin of Tessa using the data of 13 stations during 30 years in three scales: seasonal, monthly and annual.

The Regularized Spline (RS) and Tension Spline (TS) methods were used to interpolated annual Rainfall of 1984 with different weight coefficients. The RMSE of each method is dressed in Table 4.

Using the RMSE, we concluded that the Regularized Spline method is optimum with weight coefficient 0.1. for the validation of this conclusion, we interpolate the annual rainfall of 1985 by the Regularized Spline. We find a low RMSE equal to

Table 4 The RMSE of regularized and tension spline with different weight coefficients

Method	RS(0.1)	RS(0.01)	RS(0.001)	RS(0.5)	TS(1)	TS(5)	TS(10)
RMSE	**0.003**	0.004	0.007	0.011	0.018	0.021	0.023

The regularized spline of 0.1 weight coefficient have the lower RMSE so it's the best method

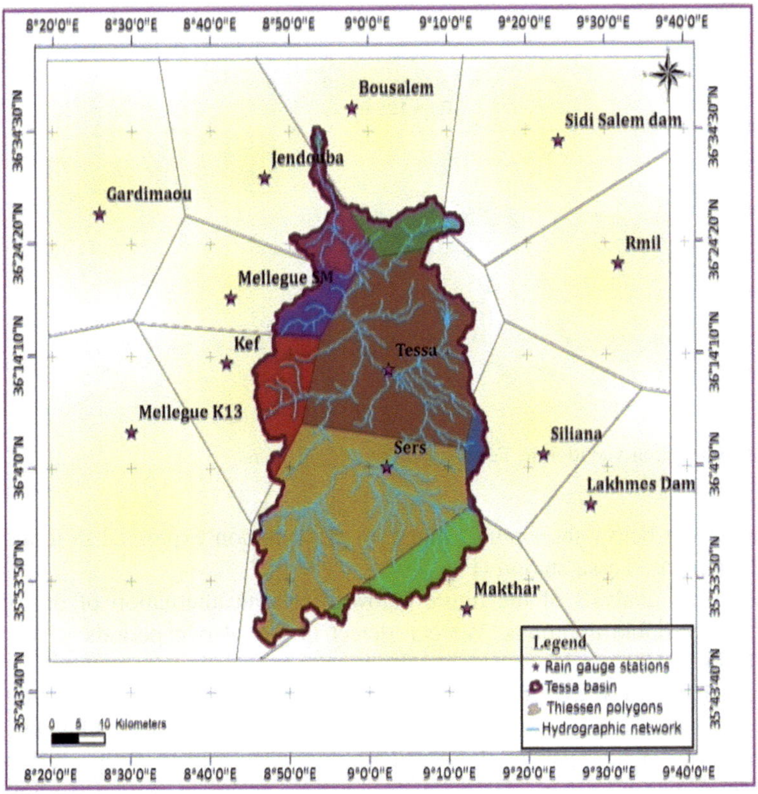

Fig. 5 Thiessen map

0.4%. So we used Thiessen polygon and Regularized Spline for the rainfall interpolation in this paper.

The area of influence generated by Thiessen polygon (Fig. 5) tool extended beyond the boundaries of the farm and therefore encompassed some of the adjoining lands of the study site.

The mean rainfall interpolate by the two methods is calculated by the equation below:

$$P_{av} = \frac{\sum_{i=1}^{n} A_i P_i}{A}$$

Where:

Variable	Spline method	Thiessen polygon
Pav (mm)	Mean rainfall in the basin	Mean rainfall in the basin
P_i (mm)	Average height h of precipitation between two isohyets i and i + 1	Recorded rainfall at the station i
A_i (ha)	Area between two isohyets i and i + 1	Polygon area
$A = \Sigma Ai$ (ha)	Basin area	Basin area
n	Total number of isohyets	Number of station

4.3.1 Average Rainfall in the Monthly Scale

The variation of the average rainfall monthly using Thiessen polygon are presented in the following graphics (Fig. 6):

The monthly average rainfall of Tessa basin, increases towards the month of January with value of 56 mm, the month of March with 48 mm and the month of September a with 41 mm. The month of July is the driest month.

Figure 7 present the isohyets map in the monthly scale using Regularized Spline interpolation.

The monthly average rainfall of Tessa extracted by spatial interpolation of isohyets is illustrated in Table 5:

We conclude that the wet months are January (55 mm) and March (45 mm) while the dry month is July (10 mm).

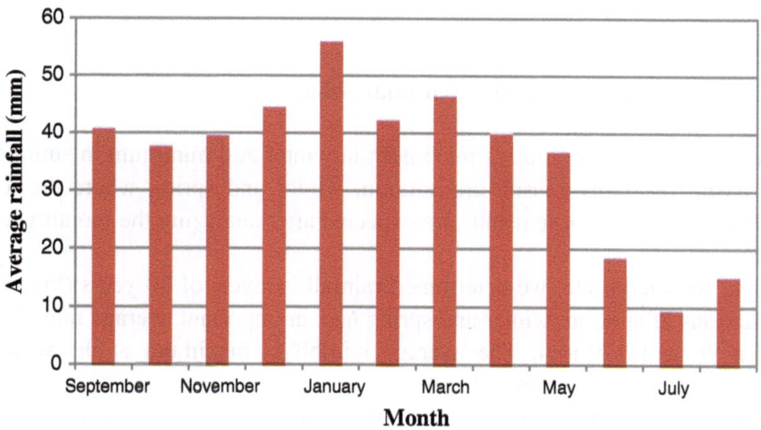

Fig. 6 Average rainfall in the monthly scale by Thiessen polygon

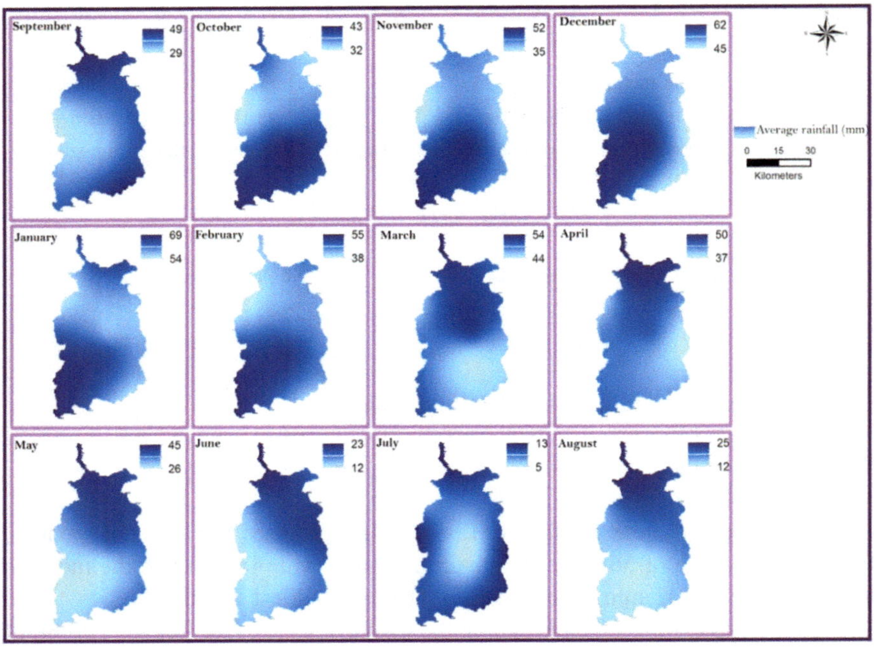

Fig. 7 Variation of average rainfall monthly

Table 5 Average monthly rainfall by regularized spline method

Month	S	O	N	D	J	F	M	A	M	J	J	A
P_{av} (mm)	42	37	38	43	**55**	40	45	39	36	18	**10**	15

The most rainy month is january (wet) and the low rainy month (dry) is July

4.3.2 Average Rainfall in the Seasonal Scale

The seasonal average rainfall is maximum in winter and minimum in summer. For Tessa basin, the rainy season are autumn, winter and spring where the average rainfall is important. These results are expected after analyzing the monthly average rain (Fig. 8).

At the seasonal scale, we determined rainfall isohyets of 30 years (Fig. 7). The three seasons of autumn, winter and spring had an important average rainfall varied between 96 and 179 mm. The average rainfall is maximum at the winter and minimum at the summer (Fig. 9).

Table 6 resumes the average rainfall of Tessa basin in each season extracted by spline method:

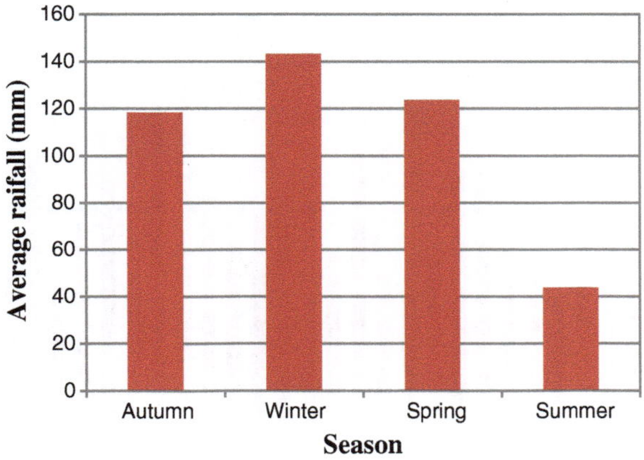

Fig. 8 Average rainfall in the seasonal scale

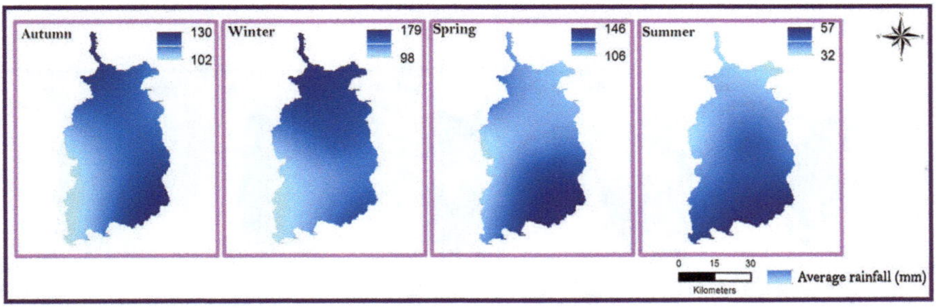

Fig. 9 Variation of average rainfall in the season scale

Table 6 Average seasonal rainfall

Season	Autumn	Winter	Spring	Summer
P_{av} (mm)	117	139	120	43

4.3.3 Average Rainfall in the Annual Scale

At the annual scale, 30 isohyets spatial distributions map are determined. As the last twelve years had the most important precipitation, we presented isohyets maps of the period 2002–2013 (Fig. 10).

In the annual scale, 11 years had an average rainfall upper the normal (428 mm). The maximum average rainfall falls in 2003. In fact, in this year the Medjerda basin (to which belongs the studied catchment) had three important inundations (Fig. 11).

The year 2002 have height average rainfall that range between 536 and 753 mm. the years 2003, 2004, 2008, 2010 and 2011 have an important annual average rainfall exceed 500 mm. we note that this years have known great inundations.

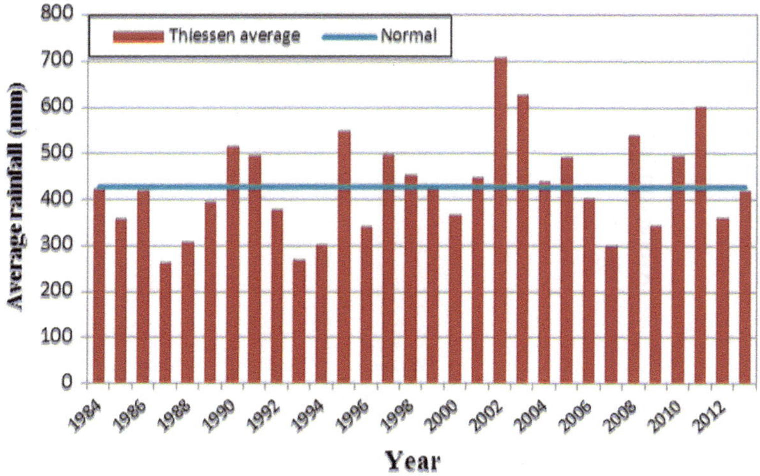

Fig. 10 Average rainfall in the annual scales

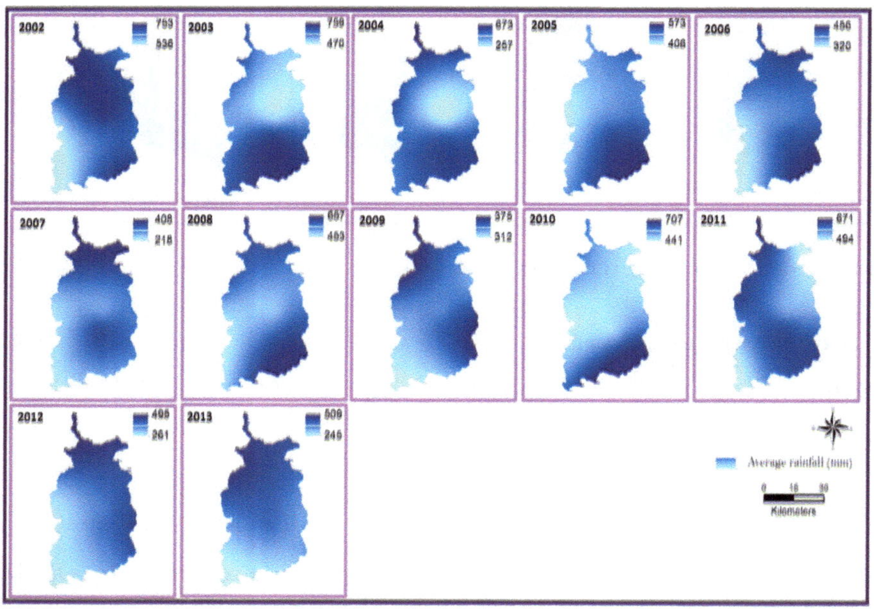

Fig. 11 Variation of annual average rainfall of the last 10 years

Figure 12 show the variation of average annual rainfall extracted from the spline method and the normal of 30 years.

We conclude from this figure, that 13 years beyond the normal (428 mm). As we noted from the maps. the year 2002 was the wettest with an average of 700 mm.

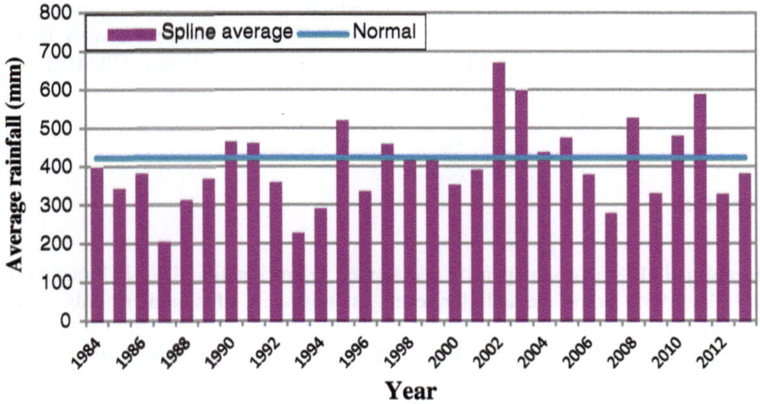

Fig. 12 Variation of annual average rainfall of the last 10 years

followed by the year 2003 with 604 mm and 2011 with 593 mm, Minimum average rainfalls fallen in the years 1987, 1993 and 2007. The average rainfall of 30 years given by the isohyet method is 418 mm.

5 Conclusion

In the literature, the results of the comparison of interpolation methods differ from one study to another. The successful performance of the methods depends on several factors, in particular, temporal and spatial resolutions of the data and the parameters of the models.

Analysis of monthly rainfall data of 13 gauged stations during 30 years, showed an irregular variation. The homogeneity tests detected some breaks at monthly, seasonal and annual scales. The precise evaluation of the spatiotemporal variability of rainfall on the watershed scale presents a complex problem because of the small number of rain gauges in most cases and because rainfall is extremely varied in space and time. The choice of interpolation method for measuring rainfall depends on the quantity of valid measures, the nature of the rain in the regions under study and the quality of the observations. The choice of method is therefore crucial.

The analysis conducted during this study confirmed that GIS is a useful tool to evaluate the spatial and temporal variation of rainfall in Tessa basin.

For the spatial interpolation of rainfall, we note that both methods of Thiessen and isohyet bring closed results (qualitatively) at the three scales. Using the average rainfall over 30 years one can determine the rainfall pattern of Tessa basin with reference to the classification of Champoux and Toutant (1988), It's a temperate continental regime.

The average rainfalls will be useful to study the hydrology behavior of the watershed Tessa and its influence on the Medjerda inundation.

Acknowledgements I would like to express my gratitude to all those who gave me the possibility to complete this article. I address my deep recognition to Professor Hamadi Habaieb (INAT) for all the efforts which it exempted during this work. I want to thank my colleague Olfa Hajji (INAT) for his help in writing and finalizing this article.

References

Alexandersson H (1986) A homogeneity test applied to precipitation data. J Climatol 6:661–675

Buishand TA (1982) Some methods for testing homogeneity of rainfall records. J Hydrol 58 (1982):11–27

Champoux A, Toutant C (1988) Éléments D'hydrologie. livre édité par Griffon D'argile Inc. Quebec

DE (1987) Department of Environment. Handling geographical information. Report of the Committee of Enquiry chaired by Lord Chorley. HMSO, London

Fiedler FR (2003) Simple, practical method for determining station weights using Thiessen polygons isohyetal maps. J Hydrol Eng 8(4):219 doi:10.1061/(ASCE)1084-0699

Gousie MB, Franklin WR (2005) Augmenting grid-based contours to improve thin plate DEM generation. Photogram Eng Remote Sens 71:69–79

ISL (2012) Barrage de TESSA- Avant Projet Détaillé. Rapport définitif. Direction Générale des Barrages et Grands TravauxHydrauliques. Direction E M E. p 41

Kaijuka E (2007) GIS and rural electricity planning in Uganda. J Clean Prod 15:203–217

Michaelides SC, Tymvios FS, Michaelidou T (2009) Spatial and temporal characteristics of the annual rainfall frequency distribution in Cyprus. Atmosph Res 94(4):606–615

Mitas L, Mitasova H, Brown WM (1997) Role of dynamic cartography in simulations of landscape processes based on multi-variate fields. Comput Geosci 23:437–446

Nicholson SE, Kim J, Hoopingarner J (1988) Atlas of African rainfall and its inter annual variability. Department of Meteorology, Florida State University Tallahassee, p 237

Pettit AN (1979) A non-parametric approach to the change-point problem. Appl Stat 28(2):126–135

Punmia BC, Ashok J, Arun J (1995) Water supply engineering (Chapter 2, p 19). In: Environmental engineering book

Schwab O, delmer D, Wiliam J, Richard K (1995) Soil and water conservation engineering, 4th edn. Wiley, New York

Spatial and Temporal Variability Analysis Using Modelled Precipitation Data in Upper Catchment of Chambal Basin

Ankit Gupta, Maya Kumari and B. Krishna Rao

Abstract Reckoning precipitation by means of customary methods has many breaks and various slips due to manual noticing which is the most unintended aspects for precipitation analysis. These errors in records rises the inconsistency and grounds uncertainty while using it in simulation modeling, sediment modeling and other such type of works. TMPA 3B42, a research product of the world's foremost satellite in precipitation study is used in this work for rainfall variability analysis with good accuracy level. The integration of GIS and Remote Sensing acts as an operative tool for extracting and analyzing this precipitation product with spatially interpolating it over region. The objective of this work is to analyze the variability in rainfall over Shipra catchment between 1998 and 2012, and extraction of precipitation data from network common form data files. The study proves that the catchment has an exceedingly variable drift of rainfall in downstream portion. Also, the work results that there is high inconsistency in the year 2008 with fall in rainfall all over the catchment while precipitation was found very less in 2000 and high in 2006 followed by the year 2011. This work also states that downstream portion of the study area leftovers in dearth rainfall condition most of the time but with a tad high runoff affecting the groundwater state in the area. Precipitation shortfall, diminishing water controlling structures and increasing land use is disturbing the water availability in the region.

Keywords Chambal basin: Shipra River · TRMM: TMPA 3B42 · Rainfall Variability: Spatial & Temporal · Interpolation: IDW

A. Gupta (✉) · B.K. Rao
ICAR—Indian Institute of Soil and Water Conservation,
Research Centre, Vasad, Anand 388306, Gujarat, India
e-mail: ankitgupta18may@gmail.com

M. Kumari
Amity School of Natural Resources and Sustainable Development,
Amity University, Sector-125, Noida 201301, Uttar Pradesh, India

O. Abdalla et al. (eds.), *Water Resources in Arid Areas: The Way Forward*,
Springer Water, DOI 10.1007/978-3-319-51856-5_5

1 Introduction

Measuring rainfall, analyzing it, and quantifying the spatial and temporal variability is very important in many linked studies most of which is hydrological and climate analysis, which may vary strongly according to topography, aspect and slope characteristics of the area (Immerzeel et al. 2012). Rainfall among all the inputs in hydrological simulation is most important and effective parameter for water resource management study (Mohssen et al. 2011). Various aspects like estimation of water resource potential in a river basin, analyzing flood problems, their pattern and magnitude, estimation of dependable yield for irrigation, formation of flood and its control measures, can be studied with the help of precipitation, even irregular frequency and intensity of it causes difficulty in management and storage plans along with climate change (Bonacci 2004; Gupta et al. 2014). Traditional methods of measuring rainfall have many gaps and errors due to manual readings which are the most indirect factors for climate studies. Error in data increases the variability and causes uncertainty while using it in simulation modeling and for future climate projection. Rain gauge stations are not able to cope with the pressure of working all day and night and recording weather data, resulting gaps of several days or many.

Apart from manual methods, satellite rainfall measurement products offer the unique opportunity to improve estimates, but while assessing this with remotely sensed data over a region (with good accuracy), ground data is required to remove bias (Cheema and Bastiaanssen 2012; Dinku et al. 2008; Hunink et al. 2014). Accurate precipitation measurement is necessary nowadays for future prediction and analysis so Tropical Rainfall Measuring Mission (TRMM), the world's foremost satellite for the study of precipitation is used in this study to achieve good accuracy level. The TRMM precipitation data record is observed well in lowland areas with ground-based measurements, whereas it is more difficult to find accuracy level in the areas where topographic effects are important (Stampoulis and Anagnostou 2012; Tian and Peters-Lidard 2010; Chen et al. 2013; Dinku et al. 2008). It is also stated by Bookhagen and Burbank (2006), Gebregiorgis and Hossain (2013) that, uncertainty in TRMM data depends principally on the geography and landscape of the study area, but performance on complex topography have found reasonable (Montero et al. 2012). Information on variation of rainfall with topography helps in providing a realistic assessment of water resources and also helps in hydrological modeling. However, in recent years, spatial variability is studied more to know the better understanding of interaction among climate, erosion and tectonics (Barros et al. 2006; Anders et al. 2006; Shrestha et al. 2012; Bookhagen 2010).

Precipitation derived from remotely sensed data sets is also a valuable hydrological data which is an alternative to conventional ground data collection methods (Engman and Gurney 1991; Neale and Cosh 2012; Bastiaanssen et al. 2000; Wagner et al. 2009). TRMM Multi-Satellite Precipitation Analysis (TMPA) products used is in this study, provides precipitation estimates at every 3-h basis, whereas near-real-time product is 3B42RT and research-grade product is 3B42 and on monthly basis at temporal resolutions on a 0.25° × 0.25° grid. This data is available from January 1998 to present. However, near-real time product is less accurate, but provides quick precipitation estimates which is most suitable for near-real-time monitoring and modeling activities (Wu et al. 2012). While the research version of this product is calibrated with gauge data and is available after two months of observation. Also, different sensor calibration and additional post-processing in the algorithm were used and is more accurate and suitable for research (Huffman et al. 2007, 2010). Over the years, the TMPA products have been widely used in various research and applications (Bitew et al. 2012; Gianotti et al. 2012; Gourley et al. 2011; Wu et al. 2012).

The integration of Geographical Information System and Remote Sensing acts as an effective tool for extracting precipitation from satellite data with spatially interpolating it over study region. The objective of this work is to extract the precipitation data from TRMM NetCDF files for fifteen years from 1998 to 2012 and analysis that over Shipra catchment of Ganga Basin, India for finding out the rainfall variability.

2 Study Area

The catchment area of the Shipra basin with an outlet at the confluence point with the Chambal River near Kalu-Kher village is the study area which extends over an area of 5567.80 km^2. It is entirely located in the Malwa plateau with two major tributaries viz., The Khan River and The Gambhir River. Catchment cover parts of Indore, Dewas, Ujjain and Ratlam districts of Madhya Pradesh state of central India and a little upstream portion in the Rajasthan state. The study area lies between the geographical extent of the north of 22.50°–23.94° latitude and east 75.40°–76.21° longitude, as shown in Fig. 1. Presently, Shipra River runs dry for a period of five to six months per year as the river has lost its perennial nature. The water of the river is used for drinking, industrial, irrigation purposes and for the Holy Bath of Faith. In present, during the summers of 2016 for world's biggest gathering called Simhastha (Maha Kumbh Mela, once in every twelve years). At this celestial time, millions of people and pilgrims arrive and converge on the banks of the Shipra River to take the holy dip.

2.1 *Climate*

The study area has a subtropical climate, consisting of six rain gauge observatory and four meteorological stations maintained by the Indian Meteorological Department (IMD) but over the years, they have been proven unsuccessful due to lack of maintenance, error and data gaps. Basin receives about 93% of total annual rainfall during the monsoon period between June and September and maximum

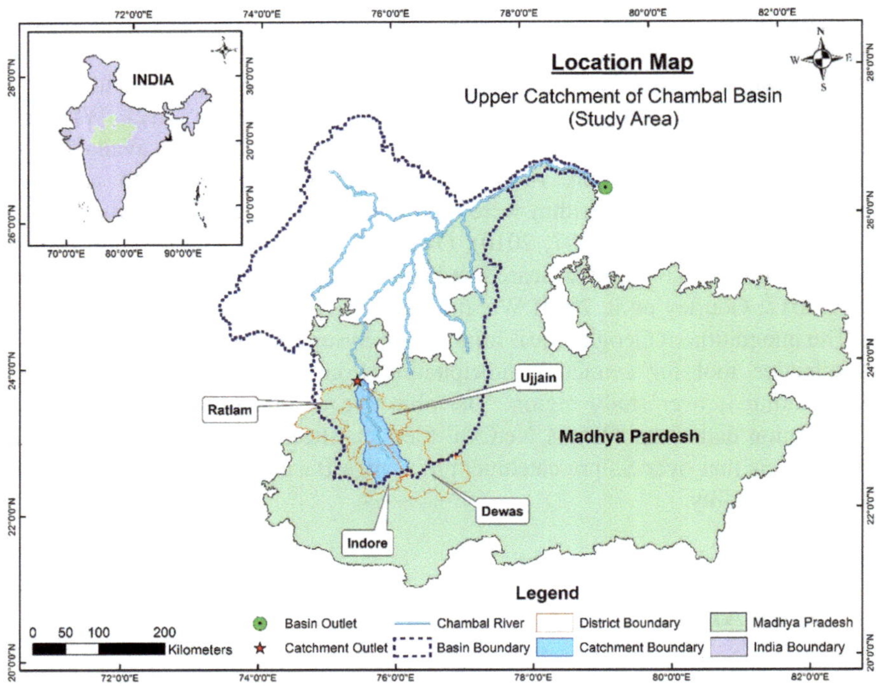

Fig. 1 Location map of study area

monthly rainfall takes place in June and August which ranges from 875 to 1126 mm. In this area, variability in rainfall has been seen during Kharif season (July to October) and Rabi season (November to February), where sometimes there is no precipitation for days and months. Accurate precipitation measurement is necessary nowadays for future prediction and analysis.

3 Materials and Method

3.1 Data and Software Used

In the present study, the digital elevation model (DEM) of Shuttle Radar Topography Mission (SRTM) of 90 m resolution has been utilized to get topographic details, catchment delineation and stream network generation shown in Fig. 2. The vertical error of the DEMs is less than 16 m (Sun et al. 2003). SRTM is currently in fact standard for elevation models in various hydrological assessments (Hunink et al. 2014). Rainfall data based on TRMM satellite products have been studied and evaluated. Its final product has 1-day temporal resolution and 25 × 25 km spatial resolution (Simpson et al. 1996). The aim of satellite product is to provide accurate worldwide tropical rain estimate by using a combination of various instruments intended entirely for rainfall observation. Moreover, some of the consequent products are calibrated with selected ground-based observation sites. Since its launch, the TRMM algorithms and products have been continuously improved, and since May 2012 version 7 is available for use (Huffman et al. 2007, 2010; Huffman and Bolvin 2012, 2013; Hunink et al. 2014). Daily Precipitation data of TMPA 3B42 research product have been used for fifteen years between 1998 and 2012 for variability analysis as it is the world's foremost satellite for the study of precipitation with good accuracy level. Spatial Analyst tool, Multidimension tool, Model builder of the ArcGIS software package has been extensively used for interpolation, data extraction and batching purposes, respectively. Precipitation data is available globally, therefore this data should be clipped for the study area and thus has been subset and downloaded using Simple Subset Wizard (SSW) of EOSDIS.

3.2 Methodology

The methodology presented here, referred to the rainfall variability analysis, and includes the coupling of precipitation data and Geostatistical approach. TRMM 3B42 research product is downloaded in NetCDF format, for 5479 days (fifteen years) and ArcPy i.e. ArcGIS Python has been used to extract the precipitation data from it. A batch model is prepared to convert these 5479 daily files into tabular form so that further analysis can be done accordingly on the daily, monthly or yearly basis. TRMM data is of $0.25° \times 0.25°$ resolution resulting in 12 sample points, in and around the catchment area after extraction by using Multidimension tool and later scripting used to merge all data in the single file.

After extracting data and assigning extracted precipitation attributes to these points using GIS, a spatial interpolation technique called Inverse Distance Weighed (IDW) was performed for all the years. This method does not make any assumptions about spatial relationships. This technique predicts the values using a linear weighted combination of a set of sample points. This interpolation method assumes that the variable being mapped decreases in weight with distance from its sample location i.e. the greater the distance, less will be the influence (Philip and Watson 1982; Tung 1983; Watson and Philip 1985; Carvalho and Woodroffe 2015). According to Chow et al. 1988, IDW in context with rainfall assumes that value at a gauged point compared to an ungauged point is inversely relative to the distance between the two points.

Another interpolation technique, Kriging, assumes that the distance or direction between the points reflects a spatial correlation which is used to explain variation. The arithmetic average is the simplest and often the least accurate method used for computation of mean precipitation (Black 1996). No such information exists and is provided about how to find out the variability between the rain gauges which gives better accuracy. There have been many comparisons of rainfall interpolation

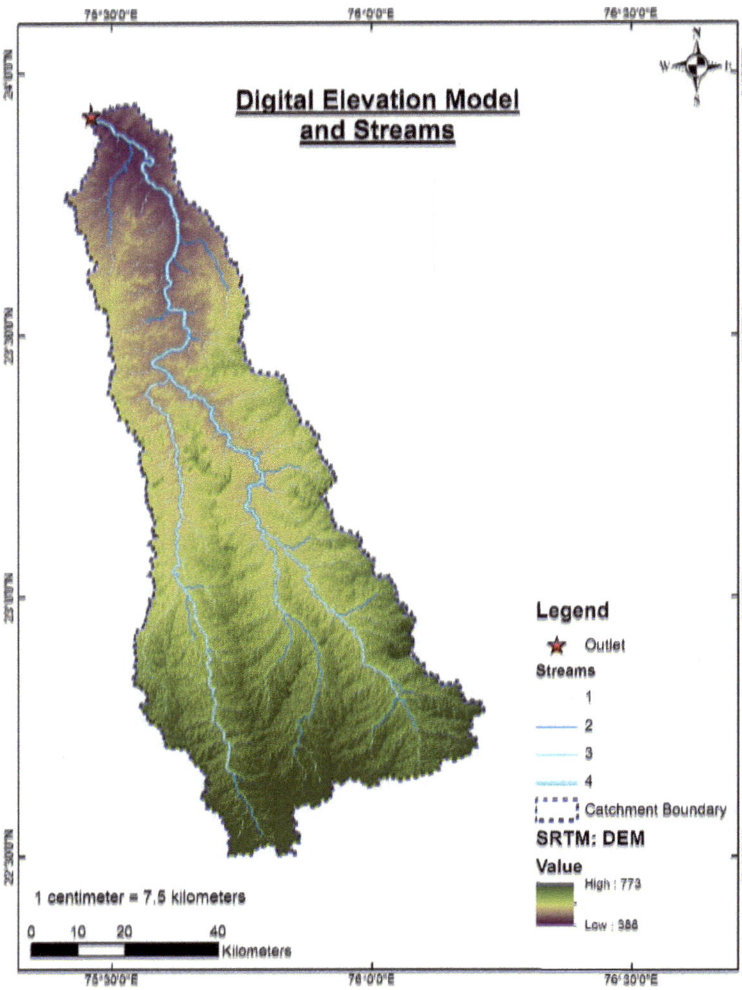

Fig. 2 SRTM—digital elevation model (DEM) and streams of Shipra catchment

method, but till date, there is no observation and exact way to demonstrate that (Shaw et al. 2011). Many of those methods, results that thin plate smoothing splines were inferior to multiquadric interpolation; Kriging is more accurate than Thiessen polygons for daily rainfall and also a good method when you have enough gauges to interpolate (Hutchinson 1995; Garcia et al. 2008; Buytaert et al. 2006). Advantages of universal and ordinary Kriging over IDW is described by Basistha et al. (2008), Kriging and IDW were the best-known methods are concluded by Li et al. (2011), whereas Dirks et al. (1998) suggested that Kriging has no clear advantage over other methods and on the other hand, Nalder and Wein (1998) found that IDW had lesser error than ordinary or universal Kriging technique of interpolation.

IDW method estimation is a linear weighted average of numerous neighboring observations which used the below-mentioned formula:

$$\hat{z}(u) = \frac{\sum_{i=1}^{n(u)} \frac{z(u_i)}{d_i^{\alpha}}}{\sum_{i=1}^{n(u)} \frac{1}{d_i^{\alpha}}}$$

Where, $n(u)$ is the number of neighboring observations, $z(u_i)$ is used to estimate the value of rainfall at an unsampled location u, d is the separation distance between the location and α is the distance weighting power, i is a step starting from 1 to n. However, as the value of α will increase, the effect of the farthest observation on the estimated value is decreased as it was discussed above that greater the distance, less will be the influence (Afanasev et al. 2014).

Taking spatially distributed rainfall considers two main steps in it, one of which is based on regression models using remotely sensed data sets and second by including variance which does not result in model outputs. The geostatistical approach may prove it to some extent includes interpolation and variance in distribution, although it is not possible to apprehend the high inconsistency in the area from common spatial interpolation approaches. The final outcome of the process is the spatial distribution and variability of rainfall over Shipra catchment of central India for fifteen years.

4 Results and Discussion

4.1 Rainfall Time-Series

As the precipitation record is not from the manual method or observation and is in digital form, totally depended on the sensors. Thus, data gaps and errors in the series are not obtained, although where there is no rainfall in a day, the measured obtained value is zero. Therefore, it is not required to detect the outliers in the datasets which is very much observed from the data extracted from the NetCDF files in tabular form. A common problem is a spatial mismatch among the study area and rainfall of one single microwave based pixel.

4.2 Spatial Interpolation

Raster surface obtained after interpolating the sample points by using an inverse distance weighted (IDW) technique results in yearly interval series of precipitation casing the entire area and allows understanding in the spatial distribution on a yearly basis. It has been noticed that downstream portion of the catchment is in deficit condition and received very less rainfall. Although the upper basin receives high rainfall followed by middle basin.

The average twelve-monthly mean for the 1998–2012 time-series is 964.87 mm for the Shipra catchment. The interpolated precipitation maps of the catchment for fifteen years are presented in Fig. 3. It has been clearly depicted that in the year 2000 the area receives very low rainfall (551.04 mm) with an average rainfall of 574.22 mm. Even its maximum rainfall (605.97 mm) is below the annual average mean rainfall followed by the years 2008 (888.70 mm), 2002 (910.99 mm) and 2005 (931.97 mm) whereas in the year 2006 the area receives heavy rainfall with average and maximum rainfall of 1438.30 and 1635.02 mm, respectively followed by the years 2011 (1504.87 mm), 2003 (1247.12 mm), 1998 (1152.45 mm) and 1999 (1186.15 mm).

It is also observed that only for one year in 1998–2012 time-series, basin receives heavy rainfall in the lower basin in 2011 and twice near average in 2001 and 2004. Rest years, downstream portion records below average. Most of the growth, development and urban areas lie in the middle and upper portion of the catchment like Ujjain and Indore and declared the drought-prone area in the year 2008–09 by the Asian Cities Climate Change Resilience Network (ACCCRN). They also state that existing water scarcity in Indore is likely to be further aggravated by rainfall variability and temperature increase in near future.

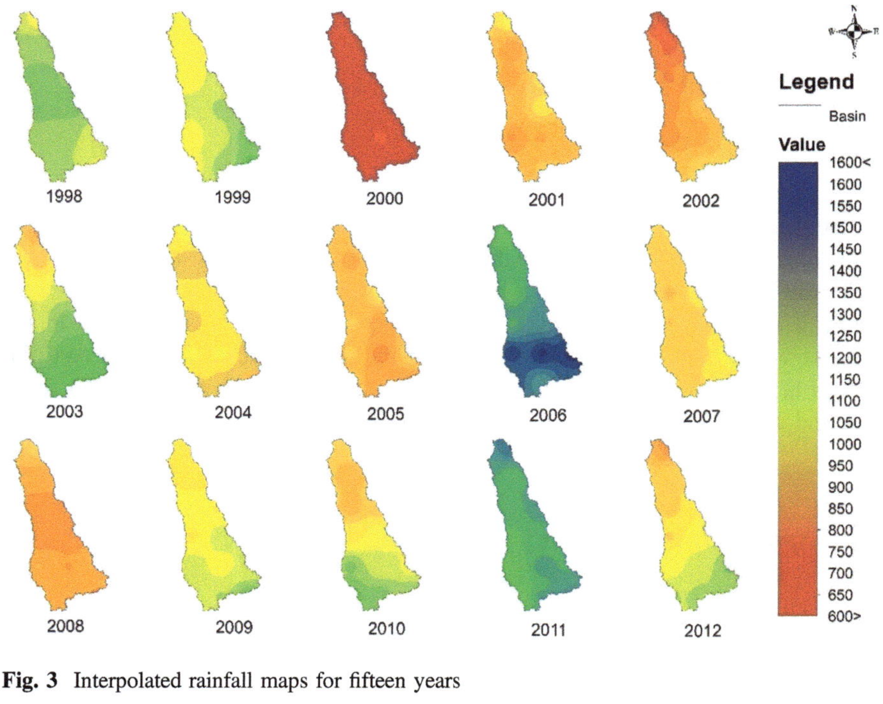

Fig. 3 Interpolated rainfall maps for fifteen years

4.3 Intra-annual Variability

The intra-annual variability can be expressed by the coefficient of variation which is the ratio of the standard deviation to the mean rainfall at each sample point. Figure 4 shows the spatial picture of the coefficient of variation, based on the entire 15-year period of annual rainfall. The map shows that in the lower region of the catchment, the time-based deviation is highest. Relatively high variability of rainfall

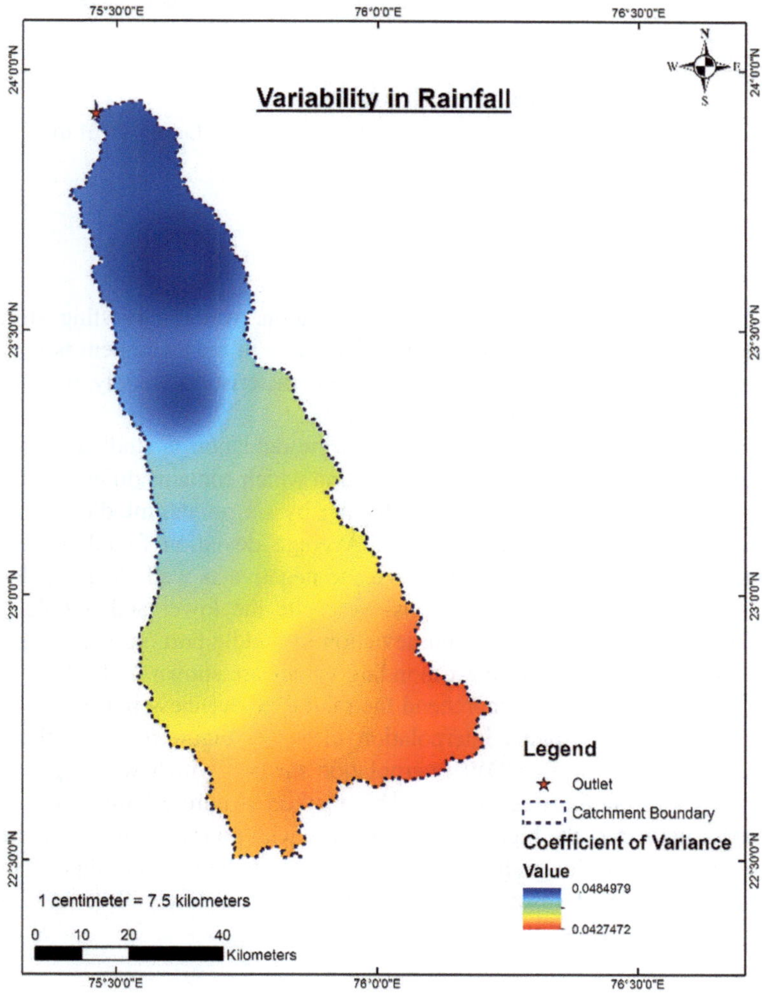

Fig. 4 Spatial demonstration of the coefficient of variation in the catchment

Table 1 Table showing covariance and deviation values over study area

S. No.	Annual average (mm)	Coefficient of variance	Deviation (mm)	Effect	Region	Average deviation (mm)
1	991.34	0.043943	26.51	+	Upper	32.91
2	999.97	0.043563	35.14	+		
3	1021.74	0.042635	56.91	+		
4	967.71	0.045015	2.88	+		
5	985.98	0.044181	21.54	+		
6	1019.70	0.042721	54.87	+		
7	929.52	0.046865	35.30	–	Middle	−18.56
8	981.84	0.044368	17.01	+		
9	909.60	0.047891	55.23	–		
10	964.11	0.045184	0.72	–		
11	898.22	0.048498	66.61	–	Lower	−61.61
12	908.19	0.047966	56.63	–		
	964.827		±35.75			

coincides with annual crop cultivation, vegetation, water harvesting structures, groundwater storage etc. Although most of the area in the catchment is very good for groundwater recharge due to the slope characteristics, mainly in upper and middle portion, thus results in lower variability.

For ease in understanding the variability, the deviation is studied in three different portions: upper, middle and lower portion which contains different number of sample points in it. Like upper portion having by six points, middle portion have four points and lower have two points. Average deviation found in the entire catchment is of 35.75 mm results positive and negative as well. Average deviation in the Upper portion calculated is positive while the lower and middle portion results in a negative value. Although deviation in middle portion is much less than that of the lower portion. All corresponding values are shown in Table 1.

It is likely to study and comprehend the rainfall unevenness in the area to some extent from common spatial interpolation methods. Figure 5 shows the annual mean rainfall based on the IDW interpolation method which was applied to the annual means at the extracted points. This method is purely based on point data which does not reflect the high spatial variability with poor resolution. On the other hand, remote sensing integration with GIS can results better with high accuracy if the data used is of high resolution. Lower the resolution, low will be the accuracy, poor variability analysis.

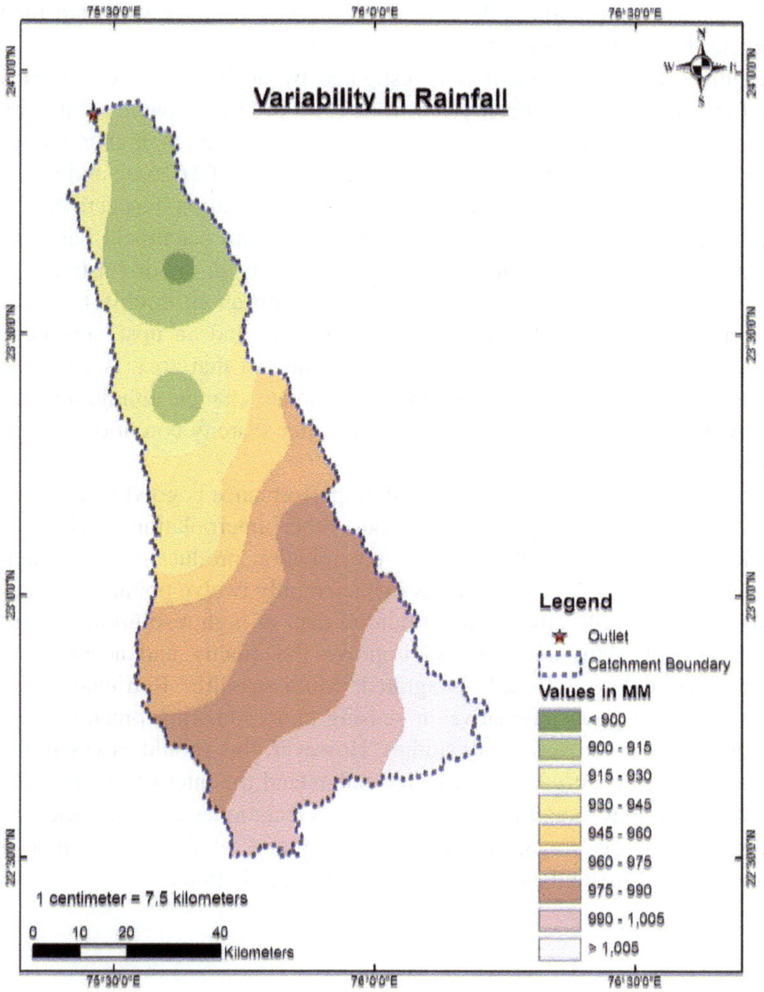

Fig. 5 Map showing variability in rainfall based on the IDW interpolation method

5 Summary and Conclusion

In the present study, an attempt has been made to understand the spatial and temporal variability of rainfall in Shipra catchment which lies in an upper Chambal basin of central India.

For analyzing the variability of rainfall, data extraction, catchment delineation and topographic detail; long-term daily TRMM 3B42 precipitation data, multidimension and spatial analyst hydrology tools of ArcGIS, digital elevation model of SRTM was used respectively. Catchment delineation is done at the outlet marked to the point of confluence of the Shipra river with the mainstream of the basin.

Surface raster interpolation is done by using inverse distance weighted method to interpolate the rainfall over the study area extent.

Variability is the catchment is investigated by analyzing 12 extracted sample points for a time span of fifteen years from 1998 to 2012. Average annual rainfall estimated was 964.827 mm for time-series, however much higher values were observed in 2006 and lesser in 2000 with a maximum of 1616.02 and 605.96 mm and a minimum of 1213.47 and 551.04 mm precipitation respectively. Overall standard deviation calculated is 43.56 mm. The Shipra catchment has a spatially temporal variable rainfall pattern with high variability in the downstream portion, although runoff is high at the confluence, also with about double the amount of precipitation in the lower region than more middle and an upstream area which inversely affect the vegetation and crop cultivation of that area. A detailed study, including geomorphic, geologic, hydrology, climate change, lithologic etc. in the area may prove helpful in water logging and water scarcity conditions in this rapid development era.

The results also shows that the procedure performance is good while using IDW interpolation techniques, however various other interpolation methods remain untouched in this work. TRMM 3B42 precipitation product offers a moderately rough resolution for such small area as extracted only twelve points, although fit for this work, but possibly other gridded rainfall data of high resolution or rain gauge observation with short distance can improve the quality and accuracy of such outputs. The newly released Integrated Multi-satellitE Retrievals for GPM (IMERG) is very good alternative in sensing and estimating precipitation across India and also for such variability studies. However, this should assessed alongside IMD gauge-based observations, to better understand the interaction among climate and erosion. Various modeling efforts, tools and techniques with the efficient use of satellite data may deal regarding better accuracy and reliability of remote sensing data, models and methods with enhanced spatial predictions.

References

Afanasev I, Volkova T, Elizaryev A, Longobardi A (2014) Analysis of interpolation methods to map the long-term annual precipitation spatial variability for the Republic of Bashkortostan, Russian Federation. WSEAS Trans Environ Dev 10:2224–3496

Anders AH, Roe GH, Hallet B, Montgomery DR, Finnegan NJ, Putkonen J (2006) Spatial patterns of precipitation and topography in the Himalaya. Spec Pap Geol Soc Am 398:39–53

Barros AP, Chiao S, Lang TJ, Burbank D, Putkonen J (2006) From weather to climate seasonal and interannual variability of storms and implications for erosion processes in the Himalaya. Spec Pap Geol Soc Am 398:17–18

Basistha A, Arya DS, Goel NK (2008) Spatial distribution of rainfall in the Indian Himalayas—a case study of Uttarakhand and region. Water Resour Manag 22:1325–1346. doi:10.1007/s11269-007-9228-2

Bastiaanssen WGM, Molden DJ, Makin IW (2000) Remote sensing for irrigated agriculture: examples from research and possible applications. Agric Water Manage 46:137–155. doi:10.1016/S0378-3774(00)00080-9

Bitew MM, Gebremichael M, Ghebremichael LT, Bayissa YA (2012) Evaluation of high-resolution satellite rainfall products through streamflow simulation in a hydrological modeling of a small mountainous watershed in Ethiopia. J Hydrometeorol 13:338–350. doi:10.1175/2011JHM1292.1

Black PE (1996) Watershed hydrology. Lewis Publishers, Boca Raton

Bonacci O (2004) On the role of hydrology in water resources management. IAHS Publ 286:88–94

Bookhagen B, Burbank DW (2006) Topography, relief, and TRMM-derived rainfall variations along the Himalaya. Geophys Res Lett 33(8):1–5. doi:10.1029/2006GL026037

Bookhagen B, Burbank DW (2010) Toward a complete Himalayan hydrological budget: Spatiotemporal distribution of snowmelt and rainfall and their impact on river discharge. J Geophys. Res 115:F03019. doi:10.1029/2009JF001426

Buytaert W, Celleri R, Willems P, Bièvre DB, Wyseure G (2006) Spatial and temporal rainfall variability in mountainous areas: a case study from the south Ecuadorian Andes. J Hydrol 329:413–421. doi:10.1016/j.jhydrol.2006.02.031

Carvalho RC, Woodroffe CD (2015) Rainfall variability in the Shoalhaven River catchment and its relation to climatic indices. Water Resour Manage 29:4963–4976. doi:10.1007/s11269-015-0198-4

Cheema MJM, Bastiaanssen WGM (2012) Local calibration of remotely sensed rainfall from the TRMM satellite for different periods and spatial scales in the Indus basin. Int J Remote Sens 33 (8):2603–2627. doi:10.1080/01431161.2011.617397

Chen Y, Ebert EE, Walsh KJE, Davidson NE (2013) Evaluation of TRMM 3B42 precipitation estimates of tropical cyclone rainfall using PACRAIN data. J Geophys Res 118(5):2184–2196. doi:10.1002/jgrd.50250

Chow VT, Maidment DR, Mays LW (1988) Applied hydrology. McGraw-Hill, New York, NY

Dinku T, Chidzambwa S, Ceccato P, Connor SJ, Ropelewski CF (2008) Validation of high-resolution satellite rainfall products over complex terrain. Int J Remote Sens 29(14): 4097–4110. doi:10.1080/01431160701772526

Dirks KN, Hay JE, Stow CD, Harris D (1998) High-resolution of rainfall on Norfolk Island, Part II: interpolation of rainfall data. J Hydrol 208:187–193. doi:10.1016/S0022-1694(98)00155-3

Engman ET, Gurney RJ (1991) Remote sensing in hydrology. Chapman and Hall Ltd., London

Garcia M, Peters-Lidard CD, Goodrich DC (2008) Spatial interpolation of precipitation in a dense gauge network for monsoon storm events in the southwestern United States. Water Resour Res 44:W05S13. doi:10.1029/2006WR005788

Gebregiorgis AS, Hossain F (2013) Understanding the dependence of satellite rainfall uncertainty on topography and climate for hydrologic model simulation. IEEE Trans Geosci Remote 51(1):704–718. doi:10.1109/TGRS.2012.2196282

Gianotti RL, Zhang D, Eltahir EAB (2012) Assessment of the regional climate model version 3 over the maritime continent using different cumulus parameterization and land surface schemes. J Clim 25:638–656. doi:10.1175/JCLI-D-11-00025.1

Gourley JJ, Hong Y, Flamig ZL, Wang J, Vergara H, Anagnostou EN (2011) Hydrologic evaluation of rainfall estimates from radar, satellite, gauge, and combinations on Ft. Cobb Basin, Oklahoma. J Hydrometeorol 12:973–988. doi:10.1175/2011JHM1287.1

Gupta A, Thakur PK, Nikam BR, Chouksey A (2014) Hydrological modelling of upper and middle Narmada basin using geospatial tools (Chapter 63, pp 663–675). In: Tiwari HL, Suresh S, Jaiswal RK (eds) Hydraulics, water resources, coastal and environmental engineering. Excellent Publishing House, Kishangarh, Vasant Kunj, New Delhi, India. ISBN 978-93-84935-04-7

Huffman GJ, Adler RF, Bolvin DT, Gu G, Nelkin EJ, Bowman KP, Hong Y, Stocker EF, Wolff DB (2007) The TRMM multi-satellite precipitation analysis: quasi-global, multi-year, combined-sensor precipitation estimates at fine scale. J Hydrometeorol 8(1):38–55. doi:10.1175/JHM560.1

Huffman GJ, Adler RF, Bolvin DT, Nelkin EJ (2010) The TRMM multisatellite precipitation analysis (TAMPA) (Chapter 1). In: Hossain F, Gebremichael M (eds) Satellite rainfall applications for surface hydrology. Springer, pp 3–22. ISBN 978-90-481-2914-0
Huffman GJ, Bolvin DT (2012) Real-time TRMM multi-satellite precipitation analysis data set documentation
Huffman GJ, Bolvin DT (2013). TRMM and other data precipitation data set documentation
Hunink JE, Immerzeel WW, Droogers P (2014) A high-resolution precipitation 2-step mapping procedure (HiP2P): development and application to a tropical mountainous area. Remote Sens Environ 140:179–188. doi:10.1016/j.rse.2013.08.036
Hutchinson MF (1995) Interpolating mean rainfall using thin plate smoothing splines. Int J Geogr Inf Syst 9(4):385–403. doi:10.1080/02693799508902045
Immerzeel WW, Pellicciotti F, Shrestha AB (2012) Glaciers as a proxy to quantify the spatial distribution of precipitation in the Hunza basin. Mt Res Dev 32(1):30–38. doi:10.1659/MRD-JOURNAL-D-11-00097.1
Ly S, Charles C, Degre A (2011) Geostatistical interpolation of daily rainfall at catchment scale: the use of several variogram models in the Ourthe and Ambleve catchments, Belgium. Hydrol Earth Syst Sci 15(7):2259–2274. doi:10.5194/hess-15-2259-2011
Mohssen M, Edwards S, Walters AS, Alqassab A (2011) The impact of El Nino and La Nina weather patterns on Canterbury water resources. In: 19th international congress on modelling and simulation, Perth, Australia
Montero-M G, Zarraluqui-S V, García-G F (2012) Evaluation of 2B31 TRMM-product rain estimates for single precipitation events over a region with complex topographic features. J Geophys Res 117(D2):D02101. doi:10.1029/2011JD016280
Nalder IA, Wein RW (1998) Spatial interpolation of climatic normals: a test of a new method in the Canadian boreal forest. Agric For Meteorol 92:211–225. doi:10.1016/S0168-1923(98)00102-6
Neale CMU, Cosh MH (2012) Remote sensing and hydrology. IAHS Red Book Series, Publ. 352, IAHS Wallingford, UK, 482 pp
Philip GM, Watson DF (1982) A precise method for determining contoured surfaces. Aust Petrol Explor Assoc J 22:205–212
Shaw EM, Beven KJ, Chappel NA, Lamb R (2011) Hydrology in practice, 4th edn. Spon, London
Shrestha D, Singh P, Nakamura K (2012) Spatiotemporal variation of rainfall over the central Himalayan region revealed by TRMM Precipitation Radar. J Geophys Res 117:D22106. doi:10.1029/2012JD018140
Simpson J, Kummerow C, Tao WK, Adler RF (1996) On the tropical rainfall measuring mission (TRMM). Meteorol Atmos Phys 60(1–3):19–36. doi:10.1007/BF01029783
Stampoulis D, Anagnostou EN (2012) Evaluation of global satellite rainfall products over Continental Europe. J Hydrometeorol 13(2):588–603. doi:10.1175/JHM-D-11-086.1
Sun G, Ranson K, Kharuk V, Kovacs K (2003) Validation of surface height from shuttle radar topography mission using shuttle laser altimeter. Remote Sens Environ 88(4):401–411. doi:10.1016/j.rse.2003.09.001
Tian Y, Peters-Lidard CD (2010) A global map of uncertainties in satellite-based precipitation measurements. Geophys Res Lett 37(24). doi:10.1029/2010GL046008
Tung YK (1983) Point rainfall estimation for a mountainous region. J Hydraul Eng 109:1386–1393
Wagner W, Verhoest NEC, Ludwig R, Tedesco M (2009) Editorial "Remote sensing in hydrological sciences". Hydrol Earth Syst Sci 13:813–817. doi:10.5194/hess-13-813-2009
Watson DF, Philip GM (1985) A refinement of inverse distance weighted interpolation. Geoprocess 2:315–327
Wu H, Adler RF, Hong Y, Tian Y, Policelli F (2012) Evaluation of global flood detection using satellite-based rainfall and a hydrologic model. J Hydrometeorol 13:1268–1284. doi:10.1175/JHM-D-11-087.1

Recent Observed Climate Change Over Oman

Said Al-Sarmi, Sultan Al-Yahyai, Juma Al-Maskari, Yassine Charabi and B.S. Choudri

Abstract This study examined trends in temperature and precipitation parameters for the Sultanate of Oman in the period between 1980 and 2013. The data set has been carefully checked for the quality control for its homogeneity with the help of low density (8 stations) and it was possible to arrive at a clear picture of climate change in Oman. The results of this study found that the general pattern of Oman's mean annual temperature is of warming, with all stations, subregions and all Oman average showing statistically significant warming at 0.05 level. Across stations of Khasab and Sohar, highest and statistically significant mean annual warming trends are found (0.5 °C decade^{-1}). Results of seasonal warming showed consistent and widespread most of the year except in wintertime. These trends are mostly higher in magnitude over the non-monsoonal area relative to the monsoonal areas in across Oman. The increase in the mean annual maximum temperature is less widespread in relation to the mean annual temperature, which has reported statistically significant increase only for 2 stations. The highest statistically significant trend of 1.6 °C decade^{-1} reported at Khasab station and for minimum temperature, all stations observed and sub-regions across Oman on an average show statistically significant warming trends, with the highest annual trends observed in the monitoring station of Sohar (1.1 °C decade^{-1}). Data monitored at stations of Saiq and Thumrait reported 0.6 °C decade^{-1}. Total precipitation of annual values show only

S. Al-Sarmi (✉) · J. Al-Maskari
Directorate General of Meteorology, Public Authority for Civil Aviation, Muscat, Oman
e-mail: s.alsarmi@gmail.com

J. Al-Maskari
e-mail: j.almaskari@met.gov.om

S. Al-Yahyai
Information Technology Department, Mazoon Electricity Company, Nizwa, Oman
e-mail: s.alyahyai@gmail.com

Y. Charabi · B.S. Choudri
Center for Environmental Studies and Research, Sultan Qaboos University, Muscat, Oman
e-mail: yassine@squ.edu.om

B.S. Choudri
e-mail: bchoudri@squ.edu.om

O. Abdalla et al. (eds.), *Water Resources in Arid Areas: The Way Forward*,
Springer Water, DOI 10.1007/978-3-319-51856-5_6

at 2 stations with statistically significant trends. Out of studied 8 stations, only two stations show negative trends (i.e. Saiq = −74.0 mm $decade^{-1}$ and Salalah = −10.8 mm $decade^{-1}$). Further, trends observed at remaining stations are negative and statistically not significant. Therefore, the results of this study offer an evidence that climate of Oman has changed during the past three decades.

Keywords Oman · Climate trends · Warming · Drying

1 Introduction

Many parts of different countries around the world, importantly developed countries have benefitted from observed climate change assessments in relation to temperature and precipitation in view of its value and importance to the benefit of society (Goddard et al. 2001). There is a high degree of agreement in the literature that the planet Earth is warming. According to the fifth assessment report of IPCC, it has been highlighted that global mean surface temperature has increased right from the late 19th century (Hartmann et al. 2013). Studies show that past three decades have been warmer at the surface of Earth than all previous decades as recorded by instruments and it is pertinent to highlight that the first decade of the 21st century has been the warmest one. On a global scale, the average of combined land and ocean surface temperature data calculated by a method of linear trend, show a warming of 0.85 (0.65–1.06) °C, in the period from 1880 to 2012. However, most of climate change studies in the past, been largely focused to North America, Europe and Australia because of data availability (Freiwan and Kadioglu 2008). A little is known about understanding trends on temperature and precipitation in the Arabian Peninsula (AP) including Oman. To some extent, this fact is true since the Gulf region covers a broad range of countries, some of which have no data or poor data availability, quality and inconsistency in reporting on observed data (Zhang et al. 2005; Kwarteng et al. 2009; Nasrallah and Balling Jr 1993). This is in addition to the fact that much of the area is desert and there are very few sites with full-fledged instruments deployed for monitoring of climate data. Oman is located in the southeastern part of Arabian Peninsula, characterized by topography such as deserts and mountain ranges. In addition, Oman host a dry environment with very high temperatures during summer months and vast area of the country covered with different climate zones. For instance, the northern part of the country is considered subtropical while the southern part is a tropical (monsoonal) type (AlSarmi and Washington 2011).

The first region-wide trend analysis of the Middle East extreme indices observed in the period from 1950 to 2003 at 52 stations covering almost 15 countries was reported by Zhang et al. (2005). The results observed from this study shows statistically significant and spatially reasonable trends in temperature indices, which clearly indicate temperature increase in the region. A study in the region, covering all dry lands, Hulme (1996), Jones and Reid (2001) reported that warming might have contributed to a reduction in the precipitation/precipitation evaporation ratio in most of these dryland regions. Further, another study by Zhang et al. (2005) conducted over the Middle East, highlighted that trends in precipitation are weak in general and do not show spatial consistency. However, in case of Oman observations carried out for the data of 1977–2003, it is reported that in general, the monthly and yearly patterns of precipitation are constant and no significant trends were observed in the study period (Kwarteng et al. 2009).

Economic and Social Commission for Western Asia (2010) highlights that Arab countries are required to show commitment and preparation with regard to emerging climate change frameworks in the international arena and these countries should be able to give more inputs to the IPCC negotiations. Additionally, climate change projections for the AP shows little agreement among the models, particularly about the sign of precipitation change. Therefore, the aim of this study was to evaluate observed temperature parameters and precipitation trends over Oman using an updated data series and uses observed data for 8 stations located within the country to analyze trends of four climate variables.

2 Data Set

2.1 Data

Oman National Meteorological Service (ONMS) provided data for 17 stations on monthly mean maximum temperature, mean temperature, mean minimum temperature and precipitation (total). All these stations have data reported for different periods; this study has selected only stations having data starting from 1980 because of issues related to reliability and consistency. The selected stations are Khasab, Sohar, Saiq, Seeb, Sur, Masirah, Thumrait and Salalah. The Station data have the following characteristics: (1) They have passed the homogeneity and (2) the reported data have also passed the quality control tests, which accounts reported data from 1980 to 2013. The seasonal definition will follow AlSarmi and Washington (2011) covering the Arabian Peninsula (AP) called as winter (DJF), spring (MA), start of summer (MJ, a premonsoon), late summer (JAS) and autumn (ON) which is a second transitional period or postmonsoon.

2.2 Quality Control

The data provided by ONMS were quality controlled in two stages: (1) a check for physically implausible data (which is a negative rainfall or maximum temperature more than minimum temperature) and (2) analysis of outliers that are identified by numerical and visual checks. Further, these data sets evaluated by comparing their values to nearest stations to find that provided data has any possible flagged values associated with real or irregular weather events. It is identified that a few physically implausible values have been set to missing.

2.3 Homogeneity Checks

According to Conrad and Pollak (1950), homogeneity in a climate time series data is defined as one where variations that are caused only by variations in climate. Therefore, it is has been considered as important to remove the in homogeneity in the reporting data. The homogeneity of data is assessed with the help of RHtestV3 software. This software uses a two-phase regression model, which can be applied to a monthly data in order to check for multiple step change points that could exist in a time series data (Wang 2003, 2008).

In the next step, step changes identified are checked against the history of each station (if available). This study identified significant change points (5% level) but most of these change points were found to be physically genuine values especially of these associated with the discontinuities took place in the period between 1997 and 1998. It is observed that these change points are linked to heating associated with the 1997/8 El Nino event. Explanations for the historical cause in the change points, the study found that only 2 stations relocation was the main reason for Khasab and Sur stations. Hence, this study used the adjusted monthly time series of these 2 stations where the change points are significant and supported by metadata. However, mean maximum temperature data for Sur station have been excluded considering identified problems in data on accuracy and adequacy.

In the case of Seeb station of Oman, an exclusive cooling trend in all seasons was detected which is irregular in comparison with its neighbouring stations of Sohar (180 km) and Sur (420 km) and the region as a whole (AlSarmi and Washington 2011). AlSarmi and Washington (2013) show clear differences between Seeb station relative to the Sohar and Sur especially in the nighttime minimum temperature indices. Higher values of nighttime temperature indices at Seeb station relative to Sohar and Sur are evident before 1994. The higher nighttime temperature values of Seeb are also evident in the warm night frequency (TN90p) and warmest nights (TNx). The variance in temperature at Seeb is also lower post 1994, the main reason for this variation could be the relocation of the station to a new site at the aerodrome about 1.5 km from the old site in 1994 and the relocation of station is not mentioned in the document metadata. Given the above probable reason, decided to exclude Seeb station from this study. Although temperature variables at Seeb station have been excluded from the analysis but the study used precipitation data reported from this station, as values of precipitation are deemed correct and homogeneous.

After quality control and homogeneity testing, all stations were retained for analysis except Seeb temperature variable. The final list of stations is summarized in Table 1 where details of station name, number, location, elevation and missing percentage (%) are listed.

Table 1 Stations name, number, location, elevation and missing percentage (%)

Station	WMO	Latitude	Longitude	Elevation (m)	Missing (%)
Salalah	41,316	17.03	54.08	22.0	0
Masirah	41,288	20.67	58.91	19.0	0.2
Saiq	41,254	23.07	57.63	1755.0	24.0
Seeb	41,256	23.58	58.28	8.4	4.4
Sur	41,268	22.53	59.47	14.0	1.5
Thumrait	41,314	17.67	54.02	467.0	17.6
Khasab	41,240	26.11	56.14	33.0	1.0
Sohar	41,246	24.47	56.63	3.6	10.5

3 Methods

3.1 Trends

Linear trends computed for variables such as mean maximum temperature, mean temperature, mean minimum temperature and data series for precipitation using the nonparametric approach proposed by Sen (1968), later this method was modified by Wang and Swail (2001) to account for time series auto correlation. This method has been considered as a robust one, which is resistant to outliers and can be used for computation when data are missing. While calculating liner trend, annual missing values are excluded from the analysis and the method has been widely used to compute trends in climate science related studies (Aguilar et al. 2009; Alexander et al. 2006; Butt et al. 2009; Zhang et al. 2005). Overall, the significance of a trend is determined using Mann-Kendall's test for data that need not conform to any particular distribution and Manton et al. (2001) adapted a similar approach. A 5% level of statistical significance has been used in the study and monthly, seasonal as well as annual trends are calculated for all the 8 stations based on stations time period.

3.2 Area Averaging

Oman regional average time series for the 4 variables were calculated following Aguilar et al. (2009) method, which avoid stations domination with high mean values by averaging time series anomalies of stations. A base reference of 1986–2008 was used to calculate the station normals and anomalies following Jones and Moberg (2003) methodology that was used minimum of 20 years of data within the span of 30-years period to calculate the normals. Therefore, there was a need for minimum of 18 years data needed to be available in this base period for the station to be used for calculations. The area average time series have been computed for 3 areas under this study: the all AP, the monsoonal area under the direct influence of the monsoon which covers 3 stations namely Masirah, Thumrait and Salalah and the non-monsoonal stations located in the North of Oman comprising of Khasab, Sohar, Saiq, Seeb and Sur stations.

4 Results and Discussion

4.1 Temperature Trends

Analysis on trends of mean, mean maximum and mean minimum temperature and precipitation are discussed. Analysis show that temperature parameter trends are consistent in sign and significance although magnitude varies. However, the precipitation trends show some changes in sign but more trends reported that are negative. Most of precipitation trends are not significant except in few stations.

4.1.1 Mean Temperature

The general pattern of Oman mean annual temperature trend is warming. All regional and subregional trends reported significant warming trends (Table 2).

Table 2 Stations and Regional annual mean temperature trends for the period 1980–2013 (All-all Arabian Peninsula, NMon- Non-monsoonal subregion, Mon-Monsoonal subregion). Trends in bold are significant at 0.05 level.

The highest statistically significant warming trends are found in Sur = 1.1 °C $decade^{-1}$ and over the NMON region 0.6 °C $decade^{-1}$. The lowest trend values are observed on the south and southeast coast of Oman (Salalah 0.1 °C $decade^{-1}$). Seasonally the warming is predominant and consistent over most of the year with the exception of winter season. The trends are higher in magnitude over the NMON region relative to the MON region.

Figure 1 shows the annual time series for all the stations and regional annual mean temperature, which indicate the gradual increase on the temperature over

Table 2 Stations and regional annual mean temperature trends (°C $decade^{-1}$)

Station/area	Trend ($decade^{-1}$)
Khasab	**0.5**
Sohar	**0.6**
Saiq	**0.3**
Sur	**1.1**
Masirah	**0.2**
Thumrait	**0.2**
Salalah	**0.1**
ALL	**0.4**
NMON	**0.6**
MON	**0.2**

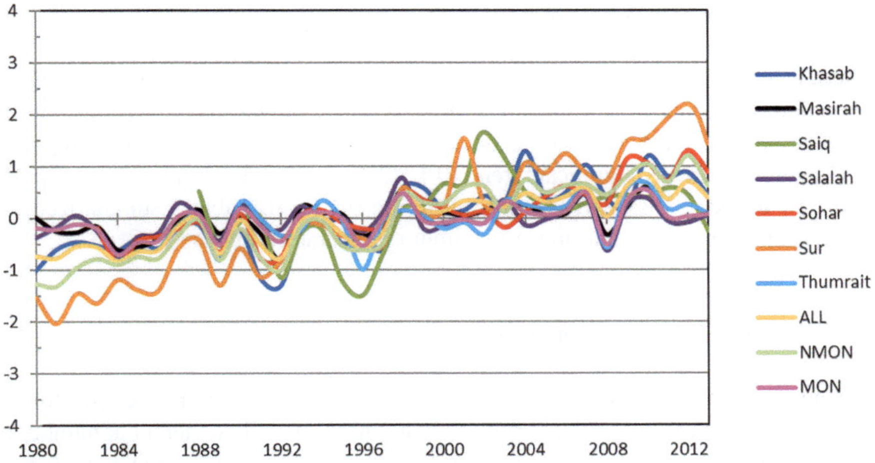

Fig. 1 All stations and regional annual mean temperature Anomlies during 1980–2013

most Oman regions especially after the year 1998 where most of stations report positive temperature anomalies. The recent significant warming over the Arabian Peninsula (AP) in the past few decades is in agreement with some of other dryland or AP region studies reported. This study shows that the non-monsoonal region located north of 20°N has experienced higher rates of warming. A similar study has been reported by Solomon et al. (2007) which indicated greater increases in the interior compared with coastal regions. A negative association between precipitation and temperatures have experienced for dry lands in which warmer (colder) years have been drier (wetter) according to Hulme (1996) and Jones and Reid (2001). Continuous negative precipitation anomalies are matched by continuous positive temperature anomalies since 1998 (AlSarmi and Washington 2011). These results generally agree with studies that have been observed in other parts of the world during the second half of the 20th century (Frich et al. 2002; Alexander et al. 2006). Almazroui et al. (2012a, b) have pointed to a decrease in rainfall over Saudi Arabia by 35.1 mm (5.5 mm) per decade during the wet (dry) season from years 1994–2009 and temperature has increased at 0.72 °C per decade in the dry season. Further, another study conducted by these authors report that maximum, mean and minimum temperatures have increased at a rate of 0.71°, 0.60°, and 0.48° per decade significantly. More recently, one of the study examined the temporal changes in climate extremes in the Arab region with regard to long-term trends (Donat et al. 2013). This study found that consistent warming trends across the region that highlights increased frequencies of warm nights and warm days, higher extreme temperature, cold nights and fewer cold days as well as shorter cold spell durations are all evident right from early 1970s.

4.1.2 Mean Maximum Temperature

There is a less broad statistically significant increase (at 0.05 level) in mean annual maximum temperature in Oman. The highest mean annual trends are observed in the Khasab, which reported 1.6 °C decade^{-1}. Thumrait over Oman's south reported the lowest statistically significant trend (0.1 °C decade^{-1}). The cooling over Salalah (−0.1 °C decade^{-1}) is nearly significant to the 0.1 level. Table 3 summarized the stations and regions annual mean maximum temperature trends.

4.1.3 Mean Minimum Temperature

It is apparent that the warming trends of the mean annual minimum temperature in Oman are statistically significant and these are more spatially coherent at number of stations in comparison with those for the annual mean and annual maximum temperature.

All stations and regions of Oman show statistically significant warming trends and the highest trend values are observed in the area of Sohar (1.1 °C decade^{-1}) while Saiq and Thumrait recorded 0.6 °C decade^{-1}. One single striking difference with the other variables is that the trend of the mean annual minimum temperature is statistically significant over all locations covering coastal, inland or mountainous areas. Similar to the mean temperature the seasonal trends of minimum temperature are clearly observed over all seasons with lower magnitude during winter. The mean annual trends of the stations and regional averages are summarised in Table 4.

Figure 2 shows the variability of the anomalies annual time series for all stations and regional annual mean temperature that is similar to the annual mean temperature show the clear increase of the minimum temperature over most Oman regions especially after the year 1998.

Table 3 Same as Table 2 but for mean maximum temperature

Station/area	Trend (decade^{-1})
Khasab	**1.6**
Sohar	0.0
Saiq	−0.2
Sur	
Masirah	0.1
Thumrait	**0.1**
Salalah	**−0.1**
ALL	**0.2**
NMON	**0.4**
MON	0.1

Table 4 Same as Table 2 but for mean minimum temperature

Station/area	Trend ($decade^{-1}$)
Khasab	0.4
Sohar	1.1
Saiq	0.6
Sur	0.3
Masirah	0.5
Thumrait	0.6
Salalah	0.4
ALL	0.6
NMON	0.7
MON	0.5

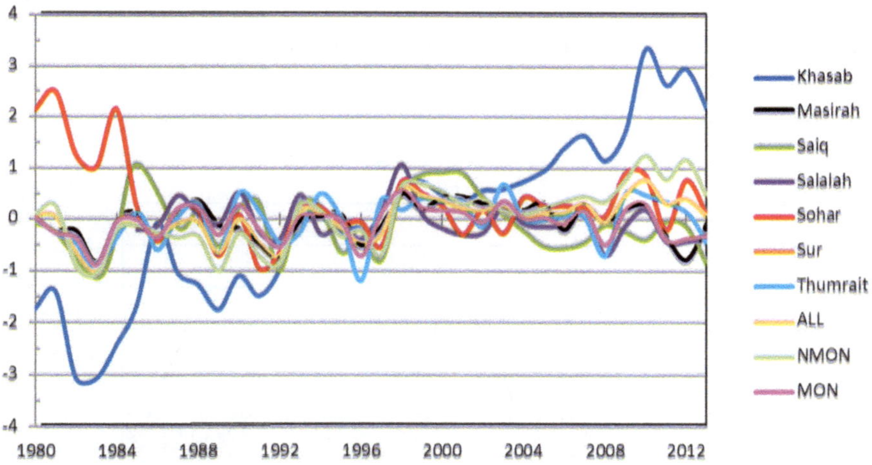

Fig. 2 Same as graph 1 but for annual mean maximum temperature

4.1.4 Precipitation Trends

Regarding trends values of total annual precipitation show that only 2 stations with statistically significant. Stations Saiq and Salalah show negative trends: (Saiq = −74.0 mm $decade^{-1}$ and Salalah = −10.8 mm $decade^{-1}$). Other than these two stations, the trends are negative and not statistically significant for remaining stations observed in Oman.

Figure 3 shows the annual anomalies time series of the precipitation for all the stations and the regions. The drying since year 1999 is unique and considered the longest period of drought since record (Fig. 4).

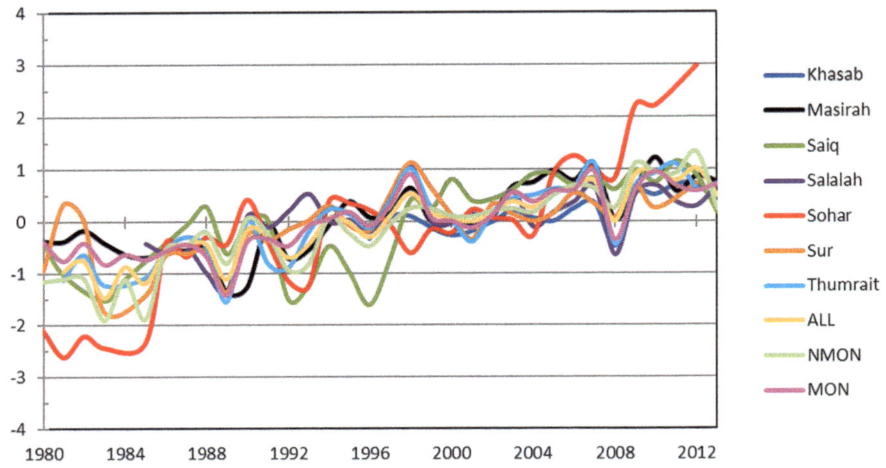

Fig. 3 Same as Fig. 1 but for annual mean minimum temperature

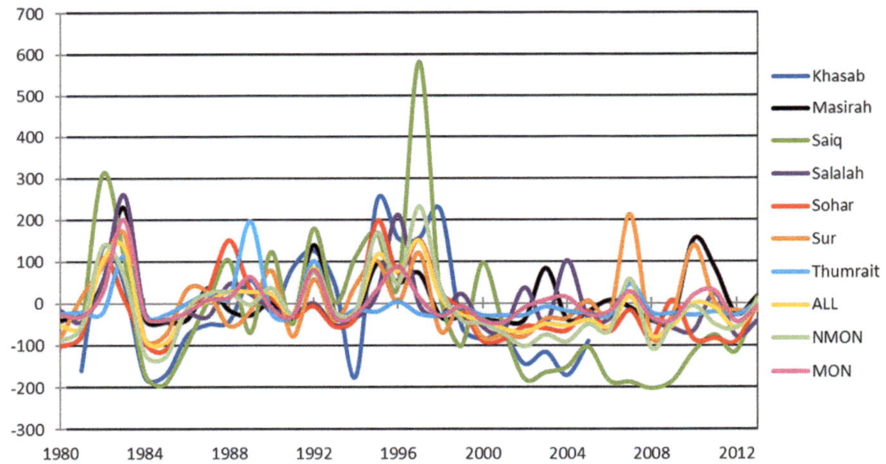

Fig. 4 Same as Fig. 1 but for total annual precipitation (mm)

5 Conclusion

Overall, the results of this study provide clear evidences that Oman climate has changed during the past three decades. The significant changes are associated mostly with the parameters of temperature than the changes in precipitation. This is in broad agreement with the recent findings obtained either over the AP or nearby regions (Almazroui et al. 2012a, b; Athar 2012; AlSarmi and Washington 2011, 2013; Donat et al. 2013; Hartmann et al. 2013; Zarenistanak et al. 2014). The trend pattern varies between the north and south of Oman where higher significant trends in temperature elements are noted in the north relative to the south. In addition, nighttime temperatures are significant at both the North and the South with higher magnitudes at the former region.

There is an indication of a regional climate jump on 1998, this jump is clearly identified in the time series of many temperature parameters and for several stations. The increase of nighttime temperature over Oman is remarkable especially during the last 2 decades. The increase is significant over both urban and rural stations, which suggest the need for a regional climate change attribution experiments. Seasonality is also an issue to consider when studying the climate changes in Oman.

References

Aguilar E, Barry AA, Brunet M, Ekang L, Fernandes A, Massoukina M, Mbah J, Mhanda A, do Nascimento DJ, Peterson TC, Umba OT, Tomou M, Zhang X (2009) Changes in temperature and precipitation extremes in western central Africa, Guinea Conakry, and Zimbabwe, 1955–2006. J Geophys Res 114:D02115. doi:10.1029/2008JD011010

Alexander LV, Zhang X, Peterson TC, Caesar J, Gleason B, Klein Tank AMG, Haylock M, Collins D, Trewin B, Rahimzadeh F, Tagipour A, Rupa Kumar K, Revadekar J, Griffiths G, Vincent L, Stephenson DB, Burn J, Aguilar E, Brunet M, Taylor M, New M, Zhai P, Rusticucci M, Vazquez-Aguirre JL (2006) Global observed changes in daily climate extremes of temperature and precipitation. J Geophys Res 111:D05109. doi:10.1029/2005JD006290

Almazroui M, Islam MN, Jones PD, Athar H, Rahman MA (2012a) Recent climate change in the Arabian Peninsula: seasonal rainfall and temperature climatology of Saudi Arabia for 1979-2009. Atmos Res doi:10.1016/j.atmosres.2012.02.013

Almazroui M, Nazrul Islam M, Athar H, Jones PD, Rahman MA (2012b) Recent climate change in the Arabian Peninsula: annual rainfall and temperature analysis of Saudi Arabia for 1978-2009. Int J Climatol 32(6):953–966. doi:10.1002/joc.3446

AlSarmi S, Washington R (2011) Recent observed climate change over the Arabian Peninsula. J Geophys Res 116:D11109. doi:10.1029/2010JD015459

AlSarmi S, Washington R (2013) Changes in climate extremes in the Arabian Peninsula: analysis of daily data. Int J Climatol n/a-n/a. doi:10.1002/joc.3772

Athar H (2012) Decadal variability of the observed daily temperature in Saudi Arabia during 1979?2008. Atmos Sci Lett 13(4):244–249. doi:10.1002/asl.390

Butt N, New M, Lizcano G, Malhi Y (2009) Spatial patterns and recent trends in cloud fraction and cloud-related diffuse radiation in Amazonia. J Geophys Res 114:D21104. doi: 10.1029/2009JD012217

Conrad V, Pollak LW (1950) Methods in Climatology. Harvard University Press: Boston, MA

Donat MG, Peterson TC, Brunet M, King AD, Almazroui M, Kolli RK, Boucherf D, Al-Mulla AY, Nour AY, Aly AA, Nada TAA, Semawi MM, Dashti HAA, Salhab TG, El Fadli KI, Muftah MK, Eida SD, Badi W, Driouech F, Rhaz KE, Abubaker MJY, Ghulam AS, Erayah AS, Mansour MB, Alabdouli WO, Dhanhani JSA, Al Shekaili MN (2013) Changes in extreme temperature and precipitation in the Arab region: long-term trends and variability related to ENSO and NAO. Int J Climatol n/a–n/a. doi:10.1002/joc.3707

Economic and Social Commission for Western Asia (2010) Regional and global priorities: climate change, food security and empowerment of women. Tech. rep. E/ESCWA/2009/C.4/5, Beirut, pp 3–8

Freiwan M, Kadioglu M (2008) Climate variability in Jordan. Int J Climatol 28(1):69–89

Frich P, Alexander LV, Della-Marta P, Gleason B, Haylock M, Tank Klein AMG, Peterson T (2002) Observed coherent changes in climatic extremes during the second half of the twentieth century. Climate Res 19(3):193–212

Goddard L, Mason SJ, Zebiak SE, Ropelewski CF, Basher R, Cane MA (2001) Current approaches to seasonal-to-interannual climate predictions. Int J Climatol 21(9):1111–1152

Hartmann D, Klein Tank A, Rusticucci M, Alexander L, Brönnimann S, Charabi Y, Dentener F, Dlugokencky E, Easterling D, Kaplan A, Soden B, Thorne P, Wild M, Zhai P (2013) Observations: atmosphere and surface. In: Stocker TF, Qin D, Plattner GK, Tignor M, Allen SK, Boschung J, Nauels A, Xia Y, Bex VMidgley PM (eds) Climate change 2013: the physical science basis. Contribution of working group I to the fifth assessment report of the intergovernmental panel on climate change. Cambridge University Press, Cambridge and New York

Hulme M (1996) Recent climatic change in the world's drylands. Geophys Res Lett 23(1):61–64

Jones PD, Reid PA (2001) Temperature trends in regions affected by increasing aridity/humidity. Geophys Res Lett 28(20):3919–3922. doi:10.1029/2001GL013840

Jones PD, Moberg A (2003) Hemispheric and large-scale surface air temperature variations: An extensive revision and an update to 2001. J Clim 16:206–223

Kwarteng AY, Dorvlo AS, Kumar GTV (2009) Analysis of a 27-year rainfall data (1977-2003) in the Sultanate of Oman. Int J Climatol 29(4):605–617

Manton MJ, Della-Marta PM, Haylock MR, Hennessy KJ, Nicholls N, Chambers LE, Collins DA, Daw G, Finet A, Gunawan D, Inape K, Isobe H, Kestin TS, Lefale P, Leyu CH, Lwin T, Maitrepierre L, Ouprasitwong N, Page CM, Pahalad J, Plummer N, Salinger MJ, Suppiah R, Tran VL, Trewin B, Tibig I, Yee D (2001) Trends in extreme daily rainfall and temperature in southeast Asia and the south Pacific: 1961-1998. Int J Climatol 21(3):269–284

Nasrallah HA, Balling RC Jr (1993) Analysis of recent climatic changes in the Arabian Gulf region. Environ Conserv 20(3):223–226

Sen PK (1968) Estimates of the regression coefficient based on Kendall's Tau. J Amer Statist Assoc 63:1379–1389

Solomon S et al (2007) Summary for policymakers of climate change 2007: The Physical Science Basis, Contribution of Working Group I to the Fourth Assessment Report of the Intergovernmental Panel on Climate Change. Cambridge Univ. Press, Cambridge, UK

Wang XL (2003) Comments on detection of undocumented changepoints: A revision of the two-phase regression model. J Clim 16:3383–3385

Wang XL (2008) Accounting for autocorrelation in detecting mean shifts in climate data series using the penalized maximal t or F test. J Appl Meteor Clim 47:2423–2444

Wang XL, Swail VR (2001) Changes of extreme wave heights in northern hemisphere oceans and related atmospheric circulation regimes. J Clim 14:2204–2221

Zarenistanak M, Dhorde AG, Kripalani RH (2014) Trend analysis and change point detection of annual and seasonal precipitation and temperature series over southwest Iran. J Earth Syst Sci 123(2):281–295

Zhang X, Aguilar E, Sensoy S, Melkonyan H, Tagiyeva U, Ahmed N, Kutaladze N, Rahimzadeh F, Taghipour A, Hantosh TH, Albert P, Semawi M, Karam Ali M, Said Al-Shabibi MH, Al-Oulan Z, Zatari T, Khelet IAD, Hamoud S, Sagir R, Demircan M, Eken M, Adiguzel M, Alexander L, Peterson TC, Wallis T (2005) Trends in middle east climate extreme indices from 1950 to 2003. J Geophys Res 110:D22104. doi:10.1029/2005JD006181

Part II
Groundwater Resources

Significance of Silica Analysis in Groundwater Studies of Domestic Shallow Wells in Parts of Jeli District, Kelantan, Malaysia

Mohammad Muqtada Ali Khan, Fatin Wahida Fadzil, Hafzan Eva Mansor, Dony Adriansyah Nazaruddin and Zameer Ahmad Shah

Abstract The present study was carried out to investigate the nature and behaviour of silica in shallow groundwater in parts of Jeli district, Malaysia to assess the groundwater and surface-water interaction, to determine the depths, residence duration of groundwater and to evaluate groundwater contamination by correlating silica with hydrochemistry and the temperatures at which groundwater has acquired present silica concentrations. The most important factors controlling Silica concentration in the groundwater is the rock type with which water comes in contact and most importantly the duration of interaction. In the present study, silica concentrations analysed in 7 samples exhibit an average value of 14 mg/l. Such silica concentrations clearly indicate the minimal impact of anthropogenic activities and small residence time of groundwater in terms of rock-water interaction. Prolonged rock-water interaction would have resulted in acquisition of high silica concentrations as the area is mostly covered by igneous rocks. Such low silica values (7–21 mg/l) also point to the influent nature of the rivers in the area where surface water has caused the smoothening of silica concentrations in groundwater. By employing silica geo-thermometry, the maximum temperature estimated is about 33 °C which under normal geothermal conditions would corresponds to a depth of 0.20 km (200 m) by considering an average heat flow of 30 °C/km. Such shallow depths (below 0.20 km) of groundwater circulation are pretty reasonable for such deficient silica concentrations.

Keywords Silica analysis · Shallow wells · Groundwater · Malaysia

---

M.M.A. Khan (✉) · F.W. Fadzil · H.E. Mansor · D.A. Nazaruddin
Department of Geoscience, Faculty of Earth Science, Universiti Malaysia Kelantan, Campus Jeli, Locked Bag No. 100, 17600 Jeli, Kelantan, Malaysia
e-mail: muqtadakhan@gmail.com

Z.A. Shah
Department of Geology, Aligarh Muslim University, Aligarh 202002, India

O. Abdalla et al. (eds.), *Water Resources in Arid Areas: The Way Forward*, Springer Water, DOI 10.1007/978-3-319-51856-5_7

1 Introduction

Seven selected groundwater samples were analysed to measure dissolved silica content representing the entire study area. The study was taken-up with intent to have an idea about the depth of circulation of groundwater and address its characteristics such as, residence time and interaction with solid phases, i.e. granular zone or bedrock. It needs to be mentioned here that although, relatively rich literature exists on relationship of silica with parameters, such as, temperature, regional heat flow values and aquifer depth, but conventionally silica analysis is not performed as part of groundwater studies (Marchand et al. 2002; Fournier 1983; Boughton and McCoy 2006; Swanberg and Morgan 1978; Geological Survey of India 1991; Khan and Umar 2010). The main aim of the present investigation is to emphasize the significance of silica analysis in various groundwater related issues in shallow domestic wells.

Groundwater acquires Silica exclusively and unequivocally through rock-water dealings (Hem 1959). SiO_2 concentration in groundwater ranges between 1 and 30 mg/l with median value being 17 mg/l (Davis 1964). Another study (Marchand et al. 2002) demonstrates that silica values of groundwater rises due to prolonged interaction with silicate rocks and that the silica levels are totally dependent on the duration during which water remains underground. Studies have also proved that silica values are higher in groundwater that have circulated deeper than waters of shallower circulations (Khan and Umar 2010).

Rock types with which groundwater interact over a longer period are the principal factors controlling its concentration levels (Davis 1964). About 70,000 non-thermal groundwater samples of USA were examined by Swansberg and Morgan (1978) using silica geothermometery and a strong correlation was found between silica geotemperatures for a given region and the known regional heat flow and this experiment brought them with the following relationship:

$$\text{Silica derived temperature in groundwater (T in °C)} = \text{mq} + \text{b}$$

In this equation, q represents the regional heat flow in m Wm^{-2}-units, and m and b are invariables having values of 0.67 °C m^2 mW^{-1} and 13.2 °C, respectively. Value of ‘b’ is dependent on the mean air temperature and ‘m’ is related to the level up to which water percolates. They estimated the depth of circulation of groundwater varied from 1.4 to 2.0 km.

2 Study Area

The study area lies between latitude 05° 42′ 49″N–05° 45′ 31″N and longitude 101° 50′ 31.62″E–101° 53′ 14″E and covers an area of about 25 km^2 (Fig. 1) and is a part of the Jeli district in the state of Kelantan in peninsular Malaysia. It is bounded on the east by Tana Merah district, on the west by Perak, on the south by

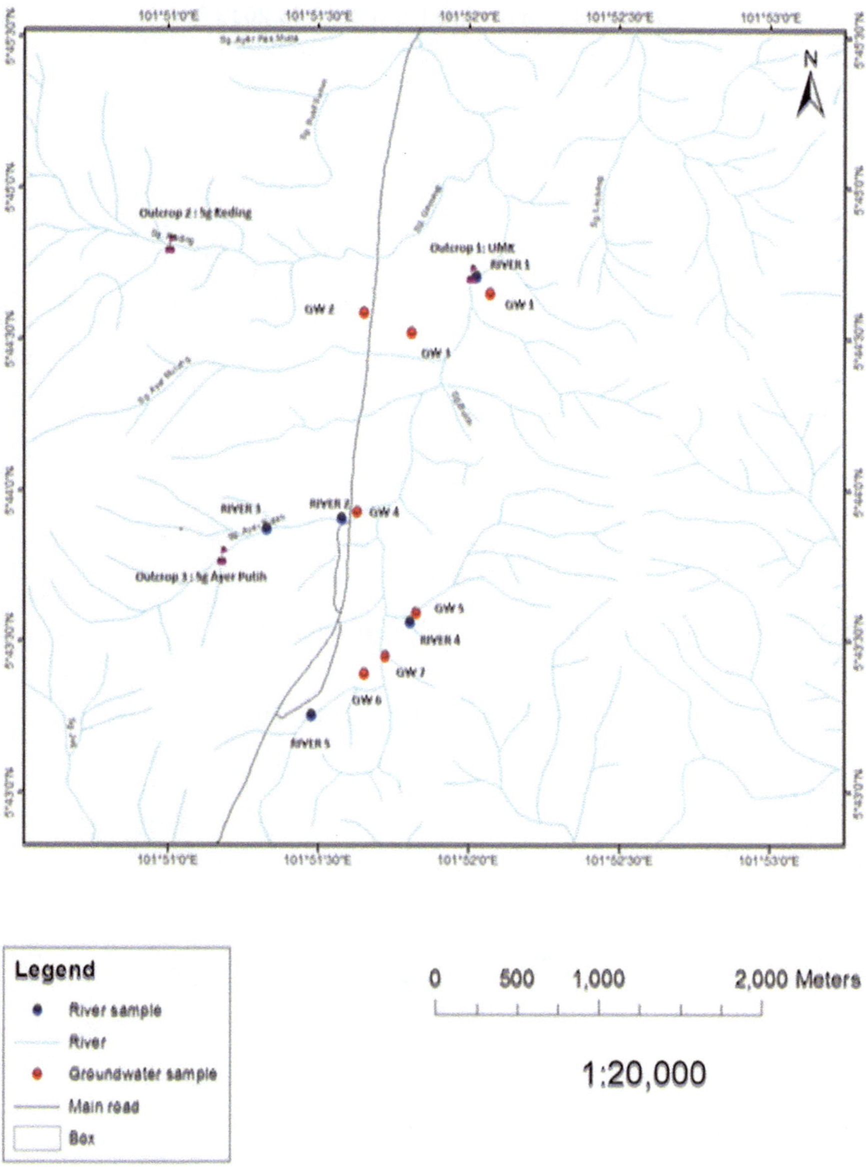

Fig. 1 Map showing location of study area

Gua Musang district and in the north by Thailand. Around 70% of study area is hilly or mountainous and covered by thick forest traversed by many rivers showing mainly dendritic and radial pattern. The altitude of the area ranges from 90 to 500 m above sea level. The area is famous for rubber plantation and orchids and majority of people are engaged in agricultural activities.

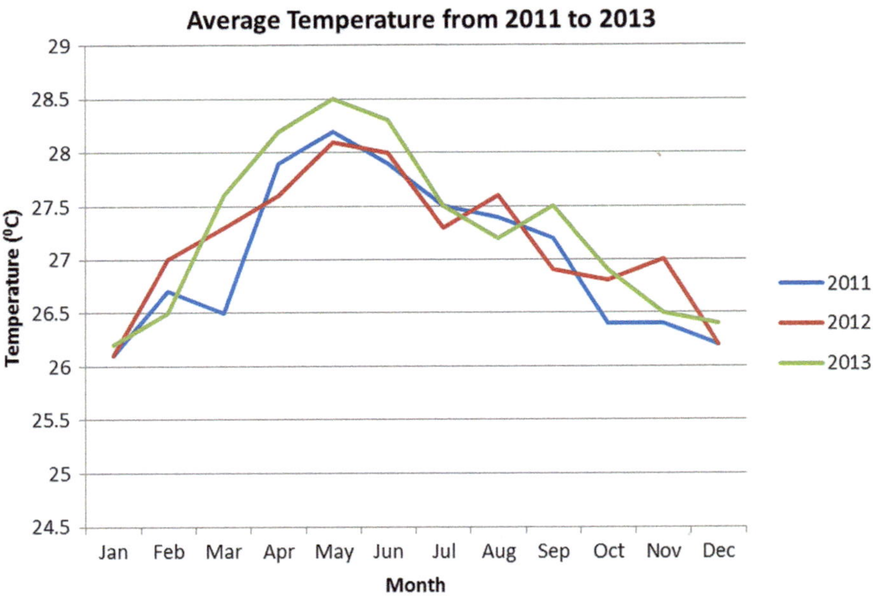

Fig. 2 Graph representing average temperature from 2011 to 2013

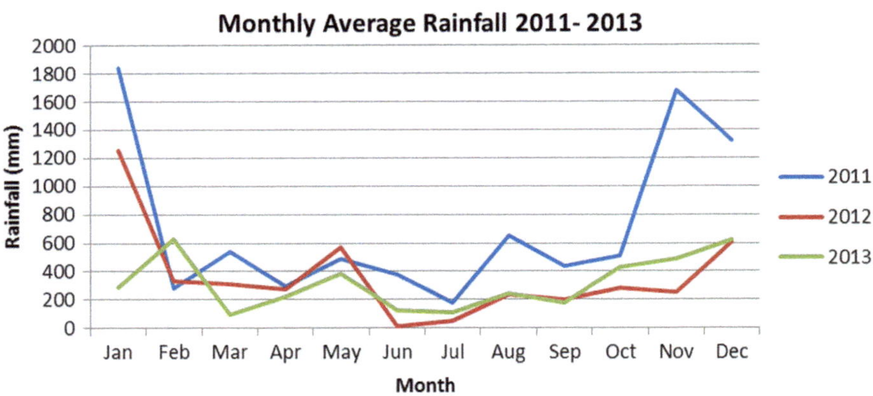

Fig. 3 Graph showing average rainfall 2011–2013

Climatically, the area experiences uniform temperature and high humidity. The temperatures vary between 21 and 32 °C throughout the year; without any drastic seasonal variation and normal temperature around 27.13 °C. Figure 2 shows the pattern of average temperature of three years. The area receives heavy rainfall with an average 8596 mm in year 2011, 4383 mm in 2012 and 3810.5 mm in 2013 (Fig. 3). This is the time when water in the rivers flow over banks and chances of floods are high.

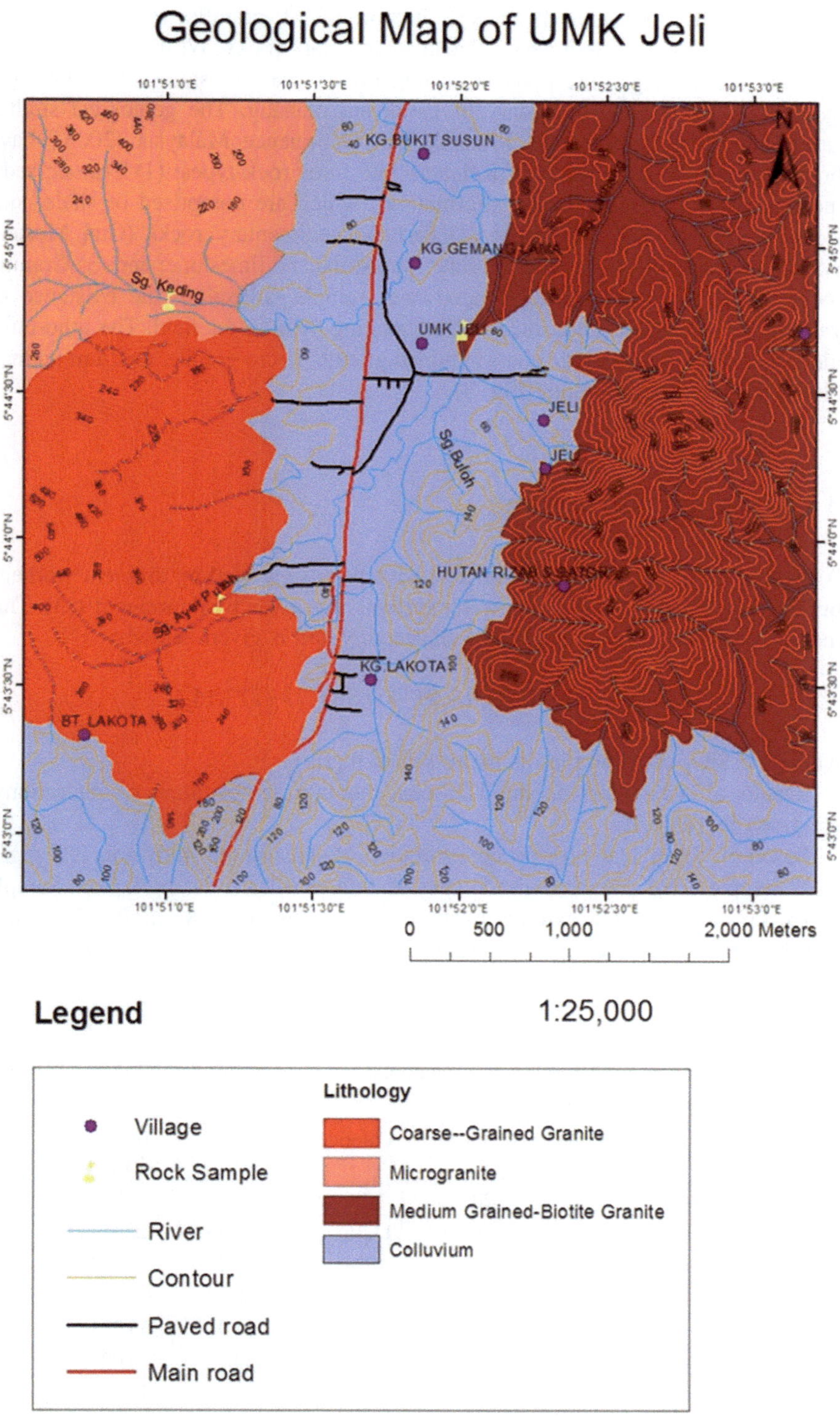

Fig. 4 Lithology map of study area

Geologically, the area is mostly comprised of granitic rocks containing numerous enclaves of Sedimentary/meta sedimentary rocks. West of the Kelantan contains the main granitic body stretching on the western side of the state and extends up to the state boundary of Perak and Pahang. The geological studies carried-out by Department of Minerals and Geoscience Malaysia (2003), have broadly classified Jeli district of Kelantan into three rock types: (1) Triassic sedimentary rocks (Gunong Rabong Formation), which are comprised of shale, siltstone, sandstone, and limestone; (2) Permian sedimentary rocks (Gua Musang Formation) which include phyllite, slate, sandstone and limestone; and (3) Granitic rocks (acid intrusive). During groundwater sample collection, the existence of various litho-units was confirmed by demarcating them in the field. The litho-units identified in the field include coarse grained granite, micro granite, medium grained biotite granite and colluvium (Fig. 4).

3 Result and Discussion

A wide range of silica geothermometers have been suggested by different workers, but the most fitted one for groundwater systems was given by Fournier (1983). This method emphasises the solubility of chalcedony and is given below:

$$t°C = \left[(1032)/\left(4.69 - \log SiO_2^-\right)\right] - 273.15$$

where the SiO_2 concentration is expressed in mg/l.

Results of silica, chloride and TDS concentration, temperatures of chalcedony, are given in Table 1.

Based on Table 1, the lowest concentration of silica in the study area is 7 mg/L from Sample GW4 and the highest concentration of silica is 21 mg/L from Sample GW5. Overall, the area exhibits average silica value of 13.86 mg/L. The silica distribution of the study area is shown in Fig. 5. The areas with lowest silica

Table 1 Silica, chloride, TDS values in groundwater and temperatures derived (Silica, Chloride and TDS values are given in mg/L)

Sample no	Sample location code	Silica	Chloride	TDS	Temperature °C chalcedony
1	GW1	20	9.94	16.68	31.367
2	GW2	19	11.36	42.45	29.379
3	GW3	9	9.94	13.23	3.099
4	GW4	7	9.94	80.44	−4.742
5	GW5	21	19.88	101.4	33.283
6	GW6	9	24.14	153.8	3.099
7	GW7	12	12.78	93.16	12.658

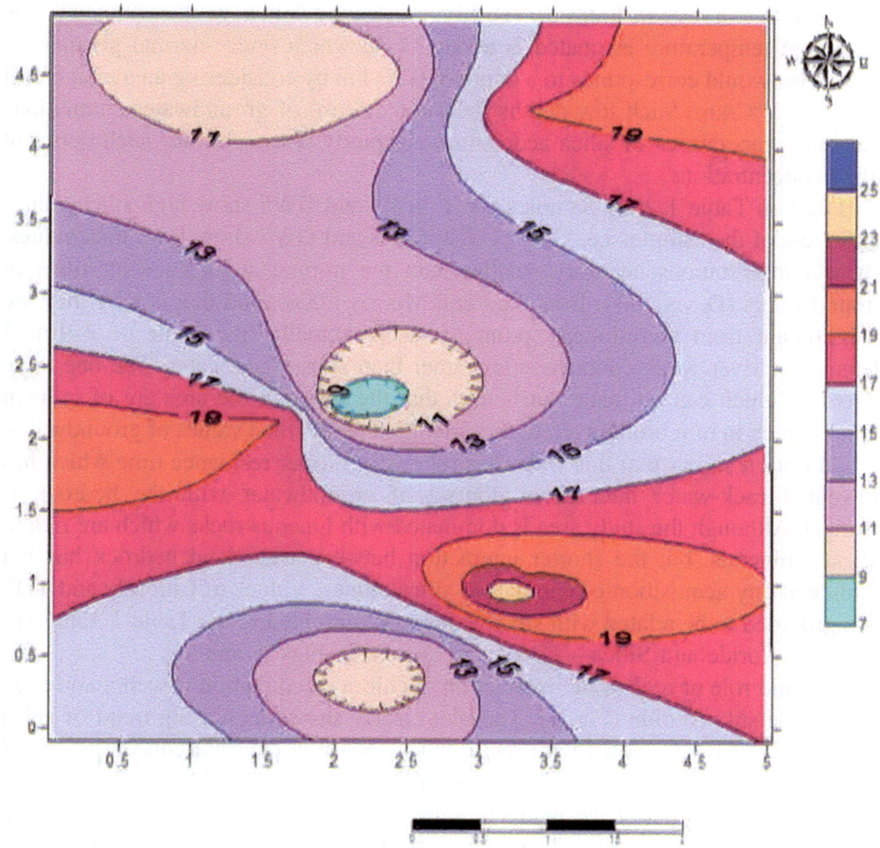

Fig. 5 Distribution map of silica in the study area in mg/L

concentrations lie in the central part while the highest values are towards north east and northwest side of study area.

The temperature of the silica is calculated using the Fournier equation. The highest temperature of the silica for groundwater as estimated by the chalcedony equation is 33.283 °C, which is from Sample GW5. The lowest temperature recorded is 3.099 °C from Sample GW3 and GW6. The concentrations of silica for these samples are only 9 mg/L. Other samples show temperatures ranging between 12.658 and 31.367 °C for Sample GW1, GW2 and GW7. Sample GW4 shows Silica value as low as 7 mg/l which is insignificant as far as the present investigation is concerned.

The temperature calculated from silica values can directly manifest the depth from which groundwater has circulated by taking geothermal gradient into consideration and excluding the ambient surface temperature of 27 °C. Presuming typical terrestrial thermal gradient of 30 °C/km, the remaining temperature will be divided by 30 °C in order to get the depth for the groundwater in kilometres

(Geological Survey of India 1991). By employing silica geothermometry, the maximum temperature estimated is about 33 °C which under normal geothermal conditions would corresponds to a depth of 0.20 km by considering an average heat flow of 30 °C/km. Such low depths (below 0.20 km) of groundwater circulations and low temperatures of silica acquisition are pretty reasonable for such deficient silica concentrations.

Based on Table 1, only Samples GW1, GW2 and GW5 show high silica values while rest of the samples i.e. GW3, GW4, GW6 and GW7 show low silica values. These concentrations are very smaller than the normal silica concentrations of groundwaters (Davis 1964; Boughton and McCoy 2006), and that is why they are insignificant from thermometry point of view. Actually, the value of 2–9 mg/l demarcates river water characteristics rather than subsurface water. The one logic assertion which can be made out here is that the rivers in the area are of influent nature which in turn brought about the smoothening of silica values of groundwater. In addition, it shows that this water has relatively shorter residence time which has prevented rock-water interaction (contact of groundwater with the bedrock is shorter). Although the study area is dominated with igneous rocks which are rich in silicate minerals, but the shorter interaction between water and bedrock has not facilitated any acquisition of anomalous silica values. Values of Chloride and TDS also provided to be related with silica values are presented in the Table 1. Graph of Silica—Chloride and Silica—TDS are presented in Figs. 6 and 7.

The mild role of rock-water interaction in chloride acquisition is well manifested in silica versus chloride (Fig. 6). The plot clearly shows increasing trend of silica concentrations is associated with more or less constant chloride concentrations of around 10 mg/l (as shown by the arrow in Fig. 6) which indicates that silica has increased only via rock-water interaction and chloride is hardly contributed through any other source to the groundwater system. Although such low chloride concentrations hardly make any sense of involvement of anthropogenic activities, but while correlating them with silica values, a very mild role of anthropogenic

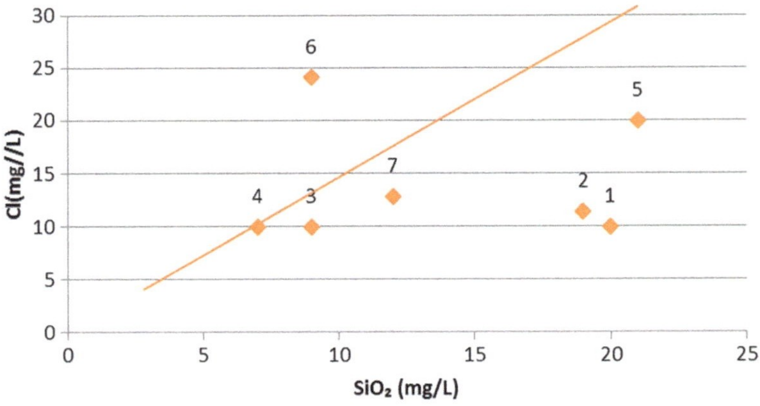

Fig. 6 SiO_2 versus Cl plot for groundwater

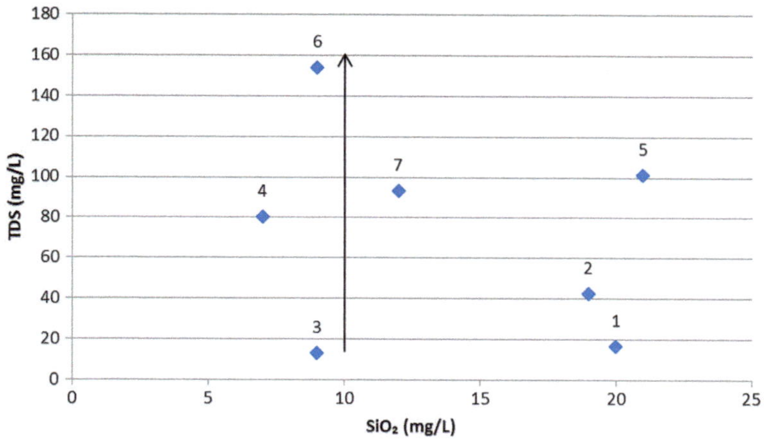

Fig. 7 SiO_2 versus TDS plot of groundwater

Table 2 Values of silica, chloride and total dissolved solids of river samples (Silica, Chloride and TDS values are given in mg/L)

Sample no	Sample location code	Silica	Chloride	TDS	Temperature °C chalcedony
8	River 1	21	9.94	39.31	33.283
9	River 2	19	9.94	41.01	29.379
10	River 3	20	9.94	34.19	31.367
11	River 4	25	9.94	37.82	40.332
12	River 5	16	8.52	31.44	22.901

activities comes into play. This role is well established by low silica values having comparatively higher chloride concentrations (in sample GW6).

Overall, the ion-acquisition in the area under investigation is dominantly through rock-water interaction. However, the role of anthropogenic activities is effective to some extent in a very few samples. This is well expressed in silica versus TDS plot where average silica value of around 10 mg/l is associated with TDS values up to 180 mg/l (Fig. 7). Very small changes in silica concentrations accompanied by sizeable fluctuations in TDS can be well attributed to the mild role of anthropogenic activities.

Moreover, rivers in the study area were also brought under this investigation and five samples were collected from them to analyse the mixing behaviour of surface water and groundwater. The results of silica, Cl and TDS concentration are shown in Table 2.

Similar interpretations were made in river water samples (Fig. 8). The negligible role of rock-water interaction in chloride acquisition is again clear from Fig. 8, where varying silica values exhibit very similar chloride concentrations. The consistent rise in the silica concentrations can be well attributed to the surface run-off

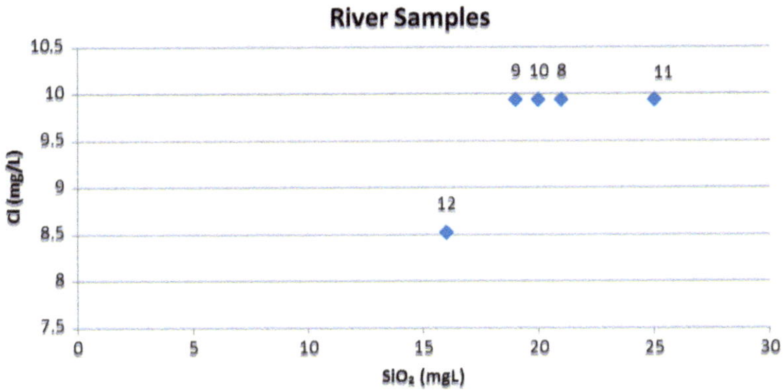

Fig. 8 SiO_2 versus Cl plot for river water

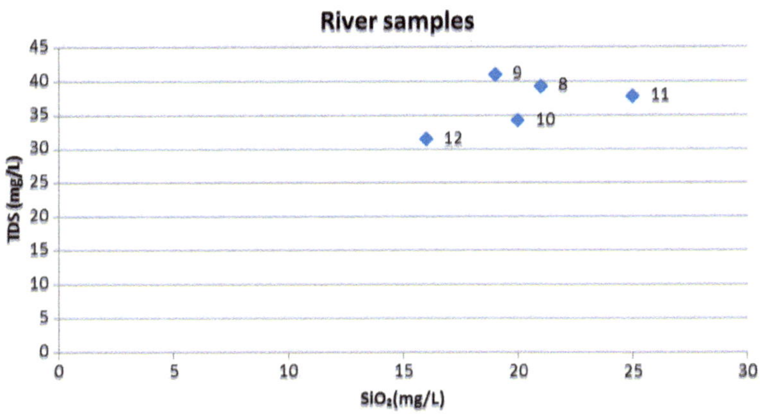

Fig. 9 SiO_2 versus TDS plot for river water

draining into the rivers in the area which is usually loaded with comparatively higher silica concentrations.

In surface waters Cl concentrations serve as path finders for delineation of anthropogenic activities. However, in the present study, similar and low Cl values associated with lower but varying silica concentrations hardly depict any such thing. The low but consistent rise in silica values can be well attributed to diatoms in the river water (Aigars et al. 2014; Conley 1997).

In Fig. 9, the Silica values have changed more significantly than TDS values which can indicate:

1. The hydrologic behaviour of the river varies from place to place.
2. The overall smaller TDS values indicate less interaction of water with the underlying rocks and banks and perhaps the continuously flowing water which does not facilitate prolonged water-rock interfaces.

4 Conclusion

The present investigation shows that SiO_2 values are very vital while assessing the depth of groundwater circulations, evolution of groundwater chemistry through various hydrogeochemical processes, geogenic or anthropogenic involvement that may influence the distribution of major ions in groundwater. Correlations of silica values with conservative ions like Cl or total solute concentration are very decisive in demarcating the sources of various chemical species and in management and conservation of groundwater resources. Keeping all these applications under consideration, silica analysis should be involved in all possible groundwater investigations.

Acknowledgements The financial assistance received from Fundamental research Grant (R/FRGS/A08.00/00644A/002/2015/000228) is highly acknowledged. The authors are also grateful to faculty of Earth Science, Universiti Malaysia Kelantan, Campus Jeli, for endowing with basic facilities to carry out the present investigation.

References

Boughton CJ and McCoy JK (2006) Hydrogeology, aquifer geochemistry, and ground-water quality in Morgan County, Scientific Investigation Report 2006-5198, U.S. Department of Interior and U.S. Geological Survey, Reston Virginia

Conley DJ (1997) Riverine contribution of biogenic silica to the oceanic silica budget. Limnol Oceanogr 42:7742–7777

Davis NS (1964) Silica in streams and groundwater. Am J Sci 262:870–891

Department of Minerals and Geoscience Malaysia (2003) Quarry resource planning for the State of Kelantan. Osborne & Chappel Sdn. Bhd, Kota Bharu, Malaysia

Fournier RO (1983) A revised and expanded silica geothermameter. Bull Geoth Res Counc 11:329

Geological Survey of India (1991) Geothermal Atlas of India, Geological Survey of India special publication No 19

Hem JD (1959) Study and interpretation of chemical characteristics of natural water. U.S. Geological Survey Water Supply paper 1473, 363 pp

Aigars J, Jurgensone I, Jansons M (2014) Dynamics of silica and phytoplankton population under altered conditions of river flow in the Daugava River, Latvia. Est J Ecol 63(4):217–231. doi:10.3176/eco.2014.4.02

Khan MMA, Umar R (2010) Significance of silica analysis in groundwater in parts of Central Ganga Plain, Uttar Pradesh, India. Curr Sci 98(9):1237–1240

Marchand D, Rayan MC, Bethune DN, Chu A (2002) Groundwater-Surface water interaction and nitrate origin in municipal water supply aquifers. Sanjose, Costa Rica

Swanberg CA and Morgan P (1978) The linear relation between temperatures based on the silica content of groundwater and regional heatflow: a new heat flow map of the United States. J Pure Appl Geophys 117(1–12)

Assessment of Global Change Impacts on Groundwater Resources in Souss-Massa Basin

Seif-Ennasr Marieme, Hirich Abdelaziz, Zine El Abidine El Morjani, Choukr-Allah Redouane, Zaaboul Rashyd, Nrhira Abdessadek, Malki Mouna, Bouchaou Lhoussaine and Beraaouz Elhassane

Abstract The Souss-Massa region is considered as the most productive and adaptive area in terms of horticulture in Morocco since it provides about 90% of fruit and vegetables exportation at national level. In the last decades this region has suffered from water scarcity problem aggravated by climate change and agricultural intensification leading to water table lowering, groundwater salinization and pumping costs increase. This study aims to assess the global change impacts on groundwater management by developing many scenarios combining cropping system, climate change, political interventions and irrigation strategies, practices and technologies. Climate change scenario suggested that the region will be subjected to an increase in temperature and decline in rainfall. As result of global change the groundwater quality is deteriorated more and more due to agricultural pollution and sea water intrusion. Many strategies and scenarios are already or will be implemented such as cropping system scenarios which are corresponding to change in production pattern especially adopting the soilless system and new crops and varieties. The political interventions and strategies which the government has

S.-E. Marieme (✉) · M. Mouna · B. Lhoussaine · B. Elhassane
Applied Geology and Geo-Environmental Laboratory,
Faculty of Sciences, Ibn Zohr University, Agadir, Morocco
e-mail: seif.ennasr.marieme@gmail.com

M. Mouna
e-mail: malki.mouna@gmail.com

B. Lhoussaine
e-mail: lbouchaou@gmail.com

B. Elhassane
e-mail: beraaouz@gmail.com

H. Abdelaziz · Z. Rashyd
International Center for Biosaline Agriculture, Dubai, United Arab Emirates
e-mail: h.aziz@biosaline.org.ae

Z. Rashyd
e-mail: r.zaaboul@biosaline.org.ae

O. Abdalla et al. (eds.), *Water Resources in Arid Areas: The Way Forward*,
Springer Water, DOI 10.1007/978-3-319-51856-5_8

implemented and willing to setup as the Public Private Partnership Projects are (i) sea desalination for irrigation in Chtouka area, (ii) modern irrigation district in El Guerdane zone, and (iii) aquifer contract. These decisions developed as political scenarios aim to reduce the global change impacts and allow an integrated water resources management for future sustainability. Scenarios related to irrigation strategies especially at field scale which are conversion of surface to drip irrigation, using weather stations network in irrigation scheduling, reusing waste water and installing discharge meters in pumping stations were taken in consideration in this assessment.

Keywords Climate change · Aquifer · Water resources · Irrigation · Scenarios · Salinity

1 Introduction

The water scarcity exacerbates water supply-demand imbalance, giving rise to some concern in water management especially for agriculture (FAO 2004). The agriculture represents one of the pillars of the Moroccan economy with 19% of the gross domestic product (GDP) (Addison et al. 2012). Nevertheless, Moroccan agricultural production is threatened by climate change impacts which aggravate the situation of water scarcity (Schilling et al. 2012).

Therefore, Morocco has launched several strategies for water use optimization in order to alleviate the impact of water scarcity. One of the biggest projects that aims to improve farmer's conditions, agricultural productivity and irrigation water supply is the National Green Plan. In spite of socio-économic development brought by this project, especially in Souss Massa region, the level of groundwater storage has seriously decreased (Choukr-Allah et al. 2012). Indeed, groundwater management is impeded by salinization, as a result of climate change and agricultural production intensification (Hsissou et al. 1999; Boutaleb et al. 2000; Bouchaou et al. 2008). Indeed, groundwater management is impeded by salinization, as a result of climate

Z.E.A. El Morjani
Polydisciplinary Faculty of Taroudant, Ibn Zohr University, Agadir, Morocco
e-mail: elmorjaniz@gmail.com

C.-A. Redouane
Agronomic and Veterinary Medicine Hassan II Institute,
Complex of Horticulture, Agadir, Morocco
e-mail: redouane53@yahoo.fr

N. Abdessadek
Souss-Massa-Drâa River Basin Agency, Agadir, Morocco
e-mail: anrhira@gmail.com

change and agricultural production intensification (ABHSMD 2008). This situation calls for a fast intervention by both public and private sector to find out an eventual solutions as well as an alternative water resource. Among the succeed stories of Private-Public-Partners (PPP) initiatives in the Souss Massa region, the irrigation district of El Guerdane zone which is managed by a private operator and has been implemented to save the citrus orchards in this zone, and the desalination project for irrigation purposes which will be built in the Chtouka zone to supply irrigation water for greenhouses growers. The government has also encouraged farmers to convert their surface irrigation system to drip irrigation by providing subsidies which can reach 100% of overall investment related to drip irrigation equipment, and to be organized within agricultural water user associations (WUAs) to facilitate their access to irrigation water and extension services.

The objective of this study is to assess the global change impacts on groundwater management by developing many scenarios combining cropping system, climate change, political interventions and irrigation strategies, practices and technologies.

2 Overview of the Souss-Massa Basin

The Souss-Massa basin, located in south- western Morocco, is one of the country's most important hydrological catchments with an area of 27,000 km^2. Elevations in the catchment range from 0 m (Atlantic Ocean) to 4168 m (Toubkal peak in the High Atlas Mountains) (Bouchaou et al. 2011). Situated between the Atlantic Ocean, the High Atlas, and the Anti-Atlas mountains. The watershed of the study area is composed of 25% of plains and 75% of mountains. The main plains are: Souss plain (4500 km^2), Chtouka plain (1260 km^2), Tiznit plain (1200 km^2) (ABHSMD 2007a). Three factors determine the semi-arid Mediterranean climate of the region, namely relief, ocean coast and the Sahara. Thus, the north of the region, dominated by Atlas, is characterized by a semi-arid to humid climate progressing towards the plain. The latter, which occupies the sunken relief of the Atlas and the basins of Souss and Massa, has an arid climate despite a wide opening on the Atlantic. Finally, the southern and southeastern part of the region that makes up the north side is covered by the Sahara Desert climate. The precipitations are very variables in space and time with a rainfall average of 200 mm/year (Bouragba et al. 2011).

The region is surrounded by two rivers, namely Souss from the north and Massa from the south, giving to the region the name Souss-Massa. Souss river take in important inflow generated from High Atlas Mountains while Massa river receive influx from the Anti Atlas Mountains (Bouchaou et al. 2011). The hydrological regime of rivers is characterized by strong seasonal and interannual irregularity. The maximum inflow occurs during the months of January, February and March, with a minimum in August (ABHSMD 2007b).

3 Drought Characterization in the Souss-Massa Basin

3.1 FAO Climatic Aridity Index

Aridity is a spatial climatic phenomenon involving low rainfall. In fact, aridity is expressed when precipitation (P) is less than potential evapotranspiration (ETP). Three arid zones, namely hyper-arid, arid and semi-arid, that could be identified using the climatic aridity index P/ETP (Sjöholm et al. 1989):

- The hyper-arid zone (arid index 0.03) comprises dry land areas without vegetation, with the exception of a few scattered shrubs.
- The arid zone (arid index 0.03–0.20) is characterized by pastoralism and no farming except with irrigation.
- The semi-arid zone (arid index 0.20–0.50) can support rain-fed agriculture with more or less sustained levels of production.

Despite the relatively high cumulative annual rainfall in some areas of the basin (North), most of plains of the region remain in the arid to semi-arid bioclimatic layer, more severe in the south where the degree of aridity is below 0.2 which indicates that this zone is arid due to Saharian influence (Fig. 1).

3.2 Standardized Precipitation Index SPI

The SPI, currently used to monitor drought, is characterized by linking between rainfall deficits and its different impacts on groundwater, water storage in reservoirs, soil moisture, snow cover and river flows which make it very recommended for drought studies.

In analyzing dried and humid period, the SPI uses only the precipitation with an optimum period of 30 years with continuous monthly precipitation values. Thus, the SPI can be calculated for different time scales (1, 3, 6, 9, 12, 24 and 48 months) for assessing short and long-term severity of drought (McKee et al. 1993).

The SPI 3 provides an indication of moisture conditions in short terms and estimated rainfall over a season. The SPI 3 provides a comparison between the total rainfall over the three-month period examined and the total rainfall for the same three months in all years for which there are records. SPI 3 months ending February compares the cumulative rainfall for December, January and February of the year examined with accumulated rainfall from December to February for all years presented in the history of observations performed with the studied station. SPI 3 indicates the climatic drought especially it provides seasonal variation of rainfall recorder mainly in winter season where rainfall is often recorded. The SPI calculated on 6 months provides an indication of trends in precipitation over a season and up to medium term. For example, an SPI of 6 months ending March would provide a very good indication rainfall amounts observed during the wet season

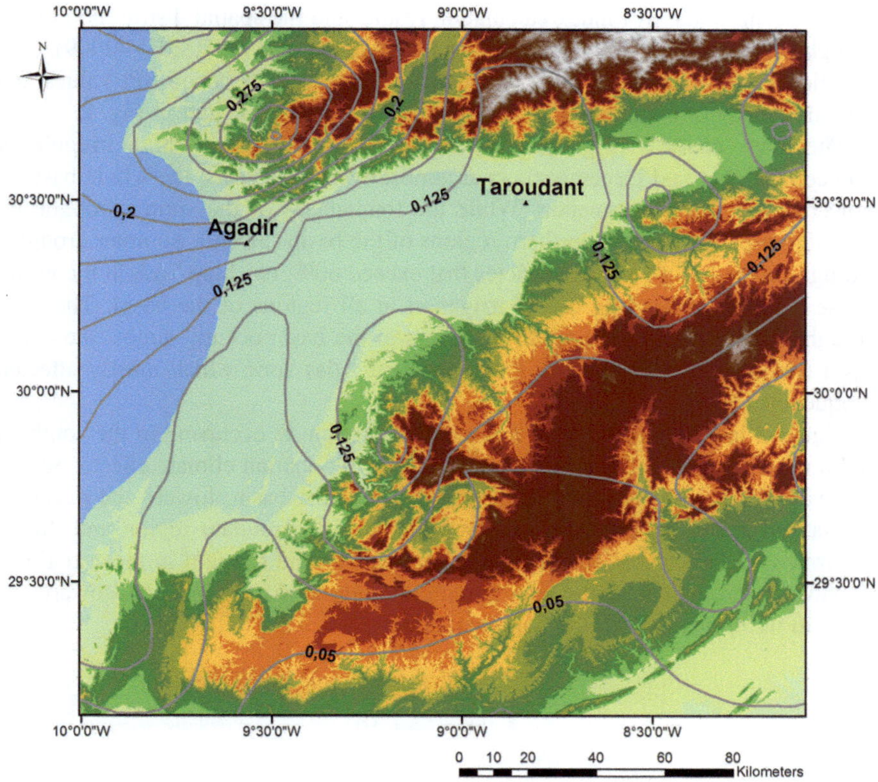

Fig. 1 FAO climatic aridity index applied on the Souss-Massa region (ABHSMD 2009)

from October to March, which is important for Mediterranean agriculture where most of crops are cultivated during this period. SPI 6 indicates the agricultural drought. SPI 12 based on 12 months establish a comparison of rainfall during a period of 12 consecutive months and those recorded on the same series of 12 consecutive months in all years for which there are records. SPI 12 indice for these time scales are usually associated with relatively long term rivers flow, reservoir levels and groundwater levels. SPI 12 indicates the hydrological drought (WMO 2012).

Drought intensity is described in categories depending on the SPI values as following (Table 1):

Table 1 SPI values and corresponding drought categories according to McKee et al. (1993)

SPI values	Drought category
0 to −0.99	Mild drought
−1.00 to −1.49	Moderate drought
−1.5 to −1.99	Severe drought
<−2.00	Extreme drought

According to the SPI index calculation results and the spatial distribution of the drought categories in Souss-Massa basin carried out by ABHSMD (2009). The variation of the sectorial drought frequencies calculated based on SPI values and their frequency of occurement of drought categories is presented in Fig. 2.

The maps analysis (Fig. 2) indicates that early drought (rainfall irregularity between October and December) occur almost every two years, the whole basin is concerned by this phenomenon. While the frequencies of the spring drought are also high especially in the northern regions of the basin (>64%). Summer droughts throughout the basin have frequencies that exceed 60%, reaching 76% in the north. These data indicate strong use of irrigation in all regions of the basin. The mild drought which affecting the water resources in the basin occupy almost the entire basin (frequency >40%) and it is the Northern Atlas zone which mostly affected (frequency >60%).

Figure 2 indicates that severe and extreme drought is occurring in the southern and northern part of the Souss-Massa basin due to Saharian climate effects. Severe and extreme drought occurred in several parts of the basin, toward the ocean in Massa and Tiznit region and in the western are in Taroudant region and this is confirmed by the critical situation of water scarcity and drought in those regions. However hydrological severe and extreme drought is mainly recorded toward the south in the frontiers with the Saharian climate.

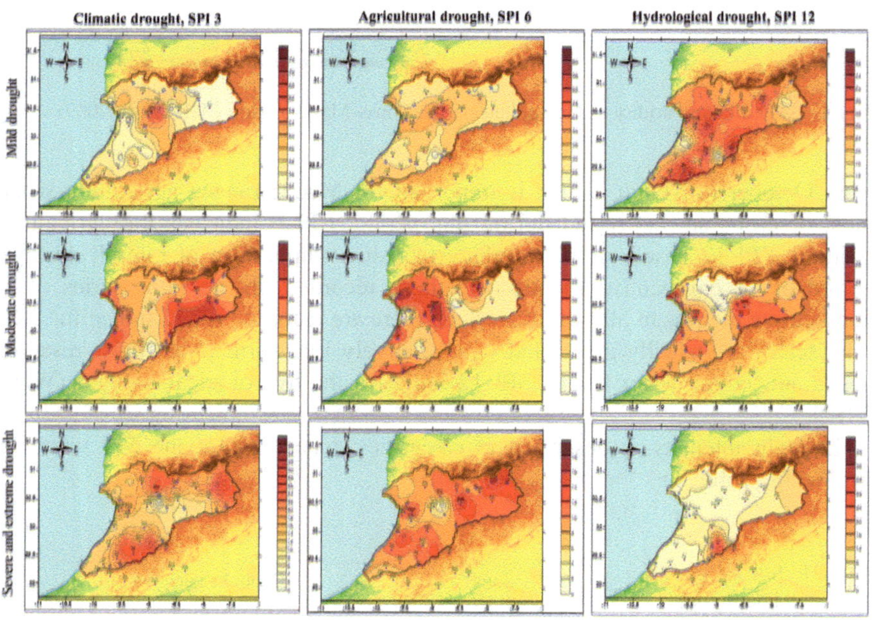

Fig. 2 Frequencies variation of drought categories for climatic, agricultural and hydrological drought (ABHSMD 2009)

4 Climate Change Projections

Two emission scenarios have been used for climate projections. RCP 4.5 (Representative concentration pathways) which is corresponding to radiative forcing of 4.5 Wm^{-2} towards 2100 with a CO_2 emission of 660 eq-CO_2 in case of emission reduction strategies were respected and Greenhouse Gas (GHG) emission stabilized towards 2100. RCP 8.5 is the worst scenario to be likely occurred especially if no measures are taken to reduce GHG emissions. This scenario will likely have a radiative forcing equal to 8.5 Wm^{-2} towards 2100 and continue to increasing after that. The CO_2 emission in this scenario will likely reach more than 1370 eq-CO_2 by 2100 (Stocker et al. 2013). The reason of choosing both scenarios is that RCP 8.5 is the likely to be occurred in case if no mitigation measures to reduce greenhouse gas concentration in the atmosphere are implemented. RCP 4.5 is among the optimistic scenarios and the most realistic in case if mitigations plans are implemented towards 2050.

A study has been carried out recently by Díaz and Pérez (2014) within the "Climatique" project (Observatory of climate change in Souss Massa Drâa region and Canary Islands) using WRF model (Weather Research and Forecasting) and ERAinterim Database (http://www.ecmwf.int/en/research/climate-reanalysis/era-interim) for historical data predicted that:

- Maximum and minimum temperatures will increase up to 1.5 and 2.5 °C during 2045–55 under RCP 4.5 and RCP 8.5 emission scenarios respectively. During 2090–2100 the temperature will likely increase up to 3 and 6 °C under RCP 4.5 and RCP 8.5 emission scenarios respectively and the changes will be more important at elevated zones. During summer, the changes will be more significant.
- Precipitations will decrease, most important changes will occur at high altitude areas. During winter, the changes will be more significant and the largest changes appear for the period 2090–2100 with a reduction of 40 and 120 mm/year under RCP 4.5 and RCP 8.5 emission scenarios respectively.
- For the winds no significant changes during 2045–55 were appreciated. In the horizon of 2090–2100 decade and for the RCP 8.5 scenario, important changes are observed at high altitude areas with an increase in the East part.

Figure 3 presents the changes in minimal and maximal temperatures, precipitations and mean wind speed during the decade 2045–2055 and 2090–2100.

Within the same observatory the results indicate that regional population in Souss Massa will have an increasing rate higher than the national average enhancing its place as an important region in terms of economic development. The agriculture will tend to use water with an efficient manner by the application of new irrigation technologies and techniques. Potable water in rural area will decrease due to rural migration to urban areas as result the demand of potable water in the urban area will increase. The industrial water demand will also increase due to the industrial development in the region. Climate change exhibited by decrease in

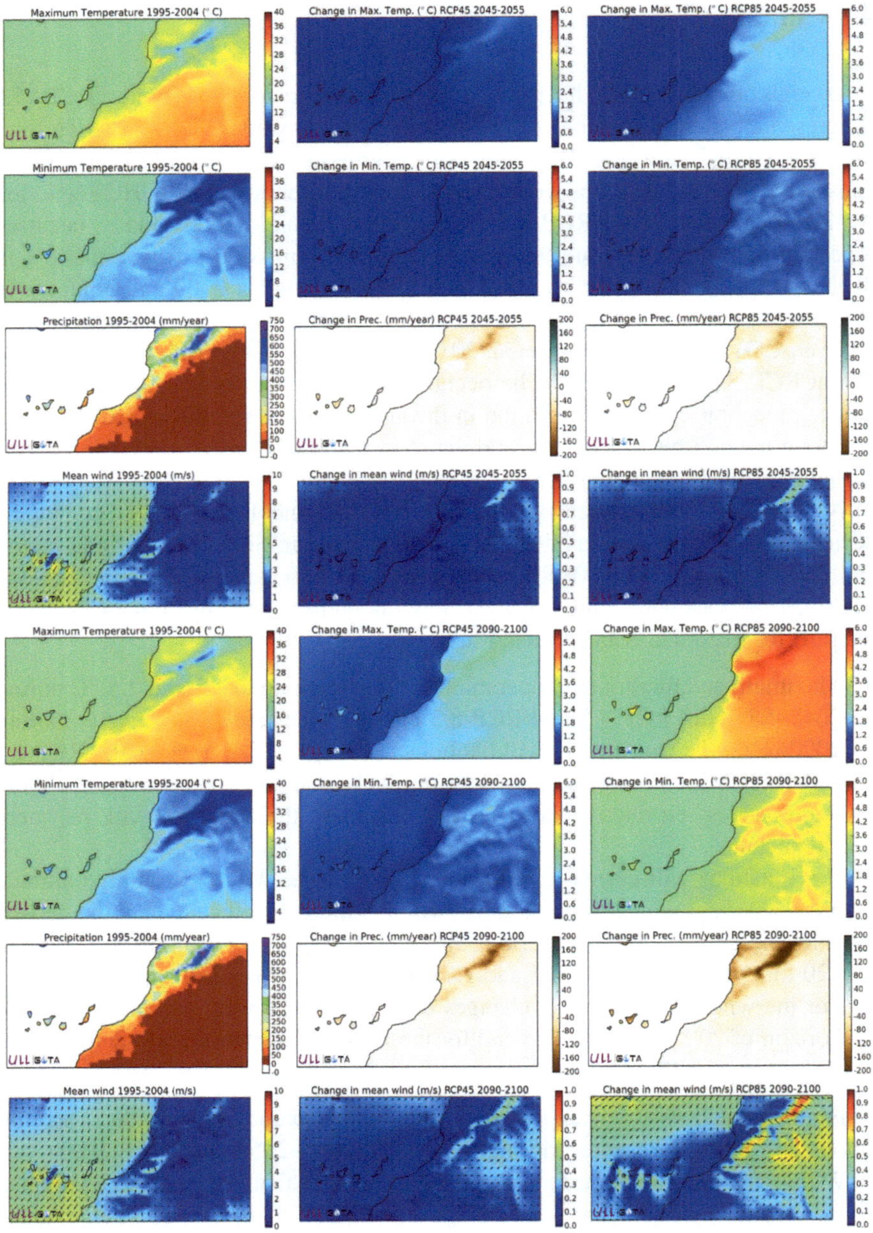

Fig. 3 Changes in minimal and maximal temperatures, precipitations and mean wind speed during the decade 2045–2055 and 2090–2100 (Díaz and Pérez 2014)

Table 2 Projection of water use in mm^3 by sector in the horizon of 2030 and 2050 (Díaz and Pérez 2014)

Sectors	2030	2050
Agriculture	1541	1826
Urban potable water	276	445
Rural potable water	30	37
Industrial water	54	68
Sanitation	31	29
Total	1932	2405

precipitations and increase of sea level will lead to an increased competition for water resources between sectors and groundwater salinization. This study has estimated that the water consumption by each sector in 2030 and 2050 will decrease as presented in Table 2.

Recently the Moroccan Agency for Solar Energy (MASEN 2014) has carried out a study on evaluating the impact of climate change on water resources in the Souss-Massa region testing three CORDEX models (CCCma, DMI, KNMI) under two emission scenarios (RCP 4.5 and RCP 8.5) using the European domain (http://www.euro-cordex.net/060378/index.php.en). The obtained results in terms of change in precipitation and temperature are presented in Figs. 4 and 5:

Projection results (Fig. 4) indicate that precipitation variation in average in the Souss-Massa basin will be:

- −14 and −29% in the period 2021–2050 for CCCma model (optimistic model) under RCP 4.5 and RCP 8.5 emission scenarios respectively
- −22 and −30% in the period 2021–2050 for KNMI model (intermediary model) under RCP 4.5 and RCP 8.5 emission scenarios respectively
- −20 and −40% in the period 2021–2050 for DMI (pessimistic model) under RCP 4.5 and RCP 8.5 emission scenarios respectively

Therefore, the projection result of the pessimistic model should be taken in consideration by the government, in case of extreme droughts or late rains in the future.

Figure 5 shows the temperature variation in % in the Souss-Massa basin for the simulated model during the period 2021–2050.

Projection in terms of temperature indicates in average a variation of:

- +1.5 and 2.5 °C for the CCCma model in the period 2021–2050 under RCP 4.5 and RCP 8.5 emission scenarios respectively.
- +1 and 1.75 °C for the KNMI model in the period 2021–2050 under RCP 4.5 and RCP 8.5 emission scenarios respectively.
- +1 and 1.5 °C for the DMI model in the period 2021–2050 under RCP 4.5 and RCP 8.5 emission scenarios respectively.

All models predicted more variation in temperature especially during the winter season.

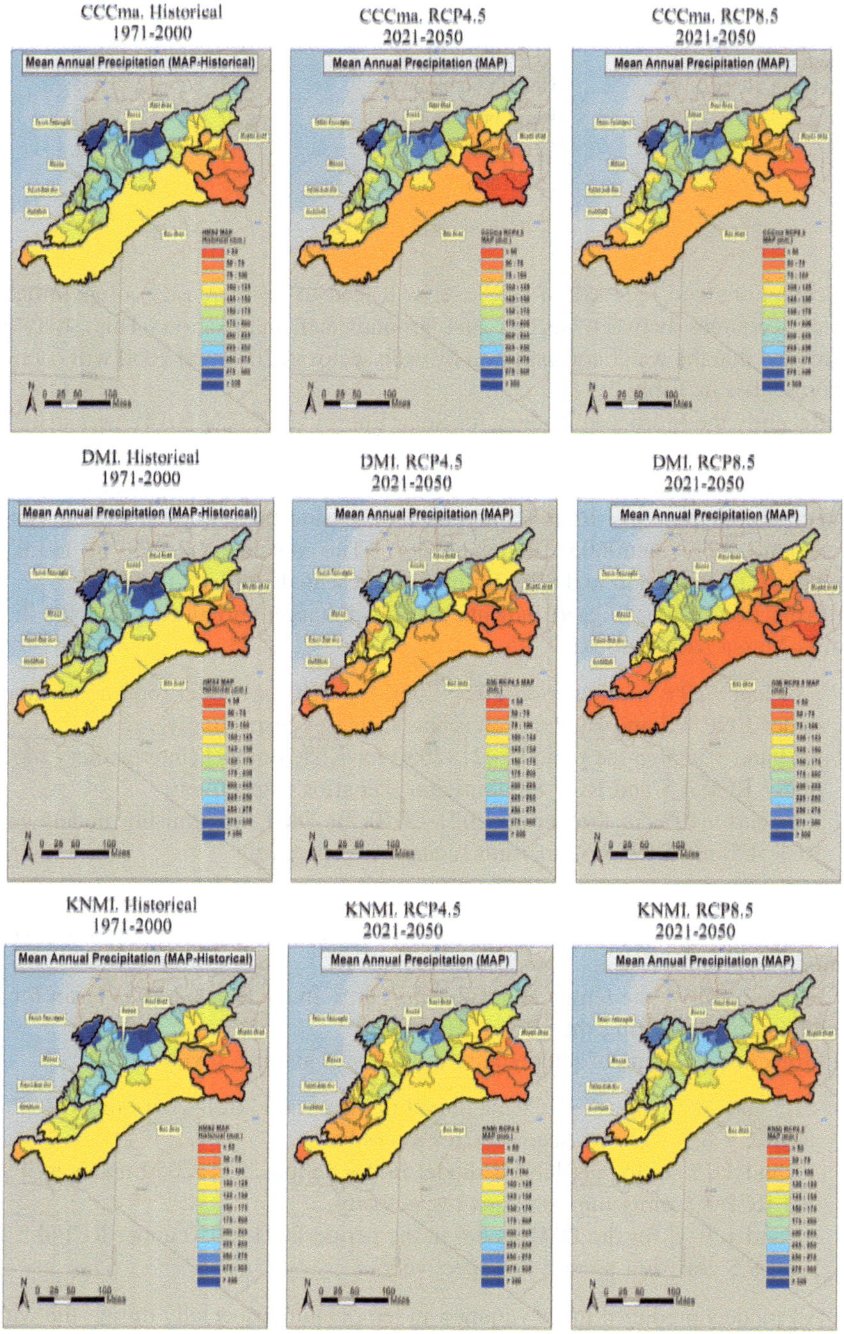

Fig. 4 Precipitation projection for CCCma, DMI and KNMI models under RCP 4.5 and RCP 8.5 emission scenarios for the period 2021–2050 (MASEN 2014)

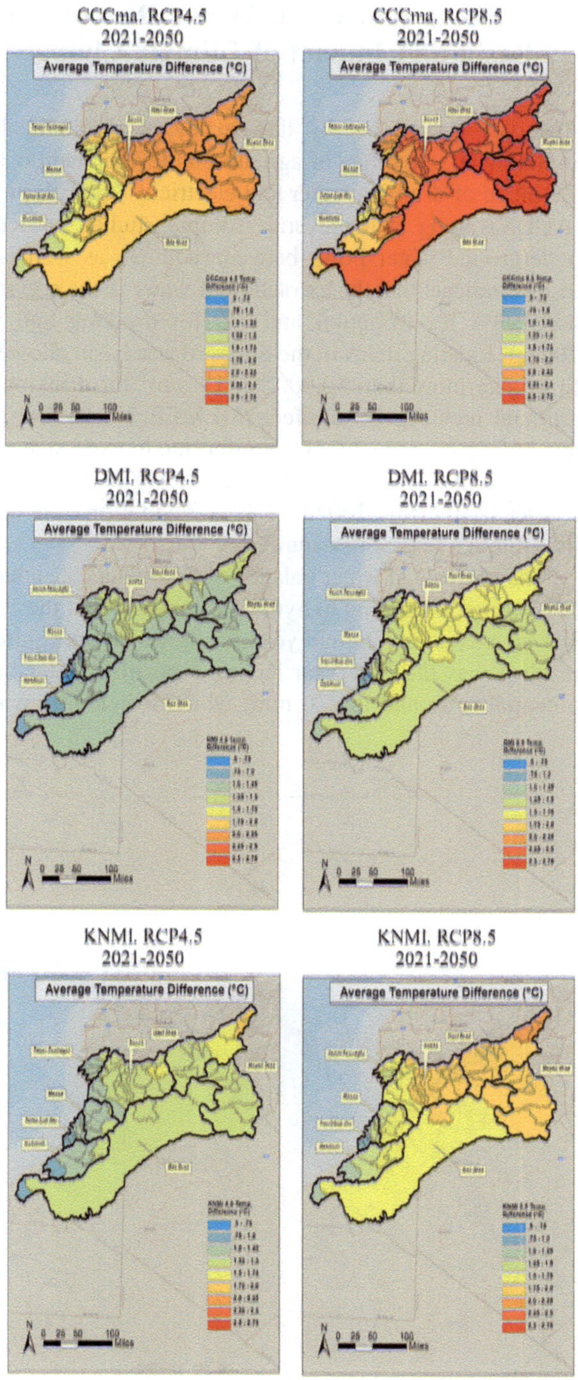

Fig. 5 Changes in temperature in the horizon of 2021–2050 in the Souss-Massa basin (MASEN 2014)

5 Water Balance and Water Demand in Souss-Massa Basin and Impact of Climate Change

A survey carried out by the agency of Hydraulic basin of Souss Massa Drâa (ABHSMD), the office of agricultural development of Souss Massa (ORMVASM), the office of drinking water and electricity (ONEE) and associations of water users (AUEA) along with several isotopic studies carried out to determine the the underground water flows between aquifers as well as losses into the ocean. The results indicate that the extraction of water in the region reached 645 mm^3, which is used 93% for irrigation and 7% for drinking and industrial water (ABHSMD 2007c). Existing dams in the Souss Massa basin allowed a storage of 800 mm^3/year to irrigate more than 50,000 ha, 12 mm^3 for drinking water and 162 mm^3 for the artificial recharge of aquifers (MEMEE 2014). The decrease of the groundwater level of Souss and Chtouka aquifer due to excessive pumping (Fig. 6) and drought varies from 100 to 370 mm^3/year for Souss aquifer and 60 mm^3 for Chtouka aquifer (ABHSMD 2007c). Tables 3 and 4 provide details on the water balance in the Chtouka and Souss aquifers.

The analysis of water balance of Souss indicates that water deficit vary from 100 to 370 mm^3 according to hydraulic conditions of the concerned year. The year 1996 was an exceptional humid year during which the aquifer recharge was important, as result it was the only year having a positive water balance. The current recharge potential is around to 270 mm^3 which is below to water withdrawal.

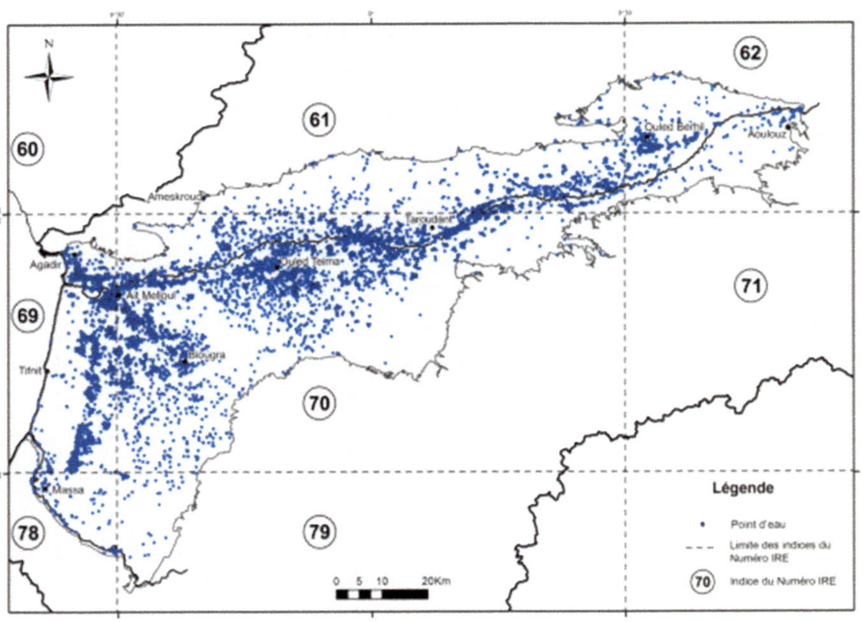

Fig. 6 Distribution of wells and boreholes in the Souss-Massa basin (ABHSMD 2007c)

Table 3 Water balance of Souss aquifer (ABHSMD 2007c)

Souss aquifer	1976	1979	1985	1994	1996	1998	2003	2007
Aquifer recharge								
Rain water infiltration	66.2	62.8	57.8	31.3	105	29.7	39.6	31
Water infiltration at rivers banks	88.7	208.5	50.2	17.3	490	31	199	160
Irrigation water infiltration	14.3	13.7	8	10.2	80	17.4	15.8	4.5
Artificial recharge	–	–	–					9
Upward leakage from deep aquifers	3	3	3	3	3	3	3	3
Supply by aquifers junction	48	48.8	43.7	46.2	192	174.9	65	60
Total input	220	337	163	108	870	256	323	268
Water withdrawal								
Underground flow to the sea	22	19.9	15	19	142	16.4	4	4
Drainage downstream Souss	8.2	60.5	0	0	0	0	0	0
Irrigation withdrawal of the traditional irrigation district	116	73.7	11.1	65.4	33.8	67.6	519	521
Irrigation withdrawal by pumping in the public and private modern districts	250.6	278.1	365.4	375	431	488		
Potable and industrial water withdrawal	8.1	9.8	16.8	18.6	30	41.9	28.7	26
Total output	405	442	408	478	637	614	551	555
Balance: Input-Output	−185	−105	−246	−370	233	−358	−228	−283

Table 4 Water balance of Chtouka aquifer (ABHSMD 2007c)

Chtouka aquifer	1972	1996	1998	2003	2007
Aquifer recharge					
Rain water infiltration	8.2	45.0	13.5	7.7	3.5
Flooding water infiltration of Oued N'Sfa river	0.3	2.0	2.2	3.0	2
Irrigation water infiltration	6.1	6.0	12.7	6.8	15.7
Upward leakage from deep aquifers	10.9	15.0	15.0	15.0	10.0
Supply by aquifers junction	–	–	–	2.6	1.7
Total input	26	68	43	35	33
Water withdrawal					
Irrigation water withdrawal	9.7	32.4	47.9	76.1	78
Potable and industrial water withdrawal	1.0	0.7	1.2	7.1	7.2
Underground flow to the sea	14.3	16.0	12.5	5.7	3
Outputs by drains and springs (Oued Massa)	5.5	7.0		3.7	2.2
Total output	31	56	62	93	90.5
Balance: Input-Output	−5	12	−18	−58	−58

Table 5 Impact of climate change on water balance of Chtouka aquifer during 2021–2050

Scenario of CC	% recharge variation	Natural recharge (mm^3/an)	Renewable potential (mm^3/an)	Water withdrawal (mm^3/an)	Water balance (mm^3/an)
Sans CC		5.5	26	85.2	−58
CCCma4.5	−40	3.3	24	85.2	−60
CCCma8.5	−54	2.6	23	85.2	−60
DMI4.5	−95	1.1	22	85.2	−62
DMI8.5	−92	2.4	23	85.2	−61
KNMI4.5	−80	1.1	22	85.2	−62
KNMI8.5	−56	2.4	23	85.2	−61

Table 5 presents the projection of tested models under RCP scenarios in the horizon of 2021–2050 in terms of impacts of CC on water balance of Chtouka and Souss aquifer using hydrological model HEC-HMS (MASEN 2014). The HEC-HMS model has been used to simulate variation of runoff and recharge using 3 climate change models (optimistic, middle and pessimistic models). The model requires several input data including precipitation, temperature, evapotranspiration, land use, soil profile…etc. The estimation of agricultural water demand was based on agricultural development strategies, cultivated and irrigated areas. Crop water requirement including evapotranspiration was estimated using Blaney–Criddle method (Blaney and Criddle 1962). Crop coefficient has been provided by FAO 56 paper (Allen et al. 1998). The estimated future crop patterns were provided by the ABHSMD (2007e).

Projections results indicate clearly a great impact of climate change on water balance of Chtouka aquifer. The water balance is likely to be reduced by 95% under DMI model and RCP 4.5 scenario which is the pessimistic model. CCCma, the optimistic model was predicted a reduction of 40% under RCP 4.5 scenario.

Compared to Chtouka aquifer the water balance of Souss aquifer will be less affected by climate change. The highest reduction is obtained under DMI model and RCP 8.5 scenario and the lowest reduction under CCCma model and RCP 4.5 scenario.

6 Water Quality Monitoring

The Hydraulic Basin of Souss Massa Drâa Agency (ABHSMD) has implemented a monitoring network of groundwater quality since 1990. The measurements are carried out once per year (ABHSMD 2007d). The monitored parameters are: Electrical Conductivity (EC), pH, N^+, K^+, Ca^{2+}, Mg^{2+}, Cl^-, HCO_3^-, SO_4^{2-}, NH_4^+, NO_2^-, NO_3^-, PO_4^{3-}, Organic Matter (OM), Fecal coliforms and total coliforms.

In general the groundwater quality in the Chtouka aquifer degrades from north to south (ABHSMD 2007d; Krimissa et al. 2004; Bouchaou et al. 2008; Görlich et al.

2015). This degradation is linked mainly to evaporitic dissolution, marine intrusion, and to agricultural activity within the perimeters of Massa and Tassila. The salinity coming from Paleozoic schist is very clear in shallow aquifer of Chtouka-mass. According to Tagma et al. (2009) the groundwater of Chtouka–Massa aquifer is highly contaminated. In fact, in 35% of the wells, nitrates concentration exceeds 50 mg/L which is the threshold of the potability. While in Souss aquifer, the percentage of affected wells is only 7% with an average of nitrates concentration of 22 mg/L. The areas where drip irrigation is concentrated seem to be more affected by nitrate pollution as result of the excess of nitrogen fertilizers application.

Due to groundwater overexploitation the coastal area of the Souss-Massa basin tends to be affected by salinization problem led by the sea water intrusion (Figs. 7 and 8). According to surveys carried out by the ABHSMD agency in the Souss-Mass basin the saline groundwater is located in Issen and the south Agadir coastal zone in Souss aquifer, Sidi Bibi, Massa, Belfâa and Aït Milk in the Chtouka aquifer, Tiznit aquifer, Ifni basin, the coastal zone of Tamraght basin and at the center of Tamri aquifer (ABHSMD 2007d). This salinity origin from marine intrusion is confirmed by specific chemical and isotopic tracers carried out within many studies in the area (Hsissou et al. 1999; Bouchaou et al. 2008; Vinson et al. 2013; Görlich et al. 2015)

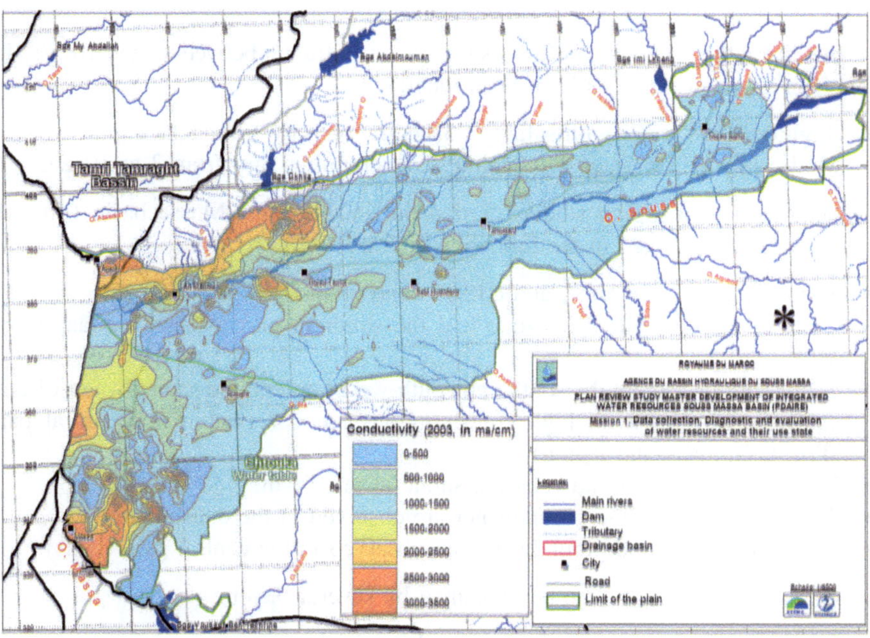

Fig. 7 Electrical Conductivity map of Souss-Massa basin. *Source* ABHSMD (2007c)

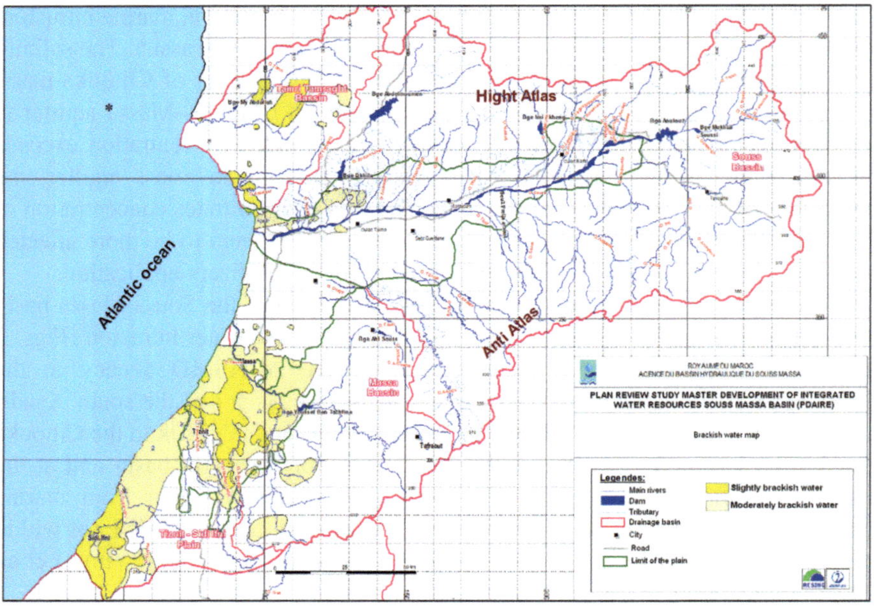

Fig. 8 Saline water map of Souss-Massa basin. *Source* ABHSMD (2007c)

In general the saline water shows salinity concentration between 1 and 10 g/L. In this interval two categories have been identified:

- Slightly saline water with salinity concentration vary from 1 to 3 g/L
- Moderately saline water with salinity concentration vary from 3 to 10 g/L

According to mapping studies carried out in the Souss-Massa basin (Fig. 8), saline water is localized in the following zones:

- Souss aquifer: Issen zone, and coastal zone in southern Agadir
- Chtouka aquifer: Sidi Bibi zone and the southern part (Massa, Belfâa et Aït Milk)
- Tiznit aquifer: except the south and south-west borders, the groundwater of this aquifer are saline with high concentrations are observed in the central part (Tiznit, Ouijjane and Aglou)
- Ifni basin: all the groundwater are saline in this basin
- Tamraght basin: saline groundwater are observed in the coastal areas
- Tamri aquifer: saline groundwater are observed in the central part

The increased salinity can be explained by several factors:

- Leaching of salty Triassic field of the Issen basin

- Leaching of evaporate fields (Jurassic and Cretaceous) of Tamri-Tamraght basin
- Movement of groundwater at shallow depth in shales in south of Chtouka and in Tiznit plain
- Groundwater flow in the metamorphic and granitic body in Ifni basin
- Sea water intrusion due to lowering of piezometric level especially in the near-coastal zones

7 Groundwater Modelling

Models are useful tools used to simulate and predict the effect of factors on receptors and they can give a quite clear vision about the current or future situation. The ModFlow model is used to simulate those changes and generate many outputs as the water table piezometric level, drawdown, water table depth, recharge, concentration… etc. in order to predict the impacts of all scenarios on quantitative and qualitative parameters of the aquifers. The ModFlow USGS 2005 from SWS package has been used to simulate changes in the aquifer from 1968 to 2003 (ABHSMD 2007f). First the model has been calibrated using data of the year 1968. The input data of the model are (Fig. 9):

- Water extraction points: wells, drains
- Groundwater losses into the sea (determined using isotopic studies)
- Climatic data: rainfall, evapotranspiration
- Land use: irrigated areas
- Rivers flows and infiltration
- The data of upward leakage through the Turonien body
- Hydraulic conductivity and storage
- Aquifer body elevation

Figure 10 shows the ModFlow model output in terms of piezometric level of Souss-Massa aquifer system as simulated in 1968 and 2003.

Therefore, the Souss-Chtouka aquifer system has suffered overall during the period 1968–2003, a considerable deficit evaluated by the model to nearly 5.53 billion m^3, equivalent to an average annual deficit of 158 mm^3/year. This deficit is mainly related to the imbalance between renewable reserves and water extraction evaluated at the end of the simulation model respectively 251 and 576 mm^3.

Nrhira (2011) has predicted three scenarios of water resources management in Souss-Massa basin towards 2020: disastrous scenario (no action), water resources safeguarding scenario and drip irrigation generalization scenario. The ModFlow model was used for scenarios simulation. Table 6 shows several indicators values for each scenario:

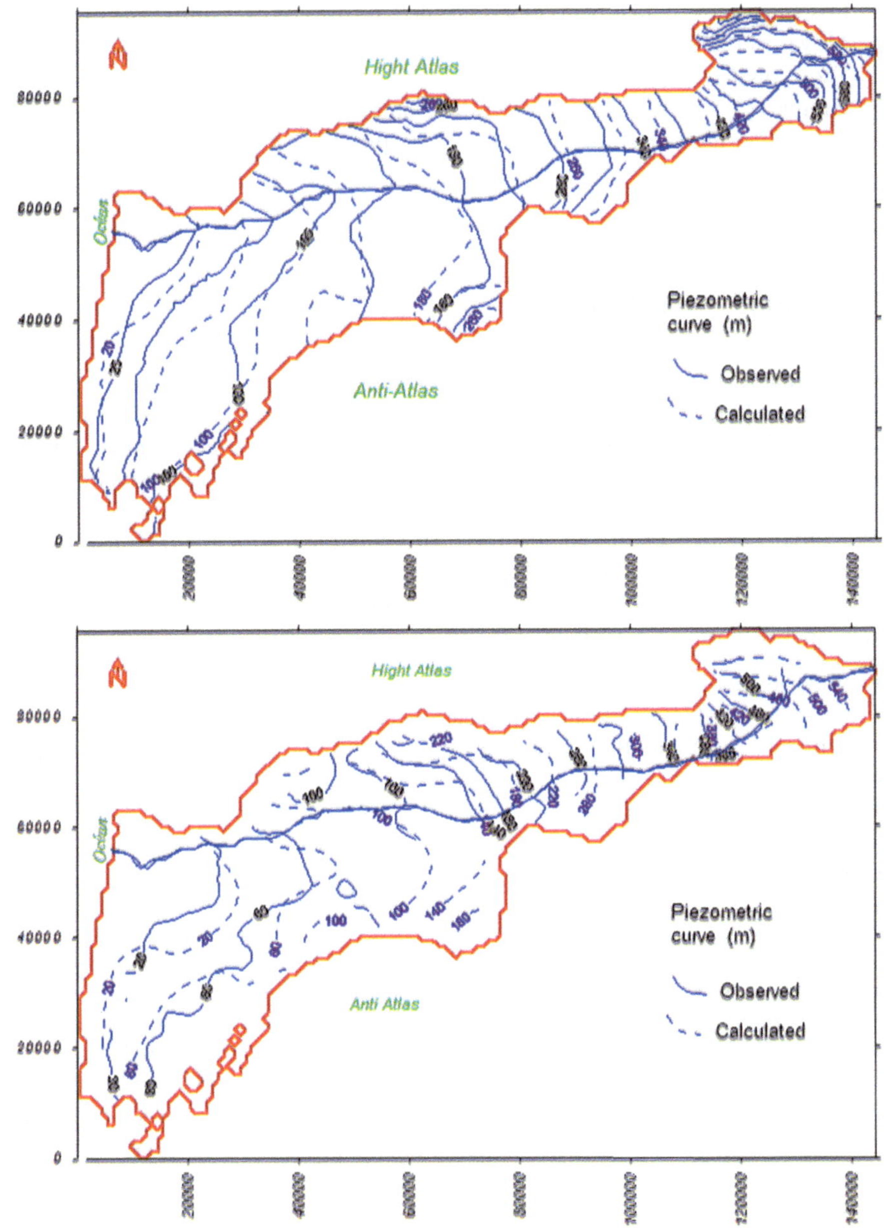

Fig. 9 Piezometric level of Souss-Chtouka aquifer system as simulated by ModFlow model in 1968 (map *above*) and 2003 (map *below*). *Source* ABHSMD (2007e)

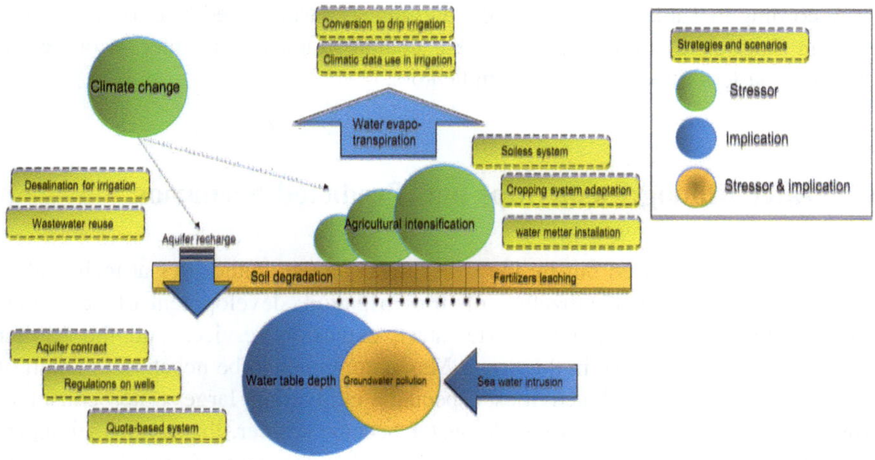

Fig. 10 Groundwater management model

Table 6 Impact of climate change on water balance of Souss aquifer during 2021–2050

Scenario of CC	% recharge variation	Natural recharge (mm^3/an)	Renewable potential (mm^3/an)	Water withdrawal (mm^3/an)	Water balance (mm^3/an)
Sans CC		191	200	547	−284
CCCma4.5	−24	145	153	547	−329
CCCma8.5	−27	140	148	547	−334
DMI4.5	−44	108	115	547	−367
DMI8.5	−48	100	107	547	−375
KNMI4.5	−43	108	115	547	−367
KNMI8.5	−24	144	152	547	−330

Table 7 Comparison of scenarios in the horizon of 2020

Indicators	Scenarios		
	Disastre	Safeguarding	Drip irrigation generalization
Maximal decrease in the piezometric level	50	20	30
Overall Aquifer deficit (mm^3)	406	27	97
Area to be dewatered (ha)	8800	1600	2100
Loss of job positions	42,000	17,000	25,000
Average of the increased cost of pumping (Dhs)	260,000	40,000	80,000
Overall cost of the piezometric decline (Mdhs)	340	114	273

According to Table 7 the water safeguarding scenario is predicted to be the most efficient strategy to minimize the global change impacts on water resources as shown to be less affecting all cited indicators.

8 Water Management Gaps and Predicted Scenarios

Despite all efforts carried out by the water management authorities at technical or political level as the generalization of drip irrigation, development of new alternative water resources (treated wastewater), extension services and the aquifer contract, the water balance in the Souss-Massa continues to be negative as result of the increased agricultural activities especially cultivating large water consumer crops as banana and forage crops. As water became scarcer, the various strategies diverged. Disinterest in farming (resilience or abandonment of crop land) or the modification of cropping patterns (reduction of irrigated surface area and changes in production systems) occurred and alternative water resources were sought, leading to massive over-scaled investment in a race to pump water from the aquifer, or to shifting land tenure.

The scenario to be achieved by the ABHSMD Integrated Water Resources Management and Development Plan (IWRM-DP) is based on restoring the balance between the supply and demand, mainly by increasing water availabilities, whether from underground, surface or non-conventional sources. This strategy confirms that on the one hand the same conceptual framework is sustained despite the problems encountered and the failure of previous IWRM-DP strategy to restore the balance, and on the other hand, that emphasis is being placed on the offer without really controlling the demand. The water management authorities in the region have established many strategies in terms of water supply and demand management. For the water supply management, the groundwater availability should be enhanced by artificial recharge of the aquifer, rain water harvesting, reuse of treated wastewater, desalination and demineralization of brackish water. In terms of demand management, it is now time to practice all water saving measures in terms of irrigation or potable water by adopting all technical means as the conversion of 50,000 ha to be irrigated using drip irrigation. Also pumping groundwater in modern public districts irrigated by surface water should be stopped in order to cover and water deficit at aquifer level.

Change in the agricultural production trends following the market conditions and water availability is considered one of the stressors and at the same time implication, stressor because the increase in agricultural activities leads to more water extraction and implication because it depends to water availability. The Cropping system scenarios are corresponding to change in production pattern especially adopting the soilless system and new crops and varieties.

The political interventions and strategies which the government has implemented and willing to setup as the Public Private Partnership projects (the project of using desalination for irrigation in Chtouka zone, modern irrigation district in El Guerdane zone, aquifer contract) in order to reduce the global change impacts are also developed as political scenarios. Scenarios related to irrigation strategies especially at field scale which are conversion of surface to drip irrigation, using weather stations network in irrigation scheduling, reusing wastewater and installing water meters in pumping stations should be also taken in consideration for future management of groundwater in Souss-Massa basin.

Figure 10 show the groundwater management model including stressors and their implications as well as strategies and scenarios implemented or to be implemented to save groundwater in Souss-Mass basin.

Table 8 presents several strategies and instruments of water supply and demand management as well as their impacts. There are many instruments for water supply and demand management, some of them are already implemented in the region and the rest are under study stage or in discussion. As well known that climate change affect and will affect the Souss-Massa region exhibited by increased temperature and reduced precipitation which make more pressure on water resources either for agriculture or for domestic or industrial use, this will decrease obviously the water availability for the next generations. The case of the Souss-Massa basin confirms these global-scale findings at the local level. The rain stations indicate growing rainfall deficits, possibly related to climate change, especially over the High Atlas Mountains, which are the main source of water for the entire hydro- system in southern Morocco. The introduction of future water availability scenarios in the management of the irrigation system (crop diversification, drip irrigation, improvement of the institutional aspects, the joint use of surface and groundwater, the use of alternative water resources) have not yet been examined. Clearly, the continued use of water in the coastal and near-coastal areas further aggravate the depletion of groundwater and degradation of its quality.

Bouchaou et al. (2011) suggested that groundwater extraction should be moved from areas with high population density, especially areas along the coast, which are more susceptible to contamination. This may affect the recharge area in the upstream and reduce the contribution to provide the downstream segment. Therefore, the artificial recharge and the improving of the irrigation schemes could help significantly to anticipate further deterioration of water resources, both quantitatively and qualitatively. Seawater desalinization may be an option in cases of crisis, however this practice is costly.

Table 8 Water management strategies and instruments as well as their expected impacts in Souss-Massa basin

Strategies	Instrument	Expected impacts			Implementation
		Increasing resources	Reduction of groundwater withdrawal	Better valorization of water	
Water supply increase	*Aquifer recharge*	Yes	No	No	Through storage dams
	Irrigation district using surface water (case of El Guerdane district)	Yes	Yes	No	El Guerdane PPP project (Amensouss) already implemented covering 10,000 ha and consist to transport 45 mm^3 water from Aoulouz dam with a production cost of about 1 billion DH
	Using desalination for irrigation (case of Chtouka desalination project)	Yes	Yes	Yes	PPP project, study stage completed, funds agreed (desalination unit capacity equal to 160,000 m^3/day)
	Reuse of treated wastewater	Yes	Yes	No	To be developed
	Saline water demineralization	Yes	Yes	No	Not used yet
	Rain water harvesting	Yes	No	No	Only for drinking water in mountain areas
	Water transport from Northern basins	Yes	No	No	Suggested
Water demand management	Conversion surface irrigation to drip irrigation	No	Yes	Yes	About 30,000 ha already converted (reduction of 2000 m^3/ha year in citrus orchards)
	Using climate, soil and plant data for irrigation scheduling	No	No	Yes	About 30 weather stations are already installed covering all the basin
	Water pricing	No	Low impact	No	Adopted in the irrigated district using surface water
	Cropping system conversion to crops that value better water	No	No	Yes	To be developed

(continued)

Table 8 (continued)

Strategies	Instrument	Expected impacts			Implementation
		Increasing resources	Reduction of groundwater withdrawal	Better valorization of water	
	Water meters installation for better control of water pumping	No	Large impact	No	In discussion
	Adopting soilless production system to reduce soil and groundwater pollution	No	No	Yes	Adopted by the large producers
	Participatory agricultural water management	No	No	Yes	Already implemented through the Agricultural water users association (AUEAs)
Policies of water demand regulation	*Aquifer contract*	No	Yes	Yes	In progress
	Control of new extraction wells and deepening of existing wells	No	Yes	No	Implemented
	Registration of existing wells and boreholes	No	No	No	Implemented
	Quota-based system	No	Yes	No	In discussion
	Control of irrigated areas	No	Yes	No	In progress
	Extension and awareness services	No	Yes	Yes	Implemented

9 Conclusion

A clear communication between all the stakeholders is on the basic conditions needed for groundwater resources management. Furthermore, the water authorities have to do more efforts to make aware the water users about the serious water problem in the region before moving to political and practical solutions by enhancing the extension services and technical assistance. As 95% of water is used for irrigation, more works should be done on water saving at field scale by improving irrigation techniques and methods and adopting less water consumer crops. Regulations could play an important role to limit the overexploitation of groundwater by the implementation of aquifer contract. Looking for new alternative water resources as desalination and treated wastewater reuse could be also a judicious choice to restore the negative groundwater balance and to save water for future generation. The worst scenario (no action) predict the disappearance of the aquifer in the 20 or 30 coming years which calls for an urgent actions and strategies by all the actors because if water is our right is also the right of the sustainability of the future population.

References

ABHSMD (2007a) Volume 1: Contexte général du PDAIRE. Etude de Révision du Plan Directeur d'Aménagement Intégré des Ressources en Eau (PDAIRE) des bassins du Souss Massa, vol 1. Agence de Bassin Hydraulique de Souss Massa Drâa, Agadir, Morocco

ABHSMD (2007b) Volume 2: Ressources en eau de surface. Etude de Révision du Plan Directeur d'Aménagement Intégré des Ressources en Eau (PDAIRE) des bassins du Souss Massa, vol 2. Agence de Bassin Hydraulique de Souss Massa Drâa, Agadir, Morocco

ABHSMD (2007c) Volume 3: Ressources en eau souterraine. Etude de Révision du Plan Directeur d'Aménagement Intégré des Ressources en Eau (PDAIRE) des bassins du Souss Massa, vol 3. Agence de Bassin Hydraulique de Souss Massa Drâa, Agadir, Morocco

ABHSMD (2007d) Volume 6: Qualité de l'eau et charge polluante. Etude de Révision du Plan Directeur d'Aménagement Intégré des Ressources en Eau (PDAIRE) des bassins du Souss Massa, vol 6. Agence de Bassin Hydraulique de Souss Massa Drâa, Agadir, Morocco

ABHSMD (2007e) Volume 10: Demande en eau agricole. Etude de Révision du Plan Directeur d'Aménagement Intégré des Ressources en Eau (PDAIRE) des bassins du Souss Massa, vol 10. Agence de Bassin Hydraulique de Souss Massa Drâa, Agadir, Morocco

ABHSMD (2007f) Volume 12: Les Modèles Hydrodynamiques, Hydrologiques et Agro-économiques. Etude de Révision du Plan Directeur d'Aménagement Intégré des Ressources en Eau (PDAIRE) des bassins du Souss Massa, vol 12. Agence de Bassin Hydraulique de Souss Massa Drâa, Agadir, Morocco

ABHSMD (2008) Situation hydrologique du bassin hydraulique de Souss Massa. Agence du Bassin Hydraulique de Souss Massa, Agadir, Morocco

ABHSMD (2009) Étude du plan de gestion de la secheresse dans le bassin de Souss Massa. vol Mission 1. Agence de Bassin Hydraulique de Souss Massa Drâa

Addison B, El Korchi T, Rosenstock J, Badran K, Baker J, Collins B (2012) Water management and conservation in rural Morocco. Worcester Polytechnic Institute

Allen RG, Pereira LS, Raes D, Smith M (1998) Crop evapotranspiration-Guidelines for computing crop water requirements-FAO Irrigation and drainage paper 56. FAO, Rome:326

Blaney HF, Criddle WD (1962) Determining consumptive use and irrigation water requirements, vol 1275. US Department of Agriculture

Bouchaou L, Michelot J, Vengosh A, Hsissou Y, Qurtobi M, Gaye C, Bullen T, Zuppi G (2008) Application of multiple isotopic and geochemical tracers for investigation of recharge, salinization, and residence time of water in the Souss-Massa aquifer, southwest of Morocco. J Hydrol 352(3):267–287

Bouchaou L, Tagma T, Boutaleb S, Hssaisoune M, El Morjani ZEA (2011) Climate change and its impacts on groundwater resources in Morocco: the case of the Souss-Massa basin. In: Treidel H, Martin-Bordes JL, Gurdak JJ (eds) Climate change effects on groundwater resources: a global synthesis of findings and recommendations. CRC Press, pp 129–144

Bouragba L, Mudry J, Bouchaou L, Hsissou Y, Krimissa M, Tagma T, Michelot J (2011) Isotopes and groundwater management strategies under semi-arid area: case of the Souss upstream basin (Morocco). Appl Radiat Isot 69(7):1084–1093

Boutaleb S, Bouchaou L, Mudry J, Hsissou Y, Chauve P (2000) Effects of lithology on quality of water resources. The case of oued Issen (Western Upper Atlas, Morocco). Hydrogeol J 8: 230–238

Choukr-Allah R, Ragab R, Rodriguez-Clemente R (eds) (2012) Integrated water resources management in the Mediterranean region: dialogue towards new strategy. Springer, Netherlands

Díaz JP, Pérez JC (2014) Projections climatiques pour la région de Souss Massa Drâa: décades 2045-55 et 2090-2100. Paper presented at the Séminaire de diffusion des résultats du projet Climatique, Agadir, Maroc

FAO (2004) Water desalination for agricultural applications. In: Beltrán JM, Koo-Oshima S (eds) FAO expert consultation on water desalination for agricultural applications, Food and Agriculture Organization, Rome, 26–27 April 2004

Görlich I, Weigand S, Beuel S, Bouchaou L, Reichert B (2015) Status analysis of sea water intrusion and groundwater balance modeling of the upper coastal aquifers in Agadir, Morocco. Grundwasser 20(1):25–37

Hsissou Y, Mudry J, Mania J, Bouchaou L, Chauve P (1999) Utilisation du rapport Br/Cl pour déterminer l'origine de la salinité des eaux souterraines: exemple de la plaine du Souss (Maroc). C R l'Académie Sci 328(6):381–386

Krimissa S, Michelot J-L, Bouchaou L, Mudry J, Hsissou Y (2004) Sur l'origine par altération du substratum schisteux de la minéralisation chlorurée des eaux d'une nappe côtière sous climat semi-aride (Chtouka-Massa, Maroc). CR Geosci 336(15):1363–1369

MASEN (2014) Etude des impacts des changements climatiques sur les ressources hydriques de la region souss-massa-draa. Moroccan agency for solar energy (MAZEN), Agadir, Morocco

McKee TB, Doesken NJ, Kleist J (1993) The relationship of drought frequency and duration to time scales. In: Proceedings of the 8th conference on applied climatology, vol 22. American Meteorological Society, Boston, MA, pp 179–183

MEMEE (2014) Les bassins hydraulique du Maroc. Ministère de l'Energie, des mines, de l'Eau et de l'Environnement, Rabat, Morocco

Nrhira A (2011) La sauvegarde des eaux souterraines dans le Bassin du Souss Massa. Paper presented at the groundwater 2011: Gestion des ressources en eau souterraine, Orléans, France, 14–16 Mar 2011

Schilling J, Freier KP, Hertig E, Scheffran J (2012) Climate change, vulnerability and adaptation in North Africa with focus on Morocco. Agric Ecosyst Environ 156:12–26. doi:10.1016/j.agee. 2012.04.021

Sjöholm H, Reynders M, Ffolliott P (1989) Arid zone forestry: a guide for field technicians. FAO conservation guide 20

Stocker T, Qin D, Plattner G, Tignor M, Allen S, Boschung J, Nauels A, Xia Y, Bex B, Midgley B (2013) IPCC, 2013: climate change 2013: the physical science basis. In: Contribution of working group I to the fifth assessment report of the intergovernmental panel on climate change

Tagma T, Hsissou Y, Bouchaou L, Bouragba L, Boutaleb S (2009) Groundwater nitrate pollution in Souss-Massa basin (south-west Morocco). Afr J Environ Sci Technol 3:301–309

Vinson DS, Tagma T, Bouchaou L, Dwyer GS, Warner NR, Vengosh A (2013) Occurrence and mobilization of radium in fresh to saline coastal groundwater inferred from geochemical and isotopic tracers (Sr, S, O, H, Ra, Rn). Appl Geochem 38:161–175

WMO (2012) Guide d'utilisation de l'indice de précipitations normalisé. World Meteorological Organisation OMM-N° 1090, Geneva, Switzerland

An Overview of Stable Isotopes in Northern Oman's Main Aquifers as an Insight into Recharge Process

Khadija Semhi, Osman Abdalla and Rashid Al Abri

Abstract This review presents a synthesis of literature published on deuterium and oxygen isotopes in groundwater collected from different aquifers in North Oman. The identification of the main hydrological processes which characterize the main aquifers in the North Oman was the main objective of several previous studies. The main focus was to identify zones and mechanisms of groundwater recharge to the different aquifers. Hajar Super group was identified as a main aquifer that receives recharge and transfer freshwater into neighboring aquifers namely ophiolite, alluvium, Hawasina and Tertiary aquifers. A mass balance was also applied in previous studies to groundwater alluvium aquifer and indicated that 2/3 of recharge occurs along the wadi channels and that 1/3 is from the subsurface. Recently it was estimated that 13–68% of groundwater in the ophiolite aquifer is coming from the HSG contrary to the previous estimates of 10%.

Keywords Recharge · Groundwater · Isotopes · Evaporation · Ophiolite · Alluvium

1 Introduction

Development and sustainability in arid areas largely depend on water resources; in particular groundwater which comprises the main source. Understanding and assessing groundwater renewability in form of natural recharge is fundamental for the development plans and sustainability. Informed decisions on abstraction rates

K. Semhi (✉) · O. Abdalla · R. Al Abri
Department of Earth Sciences, College of Science, Sultan Qaboos University,
P.O.box 36 Al-Khod 123, Muscat, Oman
e-mail: semkhad@yahoo.fr

R. Al Abri
Ministry of Regional Municipalities and Water Resources, Muscat, Oman

O. Abdalla et al. (eds.), *Water Resources in Arid Areas: The Way Forward*,
Springer Water, DOI 10.1007/978-3-319-51856-5_9

and well developments are merely based on groundwater renewability. In the view of the global climatic changes and variation in precipitation intensity and frequency, recharge estimates to groundwater basin becomes more challenging and difficult to assess. This is particularly difficult in geologically diverse environment where water is hosted in multi aquifers and water exchange between aquifers is important. Growing and overwhelming water demand in arid areas exerts pressure on groundwater causing its overexploitation which leads to its depletion, quality deterioration and seawater intrusion along coastal aquifers.

Groundwater recharge can be estimated using different tools, including chemical and isotopic techniques (Simmers 1988; Scanlon et al. 2006; Mukherjee et al. 2007). Isotope techniques have been largely used in hydrology investigations for groundwater age determination, mixing with surface waters, movement in aquifers and also for characterization of aquifers in term of porosity and transmissivity. Isotopes are also found to be very powerful tools to investigate aquifer interrelations, and infiltration through the vadose zone. In previous studies, hydrological investigations using isotopes revealed mixing and interactions between deep and shallow waters (McCarthy et al. 1992; Horst et al. 2007; Nakaya et al. 2007; Makni et al. 2014) but did not distinguish those between different types of aquifers. Among isotopes, ^{18}O and ^{2}H are the most commonly used to determine such hydrological information including the sources of recharge and groundwater mixing.

Heavy isotope over the light one (D/H and $^{18}O/^{16}O$ ratios) in precipitations vary according to elevation and distance from the ocean. The water vapor is more depleted in heavy isotopes than sea water, and because of successive precipitations the same vapor mass becomes depleted in heavy isotopes than the first precipitation events. Prior to condensation, the D/H and $^{18}O/^{16}O$ ratios in water may change by fractionation during its travel from the atmosphere to groundwater. The D/H and $^{18}O/^{16}O$ ratios in water may also sometimes change within the aquifer due to evaporation and exchange and mixing processes. The processes such as mixing between waters and interaction with rocks may change the isotope values. Such approaches are commonly used for hydrological purposes (Yeh et al. 2009; Saka et al. 2013).

Application of stable isotopes to understand the hydrological system in Oman has been carried out in previous studies such as Clark (1987, 1988), Weyhenmeyer et al. (2000, 2001, 2002), Matter et al. (2001), Fleitman et al. (2003), Strauch et al. (2014) and Abdalla et al. (2016a, b). Besides, several other studies have investigated quality of groundwater in Oman. These studies include Semhi et al. (2009), Abdalla and Al Abri (2014), Askri (2015) and Chitrakar and Sana (2015).

In the current study we collect and synthesize published δD and $\delta^{18}O$ data in groundwater of North Oman. The objectives of this review are: (1) a construction of local meteoric water line for Oman precipitations, (2) to discuss the most important hydrological processes which control groundwater in Oman (3) to define the recharge source of groundwater.

2 Materials and Methods

Within the first studies which were carried about groundwater in North Oman Weyhenmeyer et al. (2002) collected samples from more than 200 wells between 1996 and 1998. Other samples were also collected from several springs in the Jabal Akhdar Mountains and along the foothills. Within the same study, other water samples were collected during rainstorm events between 1995 and 1998 (Weyhenmeyer et al. 2002). These last samples from winter and summer precipitation reflect different origins of water vapor. For the same aims, Clark (1988) collected a large number of rainwater samples during July 1985 from Oman. During the most recent study by Abdalla et al. (2016a, b), about 200 samples were collected from Aflaj, groundwater and precipitations. The groundwater is hosted in the main types of aquifers: Limestone of Hajar Super Group (HSG) and limestone of Tertiary, ophiolite and alluvium.

Among analyzed parameters in water for both previous and recent studies: alkalinity, major and minor ions (by ion chromatography and by standard ICP (inductively coupled plasma spectrometer) methods). Analytical precision for all measurements ranged between 5 and 10%. A ion charge balance within ±5% was considered for chemical analyses.

Stable isotopes were analyzed according to standard gas mass spectrometry techniques by Laser Absorption Spectroscopy (LAS) using a Los Gatos Research (LGR) Isotopic Water Analyzer (IWA-35d-EP) Model 912-0026.

For oxygen isotope analyses, 2-ml water samples were equilibrated with a CO_2 standard gas in an Isoprep automated, online-equilibration bench for 5 h. The equilibrated CO_2 sample was then directly inlet into a VG-Prism isotope ratio mass spectrometer and analyzed for $\delta^{18}O$ of CO_2. For hydrogen isotope analyses of waters, two techniques were used. In the first, 4 ml of water were reduced to hydrogen gas by reaction with a zinc reagent in sealed Pyrex tubes at 500 °C for 30 min. At this temperature the zinc reacts quantitatively with H_2O, thus producing H_2 with an isotopic composition equal to that of the water. The samples were analyzed by isotope ratio mass spectrometry. Most samples were analyzed again by a new online-equilibration technique using platinum-coated 'Hokko beads' as a hydrogen reaction catalyst (Weyhenmeyer et al. 2002). For the analyses, 20–40 mg of Hokko beads are added to 3 ml of water and then equilibrated with a H_2 standard gas in the Isoprep equilibration. Most water samples were analyzed in duplicate with an overall analytical accuracy of ±0.1% for $\delta^{18}O$ and ±1.0% for $\delta^{2}H$ with the zinc method and ±0.6% using the online-equilibration technique.

The results are reported with respect to International Atomic Energy Association IAEA international standard Vienna standard Mean Oceanic Water VSMOW. The $^{2}H/^{1}H$ and $^{18}O/^{16}O$ ratios are expressed in delta values, δ, in units of per mil relative variation with respect to Vienna Standard Mean Ocean Water (V-SMOW). δ is defined as:

$$\delta = \left(R_{sample} - R_{standard}\right)/R_{standard}Z$$

where R_{sample} is the ratio of the heavy to the light isotope measured for the sample and $R_{standard}$ is the equivalent ratio for the standard.

3 Local Meteoric Water Line in Oman

Among studies which discussed the source of precipitation in Oman, Roberts and Wright (1993) and Kwarteng et al. (2009) suggested 4 sources of moisture to Oman as follow:

(1) Summer rainstorms that are associated with localized convective cells.
(2) Rainfall during November to April originate from North Atlantic or the Mediterranean Sea.
(3) Seasonal tropical cyclones equally occur in two periods, May to June and October to November originate from the Arabian Sea.
(4) On-shore monsoon occurs in the south from June to September.

The study of Fleitmann et al. (2003) indicated that a southern moisture source (Indian Ocean) is main source of recharge to groundwater when the monsoon rainfall belt moved northward and reached Northern Oman during the last 330,000 years.

The investigation of δD versus $\delta^{18}O$ in precipitations collected from North Oman, by Weyhenmeyer et al. (2002) has defined two local meteoric water lines (LMWL): Northern Oman Meteoric Water Line (NOMWL; $\delta D = 5\delta^{18}O + 10.7$) and Southern Oman Meteoric Water Line (SOMWL; $\delta^2H = 7.2\delta^{18}O - 1.1$) and showed differences between isotopic composition of precipitation from northern moisture sources (Mediterranean frontal systems) and isotopic composition of precipitation from southern moisture sources (Arabian Sea and Bay of Bengal) (Table 1).

Table 1 Average isotopic composition of precipitation and groundwater from different aquifers

Location	δD	$\delta^{18}O$	Reference
Precipitation	−2.18‰	−1.98‰	Abdalla et al. (2016a, b)
	−4.52	−1.69	Clark et al. (1988)
Alluvium	−8.08	−1.65	Abdalla et al. (2016a, b)
	−15.6 to −0.2	−3.6 to 0.5	Macumber et al. (1997)
	−13.04	−3.15	Weyhenmeyer et al. (2002)
Ophiolite	−5.63	−1.21	Abdalla et al. (2016a)
	−9.9 to 0.5	−0.2 to −1.2	Macumber et al. (1997)
HSG	−11.47	−2.38	Abdalla et al. (2016a)
	−14.6	−3.59	Weyhenmeyer et al. (2002)
Tertiary	−10.26	−1.63	Abdalla et al. (2016b)

Table 2 Correlation equations between δD and $\delta^{18}O$ in groundwater from different aquifers

Location	Correlation equation	
Alluvium	$\delta D = 0.13\delta^{18}O - 0.63$	Abdalla et al. (2016a, b)
	$\delta D = 3\delta^{18}O + 0.3$	Matter et al. (2001)
HSG	$\delta D = 5\delta^{18}O + 0.57$	Abdalla et al. (2016a, b)
	$\delta D = 5.1\delta^{18}O + 3$	Macumber et al. (1997)
Ophiolite	$\delta D = 3.85\delta^{18}O - 0.97$	Abdalla et al. (2016a, b)
Tertiary	$\delta D = 0.14\delta^{18}O - 0.19$	Abdalla et al. (2016a, b)
Precipitation	NOMWL; $\delta D = 5\delta^{18}O + 10.7$ SOMWL; $\delta D = 7.2\ \delta^{18}O - 1.1$	Weyhenmeyer et al. (2002)

The previously established local meteoric water line for the Batinah region by Macumber et al. (1997) is: $\delta^2H = 5.1\ \delta^{18}O + 8.0$ (Table 2).

For the same objectives Clark (1988) collected rainwater samples during July 1985 from Oman (Fig. 1). Their plot in the δD versus $\delta^{18}O$ plot defines a regression line with a slope of 2.72 and an intercept of 0.085 which is much lower than those of meteoric water lines (GMWL) (Craig 1961) due to high evaporation. The NOMWL suggested by Clark (1988) plots between the NOMWL and SOMWL established by Weyhenmeyer (2002). All these data are plotted in Fig. 1. Rainwater samples collected by Clark were only during July, whereas those collected by Weyhenmeyer span a longer period of time which may favor Weyhenmeyer (2002) data for any comparison of groundwater water or precipitation samples collected from North Oman. Clark's data also include samples from south Oman regions which explain it's diversion as the source of moisture in South Oman is merely from the Arabian Sea. The plot of Clark (1988) data between the NOMWL and SOMWL could be attributed to mixing between the precipitation from Indian Ocean and Mediterranean Sea.

Like in most of previous studies, one of the objectives of the most recent study carried about hydrological system in Oman by Abdalla et al. (2016a, b), was the determination of source of moisture using δD and $\delta^{18}O$ in precipitations combined to those in groundwater from different aquifers. Similar to Clark's data, rainwater samples were collected by Abdalla et al. (2016a, b).

Mean isotopic ratios in precipitations are −1.98 per mil for $\delta^{18}O$ and −2.18 per mil for δD (Table 1). Oman's precipitation is more depleted in $\delta^{18}O$ and δD than precipitation of previous studies in Oman, Bahrain and UAE (Rizk and Alsharhan 1999), suggesting continental and altitude effects. The plot of δD versus $\delta^{18}O$ for rainwater samples collected by Abdalla et al. (2016a, b) (Fig. 1) indicates that the samples are randomly scattered above and below the GMWL by Craig (1961). However, these samples plot along both the NOMWL and SOMWL already defined by Weyhenmeyer (2002) and therefore demonstrate that both sources can be active at the same time.

The new finding of the study of Abdalla et al. (2016a, b) compared to previous studies is that the 2 moisture sources previously defined by Weyhenmeyer (2002) characterize the rainfall in the whole North Oman. The most important contribution

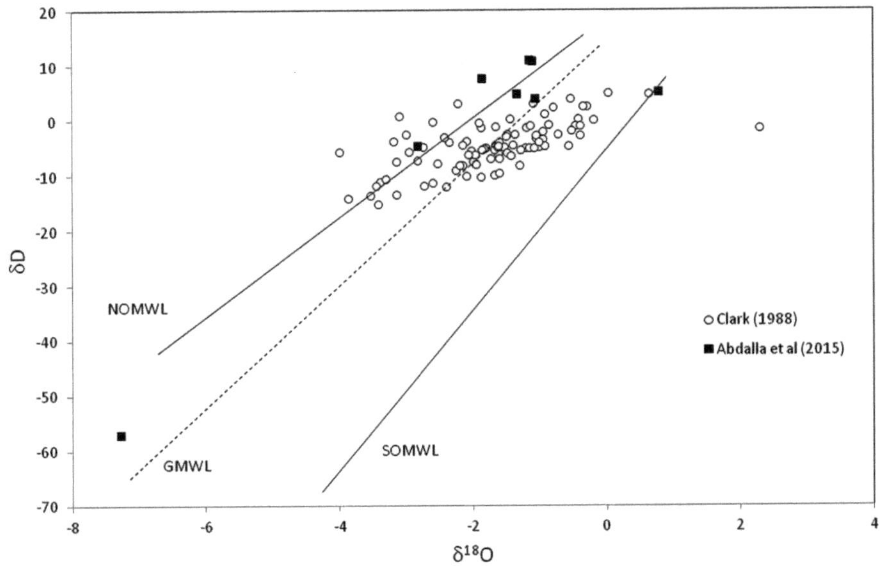

Fig. 1 δD versus δ^{18}O in precipitations collected by Clark (1988)

should be from the South, as also indicated by Fleitmann et al. (2003) and Weyhenmeyer (2002), since the most depleted precipitations in $\delta\ ^{18}$O and δD are those coinciding with SOMWL. It can be noted that the moisture source from Mediterranean Sea, has the potential to cross beyond North Oman Mountains (NOM).

4 Hydrological Processes: Evaporation, Water-Rock Interaction and Recharge

– **Water-rock interaction**

To determine the main processes which control the chemistry of groundwater, samples are plotted in the Gibbs diagram (TDS = f(Na/Na + Ca)). Groundwater chemistry is either controlled by evaporation or by water-rock interaction. The groundwater samples collected by Abdalla et al. (2016a) from the ophiolite aquifer plot in the zone of water-rock interaction (Fig. 2) except for a few samples which show the dominance of evaporation. In the opposite, groundwater from limestone of HSG aquifer is less affected by evaporation. Groundwater from limestone of Tertiary is more affected by evaporation than the other aquifers. The significance of

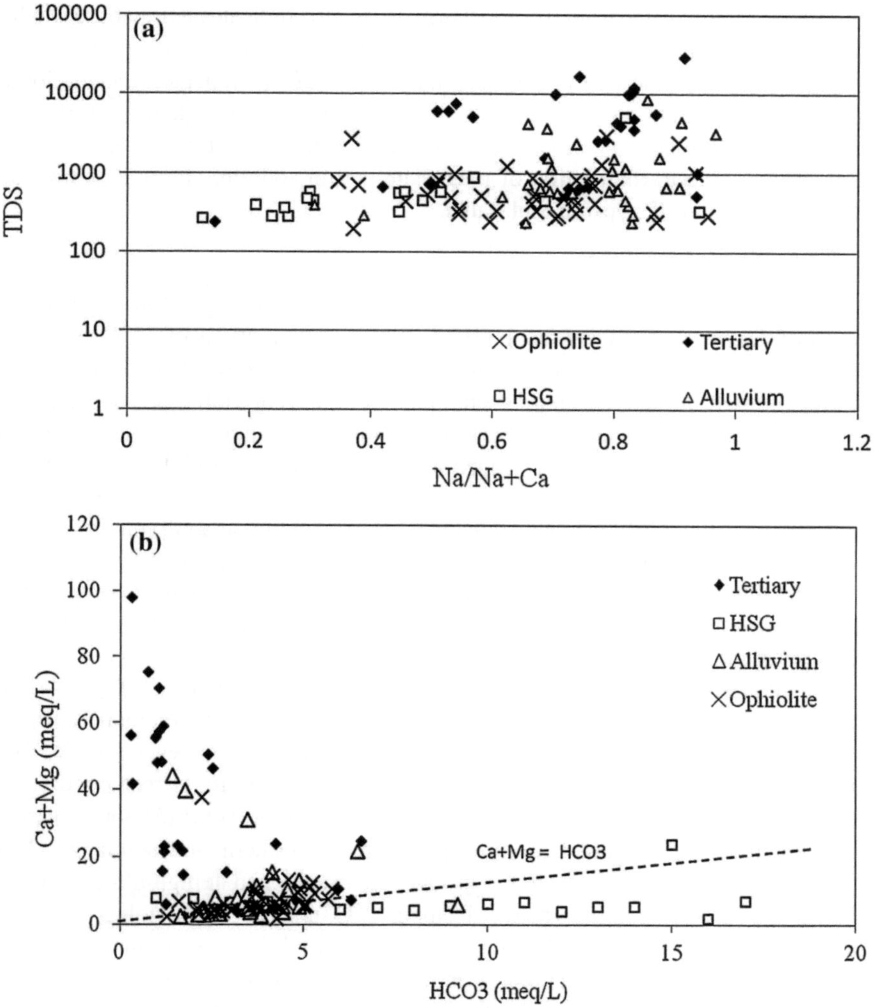

Fig. 2 Gibbs diagram for groundwater from the main aquifers in (**a**) and variations of Ca + Mg versus HCO_3 in (**b**)

mineral dissolution is well demonstrated in the previous studies (Abdalla et al. 2016a, b) by the Ca + Mg/HCO_3 ratio.

The Ca + Mg/HCO_3 ratio (Fig. 2) of groundwater from ophiolite aquifer is above 1 for most of samples. Such ratios suggest other sources of Ca and Mg such as the dissolution of silicates in the ophiolites. The dissolution of evaporites (such as gypsum, anhydrite and halite) intercalated within in the HSG (Cooper et al. 2013) seems to be the additional source of the Ca and Mg in the HSG groundwater. Groundwater from the HSG is subject to calcite dissolution and to a lesser extent dolomite dissolution in addition to evaporites. The excess of Ca + Mg over HCO_3

(Fig. 2) in groundwater from Tertiary formations also indicates extra sources other than limestone dissolution. This source can be the dissolution of evaporites.

Calculation of saturation indices by Matter et al. (2001) indicated a saturation to calcite and dolomite in groundwater of the Jabal Akhdar. Matter (2001) also reported that alluvial waters become enriched in magnesium towards the southern alluvial plain due to dissolution of magnesite and in sodium, chloride and sulfate due to dissolution of gypsum and halite.

– **Groundwater recharge**

Among the first studies which were recorded about recharge of groundwater in the North Oman, that of Clark et al. (1987, 1988). Investigations of $\delta\ ^{18}O$ and δD in the coastal plain geologically dominated by alluvium, show that localized convective rainfall in the mountains and piedmont during hot months dominates recharge to shallow coastal groundwaters. The rainfall systems of winter months are the main source of the deeper confined aquifers, recharged at higher elevations of this alluvium aquifer groundwater.

Alsharhan et al. (2001) showed that the recharge of alluvial aquifer is from wadis located at high elevations. At low elevations, recharge of this aquifer is from the Arabian Gulf. Alsharhan et al. (2001) differentiated between two levels of recharge, one at elevation of 2000 m and the other level at elevation between 600 and 1200 m on the North side of the mountains.

The hydrochemistry of groundwater samples from more than 200 wells and springs collected from coastal plain alluvial aquifer (eastern Batinah, northern Oman) between 1995 and 1998 was investigated by Weyhenmeyer (2002). This study clearly indicates that the Batinah coastal alluvial aquifer is recharged at the pediment of NOM subsequent to rain falls at the plateau of NOM (ca 3000 mamsl). Such precipitation is from two moisture sources: northern and southern). Two plumes of water are formed from the groundwater recharged at high-altitude. These plumes are depleted in the heavy isotopes ^{18}O and D when going from the mountains to the sea. Similarly Matter (2001) showed that waters in springs and wells from alluvial deposits are depleted in ^{18}O and D isotopes which suggests that recharge zone of alluvial groundwater occurs in the piedmont zone near the Jabal Akhdar mountain.

The most recent exploration of isotopes in groundwater of North Oman, has been carried out by Abdalla et al. (2016a, b) who collected several samples of groundwater from different aquifers. These aquifers include Hajar Super Group formation (HSG), ophiolite, alluvium, Hawasina and Tertiary. Like most of arid countries and due to high temperature, groundwater in these aquifers is affected by evaporation at various degrees. Groundwaters at higher elevations reflect a lesser evaporation slope and greatly show an altitude effect. The water rock interaction was also identified as a significant process. Similarly results have been demonstrated by Abdalla and Al Abri (2014) for groundwater in Oman. The $\delta D - \delta^{18}O$ relationship for all water samples from alluvium aquifer shows 3 groups of wells (Fig. 3). The first group plots on the GMWL and includes wells that are located

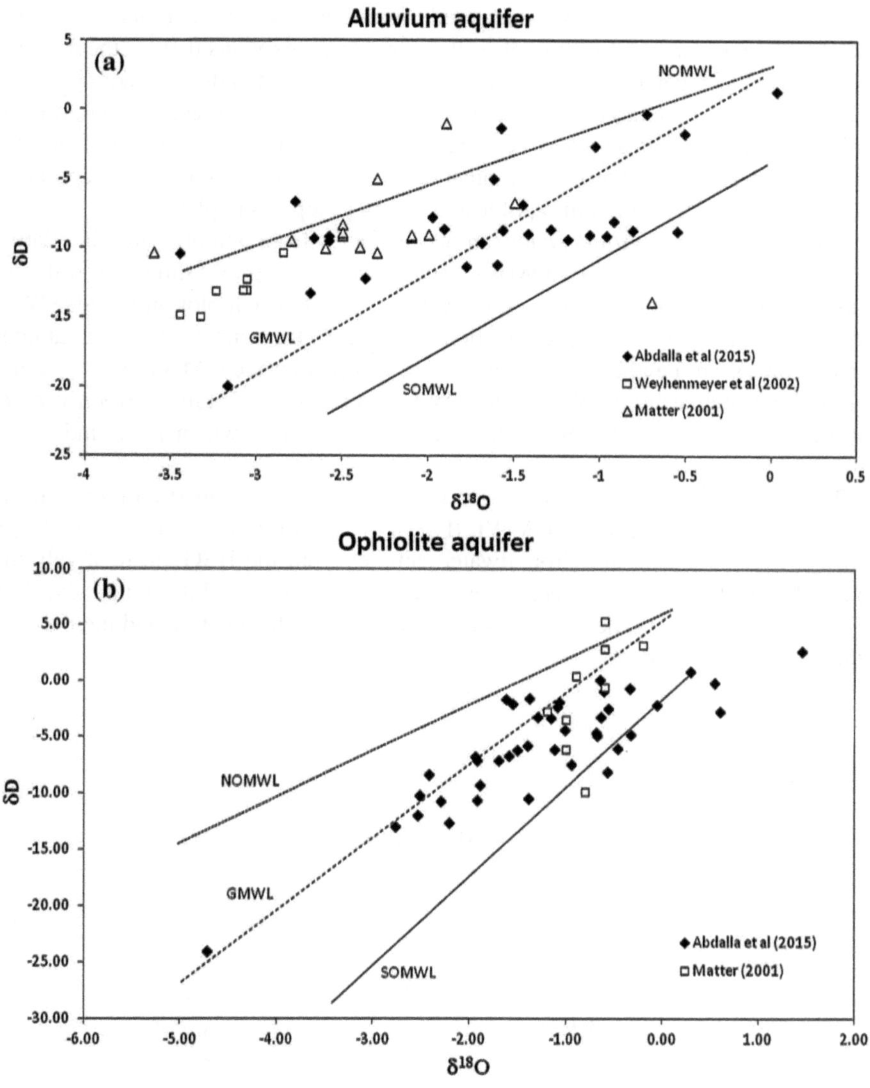

Fig. 3 δD versus δ^{18}O for **a** groundwater from alluvium aquifer and **b** groundwater from ophiolite aquifer

near the coast. The second group of wells plot parallel to GMWL which indicates their recharge from meteoric water and with little impact of evaporation on the isotopes. Most of wells of this group are located in the desert or near the coast. They don't seem very affected by evaporation since they plot parallel to GMWL which may reflect rapid infiltration in the aquifer via fractures and conduits. The third group of groundwaters from alluvium aquifer plots on a line mostly horizontal, which reflects a significant evaporation. It can be noted that the alluvium groundwater of group 2 is recharged from the Northern moisture source (Mediterranean) and the location of wells in the desert explains that, as the desert is the first area to receive rain from the Northern source. The alluvium aquifer at the coast receives the recharge from the southern source and is represented by group 1. This rainfall likely takes place during winter time, however rainfall from the southern source takes place in summer and is subject to excessive evaporation recharges in coastal alluvium aquifer and forms group 3 samples.

The distribution of groundwater samples collected from ophiolite aquifer relative to water lines (GMWL and NOMWL and SOMWL) (Fig. 3), showed a differentiation of water samples as follows: (1) water samples which plot on the SOMWL. For these samples, southern precipitation seems to be the source. (2) water samples which don't seem be affected by evaporation and plot on the GMWL and (3) water samples plot between GMWL and SOMWL and reflect a mixture between vertical infiltration not affected by evaporation and young waters which have undergone evaporation.

The $\delta D - \delta^{18}O$ relationship showed that most of wells from HSG plot near the SOMWL rather than near to NOMWL (Fig. 4) which suggests a source of recharge mostly from Indian ocean. Groundwater samples from the HSG are relatively the most depleted and less affected by evaporation. They plot along, the line ($\delta D = 5\delta^{18}O + 0.57$). The HSG rocks are located at high elevation and their waters reflect an altitude effect.

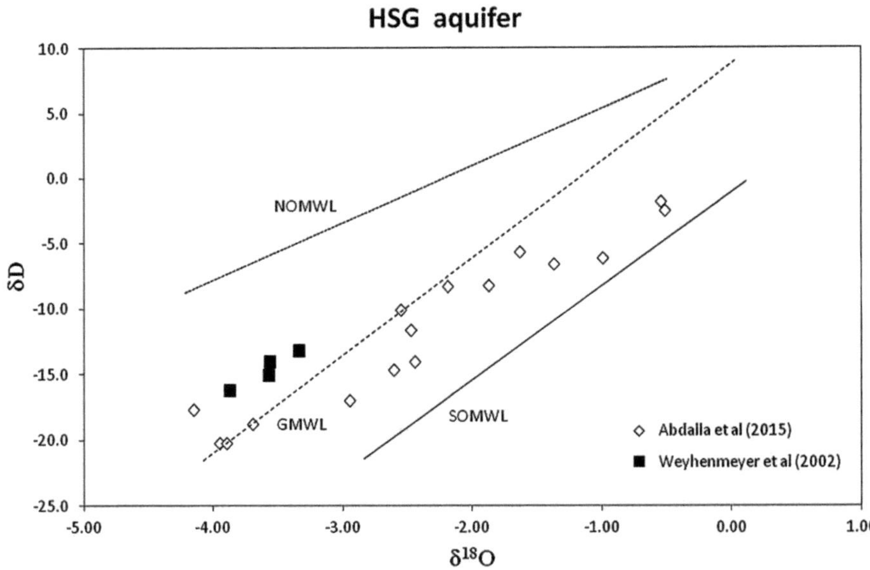

Fig. 4 δD versus $\delta^{18}O$ for groundwater from HSG aquifer

5 Modelling of the Interaction Between Aquifers

Application of isotopes in hydrological systems to quantify the contribution of different sources has been largely used in several previous studies. For Oman groundwater, either $\delta D - \delta^{18}O$ or the $^{87}Sr/^{86}Sr$ ratio were applied for such aims by Matter (2001), Weyhenmeyer (2002) and Semhi et al. (2016).

A mass balance calculated by Semhi et al. (2015) indicated that the contribution of waters from HSG to ophiolite groundwater varies between 13 and 68%. Mixing calculations by Weyhenmeyer (2002) based on strontium isotopes indicated that infiltration on the coastal plain itself accounts for less than 10% of the total groundwater recharge in the plume areas. The remaining 90% of groundwater in the plumes originate in the Jabal Akhdar mountains.

A previous numerical model was also applied but to alluvial aquifer system by (Matter 2001). This model showed that 2/3 of recharge of alluvial aquifer occurs along the wadi channels and that 1/3 is from subsurface along boundaries of Jabal Akhdar and ophiolite foothills. Semhi et al. (2016) suggested two sources of recharge to alluvium aquifer when δD and $\delta^{18}O$ are considered along with $^{87}Sr/^{86}Sr$ diversity.

6 Conclusion

Stable isotopes applied to hydrological system of Oman indicate that there are two main sources of moisture: (1) a source from the North (Mediteranean sea) and (2) a source from the South (Indian sea).

Most of previous studies dealing with stable isotopes in groundwater in Oman demonstrated the role of HSG in recharge of the other aquifers. Located at higher elevation, groundwater from HSG is depleted in D and ^{18}O and reflects a slight evaporation. In the opposite chemistry of groundwater from the other aquifers such as ophiolite and tertiary and alluvium are characterized by both evaporation and the water-rock interaction effect.

As it was discussed in this review, several studies were carried about hydrological system in Oman. In these studies the authors reported informations about recharge zone of groundwater. The seasonal dynamics of recharge to groundwater from each aquifer is poorly investigated. For this objective, it is necessary to collect groundwater samples during different periods. This investigation will be useful to evaluate the residence time and to characterize the water-rock interaction. Quantification of this interaction can be then established for different aquifers.

Acknowledgements We are thankful to the authors of previous studies where application of stable isotopes for hydrological system of Oman started. Special thanks to Dr. Clark. I.D. for his investment in characterization of groundwater in Oman.

References

Abdalla O, Al-Abri R (2014) Factors affecting groundwater chemistry in regional arid basins of variable lithology: example of Wadi Umairy, Oman. Arab J Geosci 7(7). doi:10.1007/s12517-013-0981-7

Abdalla O, Al Abri R, Semhi K, Al Hosni T, Al-Yaroubi S, Amerjeed M (2016a) Groundwater recharge to ophiolite aquifer in north Oman: constrained by stable isotopes and geochemistry. Environ Earth Sci 75:1117. doi:10.1007/s12665-016-5887-8

Abdalla O, Al Abri R, Semhi K, Al Hosni T, Al-Yaroubi S, Clark I, Amerjeed M (2016b) Hydro-chemical evolution of groundwater in a sequence of tertiary formations in Northwest Oman. Environ Earth Sci 75:1410. doi:10.1007/s12665-016-6196-y

Alsharhan AS, Rizk ZA, Nairn AEM, Bakhit DW, Alhajari SA (2001) Hydrogeology of an arid region: the Arabian Gulf and adjoining areas. ISBN: 978-0-444-50225-4

Askri B (2015) Hydrochemical processes regulating groundwater quality in the coastal plain of Al Musanaah, Sultanate of Oman. J Afr Earth Sci 106:87–98

Chitrakar P, Sana A (2015) Groundwater flow and solute transport simulation in Eastern Al Batinah coastal plain, Oman: case study. J Hydrol Eng. doi:10.1061/(ASCE)HE.1943-5584.0001284 (05015020)

Clark ID, Venscroff PRA, Fritz P (1988) Origin and age of coastal groundwaters in Northern Oman. Natuurwet Tijdschr 70:75–84 (12 figs)

Clark ID (1987) Groundwater resources in the Sultanate of Oman; their origin—circulation times—recharge processes and palaeoclimatology. Isotopic and geochemical approaches, Paris. Unpublished Ph.D. thesis, Université de Paris-Sud, Centre Orsay, France

Cooper DJW, Ali MY, Searle MP, Al-Lazki AI (2013) Salt intrusions in Jabal Qumayrah, northern Oman mountains: implications from structural and gravity investigations. GeoArabia 18 (2):141–176

Craig H (1961) Isotopic variations in meteoric waters. Science, New Series 133(3465):1702–1703

Fleitmann D, Burns SJ, Mudelsee M, Neff U, Kramers J, Manfini A, Matter A (2003) Holocene forcing of the indian monsoon recorded in a stalagmite from Southern Oman, report. Science 300

Horst A, Mahlknecht J, Merkel BJ (2007) Estimating groundwater mixing and origin in an overexploited aquifer in Guanajuato, Mexico, using stable isotopes (strontium-87, carbon-13, deuterium and oxygen-18). Isot Environ Health Stud 43(4):323–38

Kwarteng AY, Atsu S, Dorvlo AD, Ganiga T, Vijaya Kumar GT (2009) Analysis of a 27-year rainfall data (1977–2003) in the Sultanate of Oman. Int J Climatol 29:605–617

Macumber PG, Niwas JM, Al Abadi A, Seneviratne R (1997) A new isotopic water line for northern Oman. Paper presented at the third gulf water conference, ministry of water resources, Sultanate of Oman

Makni J, Brahim FB, Hassine S, Bouri S, Dhia HB (2014) Hydrogeological and mixing process of waters in deep aquifers in arid regions: South East Tunisia. Arab J Geosci 7(2):799–809

Matter J, Waber N, Loew S, Matter A (2001) Characterization of recharge areas, groundwater mixing and flow paths in Northern Oman. Results from Isotope Hydrogeology and Geochemical Modeling. American Geophysical Union, spring meeting 2001, abstract #H41A-12

Matter JM (2001) Geochemical evolution and hydrodynamics of groundwaters in the alluvial aquifer of the Dakhiliya area. Dissertation ETH Nr. 14354, Sultanate of Oman

McCarthy KA, McFarland WD, Wilkinson JM, White LD (1992) The dynamic relationship between ground water and the Columbia River: using deuterium and oxygen-18 as tracers. J Hydrol 135:1–12

Mukherjee A, Fryar AE, Rowe HD (2007) Regional scale stable isotopic signature and recharge of the deep water of the arsenic affected areas of West Bengal, India. J Hydrol 334(1–2):151–161

Nakaya S, Uesugi K, Motodate Y, Ohmiya I, Komiya H, Masuda H and Kusakabe M (2007) Spatial separation of groundwater flow paths from a multi-flow system by a simple mixing model using stable isotopes of oxygen and hydrogen as natural tracers. Water Resour Res 43 (9). doi:10.1029/2006wr005059

Rizk ZS, Alsharhan AS (1999) Application of natural isotopes for hydrogeologic investigations in United Arab Emirates. In: Proceedings of the 4th gulf water conference, Bahrain, vol 1, pp 197–228, 13–17 Feb

Roberts N, Wright HE Jr (1993) Vegetational, lake-level, and climatic history of the Near East and Southwest Asia. Global climates since the last glacial. University of Minnesota Press, Minneapolis, pp 194–220

Saka D, Tetteh T, Akiti TT, Osae S, Appenteng MK, Gibrilla A (2013) Hydrogeochemistry and isotope studies of groundwater in the Ga West Municipal Area, Ghana. Appl Water Sci 3 (3):577–588

Scanlon BR, Keese KE, Flint AL, Flint LE, Gaye CB, Edmunds WM, Simmers I (2006) Global synthesis of groundwater recharge in semiarid and arid regions. Hydrol Process 20:3335–3370

Semhi K, Abdalla O, Al Abri R, Al-Hosni T, Clark I (2016) Strontium isotopes as tool for estimation of groundwater recharge and aquifer connectivity. Groundwater Sustain Dev. doi:10.1016/j.gsd.2016.11.001 (In press)

Semhi K, Abdalla O, Al-khirbash S, Khan T, Al-Saidi S, Farooq S (2009) Mobility of rare earth elements in the system soils-plants-groundwater: a case study of an arid area (Oman). Arab J Geosci 2(2):143–150

Simmers I (ed) (1988) Estimation of natural groundwater recharge. NATO AS1 Series C, Reidel, Dordrecht, The Netherlands

Strauch G, Al-Mashaikhi KS, Bawain A, Knöller K, Friesen J, Müller T (2014) Stable H and O isotope variations reveal sources of recharge in Dhofar, Sultanate of Oman. Isot Environ Health Stud 50(4):475–490 (Isotopes in Environmental and Health Studies von Taylor & Francis)

Weyhenmeyer CE, Burns SJ, Waber HN, Aeschbach-Hertig W, Kipfer R, Loosli HH, Matter A (2000) Cool glacial temperatures and changes in moisture source recorded in Oman groundwaters. Science 287:842–845

Weyhenmeyer CE, Burns SJ, Waber, HN, Macumber PG, Matter A (2002) Isotope study of moisture sources, recharge areas and groundwater flowpaths within the Eastern Batinah coastal plain, Sultanate of Oman. Water Resour Res 38:1184. doi:10.1029/2000WR000149,2002

Weyhenmeyer CE (2002) Groundwater evolution in an arid coastal region of the Sultanate of Oman based on geochemical and isotopic tracers. Water-rock interaction volume 40 of the series water science and technology library, pp 1–38

Weyhenmeyer CE, Waber HN, Kramers J, Burns SJ, Matter A (2001) Strontium isotope ratios ($^{87}Sr/^{86}Sr$) as tracers for recharge areas, groundwater movement and mixing in an arid coastal region of the Sultanate of Oman. American Geophysical Union, fall meeting 2001, abstract #H11D-0260

Yeh HF, Lee CH, Hsu KC, Chang PH (2009) GIS for the assessment of groundwater recharge potential Zone. Env Geol 58:185–195

Satellite-Based Estimates of Groundwater Storage Changes at the Najd Aquifers in Oman

Mohamed Saber, Saif Alhinai, Ahmed Al Barwani, Ahmed AL-Saidi, Sameh A. Kantoush, Emad Habib and David M. Borrok

Abstract The Najd aquifers in Oman, are located in one of the most arid zones in the world. In such regions, there is a shortage in the water resources where groundwater is a very critical component for human life. The main aim of this contribution is to use the satellite remote sensing data of the Gravity Recovery and Climate Experiment (GRACE) along with the Global Land Data Assimilation System (GLDAS), to estimate the groundwater storage changes at the Najd aquifers. Groundwater storage changes were calculated from both GRACE/GLDAS data and from the groundwater level measurements. It was found that the estimated groundwater storage changes from GRACE and water levels coincide in their trends showing a noticeable depletion within the time period from Oct. 2002 to Sept. 2014. The spatial distribution maps of the groundwater storage changes show slightly changes from Oct. 2003 to Sept. 2010, but a significant decreasing were observed from 2010 to 2014. The groundwater storage over Najd aquifers was decreased by about 0.44 and 0.46 km^3/year as calculated from GRACE data and

M. Saber (✉)
Geology Department, Faculty of Science, Assiut University, Assiut 71516, Egypt
e-mail: mohamedmd.saber.3u@kyoto-u.ac.jp; msaber_75@yahoo.com

M. Saber · S.A. Kantoush
Disaster Prevention Research Institute, Kyoto University, Gokasho
Uji 611-0011, Japan

S. Alhinai
Water Management Systems, Special Economic Zone Authority, Duqm, Oman
e-mail: saif.alhinai@duqm.gov.om

A. Al Barwani · A. AL-Saidi
Ministry of Regional Municipalities and Water Resources, Muscat, Oman

E. Habib
Department of Civil Engineering and Institute for Coastal and Water Research, University of Louisiana at Lafayette, 42291, Lafayette, LA, USA
e-mail: habib@louisiana.edu

D.M. Borrok
School of Geosciences, University of Louisiana at Lafayette,
42291, Lafayette, LA, USA
e-mail: dborrok@louisiana.edu

O. Abdalla et al. (eds.), *Water Resources in Arid Areas: The Way Forward*,
Springer Water, DOI 10.1007/978-3-319-51856-5_10

groundwater levels, respectively. We also found that groundwater storage was affected by the strong storm events as observed in 2007 and 2010. This contribution could be helpful for the long term sustainable groundwater management in the study area and other arid regions.

Keywords Groundwater storage changes · GRACE · GLDAS · Najd aquifers · Oman

1 Introduction

The Najd aquifers are very crucial groundwater resources for Oman, therefore, an understanding of the groundwater variability is needed in order to determine the sustainability of such groundwater in this arid region. The Najd area has been categorized as one of the most aridity regions over the world, and contains about five towns and thirty villages with a population of about 21,000 residents (PAWR 1986). Generally, most of the arid regions are suffering from the scarcity of water resources. It is anticipated that the water resources in the Middle East and North Africa will be declined within the next century due to a decrease of rainfall in the range between 10 and 25%, and an increase of evaporation between 5 and 20%, which is being associated with a surge in the water consumption demand (Bates et al. 2008).

Several research studies on groundwater resources in Oman have been carried out; e.g. Clark and Fontes (1990) has applied isotopic and geochemical methods to estimate groundwater mean circulation times. Müller (2012) focused on developing a groundwater model that offers the possibility of studying potential recharge scenarios, and Herb 2011 determined the groundwater ages through radiocarbon and tritium dating assessment of the chronology of the paleotemperature record for the Dhofar groundwater, relying on noble gas measurements. Al-Mashaikhi (2011), and Al-Mashaikhi et al. (2012), focused on the groundwater chemistry and groundwater ages as well as appraisal of groundwater recharge by using hydraulics, hydrochemical and isotope evidences. As stated by Macumber et al. (1998) that extreme rainfall events can recharge the shallow aquifer system based on the study of a cyclonic storm in autumn 1992 in the Al-Wusta region in central Oman, they also concluded that the evapotranspiration losses are small, the infiltration rate is rapid, and that cyclones can produce fresh groundwater resources under the right physical conditions. However, the previous studies discussed different issues regarding groundwater, there is a noticeable gap and critical need for evaluating of the groundwater variability and changes. This could be helpful in the sustainable management of water resources in the target region using remote sensing observations. Water resources management in the study area is still a great challenge in term of population increase, economic development and urbanization expansion, therefore it is extremely important to conduct this study to evaluate and estimate of the groundwater storage changes spatially and temporally and characterize its variability.

GRACE satellite mission (Tapley et al. 2004) has been an substantial step forward in monitoring Terrestrial Water Storage (TWS) globally. Since April 2002, it has been offering monthly gravity field solutions, and has examined as an effective tool to infer groundwater storage changes by subtracting contributions from other components (Tiwari et al. 2009; Rodell et al. 2009; Famiglietti et al. 2011; Chen et al. 2014; Richey et al. 2015), the mass of ice sheets (Velicogna and Wahr 2006), snow mass (Niu et al. 2007), surface water storage (Kim et al. 2009), and also to hydrologic drought characterization (Houborg et al. 2012; Thomas et al. 2014).

The spatial variability of TWS is dependent and dominated by variations in ice and snow in polar and alpine regions, surface water in wet and tropical regions, and soil moisture in mid-latitudes (Rodell and Famiglietti 2001). Global estimates of the TWS can be determined by using the temporal variations in Earth's gravity field (Wahr et al. 2004). The accuracy of the recovered mass variations increases with increasing size of the monitored basin, thus, most of the studies utilizing GRACE TWS data are targeting large watersheds for hydrological research and applications (Wahr et al. 2004). Application of GRACE-based TWS at different spatial scales ranging from continents (Syed et al. 2008), to experimental watersheds (Tamaisiea et al. 2005) have been carried out with numerous hydrological purposes. Additionally, GRACE-based TWS has been examined to estimate hydrologic fluxes in water balance computations (Syed et al. 2010) and it has been also analyzed as an essential hydrologic state (Crowley et al. 2006; Syed et al. 2008). Several quantitative analyses showed consistency in the modelled and GRACE-based TWS estimates over large regions. The spatial resolution of GRACE-TWS grids is 1° in both latitude and longitude (around 111 km at the equator) (Landerer and Swenson 2012). The consistency of GRACE data in smaller areas was not satisfactory due to the spatial resolution restrictions (Seo et al. 2009).

The main purpose of this contribution is mainly to validate and examine the potential utility of the GRACE data to assess groundwater storage changes in arid environment aquifer. In order to achieve the main objective, we first derived the groundwater storage changes from GRACE-TWS anomalies and GLDAS data. Afterwards, the results are compared with the groundwater storage changes estimated from groundwater well levels during the time period from Oct. 2002–Sept. 2014. The impacts of extreme events on the groundwater storage changes were also discussed. Finally, validation of GRACE data to monitor the groundwater storage changes applied in arid regions of Oman. This could be helpful for the sustainable groundwater management in such highly demand water resources regions.

2 Study Area

The Najd Aquifers are situated in the Dhofar Governorate, Oman, and covers an area of about 88,000 km^2. The area is internationally bounded by the Kingdom of Saudi Arabia from the north, the Republic of Yemen from the west (Fig. 1). It is flat area penetrated by major wadis (valleys), small hills and sand dunes on the north

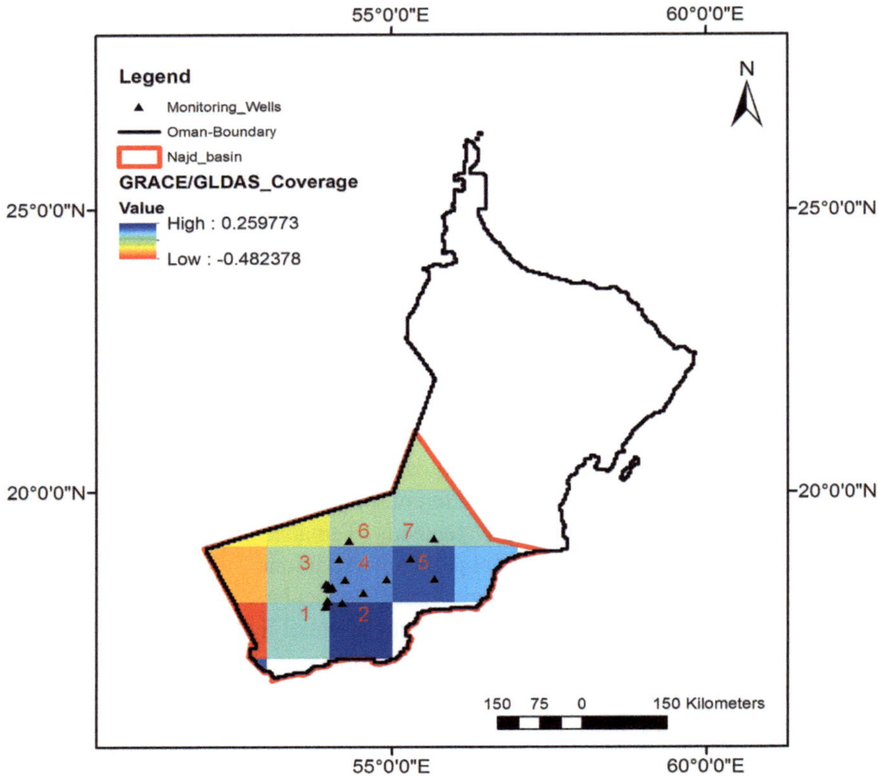

Fig. 1 Location map showing the study area of Najd aquifers in Oman, monitering groundwater wells (distribuated over GRACE/GLDAS pixels denoted from 1 to 7)

edge of the Ruba Al Khali desert (Empty Quarter desert). Vegetation is scattered and mainly consists of desert shrubs. More extensive vegetation can be found close to the Jabal chain in south (Al-Mashaikhi 2011).

Climatologically, the annual average of air temperature varies between 6.1 and 44.6 °C, with a mean of 26.2 °C. The orographic rainfall distribution controlled by the monsoon is the most responsible source of precipitation for the Dhofar area (Hildebrandt and Eltahir 2008). Rainfall amounts are generally low and variables, with annual average of about 31.2 mm, however, there some extreme events occurred in 1983, 1989, 1992, and 2007. The arid climate in the region also leads to high evaporation rates with annual averages of about 161.4 mm (Al-Mashaikhi et al. 2012).

3 Geology and Hydrogeology of the Study Area

The study region consists of alluvium deposits and geological formations of the Fars Group and the Hadhramaut Group (Roger et al. 1992). The area is influenced by several faults, which take the general direction southwest to northeast. The layers thicknesses are increased to north or northeast direction, however, these layers are thinning to the west of the study area nearby the boundary of Yemen, and increase towards north and northeast (Al-Mashaikhi 2011).

The Najd groundwater basin is separated into four aquifers categorized from the top to the bottom as A, B, C, and D as originally presented by Hydrotechnica (1985), and afterwards these categories were used by Mott MacDonald International (1991, 1994), PAWR (1986), and MWR (2000). Then, Al-Mashaikhi et al. (2012) stated that A aquifer consist of Rus, Dammam formation and other layers, but The aquifers of B, C and D are confined and comprised of Umm-Er-Radhuma (UER) formation. Some of the previous researches revealed that groundwater at Dhofar area was recharged during the humid times in the last 25,000 years (e.g. Al-Mashaikhi 2011; Clark and Fontes 1990). Al-Mashaikhi (2011) found that the monitored data reveals that water levels are declining in all aquifers, and the groundwater flow direction were determined based on the water levels contour maps showing the flow from south and southwest towards north and northeast.

4 Data Processing and Methodology

4.1 GRACE Products

GRACE is a collaborated satellite gravimetry mission between US and Germany that measures variations in the Earth's gravity field (Tapley et al. 2004). During the past decade, it has been the most frequently exploited satellite for TWS estimates. GRACE data have been utilized to assess water storage changes at global and regional scales as documented for several regions (e.g., Rodell et al. 2007a, b; Awange et al. 2008; Forootan et al. 2012, 2014).

In this paper, the scaled version of GRACE data processed and archived by Landerer and Swenson (2012) were used (http://grace.jpl.nasa.gov/data/get-data/monthly-mass-grids-land/). This enabled the calculation of GRACE TWS at ($1^\circ \times 1^\circ$) spatial resolution. TWS data are also available in monthly temporal resolution (ver. RL05 (CSR): http://grace.jpl.nasa.gov/data/get-data/monthly-mass-grids-land/), as well as the corresponding scaling factors. Additionally, the scaling factors were provided for GRACE data errors correction, for instance, the leakage errors due to signal leakage from neighboring grids, and measurement errors due to raw GRACE data processing. In order to correct TWS estimates at the target region, the GRACE pixels were multiplied by the corresponding scaling factors. Currently, the released RL05 GRACE products enhance the spatial resolution and alleviate the data errors. This assisted to analyze groundwater storage changes at regional district (Tiwari et al. 2011; Huang et al. 2015) and for watershed scale (Billah et al. 2015) with spatial areas in the range from 30 to 7000 km^2. In the present research, GRACE TWS data are used with consecutive years starting from Oct. 2002–Sept. 2014.

4.2 *Global Land Data Assimilation System (GLDAS)*

GLDAS data was introduced and sponsored by NASA Goddard Space Flight Centre (GSFC) and the National Oceanic and Atmospheric Administration (NOAA) to calculate the variations in both ocean and land mass fluxes (Rodell et al. 2002, 2004, 2007a, b, 2009). GLDAS NOAH Ver.1 (http://gdata2.sci.gsfc.nasa.gov/daac-bin/G3/gui.cgi?instance_id=GLDAS10_M) were downloaded and processed to estimate the TWS components (e.g. Soil moisture and surface runoff). GLDAS NOAH are available at spatiotemporal scales (1° pixels and monthly resolutions). The total Soil Moisture (SM) was estimated as the total summation of soil moisture values of 4 layers: (0–10, 10–40, 40–100, and 100–200 cm). The total soil moisture (SM) was then subtracted from GRACE TWS to calculate groundwater storage changes (Eq. 1).

4.3 *Groundwater Storage Anomalies Estimation from GRACE/GLDAS Data*

We utilized GRACE TWS estimates to evaluate the groundwater storage changes. The water components of GLDAS such as surface water and soil moisture storage were subtracted from the GRACE data. Since Oman can be considered as an arid country, the contribution of snow component has not been considered in the calculations. Therefore, the total water storage change on land (ΔS_{TWS}) can be represented by the following equation (Eq. 1). This equation can be re-arranged to be quantified to estimate the groundwater storage changes (Eq. 2). The 1° × 1° spatial and monthly temporal resolutions of GRACE TWS data and GLDAS data were averaged to calculate annual values within the time period Oct. 2002–Sept. 2014. Then, Soil Moisture and surface runoff anomalies were estimated by subtracting the average over the time period (Oct. 2002–Sept. 2014) from the soil moisture values for every month. The annual groundwater storage anomalies were then obtained by using Eqs. (1) and (2).

$$\Delta S_{TWS} = \Delta S_{SW} + \Delta S_{SM} + \Delta S_{GW} \tag{1}$$

$$\Delta S_{GW} = \Delta S_{LAND} - (\Delta S_{SW} + \Delta S_{SM}) \quad (2)$$

where ΔS_{TWS} is annual TWS anomlies, SW is surface water, SM is soil moisture and GW is groundwater storage. The GRACE TWS grids were multiplied by the dimensionless scaling factors grids.

4.4 Groundwater Storage Anomalies Estimates from Groundwater Levels

Monthly groundwater levels data were collected from the Ministry of Regional Municipality and Water Resources (MRMWR), Oman. The provided data were monthly data (four measurements every year) from 2002–2013, but there are some missing years (Oct. 2006–Sept. 2009) and (Oct. 2006–Sept. 2009). The monthly data then were averaged for annual level estimates (Oct. 2002–Sept. 2014). About 25 groundwater wells were selected for the availability of the data (Fig. 1). We compared the calculated GRACE groundwater storage changes from GRACE/GLDAS data and from in situ groundwater level observations in the study area to examine the potentiality of using GRACE TWS data for deriving estimates of groundwater storage changes were evaluated by. The Groundwater storage anomalies were calculated from GRACE/GLDAS data as addressed in the previous parts, and also from groundwater wells. The available wells in the study area were 25 wells distributed over only 7 GRACE/GLDAS pixels (Fig. 1), where only (3, 1, 10, 7, 2, 1, 1) wells over Pixels (1, 2, 3, 4, 5, 6, 7) respectively. The anomalies of groundwater storage were calculated by multiplying the groundwater levels anomalies by the specific yield average range of 0.3–0.5% for the Najd aquifers. The specific yield values were highly variable in the previous researches, they used specific yield to be 0.7–5.9% (GRC 2008), and 0.5% around Helat Ar Rakah (JICA 1989). According to SAWAS model, it was estimated as 1%, but in other cases, it was estimated as 2 and 10% for the volcanic and alluvium aquifers respectively (SAWAS 1996). Additionally as stated in GRC (2014), the specific yield should not exceed 1% in the aquifer.

5 Impacts of Precipitation on GRACE TWS Anomalies

Based on some availability precipitation data at the weather station of Thumrait that located at about 80 km north of the Salalah Airport, Oman, we used precipitation data within the time period from 2002 to 2009 (collected from Al-Mashaikhi 2011) to evaluate the relationship between GRACE TWS anomalies and groundwater storage anomalies with precipitation anomalies (Fig. 2a, b). It was noticed that there are impacts of extreme rainfall events on GRACE TWS and groundwater storage

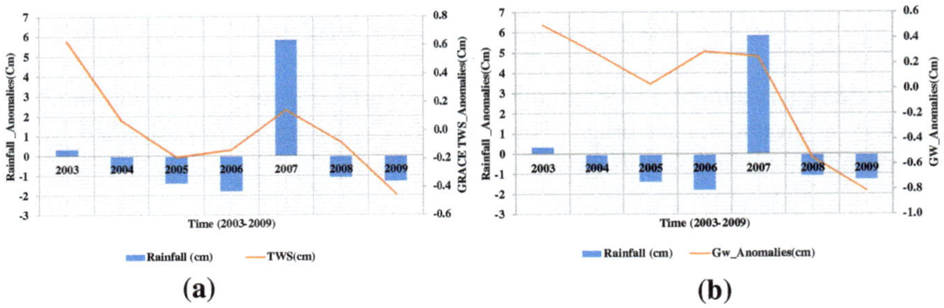

Fig. 2 Shows the relationship between rainfall anomalies and GRACE TWS anomalies (**a**) and groundwater storage anomalies within the time period from 2003 to 2009

changes as recorded in the storm events in 2007 where the TWS and groundwater were increased due to the impact of the 2007 rainfall at this station (Fig. 2).

6 Comparison of GRACE-TWS Estimates Versus In Situ Changes in Groundwater

In this study, we conducted the comparison between groundwater storage anomalies derived from GRACE/GLDAS and groundwater derived from in situ observations of groundwater levels at Najd aquifers. The study area of Najd aquifers are captured by about 15 of GRACE/GLDAS pixels (partial pixels). The available groundwater wells are distributed randomly over only 7 pixels (their locations are labeled from 1 to 7 in Fig. 1). For that reason, we compare the total average of groundwater anomalies derived from in-situ water levels with groundwater anomalies derived from GRACE/GLDAS by two scenarios. The first scenario is to compare the entire basin and the second one is to compare pixel by pixel. In the first case, the comparison shows an acceptable agreement (Fig. 3) between groundwater anomalies derived from both GRACE and water levels with coefficient of determination ($R^2 = 85$). On the other hand, the pixel to pixel comparison (Fig. 6), shows the following coefficient of determination ($R^2 = 0.81$, $R^2 = 0.66$, $R^2 = 0.7$, $R^2 = 0.65$, $R^2 = 0.67$, $R^2 = 0.73$, $R^2 = 0.47$) at pixels (1–7) respectively. It was noticed that the pixels covered by more groundwater wells exhibit good correlations such as in pixel numbers 1(3 wells), 3 (13 wells) and 4 (7 wells) (Fig. 3b, c, d), show some acceptable agreement such as in pixel 5 (Fig. 4b), but the pixels which covered by only one groundwater well exhibits non acceptable correlation such as in pixels number 2, 6, and 7 (Fig. 4a, c, d). Consequently, it is recommended to study the validation of groundwater changes estimated from GRACE and water levels to use enough coverage of groundwater wells. This will be useful for better evaluation and validation of the groundwater storage changes.

The comparison of groundwater storage anomalies estimated from GRACE TWS and in situ observations of groundwater wells shows that the

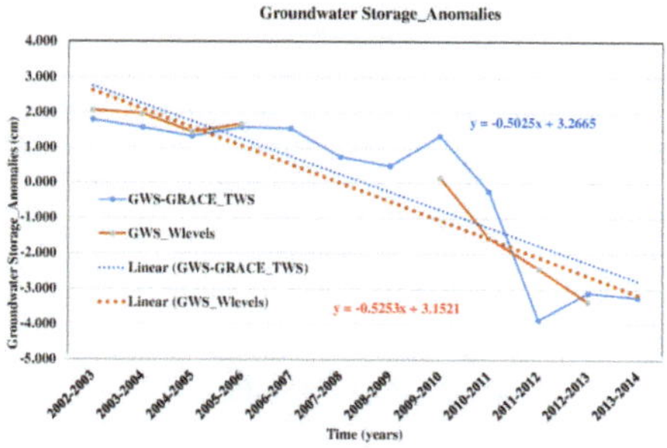

(a) The entire basin

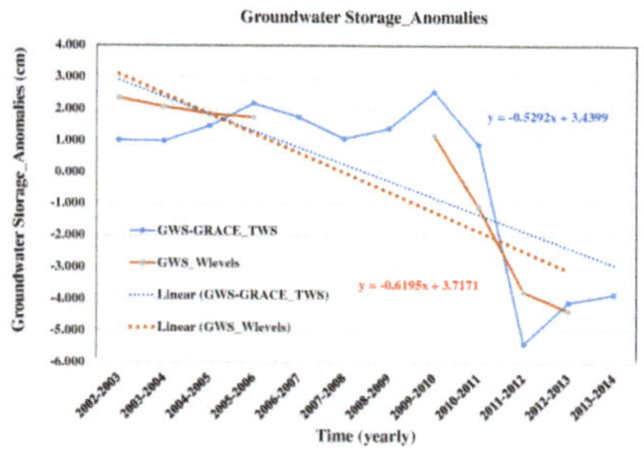

(b) Pixel 1

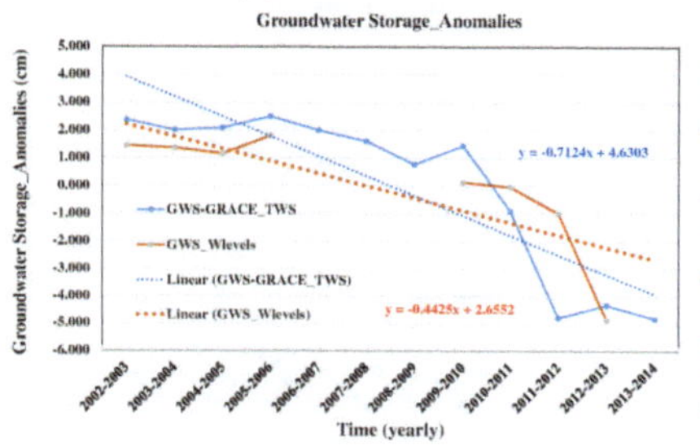

(c) Pixel 3

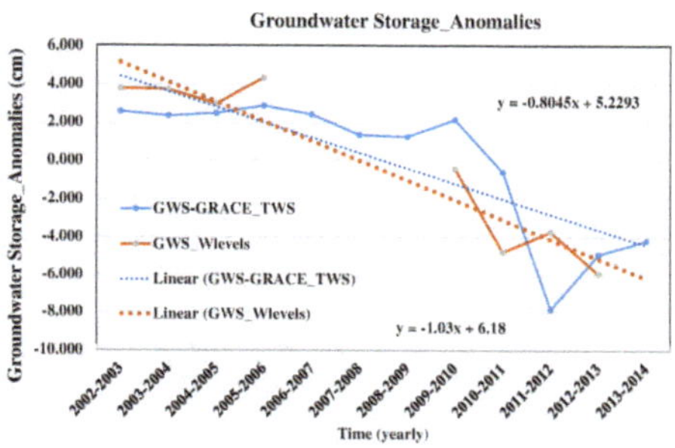

(d) Pixel 4

Fig. 3 Comparison between groundwater changes anomalies estimated from GRACE/GLDAS (cm) and groundwater changes anomalies estimated from observed groundwater levels (cm) for the time period from Oct. 2002 to Sept. 2014, at the entire basin of Najd aquifers (Pixels numbers are denoted as 1, 3, and 4)

cumulative groundwater storage volume over Najd aquifers has decreased by about 0.44 and 0.46 km^3/year respectively, during the time period of Oct. 2002–Sept. 2014 (Fig. 3a). These calculations were performed based on the linear regression for both groundwater storage changes curves derived from GRACE and water levels. The groundwater storage changes in the entire region shows slightly variation with very gentle declining rate from Oct. 2003 to Sept. 2007, then starts declining until Sept. 2009, but from Sept 2009 to Sept 2010 the groundwater storage increased. Afterwards, the storage decreased with very steep declining rate until the end time of the analysis. We found form this analysis that the groundwater storage changes have two main stages, the first stage is exhibiting slight declination in the total groundwater storage in the whole region from Oct. 2002 to Sept. 2010. Within this stage, we noticed the occurrence of two uprising peaks of groundwater storage at 2007 and 2010 (Fig. 4d), this might be due to the flash floods impacts as observed during these two years, there are extreme rainfall events, as well as strong cyclones. On June 6, 2007, Tropical Cyclone Gonu has hit the Gulf coast of Oman. At that time, the greatest storm had lost significant power and was considered as a category 1 cyclone. Also, Cyclone Phet with a Category 3 storm had hit Oman and Gulf of Oman on June 4, 2010 (http://earthobservatory.nasa.gov/IOTD/view.php?

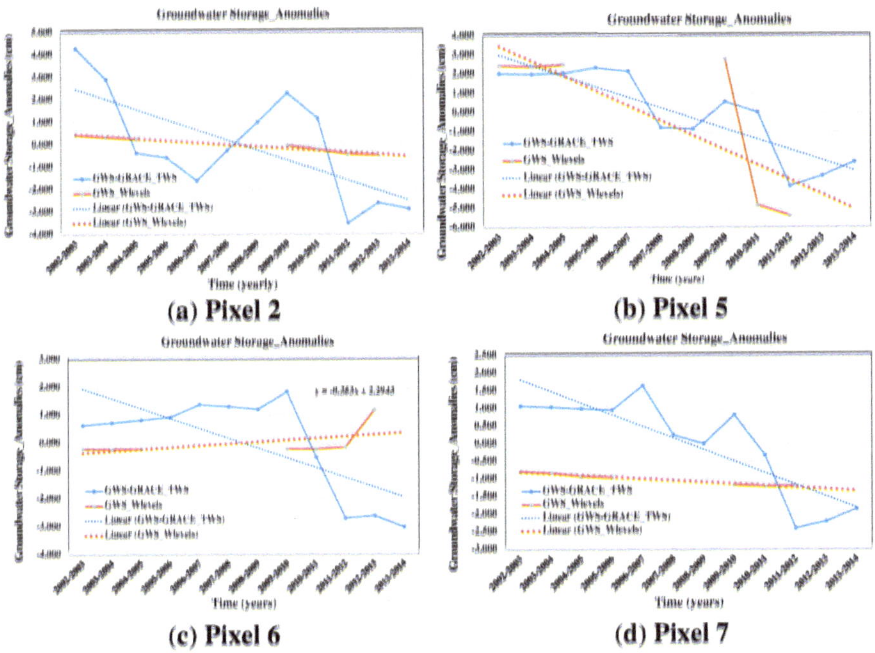

Fig. 4 Comparison between groundwater changes anomalies derived from GRACE/GLDAS (equivalent thickness in cm) and groundwater changes anomalies derived from observed groundwater levels (cm) for the period from 2002 to 2014, at pixels numbers (2, 5, 6, and 7)

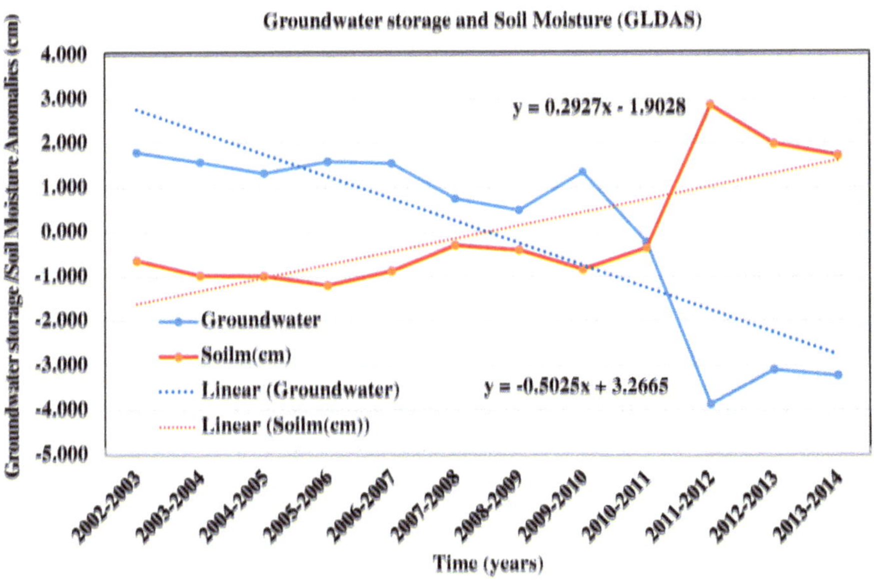

Fig. 5 The relationship between groundwater storage anomalies and soil moisture anomalies at Najd Aquifers

id=44189&eocn=image&eoci=related_image). In the second stage, we noticed that the groundwater storage changes was showing a significant declining rate indicating a great lose in the groundwater storage in this region, and then start slightly to increase in 2013. This might be due to groundwater over-drafting for the agriculture and domestic activities in the region. Additionally, Fig. 5 shows the relationship between groundwater storage and soil moisture anomalies estimated from GLDAS data. It was noticed that total water storage was significantly decreased but soil moisture was increased which might be due to increasing the agriculture activities at the region especially after 2010 as stated by GRC (2014).

7 Groundwater Distribution Maps Over Najd Aquifers

The spatial distribution maps of the groundwater changes were conducted in order to understand the spatial variability over Najd aquifers. As we noticed in the temporal analysis for groundwater storage changes in the time period from Oct. 2002 to Sept 2014, we observed that the distribution maps over the study area exhibit stability or slightly changes within the time from Oct. 2006 to Oct.

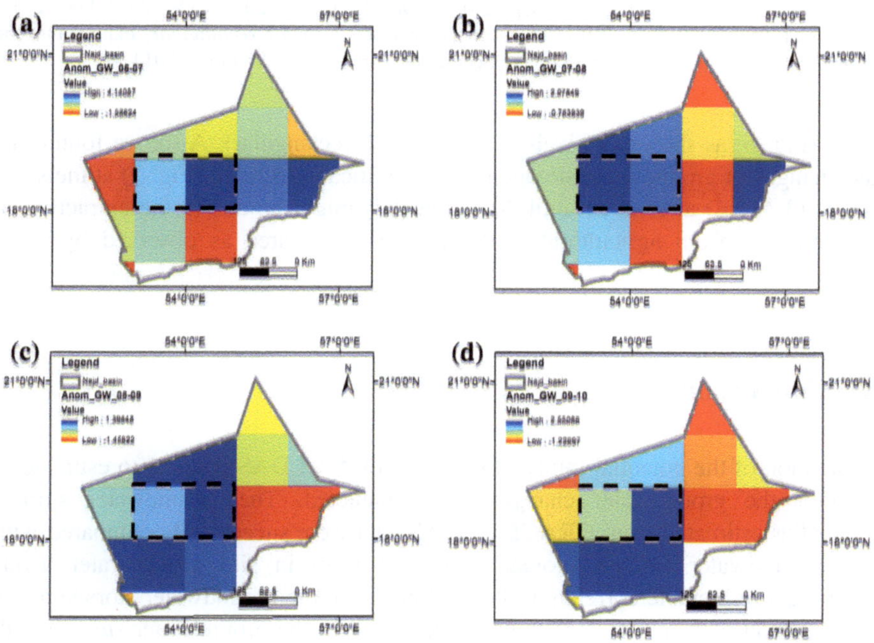

Fig. 6 Spatial distribution maps showing the changes of groundwater storage at Najd aquifers within the time periods: **a** Oct. 2006–Sept. 2007, **b** Oct. 2007–Sept. 2008, **c** Oct. 2008–Sept. 2009, and **d** Oct. 2009–Sept. 2010. The selected two pixels (*black rectangular*) exhibit that the groundwater storage is slightly increasing of keeping in steady conditions

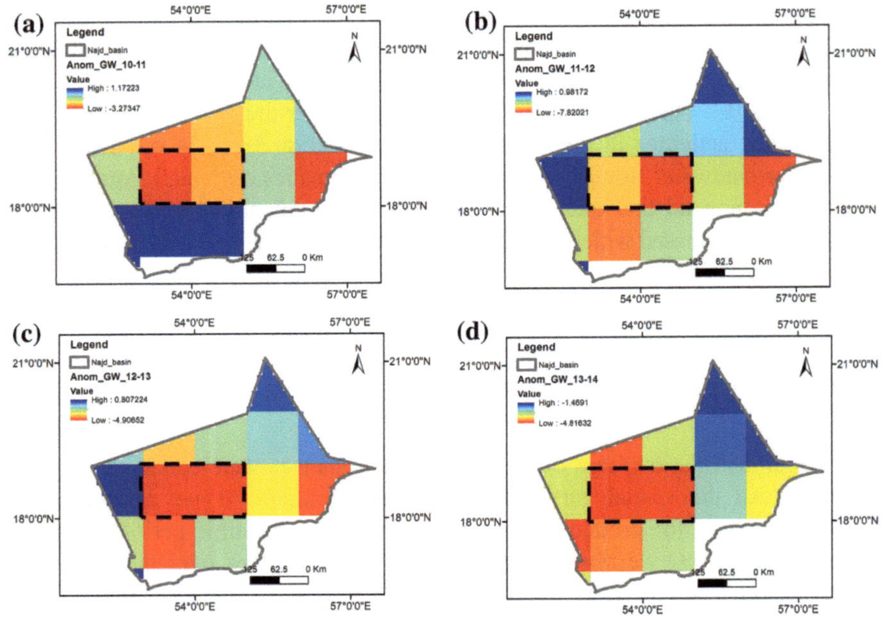

Fig. 7 Spatial distribution maps showing the changes of groundwater storage at Najd aquifers within the time periods: **a** Oct. 2010–Sept. 2011, **b** Oct. 2011–Sept. 2012, **c** Oct. 2012–Sept. 2013, and **d** Oct. Oct. 2013–Sept. 2014. The selected two pixels (*black rectangular*) exhibit that the groundwater storage is showing dramatic decreasing within after 2010 until 2014

2010 (Fig. 6) as shown in pixels 3 and 4 (black rectangular). Also, we found that the changes of groundwater storage shows significant decrease (Fig. 7) at the same pixels (3 and 4) after 2010 until 2014, and this might be due to the extraction of Groundwater from agricultural wellfield in Hanfeet area as observed by GRC (2014).

8 Conclusions

Evaluation of the potential utility of GRACE TWS/GLDAS datasets to estimate or monitor the groundwater changes was conducted. The groundwater storage anomalies estimated from GRACE/GLDAS data were successfully compared with and groundwater storage anomalies derived from in situ groundwater levels showing a reasonable correlation. It was found that the groundwater storage at the study region of Najd aquifers showing declining in the groundwater storage with the time period from Oct. 2002 to Sept. 2014. The comparison have done between the total average of ground levels with the entire groundwater anomalies derived from GRACE revealing good agreement in the decline trends with a correlation about (R^2 = 85), and the cumulative groundwater storage volume over Najd aquifers has decreased by about 0.44 km^3/year using the GRACE TWS data, and about 0.46 km^3/year from groundwater levels during the same time period. Also, the correlation between GRACE pixels and the corresponding groundwater well pixels showing that the correlation is reasonable and acceptable if the groundwater wells coverage is enough number with good distribution over the pixels (e.g. coefficient of determination (R^2 = 0.7, R^2 = 0.65) (Pixels numbers 3 (13 wells) and 4 (7 wells)) respectively. It was also observed that the groundwater storage changes in the study area has been affected by the flash floods events as noticed in 2007 and 2009. Additionally, the time series analysis and spatial distribution maps of the groundwater storage changes over the study area show slightly changes from Oct. 2003 to Sept. 2010, but after 2010 until 2014, the groundwater storage changes exhibit a significant decreasing over the region. This study provide results of the groundwater storage changes from satellite remote sensing data validated by the in situ observations of groundwater wells at the Najd aquifers, as well as the spatial and temporal variability data which could be helpful for the future sustainable groundwater management.

Acknowledgements We acknowledge the Ministry of Regional Municipality and Water Resources (MRMWR), Oman for providing the observation data of groundwater wells. The GRACE and GLDAS team are also acknowledged for processing the remote sensing data and for making it available freely for everyone.

Refrences

Al-Mashaikhi K (2011) Evaluation of groundwater recharge in Najd aquifers using hydraulics, hydrochemical and isotope evidences. PhD thesis, Jena: Friedrich Schiller University

Al-Mashaikhi K, Oswald S, Attinger S, Buchel G, Kno¨ller K, Strauch G (2012) Evaluation of groundwater dynamics and quality in the Najd aquifers located in the Sultanate of Oman. Environ Earth Sci 1195–1211

Awange JL, Sharifi MA, Ogonda G, Wickert J, Grafarend EW, Omulo MA (2008) The falling Lake Victoria water level: GRACE, TRIMM and CHAMP satellite analysis of the Lake Basin. Water Resour Manag 22:775–796. doi:10.1007/s11269-007-9191-y

Bates BC, Kundzewicz ZW, Wu S, Palutikof JP (eds) (2008) Climate change and water. Technical paper of the Intergovernmental Panel on Climate Change. IPCC Secretariat, Geneva, Switzerland, 210, 146

Billah MM, Goodall JL, Narayan U, Reager JT, Lakshmi V, Famiglietti JS (2015) A methodology for evaluating evapotranspiration estimates at the watershed-scale using GRACE. J Hydrol 523:574–586

Chen JL, Li J, Zhang Z, Ni S (2014) Long-term groundwater variations in Northwest India from satellite gravity measurements. Global Planet Change 116:(130–138) doi:10.1016/j.gloplacha. 2014.02.007

Clark ID, Fontes JC (1990) Paleoclimatic reconstruction in northern Oman based on carbonates from hyperalkaline groundwaters. Quatern Res 33(3):320–336

Crowley JW, Mitrovica JX, Bailey RC, Tamisiea ME, Davis JL (2006) Land water storage within the Congo Basin inferred from GRACE satellite gravity data. Geophys Res Lett 33:L19402

Famiglietti JS, Lo M, Ho SL, Bethune J, Anderson KJ, Syed TH, Swenson SC, de Linage CR, Rodell M (2011) Satellites measure recent rates of groundwater depletion in California's Central Valley. Geophys Res Lett 38:L03403. doi:10.1029/2010GL046442

Forootan E, Awange J, Kusche J, Heck B, Eicker A (2012) Independent patterns of water mass anomalies over Australia from satellite data and models. Remote Sens Environ 124:4273. doi:10.1016/j.rse.2012.05.023

Forootan E, Rietbroek R, Kusche J, Sharifi M, Awange J, Schmidt M, Omondi P, Famiglietti J (2014) Separation of large scale water storage patterns over Iran using GRACE, altimetry and hydrological data. Remote Sens Environ 140:580–595. doi:10.1016/j.rse.2013.09.025

GLDAS: http://gdata2.sci.gsfc.nasa.gov/daac-bin/G3/gui.cgi?instance_id=GLDAS10_M. Accessed on 28 Nov 2015

GRACE Data: http://grace.jpl.nasa.gov/data/get-data/monthly-mass-grids-land/. Accessed on 8 Dec 2015

GRC (2008) Drilling & aquifer testing in the Dhofar Governorate. MRMWR, Sultanate of Oman

GRC (2014) Update of the Nejd Groundwater Model"Consultancy Services for Provision of Technical Support for the Drilling, Aquifer Testing and Modelling at Nejd Area in Dhofar Governorate" FINAL REPORT

Herb C (2011) Paleoclimate study based on noble gases and other environmental tracers in groundwater in Dhofar (Southern Oman). Master's thesis, Institute of Environmental Physics, University of Heidelberg

Hildebrandt, Eltahir, EAB (2008) Using a horizontal precipitation model to investigate the role of turbulent cloud deposition in survival of a seasonal cloud forest in Dhofar. J Geophys Res 113: G04028

Houborg R, Rodell M, Li B, Reichle R, Zaitchik BF (2012) Drought indicators based on model-assimilated Gravity Recovery and Climate Experiment (GRACE) terrestrial water storage observations. Water Resour Res. doi:10.1029/2011WR011291

Huang Z, Pan Y, Gong H, Yeh PJF, Li X, Zhou D, Zhao W (2015) Subregional-scale groundwater depletion detected by GRACE for both shallow and deep aquifers in North China Plain. Geophys Res Lett 42:1791–1799

Hydrotechnica (1985) Thumrait military camp, reappraisal of groundwater supply, Report Nr. 6.324/R1, Sultanate of Oman Intergovernmental Panel on Climate Change (IPCC) (2008) Fourth assessment Report: climate change. The World Bank

Japanese International Corporation Agency (1989) The study on the agricultural development project in the Najd Region. Ministry of Agriculture and Fisheries, Sultanate of Oman

Kim H, Yeh PJF, Oki T, Kanae S (2009) Role of rivers in the seasonal variations of terrestrial water storage over global basins. Geophys Res Lett doi:10.1029/2009GL039006

Landerer FW, Swenson SC (2012) Accuracy of scaled GRACE terrestrial water storage estimates. Water Resour Res 48–4. doi:10.1029/2011WR011453

MacDonald M (1991) Detailed investigations for development of up to 1000 ha of irrigated land: Nejd Region. Hydrogeology Interim Report (Final), Ministry of Water Resources, Sultanate of Oman

MacDonald M (1994) Detailed investigations for development of up to 1000 ha of irrigated land: Negd Region. Final Mathematical Modelling Report (Draft), Ministry of Water Resources, Sultanate of Oman

Macumber PG, Al Abri R, Al Akhzami S (1998) Hydrochemical facies in the groundwater of central and southern Oman. Quaternary Deserts and Climatic Change, Balkema Rottendam, pp 511–520

Müller T (2012) Recharge and residence times in an arid area aquifer. PhD thesis, TU Dresden, Dresden

MWR (2000) Water resources assessment report for Najd area, Oman. Unpublished Report, Ministry of Water Resources, Sultanate of Oman

NASA Earth observatory: http://earthobservatory.nasa.gov/IOTD/view.php?id=44189&eocn=image&eoci=related_image. Accessed on 28 Dec 2015

Niu GY, Seo KW, Yang ZL, Wilson C, Su H, Chen J, Rodell M (2007) Retrieving snow mass from GRACE terrestrial water storage change with a land surface model. Geophys Res Lett 2007. doi:10.1029/2007GL030413

Public Authority for Water Resources (PAWR) (1986) Regional hydrogeological evaluation of the Najd, Open File Rep CCEWR, Muscat, Sultanate of Oman. Open File Rep CCEWR, Muscat, Sultanate of Oman, 48–86

Richey AS, Thomas BF, Lo M-H, Reager JT, Famiglietti JS, Voss K, Swenson S, Rodell M (2015) Quantifying renewable groundwater stress with GRACE. Water Resour Res. doi:10.1002/2015WR017349

Rodell M, Famiglietti JS (2001) An analysis of terrestrial water storage variations in Illinois with implications for Gravity Recovery and Climate Experiment (GRACE). Water Resour Res 37:1327–1339

Rodell M, Famiglietti J (2002) The potential for satellite-based monitoring of groundwater storage changes using GRACE: The High Plains aquifer, Central US. J Hydrol 263:245–256

Rodell M, Houser PR, Jambor U, Gottschalck J, Mitchell K, Meng CJ, Arsenault K, Cosgrove B, Radakovich J, Bosilovich M (2004) The global land data assimilation system. Bull Am Meteorol Soc 85. doi:10.1175/BAMS-85-3-381

Rodell M, Chen J, Kato H, Famiglietti JS, Nigro J, Wilson CR (2007a) Estimating groundwater storage changes in the Mississippi River Basin (USA) using GRACE. Hydrogeol J15:159–166. doi:10.1007/s10040-006-0103-7

Rodell M, Chen J, Kato H, Famiglietti JS, Nigro J, Wilson CR (2007b) Estimating groundwater storage changes in the Mississippi River basin (USA) using GRACE. Hydrogeol J 15:159–166

Rodell M, Velicogna I, Famiglietti JS (2009) Satellite-based estimates of groundwater depletion in India. Nature 460:999–1002

Roger J, Platel JP, Bourdillon-de-Crissac C, Cavelier C (1992) Geological of Dhofar (geology and geodynamic evolution during the Mesozoic and Cenozoic). Ministry of Petroleum and Minerals, Sultanate of Oman

SAWAS (1996) Availability and conceptualization of groundwater flow system, National Water Supply Authority (NWRA), Sana'a-Yemen, Technical Report5 (Unpublished Report)

Seo KW, Waliser DE, Tian B, Famiglietti JS, Syed TH (2009) Evaluation of global land-to-ocean fresh water discharge and evapotranspiration using space-based observations. J Hydrol 373:508–515

Syed TH, Famiglietti JS, Rodell M, Chen J, Wilson CR (2008) Analysis of terrestrial water storage changes from GRACE and GLDAS. Water Resour Res 44:W02433. doi:10.1029/2006WR005779

Syed TH, Famiglietti JS, Chambers DP, Willis JK, Hilburn K (2010) Satellite based global ocean mass balance reveals water cycle acceleration and increasing continental freshwater discharge. Proc Natl Acad Sci U S A 107:17,916–17,921. doi:10.1073/pnas.1003292107

Tamaisie E, Leuliette EW, Davis JL, Mitrovica JX (2005) Constraining hydrological and cryospheric massflux in southeastern Alaska using space-based gravity measurements. Geophys Res Lett 32:L20501. doi:10.1029/2005GL023961

Tapley BD, Bettadpur S, Ries JC, Thompson PF, Watkins MM (2004) GRACE measurements of mass variability in the Earth system. Science 305:503–505

Thomas AC, Reager JT, Famiglietti JS, Rodell MA (2014) GRACE-based water storage deficit approach for hydrological drought characterization. Geophys Res Lett 41:1537–1545

Tiwari VM, Wahr J, Swenson S (2009) Dwindling groundwater resources in northern India, from satellite gravity observations. Geophys Res Lett 36:L18401. doi:10.1029/2009GL039401

Tiwari V, Wahr J, Swenson S, Rao A, Singh B, Sudarshan G (2011) Land water storage variation over Southern India from space gravimetry. Curr Sci 101:536–540

Velicogna I, Wahr J (2006) Measurements of time-variable gravity show mass loss in Antarctica. Science 311:1754–1756

Wahr J, Swenson S, Zlotnicki V, Velicogna I (2004) Time-variable gravity from GRACE: first results. Geophys Res Lett 31:L11501. doi:10.1029/2004GL019779

Part III
Water Resources Management

Geological Assessment of Water-Based Tourism Sites in Jeli District, Kelantan, Malaysia

Dony Adriansyah Nazaruddin, Muhammad Muqtada Ali Khan, Sofea Rasheeqa Fazil, Zurfarahin Zulkarnain and Kausilia Raman

Abstract Water-based tourism is an important part of nature-based tourism, and its features vary from one site to others. Jeli district in the state of Kelantan, Malaysia, is rich in water resources with attractive natural and geological/geomorphological features such as lake, river, waterfall, and hot spring. This study discusses the geological assessment of potential water-based tourism sites in the district, by selecting four (4) destinations i.e. Pergau Lake, Jeli Hot Spring, Rual River, and Lata Renyok Waterfall as case studies. Geological mapping and characterisation of these sites have been conducted in 2014. These sites contain some interesting geological/geomorphological features such as granitic mountainous and hilly areas (in Pergau Lake), exotic hot water spring in a small valley (in Jeli Hot Spring), beautiful landscape and features of a river valley (in Rual River), and wonderful cascading waterfall in the alternating bright and dark-coloured rocks (in Lata Renyok Waterfall). The geological assessment of these sites was carried out through qualitative and quantitative approaches based on some geodiversity values: scientific, educational, aesthetic, recreational, and some other values. To support the assessment, the in situ water quality analysis was conducted in mid of 2014 (June–July 2014) to show that these sites have different water quality classes based on Malaysia's Interim National Water Quality Standard (INWQS) i.e. Class I for Lata Renyok Waterfall, Class III for Rual River, and Class IV for Pergau Lake and Jeli Hot Spring. All localities are potential water-based tourism attractions in the district which have some specific tourism and recreational activities e.g. swimming and bathing (in Lata Renyok Waterfall and Rual River), fishing (in Lata Renyok Waterfall, Rual River, and Pergau Lake), rafting, boating, and kayaking/canoeing (in Pergau Lake), and hot water therapy and skin treatment (in Jeli Hot Spring). It is recommended that all these localities should be conserved and developed properly as models of sustainable water-based tourism destinations in Malaysia.

D.A. Nazaruddin (✉) · M.M.A. Khan · S.R. Fazil · Z. Zulkarnain · K. Raman
Geoscience Programme, Faculty of Earth Science, Universiti Malaysia Kelantan,
UMK Jeli Campus, Locked Bag No. 100, 17600 Jeli, Kelantan, Malaysia
e-mail: dony@umk.edu.my

O. Abdalla et al. (eds.), *Water Resources in Arid Areas: The Way Forward*,
Springer Water, DOI 10.1007/978-3-319-51856-5_11

Keywords Geological assessment · Water-based tourism · Geodiversity values · Water quality analysis · Jeli district

1 Introduction

Water is important for human life, including for the purpose of recreational and medicinal benefits mainly for people who live in the city or urban areas. From time to time until nowadays, many people like picnicking beside creeks, rivers, lakes, thermal springs, and the seaside (Jennings 2006). Jennings (2006) related "water-based tourism" to any tourism or recreational activity conducted in or in relation to water bodies, such as lakes, dams, rivers, oceans, etc. According to Eurostat (2009), water is an important resource for tourism which may contribute to the attractiveness of a destination. In other words, tourism needs the attractions for water bodies (rivers, waterfalls, lakes, hot springs, sea sides, etc.). Due to tourism development, people are encouraged to maintain the water quality of the areas. Local authorities and organizations responsible for water management should consider impacts of tourism activities on the quality of water in tourism areas.

There are so many water-based sites all over the world which are conserved and developed for tourism development, such as Niagara Falls in US, Great Barrier Reef in Australia, Mediterranean Sea, ocean and beaches in the Maldives, icebergs in North Atlantic Ocean near Greenland, etc. In the South East Asia region, Bunaken National Sea Park and Toba Lake in Indonesia, Patayya Beach in Thailand, Ha Long Bay and Mekong Delta in Vietnam are some famous water-based tourism sites.

In Malaysia, some water-based sites have also been conserved and developed into tourism destinations in some of its states, such as Kenyir Lake, Perhentian Island, Tembakah Waterfall, La Hot Spring (in Terengganu), Tioman Island (in Pahang), Gunung Ledang Waterfall and Desaru Beach (in Johor), Pangkor Island and Sungai Klah Hot Spring (in Perak), Melaka River (in Melaka), some parts of Langkawi Islands, such as Dayang Bunting Lake, Kilim River, etc. However, water-based tourism sites in the state of Kelantan have not been well documented, thus this paper attempts to reveal some potential water-based tourism spots in Kelantan, especially sites in Jeli district as the case samples. This study presents the geological assessment of some potential water-based tourism sites in Jeli district i.e. Pergau Lake, Jeli Hot Spring, Rual River, and Lata Renyok Waterfall. These sites were selected based on their most representativeness and attractiveness.

2 Study Area

Jeli district in the state of Kelantan, Malaysia, has abundance of water resources with interesting geological and geomorphological features such as lake, river, waterfall, and hot spring. These resources and features make the area more interesting and can become water-based tourism attractions. The district, which becomes the study area, lies in the west of Kelantan and borders with Thailand in the north, with the state of Perak in the west, with Tanah Merah and Kuala Krai districts in the east, and Gua Musang district in the south. The district is situated between latitude 5° 20′N and 5° 45′N and longitude 101° 40′E and 101° 55′E (Fig. 1). This district covers an area of around 129,680.26 ha and is administratively divided into 3 sub-districts: Jeli, Batu Melintang, and Kuala Balah (MDJ 2011).

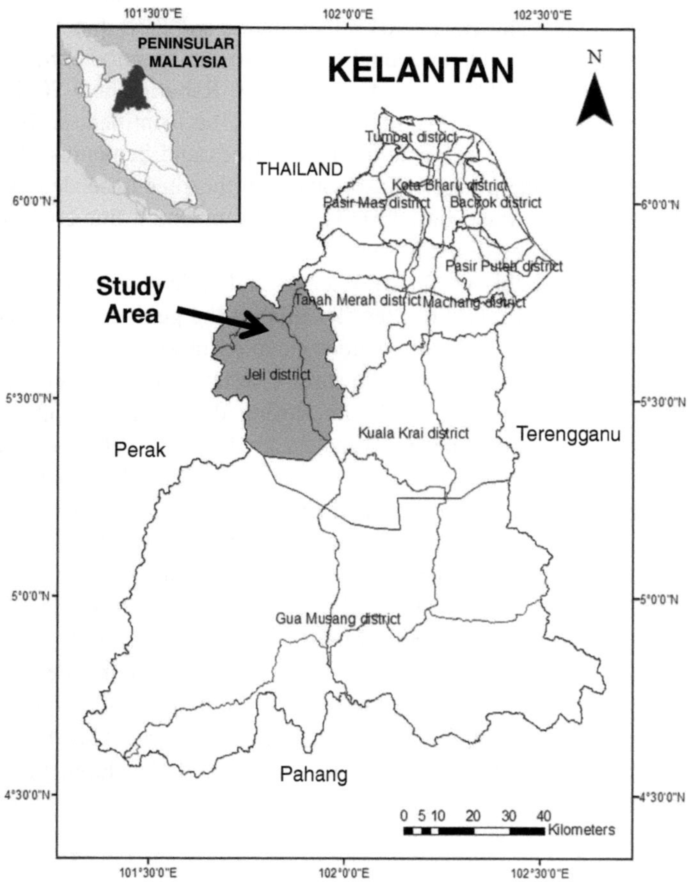

Fig. 1 Location map of the study area

3 Materials and Methods

Materials used in this study included base map (which was produced by the Department of Survey and Mapping Malaysia in 2009) and photographs. Meanwhile, some methods have been applied to achieve the objectives of this study. Geological mapping and characterisation were carried out to observe and describe the sites in details. Geological assessment was conducted in two approaches: qualitative assessment and quantitative assessment. In this study, the qualitative assessment was conducted by determining geodiversity values of each site, mainly scientific/research (important for dissemination of knowledge on geological features/records and Earth's history) and educational/training (to educate and train people who are working and studying in the field of Earth science, e.g. professional geoscientists, university students, as well as school students and the general public), aesthetic (related to beautiful features), recreational (suitable for various recreational activities), cultural (associated with the local culture), economic (the financial aspect of the resources based on their nature and their important role in the economic development), and functional (related to the functions and uses of the resources) values (Gray 2004, 2005; Komoo 2003; Raharimahefa 2012). In addition, the significance levels of the sites should also be identified, such as international, national, state-wide, regional, and local (Brocx and Semeniuk 2007). In the quantitative approach, the numerical assessment of the sites is needed to detect the potential of the sites for water-based tourism development. Some criteria based on the geodiversity values (modified from models of de Lima et al. 2010; Kirchner and Kubalikova 2011; Kubalikova 2013; Wagner and Orlewicz-Musial 2014) were established to assess and rank the sites. To support the geological assessment, the in situ water quality measurements were conducted in all sites to examine some physical and chemical parameters of the surface water in all these sites conducting in the mid of 2014 (June–July 2014), i.e. during the dry season in the state of Kelantan. Some physical parameters were measured such as *pH*, *temperature*, *total dissolved solid* (*TDS*), and *salinity* by using YSI 556 Multiparameter. Some chemical parameters i.e. *conductivity* and *dissolved oxygen* (*DO*) were also measured by using the same instrument. Furthermore, the average values for each parameter gained from these measurements in each site were compared to the 'Interim National Water Quality Standard (INWQS) for Malaysia' and then referred to 'Water Classes and Uses' (DOE Malaysia 2011; Appendix). Based on the geological assessment and water quality analysis, some recommended water-based activities are recommended for further water-based tourism development. In general evaluation, the SWOT (Strength, Weakness, Opportunity, and Threat) analysis was used to all these localities for their conservation and development in term of water-based tourism.

4 General Geology

Being located at the foot of the Main Range of Peninsular Malaysia, Jeli district can generally be divided into three types of landscape: (1) Mountainous areas, which extend in the west and north of the district and consist, among others, of mountain ridges and mountain valleys; (2) Hilly areas, which forms isolated hills and elongated hills, are distributed at the foot of the mountain range; and (3) Plain areas, which forms in the central and east of the district.

Geologically, the district is composed mostly of granitic rocks with several sedimentary/metasedimentary rock enclaves. There are three rock types can be found in the district (based on the geology of Kelantan by Department of Minerals and Geoscience Malaysia 2003): (1) Permian sedimentary rocks (Gua Musang Formation) which consists of phyllite, slate, sandstone and limestone; (2) Triassic sedimentary rocks (Gunong Rabong Formation), which consists of shale, silstone, sandstone, and limestone; and (3) The Main Range Granitic rocks (acid intrusives) which is generally of Middle Triassic age and located roughly in the west of the district stretching up to the state boundary of Perak and Pahang.

In term of structural geology, there are some regional and local structures which influence this area. Regional structures include faulting and folding, meanwhile local ones include faulting, folding, and jointing in the sedimentary rocks, and faulting and jointing in the granitic rocks (Department of Minerals and Geoscience Malaysia 2003). The dominant local structures in the district are along northwest-southeast and northeast-southwest directions (Fig. 2).

5 Potential Water-Based Tourism Sites

Some water bodies in Jeli district were selected as representatives of potential water-based tourism sites: Pergau Lake, Jeli Hot Spring, Rual River, and Lata Renyok Waterfall. Detailed geological mapping and characterisation of these sites, and supported by information from the previous literatures, are necessary to assess potential water-based tourism sites in the study area. Some values, such as scientific, educational, aesthetic, recreational, cultural as well as economic and functional values, are important for consideration of the sites. The biological aspects (animal and plant life) of the sites will be added to support the description. Figure 2 also shows the location map of these localities. Descriptions of each of these geological sites are as follows:

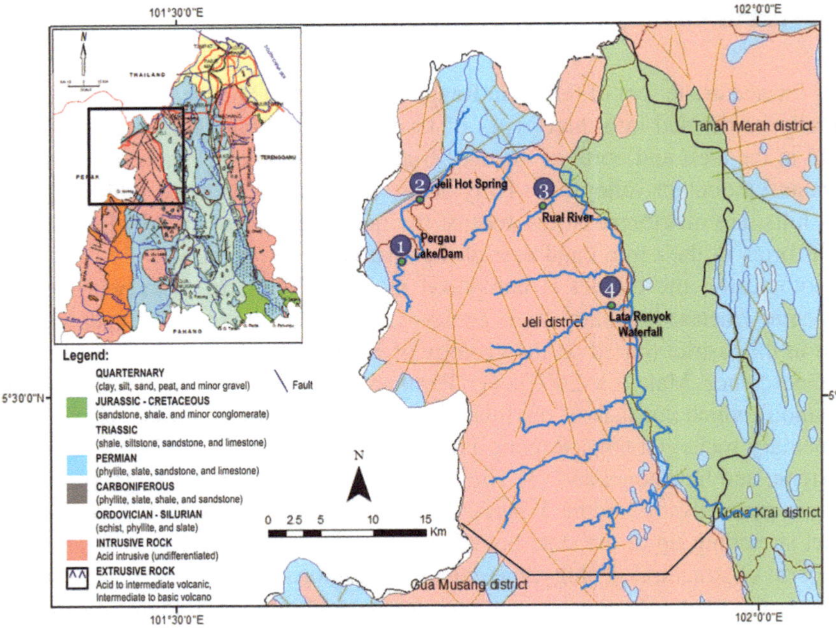

Fig. 2 General geology map and location of some potential water-based tourism sites in Jeli district, Kelantan, Malaysia. *Insert* General geology map of Kelantan. *Source* Department of Minerals and Geoscience Malaysia (2003)

5.1 Pergau Lake

This man-made lake or dam was built in 1990 as the catchment area of Pergau River to generate hydroelectric power (managed by the national electricity company). The lake (Fig. 3) is located in the coordinate of 5° 37′ 52.3″N, 101° 41′ 28.3″E, or around 30 km from Jeli town, on the East-West Highway, in Batu Melintang sub-district. It is surrounded by the tropical rainforest (Gunung Basor Forest Reserve) in the mountainous and hilly areas. It exposes some interesting natural and geological/geomorphological features. Geologically, the area is composed of different types of granitic rocks. Because of the beauty of the lake and its environments, its function has been developed into a tourism site that offers a comfortable area with various facilities and infrastructures, such as jetty, boats, chalets, and dorms. The lake possesses several small islands scattered within the lake. The main activities here are fishing of fresh water fish and netting shells. Some other water-based activities are rafting, boating, and kayaking (canoeing) (Adriansyah et al. 2015).

Fig. 3 Pergau Lake and its features. **a** A quiet and comfortable environments of the lake; **b** The jetty; **c** The dam; **d** One of granitic outcrops in the lake area; **e** Some chalets provided for visitors; **f** A bridge connecting between two islands in the middle of the lake; **g** Fishing is the favourite activity by local people and visitors; **h** Boating is another water-based activity in the lake

5.2 Jeli Hot Spring

The one and only hot spring in Jeli district is located in Kampung Bendang Lawa, Batu Melintang sub-district (coordinate: 5° 39′ 58.5″N, 101° 42′ 51.1″E). It manifests in a small valley of the Helai River and surrounded by a green hilly area, around 1 km from the East-West Highway. According to Ismail and Ashari (1998), the occurrence of hot springs in Kelantan is generally controlled by the geological structure and associated with the granite intrusion and magmatic activity in Peninsular Malaysia. Although the origin of this hot spring is still not clear, it can be known from the preliminary study that this hot spring is associated with the intrusion of the Main Range. The hot intrusion heats groundwater at relative shallow depths. The hot water is produced from the sedimentary/metasedimentary rocks and fractured granitic rocks. Adriansyah et al. (2015) explained that the temperature of this hot water is min. 42 °C (in the rainy season) to max. 50 °C (in the dry season). The hot water is accommodated in three adjacent ponds (Fig. 4). The main activity here is the hot water therapy that might give freshness to the visitor's body. In addition, the sulphure content in the water is suggested to be used for some skin problems. Some infrastructures have been provided for visitors in the area such as roofed huts with tables and chairs.

Fig. 4 Jeli Hot Spring site. **a** The attractive scene of the hot spring; **b** The main vent of the hot spring; **c** The hot water therapy is the visitors' main activity in the area

5.3 Rual River

This river is one of tributaries of Pergau River, the main river in Jeli district. The river drains from Kampung Sungai Rual to Kampung Baru Sungai Rual in Batu Melintang sub-district (coordinate: 5° 39′ 0″N to 5° 41′ 25.0″N, 101° 47′ 30.0″E to 101° 51′ 0″E) before finally it joins the Pergau River. There are some geomorphological features can be observed along the river such as cascades and rapids, and some geological features such as faults, joints, and magma mixing evidences (Fig. 5). The river is composed dominantly of granitic rocks. Since the river is surrounded by the green hilly areas, this makes the site has a magnificent view and become more interesting to be explored. The river is a suitable place for swimming and fishing to visitors who mainly come from the nearby areas. Other than natural features, the river area, mainly in Kampung Sungai Rual, has another unique cultural attraction which is a settlement area for indigenous people of Jahai tribe (Adriansyah et al. 2015).

5.4 Lata Renyok Waterfall

This waterfall, which is also known as Seri Bongor waterfall, is believed to be the most beautiful waterfall in Jeli district (Fig. 6). It is located in Kampung Renyok, Kuala Balah sub-district, nearby Jeli-Dabong Road (coordinate: 5° 34′ 47.1″N, 101° 52′ 28.5″E). This waterfall is a part of the Renyok River, one of tributaries of Pergau River. This waterfall, which has around 50 m long and 10 m high, is surrounded by the rainforest (Gunung Basor Forest Reserve) in the hilly area. The combination of its environments and the bright and dark-coloured rocks has made this waterfall more attractive. As a part of the Stong Migmatite Complex, this site

Fig. 5 Rual River. **a** The main cascade of the river; **b** One of the rapids; **c** A panoramic view of the river can be seen from a hill; **d** A magma mixing evidence with three different granitic rocks; **e** The river has some parts which are suitable for swimming

represents a rock cluster which is unique and rarely found in Malaysia. It comprises leucogranite with the metasedimentary rock enclaves (Unjah et al. 2002a, b). Adriansyah et al. (2015) stated that there are some geological features can be observed here such as systematic fractures/cracks (with the WE trending which is parallel to the river), some mini folds, boudinage structure, and veins (mainly quartz veins). Meanwhile, other than waterfall itself, there are some geomorphological features occurred in the site, such as stream/channel, rapids, cascade, ponds, and river terrace. The waterfall is also a place of a mini hydroelectric station (managed by the national electricity company) to generate and supply electricity to nearby residents. The site, which is quite popular among the local people due to its beauty, has equipped with several facilities such as multi-purpose hall, praying room, public toilets, gazebos, and parking lots. The main activities here are swimming, fishing, and photo shooting.

In this study, these potential water-based tourism sites can be classified geologically based on several categories: geodiversity (Gray 2005), scope (Brocx and Semeniuk 2007; Predrag and Mirela 2010), and scale (Brocx and Semeniuk 2007). There are eight elements of geodiversity as divided by Gray (2005): rocks, minerals, fossils, landforms, landscapes, processes, soils, and other georesources. The sites can be categorized into some scopes, such as geomorphological site, speleological site, mineralogical site, petrological site, stratigraphic site, structural site,

Fig. 6 Lata Renyok Waterfall. **a** This site is composed of leucogranite and metasedimentary rock enclaves; **b** The rocks have many systematic fractures and a main channel; **c** A boudin; **d** Some facilities in the location, such as multi-purpose hall, praying room, and toilet; **e** The mini station for hydroelectric power generation; **f** Fishing is among the main activities in the area

hydrological/hydrogeological site, and so on (Brocx and Semeniuk 2007; Predrag and Mirela 2010). The scales consist of regional scale (or megascale; covering an area of around 100 km × 100 km), large scale (or macroscale; covering an area of around 10 km × 10 km), medium scale (or mesoscale; covering an area of around 1 km × 1 km), small scale (or microscale; covering an area of around 10–100 m × 10–100 m), fine scale (or leptoscale; covering an area of around 1 m × 1 m), and very fine scale (covering an area of around 1 mm × 1 mm) (Brocx and Semeniuk 2007).

Most of the sites in the study area have geodiversity in landform/landscape and rock as well as process/phenomenon. They range into geomorphological site, petrological site, structural site, and of course hydrological/hydrogeological site, with the small to medium scales. Table 1 presents the characteristics of four potential water-based tourism sites in the district.

Table 1 Characteristics of some water-based tourism sites in Jeli district, Kelantan, Malaysia

Subject	Pergau Lake	Jeli Hot Spring	Rual River	Lata Renyok Waterfall
Location and coordinates	Mile 16 of the East-West Highway, Batu Melintang sub-district (5° 37′ 52.3″N, 101° 41′ 28.3″E)	Kampung Bendang Lawa, Batu Melintang sub-district (5° 39′ 58.5″N, 101° 42′ 51.1″E)	Kampung Sungai Rual-Kampung Baru Sungai Rual, Batu Melintang sub-district (5° 39′ 0″N to 5° 41′ 25″N, 101° 47′ 30″E to 101° 51′ 0″E)	Kampung Renyok, Kuala Balah sub-district (5° 34′ 47.1″N, 101° 52′ 28.5″E)
Main geological/geomorphological feature	A man-made lake with various granitic rocks	Hot spring	River valley with some features, e.g. cascades, magma mixing evidence	Waterfall
Other features	Surrounding mountainous and hilly areas, dam for hydroelectric power generation, and several islands in the middle of the lake	Helai River	Hilly areas, granitic rocks, fault, joints, village of aboriginal people	Granitic rocks with the enclaves of metasedimentary rocks, joints/fractures, mini folds, boudins, cascades, ponds, river terrace
Geodiversity (Gray 2005)	Rock, landforms/landscape	Process/phenomenon	Rock, landforms/landscape, process	Rock, landforms/landscape, process
Scope (Brocx and Semeniuk 2007; Predrag and Mirela 2010)	Hydrological site, geomorphological site petrological site	Hydrological/hydrogeological site	Hydrological site, geomorphological site, petrological site, structural site	Hydrological site, geomorphological site, petrological site, structural site
Scale (Brocx and Semeniuk 2007)	Medium scale	Small scale	Medium scale	Small scale

6 Geological Assessment

There are two main approaches for the geological assessment of the potential water-based tourism sites in the study area: the first one is based on the qualitative assessment and the second one is the quantitative assessment. The qualitative assessment has an advantage of relative objectivity, however, it has a difficulty to rank the sites. Meanwhile, the quantitative assessment can rank the sites by their scores, although it has possibly certain criteria which are difficult to be measured and could be the source of disagreement. Another problem in quantitative assessment is its subjectivity in assessment process (Kubalikova 2013).

For the qualitative assessment, most of the sites in this study have scientific, educational, aesthetic, recreational, and functional values, and only certain sites have cultural and economic values. These sites have levels of significance of state-wide (Pergau Lake and Lata Renyok Waterfall) and local (Jeli Hot Spring and Rual River). Table 2 shows the qualitative geological assessment of potential water-based tourism sites in the study area on the basis of geodiversity values and levels of significance.

Meanwhile in the quantitative assessment, some criteria were established to assess and rank the sites. Each criterion was divided into 0–4 grading scale. All these criteria have different weightages in order to assess the sites better (Table 3). From 100% weightage in this assessment, the scientific/research value has a significant weightage of 20% which consists of 5 points of criteria i.e. representativeness (3%), integrity (3%), rarity (3%), diversity of geological/geomorphological features (7%), and degree of scientific knowledge (4%). The educational/training value has also 20% weightage with 3 criteria i.e. exemplarity and pedagogical uses (6%), existing educational products (7%), and actual uses of a site for educational/training purposes (7%). Meanwhile, the aesthetic value has 10% weightage which consists of 2 important criteria i.e. beauty and colour of the site (5%) and global aesthetic value (5%). The recreational value has the highest weightage in this assessment with 35% proportion since this value reflects the site's suitability for tourism development. The criteria for this value are accessibility (5%), visibility (5%), logistics and facilities/infrastructures (7%), vulnerability (3%), monumentality (3%), recreational activities/uses (7%), and existing promotion of the site (5%). The cultural value, which consists of only 1 criterion i.e. presence of cultural/historical/archaeological aspects (7%), is also an important point for the tourism attraction of an area. The rest of the weightage is from the economic value, which is represented by only 1 criterion i.e. presence of businesses and local products (5%), and the functional value, which consists also 1 criterion i.e. current uses of the site for certain purposes (3%).

From the quantitative assessment among these sites, Pergau Lake is the ranking 1 with the total value of 2.6 out of maximum 4 marks (65%) and Lata Renyok Waterfall is the ranking 2 with the total value of 2.3 (57.5%). Meanwhile, the ranking 3 and ranking 4 are Rual River and Jeli Hot Spring with total value of 1.99 (49.75%) and 1.75 (43.75%) respectively. The results of the quantitative geological assessment of the potential water-based tourism sites in the study area are presented in Table 4.

Table 2 The qualitative geological assessment of potential water-based tourism sites in Jeli district, Kelantan, Malaysia

Site	Scientific value	Educational value	Aesthetic value	Recreational value	Cultural value	Economic value	Functional value	Level of significance (ranking)
Pergau Lake	Various granitic rocks; landforms/landscape in the area; water quality	Suitable for educational purposes, e.g. fieldworks, motivation camps, and tours	Beautiful landscape in the lake area surrounded by mountainous and hilly areas	Fishing, rafting, boating, kayaking/canoeing	–	Generating income by renting chalets and dorms; fishing and netting activities	Hydroelectric power generation	State
Jeli Hot Spring	Hot spring formation; landform; geothermal system; structural control; water quality	Suitable for educational purposes, e.g. fieldworks and tours	A small channel surrounded by hilly areas	Enjoy the fresh hot water	–	–	Skin treatment and hot spring therapy	Local
Rual River	River landforms/landscape; fault and other structural features; magma mixing; water quality	Suitable for educational purposes, e.g. fieldworks and tours	Interesting landscape of the river surrounded by hilly areas	Swimming/bathing, fishing	Settlement of indigenous people (Jahai tribe)	–	–	Local
Lata Renyok Waterfall	River landforms/landscape; unique outcrops of granitic and metasedimentary rocks; fractures and other structural features; water quality	Suitable for educational purposes, e.g. fieldworks and tours	Wonderful landscape of waterfall with the river terrace and surrounded by hilly areas	Swimming/bathing, fishing	–	–	Hydroelectric power generation	State

Table 3 The proposed criteria for the quantitative geological assessment of potential water-based tourism sites in Jeli district, Kelantan, Malaysia (modified from de Lima et al. 2010; Kirchner and Kubalikova 2011; Kubalikova 2013; Wagner and Orlewicz-Musial 2014)

No.	Criteria	Weightage (%)	Grading
1. Scientific/research value (20%)			
1.a	Representativeness	3	**0**—very low; **1**—low; **2**—medium; **3**—high; **4**—very high
1.b	Integrity	3	**0**—totally destroyed site; **1**—destroyed site but with the rest of geological/geomorphological feature(s); **2**—disturbed site but with the rest of the main geological/geomorphological feature(s); **3**—only low destruction of the site; **4**—site without any destruction
1.c	Rarity	3	**0**–**5** or more than 5 similar sites; **1**–**4** similar sites; **2**–**3** similar sites; **3**–**2** similar sites; **4**—the only site within the study area
1.d	Diversity of geological/geomorphological features	7	**0**—only 1 visible feature; **1**–**2** visible features; **2**–**3** visible features; **3**–**4** visible features; **4**–**5** or more than 5 visible features
1.e	Degree of scientific knowledge	4	**0**—unknown site; **1**—start to be studied; **2**—scientific paper(s) on national level; **3**—scientific paper(s) on international level; **4**—high knowledge of the site, many studies about the site in national and international level
2. Educational/training value (20%)			
2.a	Exemplarity and pedagogical uses	6	**0**—very low exemplarity and pedagogical use(s); **1**—low exemplarity and pedagogical use(s); **2**—existing exemplarity, but with limited pedagogical use(s); **3**—high exemplarity and pedagogical use(s); **4**—very high exemplarity and pedagogical use(s)
2.b	Existing educational products	7	**0**—none; **1**—less products; **2**—some but with low quality products; **3**—some with high quality products; **4**—complete and high quality products
2.c	Actual uses of a site for educational/training purposes	7	**0**—none; **1**—educative use(s) for general public (e.g. tours); **2**—educative use(s) for school children and general public (e.g. tours); **3**—educative use(s) for university students, school children, and general public (e.g. fieldworks, tours); **4**—educative use(s) for professional geologists, university students, school children, and general public (e.g. scientific expeditions, excursions, fieldworks, tours)
3. Aesthetic value (10%)			
3.a	Beauty and colour of the site	5	**0**—not beautiful and not colourful site; **1**—not beautiful but colourful site; **2**—a quite beautiful and colourful site; **3**—beautiful but not colourful site; **4**—beautiful and colourful site

(continued)

Table 3 (continued)

No.	Criteria	Weightage (%)	Grading
3.b	Global aesthetic value	5	**0**—very low; **1**—low; **2**—medium; **3**—high; **4**—very high
4. Recreational value (35%)			
4.a	Accessibility	5	**0**—very bad; **1**—bad; **2**—medium; **3**—good; **4**—excellent
4.b	Visibility	5	**0**—very bad; **1**—bad; **2**—medium; **3**—good; **4**—excellent
4.c	Logistics and facilities/infrastructures	7	**0**—none; **1**—a few; **2**—some; **3**—complete; **4**—complete and excellent
4.d	Vulnerability	3	**0**—possible risk affecting the site; **1**—possible risk affecting all geological features; **2**—possible risk affecting main geological feature(s); **3**—possible risk only affecting secondary geological feature(s); **4**—no risk of damage
4.e	Monumentality	3	**0**—not iconic; **1**—potential icon in the district level; **2**—potential icon in the state level; **3**—potential icon in the national level; **4**—potential icon in the international level
4.f	Recreational activities/uses	7	**0**—none; **1**—only 1 recreational activity/use; **2**–**2** recreational activities/uses; **3**–**3** recreational activities/uses; **4**–**4** or more than 4 recreational activities/uses
4.g	Existing promotion of the site	5	**0**—one; **1**—less promotion; **2**—adequate promotion; **3**—good promotion; **4**—emblematic site for the district or region
5. Cultural value (7%)			
5.a	Presence of cultural/historical/archaeological aspects	7	**0**—none; **1**—existing cultural feature(s) without the relation with any geological feature; **2**—existing cultural feature (s) with the relation with geological feature(s); **3**—existing cultural feature(s) with the strong relation with geological feature(s); **4**—existing cultural feature(s) with the strong relation with the site and its geological feature(s)
6. Economic value (5%)			
6.a	Presence of businesses and local products	5	**0**—none; **1**—less businesses but without any local product; **2**—less businesses but with local product(s); **3**—some existing businesses with local product(s); **4**—emblematic site for the certain business(es) and local product (s)
7. Functional value (3%)			
7.a	Current uses of the site for certain purposes	3	**0**—none; **1**—existing with one important use of the site; **2**—existing with some important uses of the site; **3**—existing with one very important use of the site; **4**—existing with some very important uses of the site

Table 4 The quantitative geological assessment of potential water-based tourism sites in Jeli district, Kelantan, Malaysia

No.	Criteria	Weightage	Pergau Lake		Jeli Hot Spring		Rual River		Lata Renyok Waterfall	
			Value	W. Value[a]	Value	W. Value[a]	Value	W. Value[a]	Value	W. Value[a]
1. Scientific/research value (20%)										
1.a	Representativeness	0.03	4	0.12	4	0.12	2	0.06	3	0.09
1.b	Integrity	0.03	4	0.12	3	0.09	3	0.09	4	0.12
1.c	Rarity	0.03	4	0.12	4	0.12	0	0	0	0
1.d	Diversity of geological/ geomorphological features	0.07	2	0.14	3	0.21	4	0.28	4	0.28
1.e	Degree of scientific knowledge	0.04	3	0.12	3	0.12	3	0.12	3	0.12
2. Educational/training value (20%)										
2.a	Exemplarity and pedagogical uses	0.06	2	0.12	2	0.12	2	0.12	2	0.12
2.b	Existing educational products	0.07	2	0.14	2	0.14	0	0	2	0.14
2.c	Actual uses of a site for educational/training purposes	0.07	3	0.21	3	0.21	3	0.21	4	0.28
3. Aesthetic value (10%)										
3.a	Beauty and colour of the site	0.05	3	0.15	0	0	3	0.15	4	0.2
3.b	Global aesthetic value	0.05	4	0.2	1	0.05	3	0.15	4	0.2
4. Recreational value (35%)										
4.a	Accessibility	0.05	3	0.15	1	0.05	3	0.15	2	0.1
4.b	Visibility	0.05	3	0.15	3	0.15	2	0.1	3	0.15
4.c	Logistics and facilities/ infrastructures	0.07	2	0.14	1	0.07	0	0	1	0.07
4.d	Vulnerability	0.03	4	0.12	3	0.09	3	0.09	4	0.12
4.e	Monumentality	0.03	1	0.03	1	0.03	0	0	1	0.03

(continued)

Table 4 (continued)

No.	Criteria	Weightage	Pergau Lake		Jeli Hot Spring		Rual River		Lata Renyok Waterfall	
			Value	W. Value[a]	Value	W. Value[a]	Value	W. Value[a]	Value	W. Value[a]
4.f	Recreational activities/uses	0.07	4	0.28	1	0.07	2	0.14	2	0.14
4.g	Existing promotion of the site	0.05	1	0.05	1	0.05	1	0.05	1	0.05
5. Cultural value (7%)										
5.a	Presence of cultural/historical/ archaeological aspects	0.07	0	0	0	0	4	0.28	0	0
6. Economic value (5%)										
6.a	Presence of businesses and local products	0.05	3	0.15	0	0	0	0	0	0
7. Functional value (3%)										
7.a	Current uses of the site for certain purposes	0.03	3	0.09	2	0.06	0	0	3	0.09
Sum (max. 4)				2.6		1.75		1.99		2.3
Percentage (max. 100%) (%)				65		43.75		49.75		57.5
Ranking				1		4		3		2

[a]W. Value (weighted value) = Weightage × Value

Table 5 Average in situ water quality measurements (mean value) of potential water-based tourism sites in Jeli District, Kelantan, Malaysia

Sites	GPS Coordinates	pH	Temp. (°C)	TDS (mg/l)	Salinity (ppt)	Conductivity (μS/cm)	DO (mg/l)	Class
Pergau Lake	05° 37′ 52.3″N, 101° 41′ 28.3″E	8.20	28.12	18	0.01	30	1.95	IV
Jeli Hot Spring	05° 39′ 58.5″N, 101° 42′ 51.1″E	8.28	49.09	230	0.16	510	1.05	IV
Rual River	5° 39′ 0″N to 5° 41′ 25″N, 101° 47′ 30″E to 101° 51′ 0″E	5.38	25.85	28.5	0.015	43	3.68	III
Lata Renyok Waterfall	5° 34′ 47.1″N, 101° 52′ 28.5″E	6.77	25.47	17	0.01	26	7.53	I

Pergau Lake received the highest value in this assessment since it has the highest recreational value, which is very important in tourism development. Besides, it has also good marks in scientific, aesthetic, economic, and functional values. Meanwhile, Jeli Hot Spring received the lowest value due to its lowest marks in recreational and aesthetic values. It has also no mark in cultural and economic values.

Pergau Lake, Rual River, and Lata Renyok Waterfall contribute as the potential recharge of the area, so that the area has no problem in term of water resources sustainability. Most population in this area depend on the surface water, instead of groundwater.

7 Water Quality Analysis

For the in situ water quality analysis, the measurements have been conducted in several measurement points for all sites depending on their coverage, i.e. 4 points in Pergau Lake, 3 points in Jeli Hot Spring, 4 points in Rual River, and 3 points in Lata Renyok Waterfall.

Based on the measured parameters, some parameters are significant to determine the water classes and uses, they are pH and DO. Meanwhile, other parameters such as temperature, TDS, salinity, and conductivity give insignificant readings. According to the measurements, the water quality of the Pergau Lake and the Jeli Hot Spring were considered as the category of "Class IV" which means that the water in these two sites can still be utilized for human uses (body contact still allowed), but it should not be used for human consumption. The Rual River was interpreted as the category of "Class III" in water quality which means that the water in this site can be used as water supply for human uses with the extensive treatment requirement. The Lata Renyok Waterfall was categorized as "Class I" which means that the site is suitable for conservation of the natural environment as well as for the water supply and the fishery of very sensitive aquatic species. The results of the average of these in situ measurements were summarised in Table 5.

Table 6 Some recommended water-based tourism activities in Jeli district, Kelantan, Malaysia

Sites	Activities					
	Swimming and bathing	Fishing	Rafting	Boating	Kayaking/canoeing	Hot water therapy and treatment
Pergau Lake	–	✓	✓	✓	✓	✓
Jeli Hot Spring	–	–	–	–	–	✓
Rual River	✓	✓	–	–	–	–
Lata Renyok Waterfall	✓	✓	–	–	–	–

✓ recommended; – not recommended

8 Water-Based Tourism Activities

One of the utilization of surface water bodies is for the development of the so-called water-based tourism. Through this kind of tourism, we hope a better understanding of the water bodies and their features can be achieved so that their existence and occurrences can be acknowledged.

Jeli district possesses many attractive water-based sites which should be conserved and developed to stimulate the tourism sector of the district. Based on the geological assessment of the sites and supported by their water quality analysis, some water-based activities are possible to be conducted in the sites such as swimming and bathing, fishing, rafting, boating, kayaking/canoeing, and hot water therapy/skin treatment. Table 6 shows the recommended activities in potential water-based tourism sites of Jeli district.

9 Evaluation

SWOT analysis was used in this study to evaluate the strengths, weaknesses, opportunities, and threats of all these localities in term of water-based tourism development. Table 7 shows SWOT analysis of each site for this purpose.

Based on the geological assessment and evaluation, it is recommended that all these sites should always be studied scientifically and developed properly for educational activities with some educational products, such as map, webpage, information panel etc. These sites should also be provided with proper accessibility, logistics and facilities/infrastructures, and active promotion in order to develop water-based tourism in Jeli district. In addition, the presence of local businesses and local products will also become attractions of the sites.

Table 7 SWOT analysis of potential water-based tourism sites in Jeli district, Kelantan, Malaysia

Site	Strength	Weaknesses	Opportunities	Threats
Pergau Lake	Good potential for research, educational, and recreational activities; high aesthetic value; good economic and functional value; good accessibility; rich in geodiversity and biodiversity	No cultural value; lack of promotion	Development of water-based activities e.g. fishing, rafting, boating, and kayaking; more active promotion; improvement of facilities; cooperation between local authority, universities, and communities	Littering and rubbish deposition; this lake is quite deep and can be dangerous to visitors
Jeli Hot Spring	Good potential for research, educational, and recreational activities; good functional value for hot water therapy and treatments	Low aesthetic and recreational values; no cultural and economic values; bad accessibility; lack of logistics and facilities; lack of promotion; poor management and protection	Suitable place for hot water therapy and treatments; proper management and development will make this site more beautiful; improvement of accessibility and facilities; more active promotion; cooperation between local authority, universities, and communities	Littering and rubbish deposition; human—wildlife conflict since it is located in the remote area
Rual River	Good potential for research, educational, and recreational activities; good aesthetic value; high cultural value of aboriginal people (Jahai tribe); good accessibility; relatively high geodiversity	No economic and functional values; bad logistics and facilities; lack of promotion	Interesting river for swimming, bathing, and fishing; providing facilities/infrastructures; more active promotion; cooperation between local authority, universities, and communities	Littering and rubbish deposition; inhabitants in the settlements surrounding the river; instability of slopes surrounding the river
Lata Renyok Waterfall	Good potential for research, educational, and recreational activities; very high aesthetic value; good functional value; rich in geodiversity	No cultural and economic values; poor logistics and facilities; lack of promotion; bad management and protection	Suitable site for swimming, bathing, and fishing; improvement of facilities/infrastructures; more active promotion; cooperation between local authority, universities, and communities	Littering and rubbish deposition; some spots in the site are quite deep and can be dangerous to visitors

For the conservation and development efforts of the sites and features, the tourism programmes and activities should be organized in the framework of the protection of the areas and promotion of the sustainable development. Tourism should also be able to encourage and empower local communities, mainly to improve their welfare. In addition, it is suggested to create a collaboration among the stakeholders to conduct this planning.

10 Conclusion

The geological assessment has been conducted in some potential water-based tourism sites in Jeli district: Pergau Lake, Jeli Hot Spring, Rual River, and Lata Renyok Waterfall. Detailed mapping and characterisation showed that these sites have significant geodiversity. Although they are dominated by granitic rocks, they occurred in beautiful landform/landscape features, such as valley, lake, and waterfall, where the water bodies exist, and surrounded by mountainous and hilly areas. In term of scale, these sites have small- to medium-sized areas (more than 10 m and less than 10 km). The qualitative assessment indicated that most of the sites have scientific, educational, aesthetic, recreational, and functional values, and only certain sites have cultural and economic values. These sites have significance level of local to state. From the quantitative assessment, Pergau Lake has the highest rank (Ranking 1) due to its highest recreational value, which is very important in tourism development, and good marks in scientific, aesthetic, economic, and functional values. Lata Renyok Waterfall and Rual River have got the second and the third ranks (Ranking 2 and 3) respectively. Meanwhile, Jeli Hot Spring has the lowest rank since it has the lowest marks in recreational and aesthetic values, and no mark in cultural and economic values. The in situ water quality analysis shows that these sites have different water quality classes based on the INWQS for Malaysia i.e. Class I for Lata Renyok Waterfall, Class III for Rual River, and Class IV for Pergau Lake and Jeli Hot Spring. All these localities have some specific tourism and recreational activities for further development, such as swimming and bathing, fishing, rafting, boating, kayaking/canoeing, and hot water therapy/skin treatment. Scientific, educational, and tourism activities in the study area should be conducted routinely for the establishment of these sites as the models of sustainable water-based tourism sites in Malaysia.

Acknowledgements Authors would like to thank the local authority (Jeli District Council), local communities as well as visitors in the sites for their contribution to this study.

Appendix

Interim National Water Quality Standard (INWQS) for Malaysia (DOE Malaysia 2011)

Parameter	Unit	Class					
		I	IIA	IIB	III	IV	V
Ammonical nitrogen	mg/l	0.1	0.3	0.3	0.9	2.7	>2.7
Biochemical oxygen demand	mg/l	1	3	3	6	12	>12
Chemical oxygen demand	mg/l	10	25	25	50	100	>100
Dissolved oxygen	mg/l	7	5–7	5–7	3–5	<3	<1
pH	–	6.5–8.5	6–9	6–9	5–9	5–9	–
Colour	TCU	15	150	150	–	–	–
Electrical conductivity[a]	μS/cm	1000	1000	–	–	6000	–
Floatables	–	N	N	N	–	–	–
Odour	–	N	N	N	–	–	–
Salinity	%	0.5	1	–	–	2	–
Taste	–	N	N	N	–	–	–
Total dissolved solid	mg/l	500	1000	–	–	4000	–
Total suspended solid	mg/l	25	50	50	150	300	300
Temperature	°C	–	Normal + 2 °C	–	Normal + 2 °C	–	–
Turbidity	NTU	5	50	50	–	–	–
Faecal coliform[b]	Count/100 ml	10	100	400	5000 (20,000)[c]	5000 (20,000)[c]	–
Total coliform	Count/100 ml	100	5000	5000	50,000	50,000	>50,000

Note

N No visible floatable materials or debris, no objectional odour or no objectional taste

[a]Related parameters, only one recommended for use

[b]Geometric mean

[c]Maximum not to be exceeded

Water classes and uses (DOE Malaysia 2011)

Class	Uses
Class I	Conservation of natural environment Water Supply I—Practically no treatment necessary Fishery I—Very sensitive aquatic species
Class IIA	Water Supply II—Conventional treatment required Fishery II—Sensitive aquatic species
Class IIB	Recreational use with body contact
Class III	Water Supply III—Extensive treatment required Fishery III—Common, of economic value and tolerant species; livestock drinking
Class IV	Irrigation
Class V	None of the above

References

Adriansyah D, Busu I, Eva H, Muqtada M (2015) Geoheritage as the basis of geotourism development: a case study in Jeli district, Kelantan, Malaysia. Geo J Tourism Geosites 15: 25–43

Brocx M, Semeniuk V (2007) Geoheritage and geoconservation—history, definition, scope, and scale. J R Soc West Aust 90:53–80

Department of Environment (DOE) Malaysia (2011) Malaysia environmental quality report 2011. Ministry of Natural Resources and Environment, Kuala Lumpur

Department of Minerals and Geoscience Malaysia (2003) Quarry resource planning for the state of Kelantan. Osborne & Chappel Sdn, Bhd

de Lima FF, Brilha JB, Salamuni E (2010) Inventorying geological heritage in large territories: a methodological proposal applied to Brazil. Geoheritage. Springer, Berlin. doi:10.1007/s12371-010-0014-9

Eurostat (2009) MEDSAT II: 'Water and tourism' pilot study. Eurostat & Euromed, 37 p

Gray M (2004) Geodiversity: valuing and conserving abiotic nature. Wiley, Chichester, p 434

Gray M (2005) Geodiversity and geoconservation: what, why, and how? The George Wright Forum, pp 4–12

Ismail A, Ashari MA (1998) Kajian kualitatif mata air panas di negeri Kelantan. Jabatan Penyiasatan Kajibumi Malaysia

Jennings G (2006) Water-based tourism, sport, leisure, and recreation experiences. In: Jennings G (ed) Water-based tourism, sport, leisure, and recreation experiences. Elsevier Ltd, 260 p. ISBN 978-0-7506-6181-2

Kirchner K, Kubalikova L (2011) Evaluation of geoheritage in the western part of Podyji National Park, Czech Republic. Revista de Geomorfologie 13:51–58

Komoo I (2003) Conservation geology: protecting hidden treasures of Malaysia. ASM Inaugural Lecture, LESTARI UKM, 51 p

Kubalikova L (2013) Geomorphosite assessment for geotourism purposes. Czech J Tourism 2 (2):80–104. doi:10.2478/cjot-2013-0005

MDJ (Majlis Daerah Jeli) (2011) Draft rancangan tempatan Jeli 2020 (ringkasan eksekutif). Unit Rancangan Pembangunan Kota Bharu, Jabatan Perancangan Bandar dan Desa Semenanjung Malaysia, 38 p

Predrag D, Mirela D (2010) Inventory of geoheritage sites—the base of geotourism development in Montenegro. Geogr Pannonica 14(4):126–132

Raharimahefa T (2012) Geoconservation and geodiversity for sustainable development in Madagascar. Madagascar Conserv Dev 7(3):126–134. doi:10.4314/mcd.v7i3.5

Unjah T, Komoo I, Mohamed H (2002a) Kepelbagaian fitur geomorfologi Kompleks Migmatit Stong, Kelantan. In: Komoo I, Leman MS (eds) Geological heritage of Malaysia (research and development of the geoheritage). LESTARI UMK, Bangi, pp 283–296

Unjah T, Komoo I, Mohamed H (2002b) Landskap geologi Kompleks Migmatit Stong Kelantan. In: Annual geological conference 2002. Geological Society of Malaysia, pp 201–205

Wagner A, Orlewicz-Musial M (2014) Functions and dysfunctions of tourism and recreation and how they influence aquatic environments. Pol J Environ Stud 23(3):1045–1050

Mapping and Modelling of Areas at Risk of Erosion: Case of Aures Center (Algeria)

Benmessaoud Hassen, Laggoun Soufiane and Chafai Chaouki

Abstract At the Mediterranean region, the problems related to the agriculture, the defense of the soil and watershed management, are one of the most significant environmental problems. To solve this problem it is necessary to study, analyze and understand the impact of the erosive action in these semi-arid areas. There are different methods that estimate soil erosion at the plot or watershed level; these methods range from simple to complex. For this work, the analysis has focused on modelling of water erosion by the RUSLE model (Revised Universal Soil Loss Equation) consisting of a modified version of the universal soil loss equation of (USLE) originally developed by Wischmeier and Smith in Predicting rainfail erosion losses—a guide to conservation planning. Agricultural handbook No. 537, USDA, Washington, 1978. A methodology of assessing the various risk indices of erosion around the geographic information system (GIS-Arc Gis 9.3). The RUSLE equation was applied by overlaying, using ArcGIS software, the various appropriate thematic maps for each factor. According to the RUSLE model, over 70% of the study areas are subject to different soil loss levels: low (<5) to very high (50–200). The average loss by water erosion across all study sites is about 8.73 t/ha/year. Total annual losses for the six selected townships are of the order of 33,995.4 t/ha/year. In light of these figures, we can conclude that the unequal distribution of areas with potential risk of erosion in the Aurès center results from the variability of the characteristics of the factors involved in the process and the model used. These results will be used for territorial planning, in reference to the project to reduce the vulnerability of these areas.

Keywords Cartography · Modelling · Quantification · Risk of erosion · Aurès

B. Hassen (✉) · L. Soufiane
Laboratory Natural Risks and Planning, Faculty of Nature and Life Sciences,
University of Batna, 02, Fesdis, 05077 Batna, Algeria
e-mail: ha123_m123@yahoo.fr

C. Chaouki
Ecole National Des Forêts, Batna, Algeria
e-mail: chafaichaouki@gmail.com

O. Abdalla et al. (eds.), *Water Resources in Arid Areas: The Way Forward*,
Springer Water, DOI 10.1007/978-3-319-51856-5_12

1 Introduction

Recent studies on the vulnerability to climate change in the Mediterranean region indicate a trend towards increased aridity accelerating water erosion (Berkane and Yahiaou 2007; Al Ali 2007; Souadi 2011). Water erosion is a natural phenomenon that evolves in line with human evolution and climate severity. This phenomenon is defined as the process of detachment of soil particles by the effect of precipitation and runoff, transport and along the path (Foster and Meyer 1972).

Losses of topsoil by erosion have immediate and long-term effects. These effects result in a reduction of soil quality and performance rather weak cultures.

Algeria is one of the most countries endangered in the world by erosion. Water erosion varies between 2000 and 4000 t/km^2/year (Pierre et al 2008), Land degradation was estimated in the Aurès central region in 1955 (according to the hydraulics science archives) between 200 and 500 tons of solid elements for the watershed of Oued-EI Abiod upstream dam Foum-EI-Guerza (Abdessemed 1984).

As such, the Aurès central region which is characterized by very rugged terrain (slopes greater than 25%, variants elevations), a dense river network, semi-arid and degraded vegetation cover and a soft bedrock lithology (marl—limestone), is an alarming example of soil degradation by water erosion which may be catastrophic (Benmessaoud 2010).

The assessment of soil erosion risks in the Aurès central region required mapping and spatialization of many factors involved in the erosive process: the aggressiveness of rainfall, slope and slope length, soil erodibility, vegetation cover and agricultural practices. Each factor has a different behavior from a zone to another. It led as well to a multitude of data mapping, store, organize and process rationally.

In this research we propose a methodology based on the universal soil loss equation (RUSLE) Wischmeier and Smith (1978), Wall et al. (2002) to identify areas at risk of erosion and to calculate soil loss, this methodology is based essentially on the use of the tools of Geomatics based on remote sensing data, field observations using GPS and the use of Geographic Information System (GIS) for analysis and modelling of erosion process.

2 Presentation of the Study Area

Aurès constitute a geographical entity located in the east of the Saharan Atlas. This set of chains of very steep mountains is heavily exposed to water erosion phenomenon especially in its central part.

Geographically, the study area is located to the south of the Aurès (Fig. 1) between the meridians 04° 6′ 45″ and 6° 38′ 59″ East and parallels 35° 24′ 49″ and 34° 58′ 30″ North.

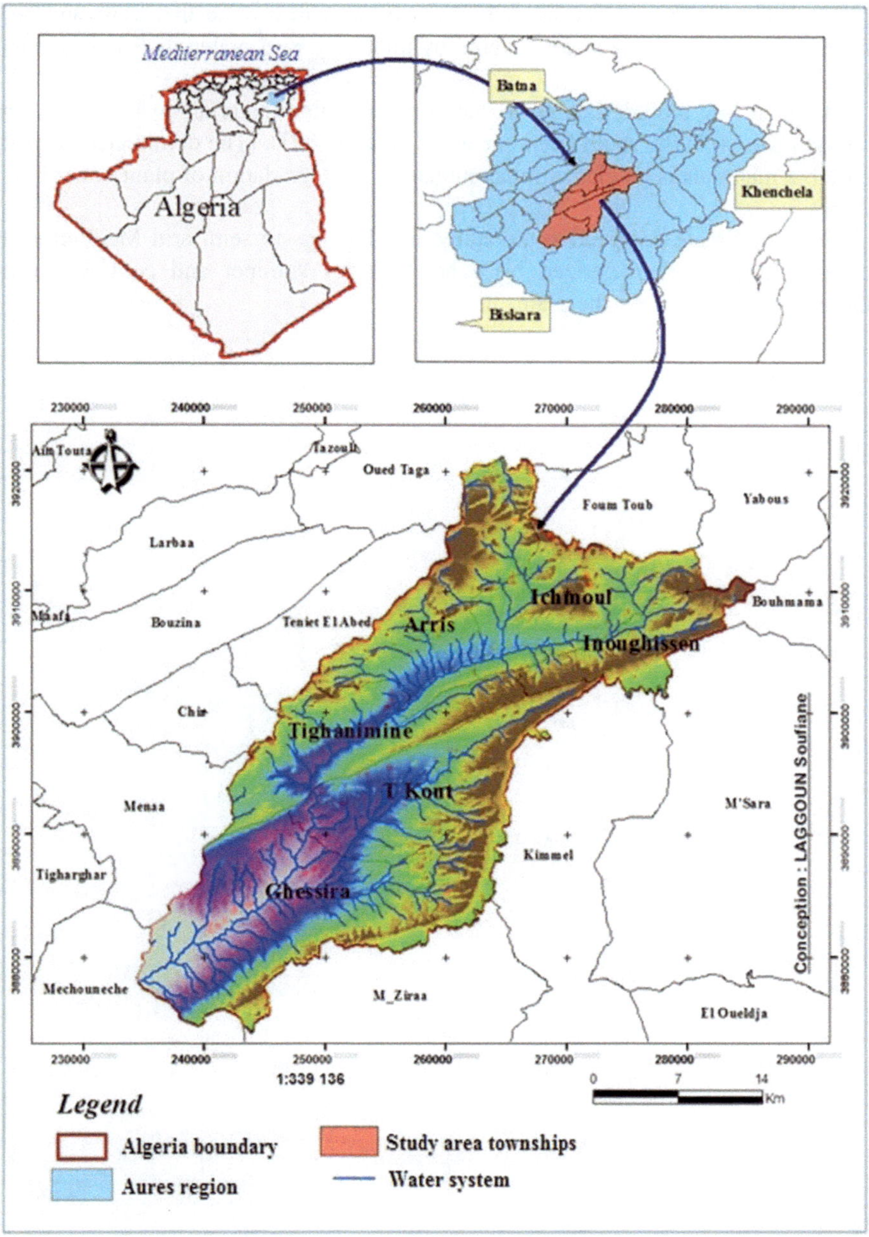

Fig. 1 Location map of the Aurès central region

Geologically, the area is divided into two main areas: The limestone and marl formations. The high areas are covered by relatively hard materials (sandstone and limestone).

Areas at low altitude belong to the domain of fine materials in the north, and coarse material in the south (Berkane and Yahiaou 2007). (The dominance of easily erodible materials is an element that enhances the degradation of plant cover in the Aurès.

From a climate perspective the study area belongs to semi arid Mediterranean climate (Fig. 2), characterized by a hot and dry summer and cool winter on mountains and fresh on the plain.

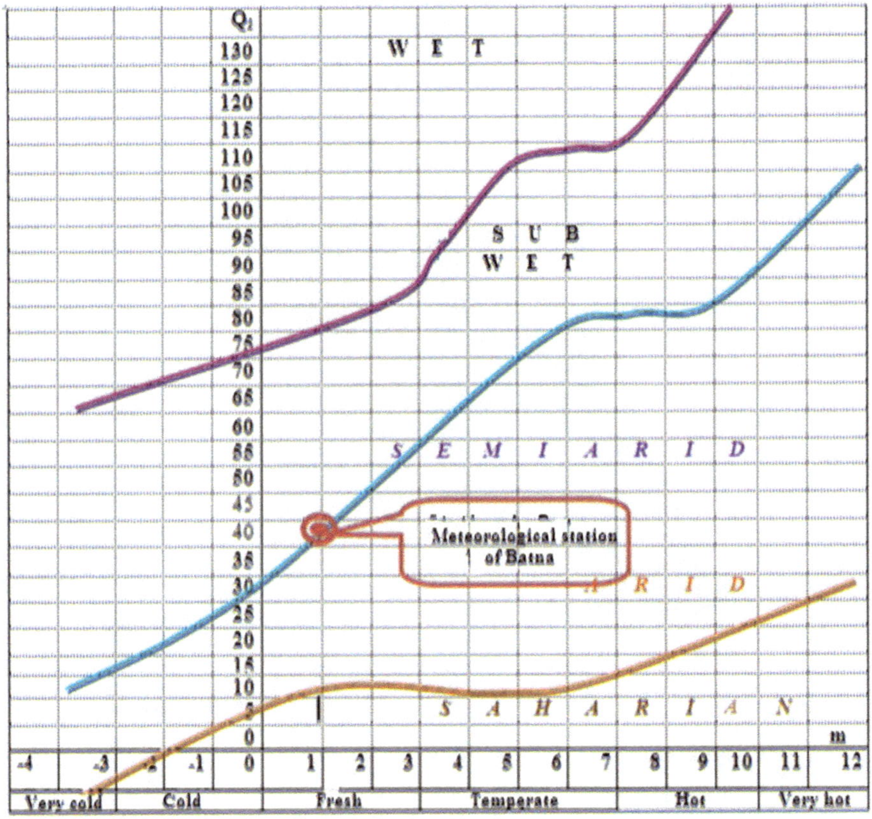

Fig. 2 Rain-thermal climatogram of EMBERGER for the study area

3 Methodological Approaches

Several methods have been oriented to models incorporating erosion factors; the most widely used model is the Universal Soil Loss Equation (USLE) and its revised form (RUSLE). The assessment of the potential risk of soil erosion in the Watershed of Aurès center is based on the modelling of erosion factors (Wall et al. 2002; Hadir 2010).

In addition, the model chosen in this work does not take into account the distribution of the vegetation cover (Leguédois 2003; El Garouani 2008) which represents a major failure since it can be crucial to assess the risk of erosion in the study area. To overcome this drawback and to improve the model, we decided to add the factor "vegetation cover". The completion of the land use map from the supervised classification of satellite imagery Landsat-8/LDCM, taken in 23 June 2013, covering the study area, facilitated the establishment of the vegetation cover map, which helped to define more precisely the most erodible areas. The digitization of topographic, geological and soil maps, analysis, combining data and modeling were carried out using the Geographic Information System (GIS).

3.1 *Field Data Collection*

Missions of field recognition have been conducted throughout the study area (06 townships). These missions aimed to make sampling points of field to compare and better understand the topographic data and satellite imagery maps.

The work was to identify the field, judiciously chosen points (Units and homogeneous or mixed type of landscape sites, erosion control works, accurately on land use, etc.). Specifically through GPS, to describe the site and take digital pictures georeferenced (800 photos).

Also with a GPS, moving in the six townships was to fill data sheets points for checking the ground truth. A total of 69 control points with these data sheets have been met.

These points sampled are distributed in a way to discover the variability in the study area (Fig. 3).

3.2 *Database and Management of Different Layers of the Model*

The mapping of soil loss in the central Aurès region based on the combination of the quantitative factors using GIS: Erosivity Index, soil erodibility index, Topographic Factor, Farming Factor, Conservation Factor.

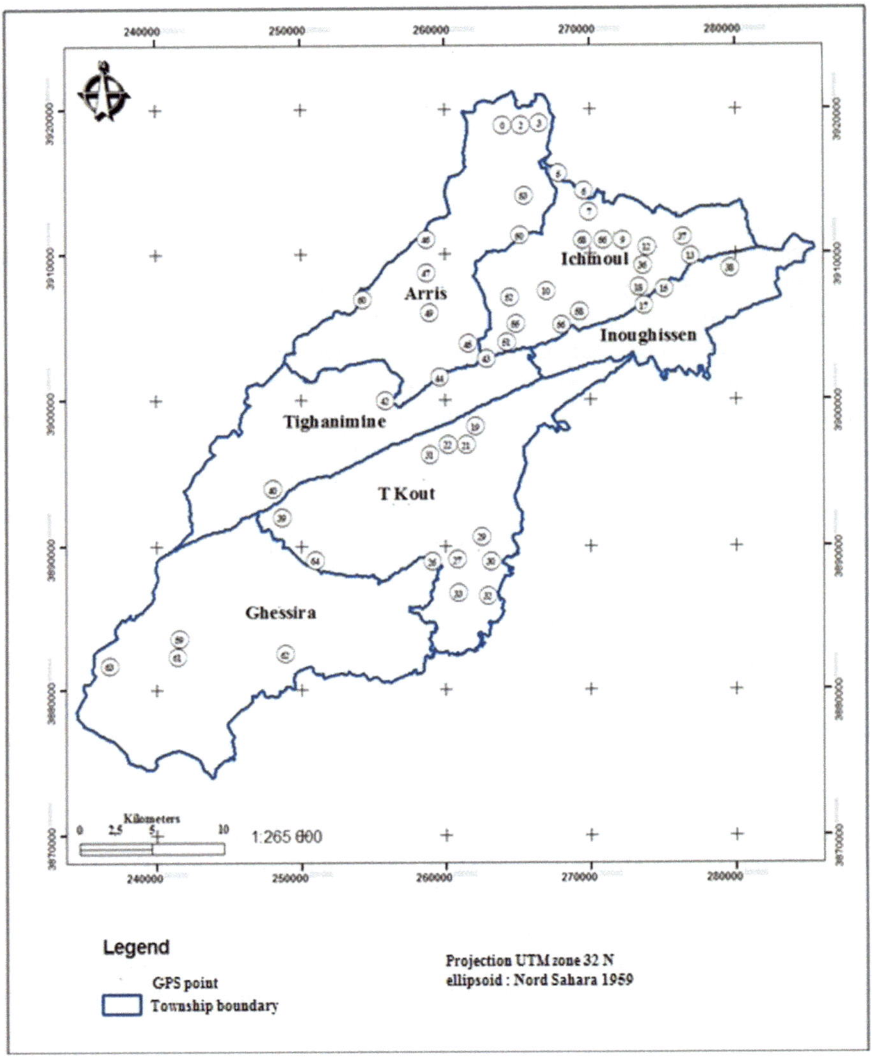

Fig. 3 Map of GPS points sampling

The geo-referenced digital data obtained by digitizing were integrated in a geographical information system, structured in several layers for Topographic Information (level curves, digital elevation model (DEM), slope map, aspect map, Hydrographical network…), soil (soil map and soil erodibility map), rainfall (rainfall data) geology (lithology) satellite image (land use).

4 Results and Discussion

The results presented detailed the spatial numerical values of the five factors of the RUSLE equation: factor of rainfall erosivity (R), slope length and inclination factor (LS), soil erodibility factor (K), vegetation cover factor (C), and erosion control practices factor (P), and also the estimated annual rate of soil loss (A) expressed in t/ha/year, for each point and for all the Aures center region.

4.1 The Rainfall Erosivity Factor (R)

After interpolation by Kriging method, we obtain the distribution map of average annual rainfall in the 16 stations studied. The average annual rainfall varies from 142.537 to 452.984 (mm). The spatial distribution of these values is like a "V" with the minimum values occupies the base in the South, while the North West and South East regions are maximum values of precipitation.

The lowest values follow the main wadi "Oued Labiod." The maxima are reached the top of the mountains. The wettest areas corresponding to the higher elevation areas (Fig. 4).

4.2 The Topographic Factor (LS)

Applications on Global Mapper, ArcMap and ArcToolbox, allowed us to perform all the tasks of mapping such as visualization, digitization, data entry and analysis of geographic data. Thus, a list of "basic" maps has been established:

- Digital Elevation Model map (DEM);
- Map of altitude slices (TIN);
- Map of slopes;
- Map of aspects;
- Relief map (shade).

The LS factor characterizes the effect of watershed topography on soil erosion depending on the length (L) and inclination of the slope (S), the values of this factor were obtained through of a digital elevation model generated after the digitization of topographic maps (Fig. 5).

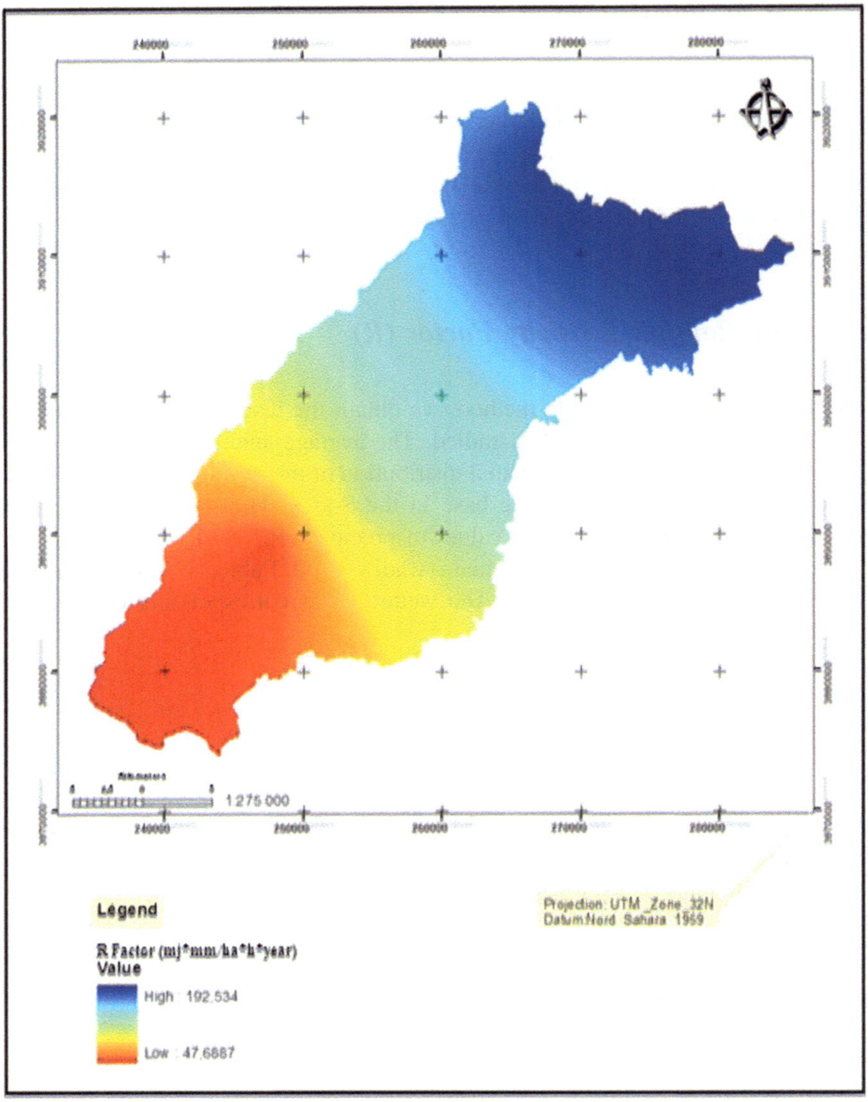

Fig. 4 Spatial distribution map of the rainfall erosivity factor (R)

4.3 The K-Factor (Soil Erodibility)

The K factor was calculated based on lithological classes that have been assigned to each polygon of the geological map. Briefly, five different K values were initially defined for each lithology class. The interpolation of the factor K results shows spatial distribution of classes between 0.0065 and 0.079 (t/ha h/ha/MJ mm) (Fig. 6).

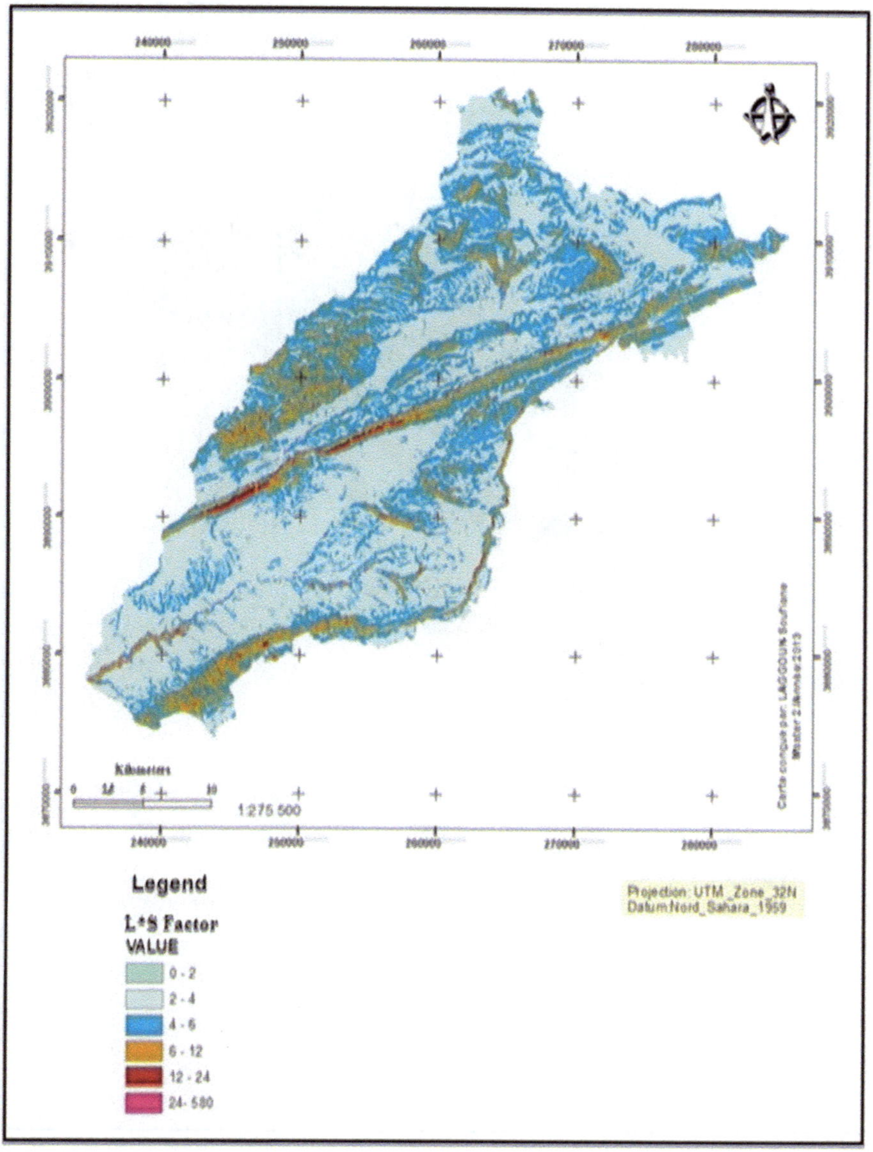

Fig. 5 Spatial distribution grid map of the factor "LS"

4.4 *Land Use Factor (C)*

The vegetation cover factor estimation based mainly on the land use map produced according to the treatments of classification on the satellite image Landsat-8/LDCM, acquired on June 23, 2013. The geographical distribution of the

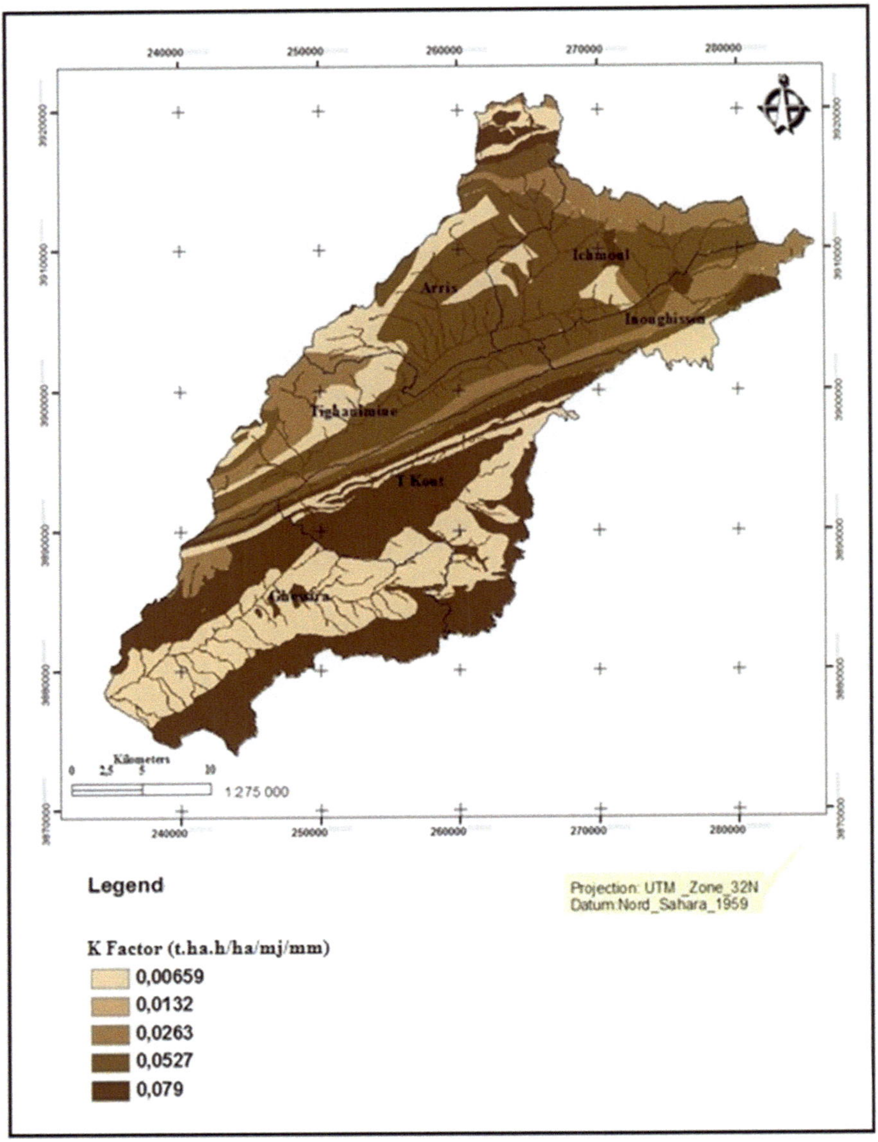

Fig. 6 Spatial distribution grid map of the factor (K)

factor (C) indirectly reflects the land use dynamics according to the physiographic units. The factor (C) values were allocated for each type of land use obtained from the classification of satellite image. Five different values of C were initially defined for each type of land use (Fig. 7).

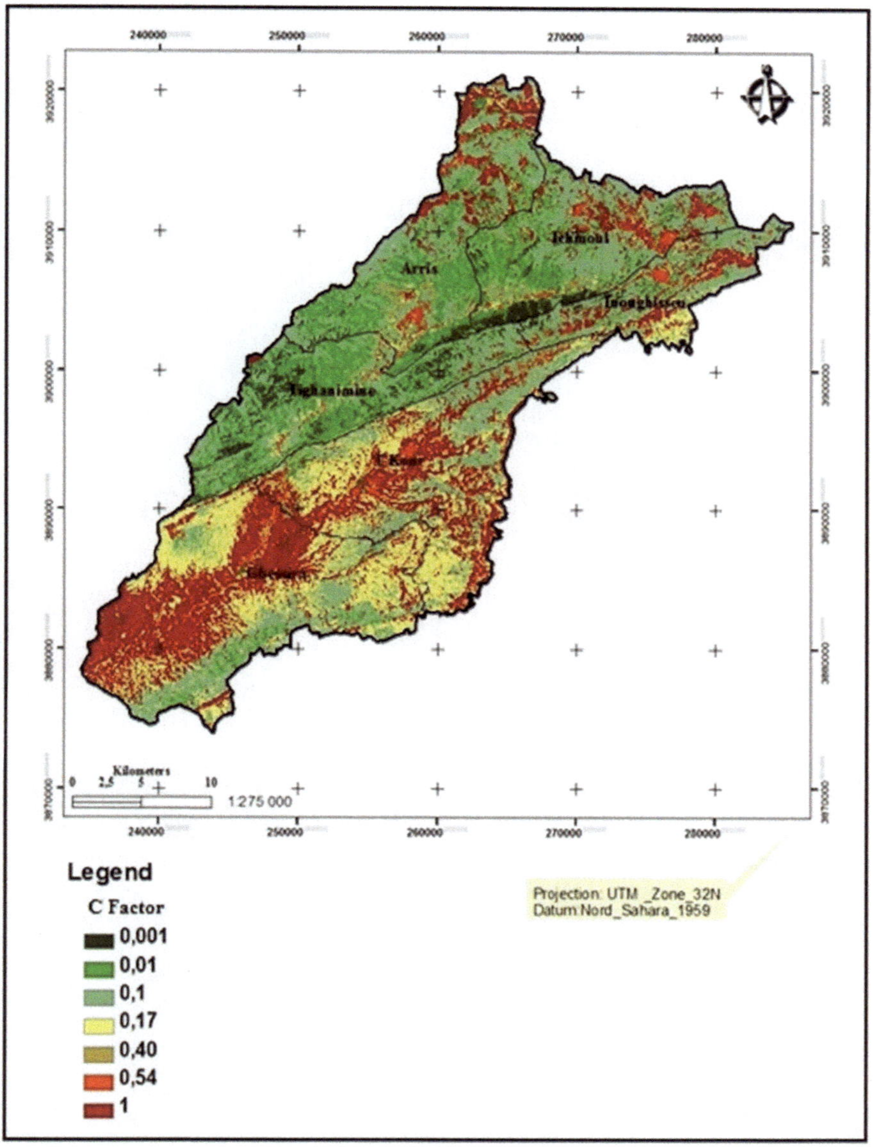

Fig. 7 Spatial distribution grid map of factor (C)

4.5 *Erosion Control Practices Factor (P)*

The erosion control practices factor (P) reflects practices that reduce the quantity of runoff and its speed, thus reducing the effects of water erosion. The values of (P) vary the practice and also according to the slope of the study area (Fig. 8).

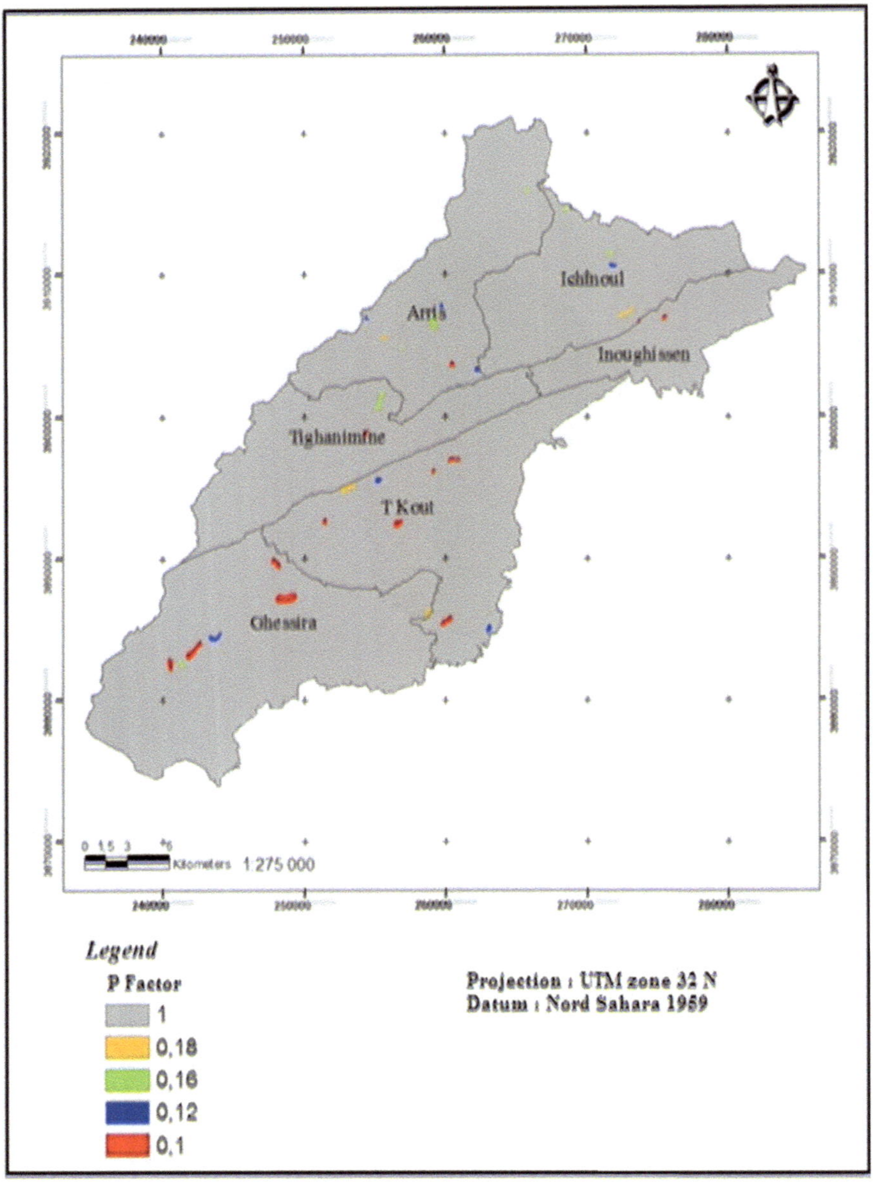

Fig. 8 Spatial distribution map of the factor (P)

Soil protection and restoration amenities that may be inert (benches, cords dry stone, etc.) or biological treatments (vegetation) that is a very effective means of erosion control. Developed areas with cords of dry stones which are improvements made on the slopes in contour.

Changes in the value of the factor (p) in these areas, in addition to that it is defined by the type of development used (dry stone cords), it depends on the values of the slope.

4.6 Synthetic Map of Soil Losses by Water Erosion

This section presents the cartographic purpose of the study because it summarizes the information maps of the factors (R, LS, K, C, P) previously presented as a single map. This map will be a valuable tool in the planning of erosion control.

According to the RUSLE model, the maps crossing of the main factors of water erosion calculated under ArcGis 9.3 with the Spatial Analyst tool (Raster Calculator) and rasterized in the step of the basic DEM pixel (30 m resolution); provides a quantitative map of soil loss at any point in the study area. The application of the formula Wischmeier and Smith (1978), taking into account the numerical values of the five factors, gives soil loss for each point by the unit t/ha/year.

For the purposes of legibility of the map, different homogeneous units were grouped into classes.

- Areas suffer of low soil losses (<5).
- Areas suffer of moderate soil losses (5–25).
- The Remaining areas suffer of soil losses ranging from average (25–50) to very high (50–200).

4.6.1 Soil Loss Rate in (t/ha/year) in the Center of the Aures

The estimation of soil loss rate is made possible by the existing interaction between factors R, K, LS, C and P. The unequal distribution of potential areas of erosion risk in the study area result the variability of the characteristics of the factors involved in the erosion process. The average loss by water erosion across all study sites is about 8.73 t/ha/year. Total annual losses for 6 towns are retained 33,995.4 t/ha/year (Figs. 9 and 10).

4.6.2 Soil Loss Rate in (t/ha/year) for Each Type of Land Use

Average soil losses differ from one zone to another, according to the influence of different factors controlling erosion. Through comparisons with maps of different factors and geological data (Figs. 11, 12 and 13).

Attributes of LIMITE ZONE d'etude avec pert

OID	NAME_2	ZON	COUNT	AREA	MIN	MAX	RANGE	MEAN	STD	SUM
0	Arris	1	168305	1,51474E+08	0	2404,45	2404,45	35,7533	48,5689	6017450
1	Ghessira	2	264712	2,38241E+08	0	17808,5	17808,5	30,0696	145,397	7959790
2	Ichmoul	3	137929	1,24136E+08	0	3748,16	3748,16	32,1359	39,2634	4432480
3	Inoughissen	4	80237	7,22133E+07	0	1426,3	1426,3	46,8435	56,2103	3758580
4	T Kout	5	207436	1,86692E+08	0	21148,4	21148,4	61,704	96,5156	1,27996E+0
5	Tighanimine	6	142812	1,28531E+08	0,043456	33595,4	33595,3	21,5816	188,961	3082110

Record: 0 Show: All Selected Records (0 out of 6 Selected) Options

Fig. 9 Attribute table of the statistical analyzes mean, maximum and minimum of soil loss in different studied towns

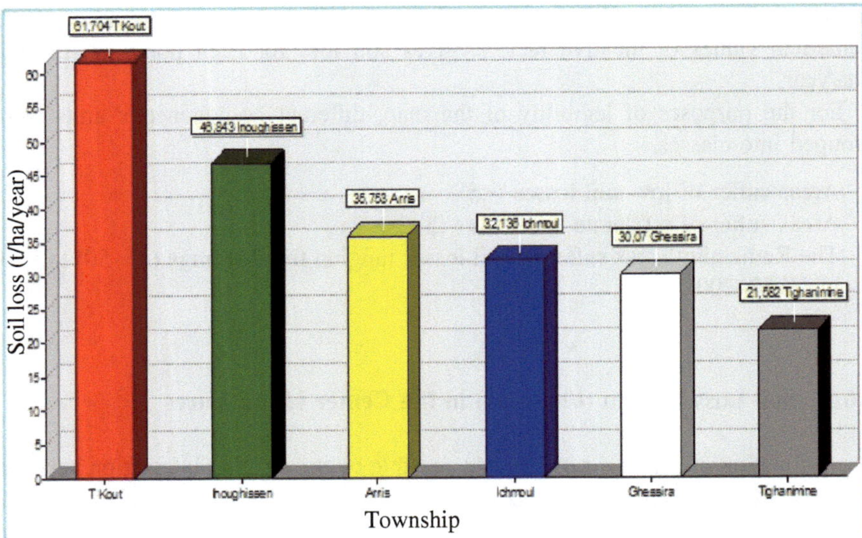

Fig. 10 Mean soil loss for each central Aurès township

Soil loss map has low erosion <5 t/ha/year in the south towns of Ghassira and West Tighanimine (very low slopes zones 0–3%), and moderate and medium erosion 5–25 t/ha/year at the center of the study area, interpreted by the presence of dense forests of Aleppo pine in the mountain (Jebel zaouai) and also the presence of calcareous marl formations. The high and very high erosion 50–200 t/ha/year

Attributes of type d'occupation avec perte en sol2

OID	TYPES D OC	ZONE CODE	COUNT	AREA	MIN	MAX	RANGE	MEAN	STD	SUM
0		1	429	386100	0	0	0	0	0	0
1	Foret dense de Pin d'Alep	2	30907	2,78163E+07	0,131387	264,013	263,882	9,64686	10,0763	298156
2	Foret claire de Pind'Alep	3	130806	1,17725E+08	0,0434562	1667,25	1667,21	16,2854	20,6044	2130230
3	Maquis de Chêne Vert et Genevrier Oxycèdre	4	100840	9,0756E+07	0,118752	6745,58	6745,46	30,5367	39,6902	3079320
4	Maquis de Chêne Vert et Genevrier Phenicie	5	260223	2,34201E+08	0,0579754	9635,97	9635,91	35,2991	61,7715	9185650
5	Arboriculture et Culture Maraichères	6	28583	2,57247E+07	0,0365836	5492,78	5492,74	40,4258	88,8677	1155490
6	Cerealiculture et Jachères	7	20724	1,86516E+07	0,155861	13826,2	13826	36,1139	136,844	748425
7	Parcours	8	148740	1,33866E+08	0,0311587	15869,1	15869,1	38,1139	117,984	5669060
8	Terrain Rocheux	9	122747	1,10472E+08	0,0385114	17194,4	17194,3	62,3545	139,344	7653830
9	Sol nu	10	109191	9,82719E+07	0,00693	17808,5	17808,5	34,3855	136,886	3754590
10	Cours d'eau sec	11	2209	1988100	0,0485693	14960,7	14960,7	31,965	331,558	70610,8
11	Agglomération	12	45267	4,07403E+07	0,258557	33595,4	33595,1	94,3763	325,411	4272130
12	Ombre	13	766	689400	0,468707	191,004	190,535	42,5417	34,7259	32586,9

Record: 0 Show: All Selected Records (0 out of 13 Selected) Options

Fig. 11 Attribute table of statistical analyzes mean, maximum and minimum of soil loss for each type of land use

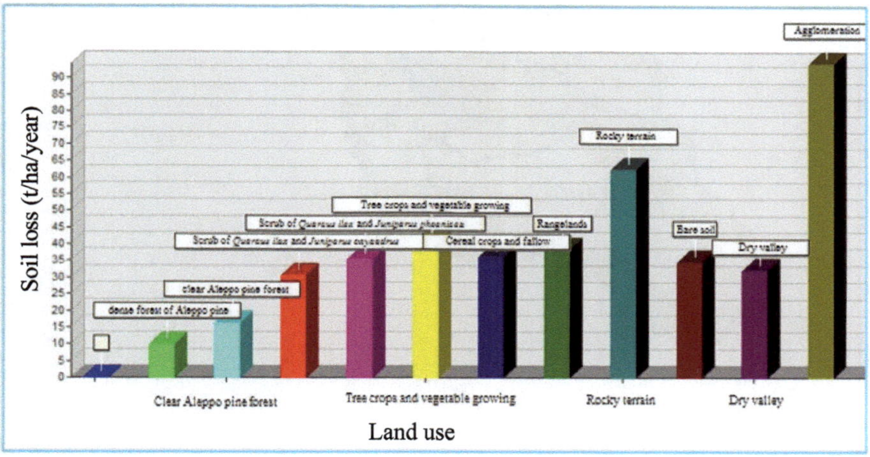

Fig. 12 Soil loss mean for each land use category

occupies only limited areas and concentrated mainly in the centers of villages Arris and T'kout, areas of bare land located on slopes relatively strong more than 25% and a high and rugged terrain.

The erosion is very intense in the Oued Labiod Valley and also in different Oueds area of Oued chenaoura, Oued Ichemoul and Oued Ghassira because of the presence of alluvium.

This study shows that land degradation by water erosion varies from 9.94 t/ha/year on a fully wooded region at 94.37 on bare ground. In general the highest values are located in the east of the study area, around north and central Tkout and eastern Inougissen.

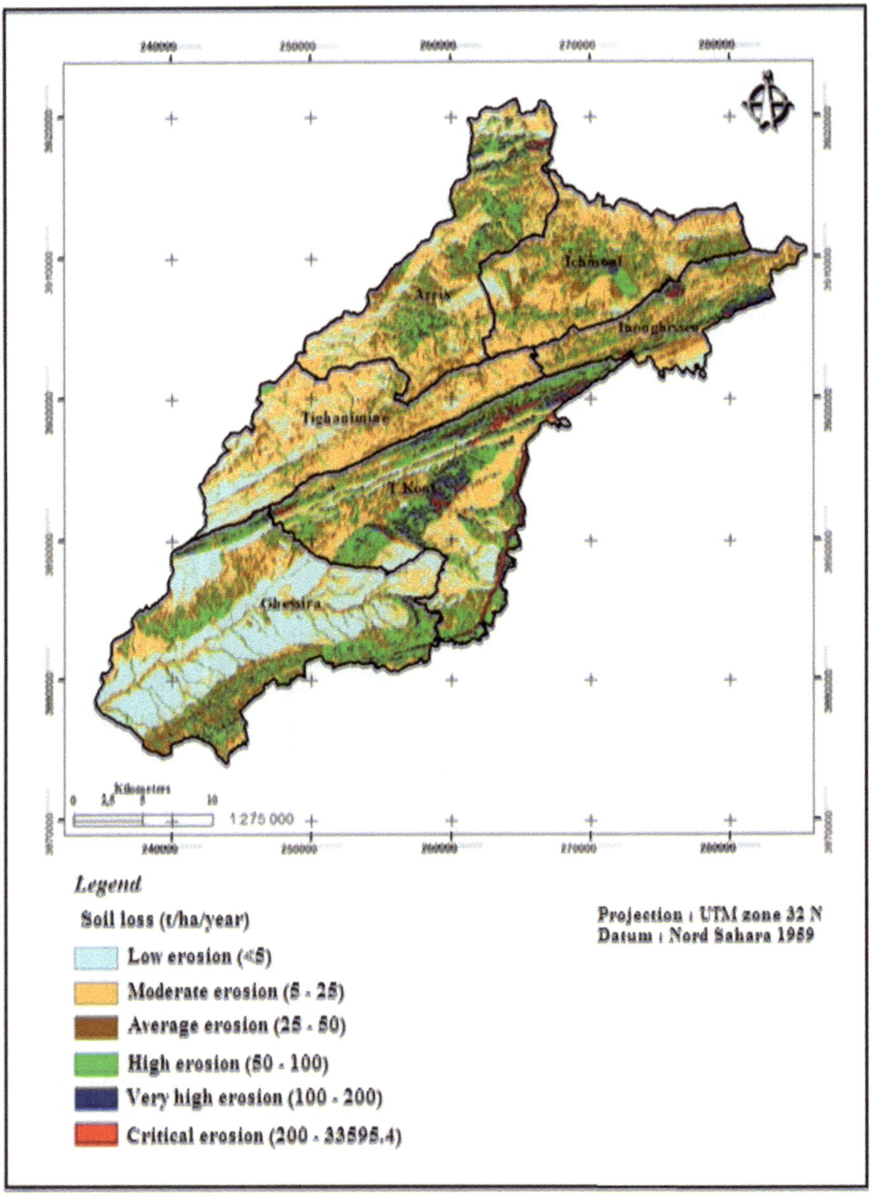

Fig. 13 Soil loss synthesis map of the study area

5 Conclusion

In light of these figures, we can conclude that the unequal distribution of areas with potential risk of erosion in the central Aures results from the variability of the characteristics of the factors involved in the process and the model used. First, the RUSLE model gives a general annual estimate of the soil loss expected.

In addition, some important factors influencing soil erosion are neglected, as the soil physical analysis also the case for estimating the R factor, based on an approximated relationship that does not take into consideration the rain intensities in 30 min and with extrapolation of the data obtained throughout the study area.

Despite these differences, the methods outlined in this study were able to give approvable information on the risk of water erosion. A more accurate estimate can be achieved for these areas using more detailed data, more adequate erosion models and more extended field studies.

References

Abdessemed K (1984) Les problèmes de la dégradation des formations végétales dans l'Aurès (Algérie). Première partie : La dégradation, ses origines et ses conséquences. Forêt méditerranéenne, Tome VI, N° 1, pp 19–26

Al Ali Y (2007) Les aménagements de conservation des eaux et des sols en banquettes : Analyse, fonctionnement et essai de modélisation en milieu méditerranéen (El-Gouazine,Tunisie Centrale). Thèse de doctorat de l'université Montpellier II, 178 pp

Benmessaoud H (2010) Etude de la vulnérabilité a la désertification par des méthodes quantitatives numériques dans le massif des Aurès (Algérie). Thèse de doctorat en Aménagement du territoire, Université El Hadj Lakhder Batna, 198 pp

Berkane A, Yahiaou A (2007) L'érosion dans les Aurès, Sécheresse vol 18, n° 3, pp 213–216

El Garouani A, Chen H, Lewis L, Triback A, Abahrour M (2008) Cartographie de l'utilisation du sol et de l'érosion nette à partir d'images satellitaires et du SIG IDRISI au Nord-Est du Maroc, Télédétection, vol 8, n° 3, p 193–201

Foster GR, Mayer LD (1972) A closed-form erosion equation for upland areas. In: Shen HW (ed) Sedimentation: symposium to honor Professor H. A. Einstein, Fort Collins, Colorado, p 12.1–12.19

Hadir S (2010) Modélisation du ruissellement et de l'érosion par le modèle STREAM dans le bassin versant de l'Oued Saboun. Mémoire de fin d'étude pour le titre d'Ingénieur Agronome, Institut Agronomique Et Vétérinaire Hassan Ii - Rabat

Leguédois S (2003) Mécanismes de l'érosion diffuse des sols Modélisation du transfert et de l'évolution granulométrique des fragments de terre érodés. Thèse présentée à l'université d'Orleans, 176 pp

Pierre R, Tichoux H, Langlois J, Fauchere R, Parent G, Trudeau F (2008) Étude de la protection des bassins-versants des barrages Foum El Gherza, Fontaine des Gazelles, Foum El Gueiss, Babar, Koudiat Medouar, K'sob et Ain Zada, Phase III. Identification et évaluation du degré d'érosion, Version définitive (TECSULT international)

Souadi Y (2011) L'érosion hydrique au Maghreb Etude d'un cas: le bassin versant de l'oued Barbara (Tunisie septentrionale), Université du Québec à Montréal

Wall GJ, Coote DR, Pringle EA, Shelton IJ (2002) RUSLE-CAN - Équation universelle révisée des pertes de sol pour application au Canada. Manuel pour l'évaluation des pertes de sol causées par l'érosion hydrique au Canada. Direction générale de la recherche, Agriculture et Agroalimentaire Canada, No de la contribution AAC2244F, 117 pp

Wischmeier WH, Smith DD (1978) Predicting rainfail erosion losses—a guide to conservation planning. Agricultural handbook No. 537, USDA, Washington

Evaluation of Water Erosion Risk in Tunisian Semi Arid Area

Olfa Hajji, Sahar Abidi, Taoufik Hermassi and Ikram Mekni

Abstract The Water erosion phenomenon is answered in central Tunisia, particularly in our study area "El Gouazine watershed". In fact, this site contains several favorable conditions for its release because of the semi-arid Mediterranean climate and its physical characteristics and specific anthropogenic. The objective of this final project study is the determination and application of a methodology for quantitative assessment of water erosion and the development of erosion map using Geographic Information Systems (GIS) in the watershed of El Gouazine to properly plan the management actions to be taken and to protect priority areas risk vis-à-vis erosion. Application of USLE, MUSLE, RUSLE and FAO/GIS approach's combines the main factors of erosion adapted according to the Tunisian conditions based on the principle combination of its main factors related to the natural environment. To this end, the layering on rainfall, soils, topography, vegetation cover provides synthetic card distribution rate to erosion in t/ha/year. From results, we can conclude that the hilly lake is characterized by low erosion with siltation rate about 1.8 t/ha/year. The evolution of water and soil conservation works has led to a remarkable reduction in soil loss about 50% at the El Gouazine basin which proves the effectiveness of erosion developments since 1996. In the end, a comparative analysis of the four methods determines the validity of RUSLE and USLE in our watershed unlike MUSLE and FAO methods. Furthermore, the results obtained identify that RUSLE model, adapted to the watersheds conditions, provides the best results of estimation.

O. Hajji (✉) · S. Abidi
Rural Engineering Water and Forest Department,
National Agronomy Institute, Tunis, Tunisia
e-mail: olfa.hajji@yahoo.fr

T. Hermassi
National Research Institute of Rural Engineering,
Water and Forests, Tunis, Tunisia

I. Mekni
Higher Engineering and Rural Equipment School,
Hydraulics and Environment, Medjez El Bab, Tunisia

O. Abdalla et al. (eds.), *Water Resources in Arid Areas: The Way Forward*,
Springer Water, DOI 10.1007/978-3-319-51856-5_13

Keywords USLE · MUSLE · RUSLE · FAO · Water erosion · Tunisia · Watershed El gouazine

1 Introduction

The global climate changes and anthropogenic pressures on the circles have deepened the concepts of management, protection, and sustainability of ecosystems to high degrees currently. It is in this general framework that is located the whole of this study, aimed to advance the role of the remote sensing and the dynamics of the new sensors now accessible to the scientific community in order to better characterize and predict some natural hazards widespread in the Mediterranean region, to quote the water erosion.

Recent studies on the vulnerability to climate change in the Mediterranean region indicate a tendency to an increase of the aridity which accelerated water erosion (De Ploey et al. 1991; Joftic et al. 1992; Shaban and Khawlie 1998; Bou Kheir et al. 2001). Soil erosion due to rain and runoff is a widespread phenomenon in the various countries of the Mediterranean region. This phenomenon continues to take considerable proportions in particular on the slopes because of the torrential rains, the high land vulnerability (soft rock, fragile soils, steep slopes and vegetation cover often degraded), overgrazing and the adverse impact of human activities: deforestation, fires, bad conduct of agricultural work, urbanism chaotic, quarrying, etc.

According to studies by FAO (1990), the situation continues to deteriorate soil by water erosion: in Greece 35%, in Morocco 40% and in Turkey 50% (Celik et al. 1996). In Tunisia, 45% of the country area is threatened by erosion (Boussema 1996) and in Algeria 45% of tellian areas either 12 million hectares (Chebbani et al. 1999).

In Tunisia, during the past two decades, the hilly lakes occupy a major place in the national strategies for the Water and Soils Conservation (WSC). In addition to their role of the environment protection, the hilly lakes appear as local reserves of water available for agriculture. Nevertheless, these hydraulic infrastructures are quite sensitive to sedimentation because of solid inputs. Water erosion affects almost 3 million hectares of agricultural soils in Tunisia, and constitutes a threat to the sustainability of reservoirs intended to mobilize the surface waters; or the dominant economic activity remains agriculture. Therefore, the methods of control are necessary, in order to ensure a sustainable management of soils and secure the agricultural productions. Several studies have been made in the framework of the Water and Soils Conservation (WSC), by resorting to the modeling of the erosion which has become a necessity for the hydrologists in order to be able to limit the areas to major risk and seek appropriate solutions. The modeling of erosion has begun by the empirical models as the universal soil loss equation (USLE), the revised version (RUSLE), its amended version (MUSLE) and the method of Food and Agriculture Organization (FAO). In addition, erosion mapping represents a fundamental tool to know the distribution and the geographical extent of the phenomenon. Thus, the combination of empirical models existing with the new techniques of specialization such as the geographical information systems (GIS) have helped to reduce the financial costs and gain the time of mapping risks and by consequence, the intervention focused efficient.

In this context, the present paper aims in the using of USLE, RUSLE, MUSLE and FAO, integrated under a Geographic Information System (GIS) in order to quantify and map the water erosion at the catchment level of El Gouazine.

2 Material and Methodology

2.1 Reminders

After Tahiri et al. (2014), multiple equations have been established to link the factors of erosion between them and quantify the losses of soil: Universal Soil Loss Equation (USLE) of Wischmeier and Smith (1961), Water Erosion Prediction Project (WEPP) of Flanagan and Nearing (1995), Soil and Water Assessment Tool (SWAT) of Arnold et al. (1998), European Soil Erosion Model (EUROSEM) of Morgan et al. (1998). Then, other models have been based on the equation of universal soil loss (USLE) of Wischmeier and Smith (1961) and its modified versions (MUSLE) and revised (RUSLE). RUSLE is an empirical model of the revised USLE equation of Wischmeier and Smith (1961). It is designed for use at the scale of the plot. RUSLE model allows you to predict the average annual rate of soil erosion of a site for scenarios involving systems of culture, management techniques and practices to control erosion (1990). RUSLE calculates the average erosion expected annual on the slopes by multiplying several factors came together: the aggressiveness of acid (R), the erodibility soil (K), the slope and its length (LS), vegetation cover (C), and the conservation tillage practices (P).

The values of these factors are determined from measurements of field and laboratory (Renard et al. 1997). MUSLE is a modified model of USLE (Renard et al. 1991). It has been used to estimate the amount of sediment from one storm (Williams and Berndt 1976). This model has replaced the factor of climatic erosivity of rainfall (R) by the maximum flow snapshot and the total volume of water runoff in order to predict the erosion of soil for an event of water erosion. The sedimentation model is based on the results of RUSLE model to calculate the balance sheet of erosion in each field elementary. It uses of homogeneous polygons resulting from the calculation of RUSLE model to assess the net movement of soil (erosion or deposition) in plots or sub-watersheds (Lewis et al. 2005). It also requires the digital terrain model (MNT) and represents an extension of the incorporation of RUSLE in a framework of GIS (Lewis et al. 2005). In addition, the method of Food and Agriculture Organization (FAO) which has been developed in 1979. This methodology is a generalization of USLE which depends on the following characteristics: the climatic erosivity of rainfall, the nature of soil, the slope and the occupation of land.

2.2 *El Gouazine Basin*

The catchment area of Lake El Gouazine has an area of 17.08 km^2. It belongs to the catchment area of Nebhana in semi-arid Tunisia central (Fig. 1). It belongs to the syncline and is constituted by marls of intercalated calcareous bars the Lutetien Bartonien. These bars lime have a strong dip. The upstream portion of the watershed is overcome by calcareous crusts sub-horizontal. The interfluves floors calci-magnesic to limestone crust more or less endured; whereas the valleys floors little advanced.

The occupation of land varies from a covered semi-forest plots in the totally devoted to agricultural activity: 32.1% for the cereal crops in alternations with of fallow land, 35% for degraded forest and 11.02% the dense forests, 2.1% for the journey, 4.12% for the scrublands (carob, lentisks) and the grassy steppe to Alfa, 3.44% for the arboriculture (olive trees and almond trees), 3% for the gullies deep and 4% for the mixed units.

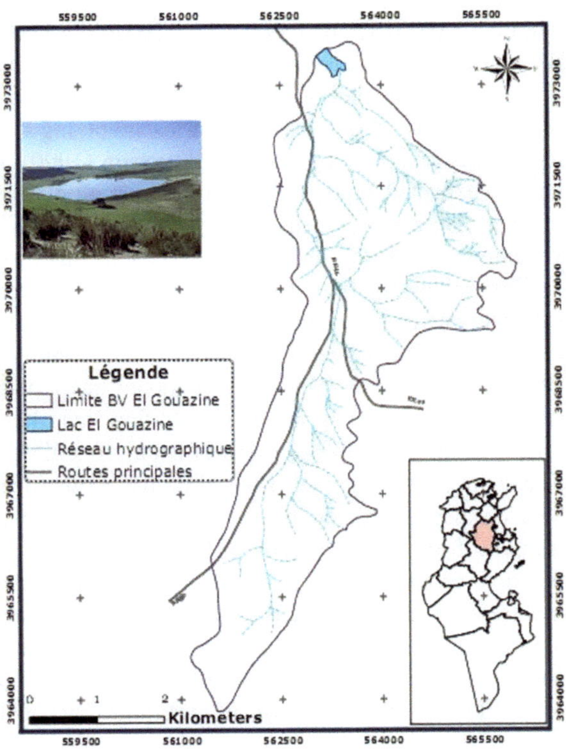

Fig. 1 Location map of the study area

Table 1 Hydrologic characteristics of El Gouazine

Characteristics of the catchment	
Aria (ha)	1708
Perimeter P (Km)	26.69
Index of compactness Ic	1.8
Length of the rectangle L (km)	11.91
Width of the rectangle L (km)	1.41
Maximum altitude Hmax (m)	590
Minimum altitude Hmin (m)	383
Index of slope Ig (m/km)	11.75
Specific rapids Dsp (m)	48.57
Characteristics of the reservoir	
Year of construction	1990
Volume to the spill Vi (m^3)	237,030
The surface spill As (ha)	9.597
Report Vi/As	2.47
Height of the dike (m)	10.63
Length of the dike (m)	232
Nature of the weir	Concrete, trapezoid
Spout height (m)	8.28
Width of the spillway (m)	20.6

A small dam was built at the outlet of the basin in 1990; the bed of the river and the hilly lake occupy 1.6% of the surface area. 40% of the area has been landscaped in front benches contours in 1996 and 1997. This development relates to the land of cereal crops and orchards (Mansouri 2001).

The climate is Mediterranean type semi-arid with a hot season of summer and a fresh season of winter. The continentally and terrain accentuate the contrasts thermal and the drought of summer. The rainfall is on average 400 mm but is experiencing a very high variability space and time. During the observation period (1994–2004) the annual rainfall varies from 253 to 577 mm.

The coordinates of the lake are 35°54′30″ in the North, 9°42′13″ in the Est. The altitude varies from 383 to 586 m. The index of slope is 11.75 m/km.

The main hydrologic characteristics of the watershed and from its reservoir downstream are listed in Table 1.

3 Description of Erosion Factors

To evaluate water erosion in El Gouazine catchment, empirical models of erosion prediction has been established which are the universal soil loss equation (USLE), the amended universal soil loss equation (MUSLE), the revised universal soil loss equation (RUSLE) developed by Fox et al. (1997) and the FAO equation.

These models have been integrated under a geographic information system (GIS) in order to quantify and map the risk of water erosion at the watershed level cited in the smokescreen. The digitalization of maps, the analysis, the combination of data and modeling has been carried out using GIS.

3.1 *Factor of Rainfall Aggressiveness (R)*

The index of aggressiveness takes into account the interactions between the height, the intensity and duration of the rains on the solid transport over a long period (Brown and Foster 1987). This climate index is calculated, for a heavy downpour, and is cumulative per episode, per month, or per season.

The climatic erosivity of rainfall is considered constant for the whole river basin that is to say it has not used a layer of this factor under a GIS, but it was considered a single value.

3.1.1 Factor of Rainfall Erosivity R According to USLE

This index takes into account the three conditions: energy, intensity of peak and duration of rainfall. Thus the index of rainfall aggressiveness is calculated by this equation where:

$$R = E_T * I_{30}$$

- I_{30} = maximum intensity of the rain in 30 min of duration for the downpour considered (mm h^{-1}).
- ET = kinetic energy total of rain (MJ ha^{-1} mm^{-1}).

The kinetic energy of the rainfall is given by the following formula:

$$E_T = 210 + 89\log(I_{30})$$

3.1.2 Factor of Rainfall Erosivity R According to MUSLE

This factor results from the following equation:

$$R = 11.8(V \cdot Q_p)^{0.56}$$

With V: the volume of runoff water (m^3) and Q_p: peak flow (m^3/s).

3.1.3 Factor of Rainfall Erosivity R According to RUSLE

- **R monthly** (Arnoldous, 1980): R is called factor rain or index of rain erosivity. The factor R can be defined as the potential capacity of the rain to produce erosion. This potential capacity is often attributed to its physical characteristics to know the quantity, intensity, the dimension of the rain drops, the distribution size of these drops and the speed of the fall which are connected between them.

In fact, several methods are presented for the determination of R factor in which one cited in Table 2.

Or:

- F: index of Fournier.
- P: annual precipitation.
- MAR: precipitation annual average.
- R: factor of rainfall erosivity.

In reality, it has not used a layer of this factor under a GIS, but it was considered a single value for the entire watershed.

The factor of rainfall erosivity is calculated from several formulas proposed by Wischmeier and Smith (1961), and then by the method of Arnoldus (Cormary and Masson 1964) based on the monthly precipitation or of the index of Fournier.

To calculate this factor, we used the data of the monthly rain, and the annual rainfall by using the formula of Arnoldus (Cormary and Masson 1964) which is presented in the form:

$$\log R = 1.47 * \log\left(\frac{Pi^2}{P}\right) + 1.29$$

Table 2 Factor of rainfall erosivity (R)

Case	Reference	R and P or F relationship
Case I	Arnoldus-linear (1980)	R = (4.17 F − 152)/17.02
Case II	Arnoldus (1980)	R = 4.17 F − 152
Case III	Yu and Rosewall (1996)	R = 3.82 F1.41
Case IV	Arnoldus (1977)	R = 0.302 F1.93
Case V	Renard and Freimund (1994)	R = 0.739 F1.847
Case VI	Renard and Freimund (1994)	R = 0.0483 P1.61
Case VII	Roose (1994)	R = P * 0.5
Case VIII	Kassam et al. (1992)	R = 117.6 (1.00105^{MAR}) for MAR < 2000 mm
Acc IX	Singh et al. (1981)	R = 79 + 0.363 F

- R: rainfall erosivity in (MJ/ha mm/h)
- Pi: rain's monthly i (mm)
- P: annual rainfall (mm).

The data used in this work are from the direction of water resources (1994–2009).

– **R (15 min) and R (30 min) of RUSLE**

This factor is calculated using the data from the instantaneous rainfall which gives the values of cumulative rains every 15 min in the same manner for R(30) which presents the values of cumulative rains every 30 min. The data used are from the database DGACTA/IRD.

This index takes into account the three conditions: energy, intensity of peak and duration of rain. E is a function of the kinetic energy of the rain 'E' and the maximum intensity of rain in duration of 15 min 'I_{15}' or 30 min at 'I_{30}'.

The R value adopted in the equation of RUSLE is the average of those subpoenaed during a hydrologic year during a multi-year period representative:

$$R = (EI)_{15} = \sum_{j=i}^{n} I_{15j} \left[\sum_{i=1}^{m} (\Delta hij, i) E_{i,j} \right]$$

Has each not of time, measure the height rushed and the kinetic energy (E_i) generated in (MJ/ha mm) depending on the intensity of precipitation (I) in (mm/h). In the equation of RUSLE, the kinetic energy, for each interval of uniform intensity in the downpour, is given by the following formula:

$$E_i = 0.29 * \left[1 - 0.72 \mathrm{e}^{(-0.05I)} \right]$$

With:

- *I:* Intensity of rain in mm/h.
- *Ei*: kinetic energy of a amounted-phase.
- *Δhi*: height of the amounted phase in mm (Table 3).

Table 3 Factor soil erodibility also occur throughout K

Pedological units	Index K (RUSLE)	Index K (USLE/MUSLE)
Complex of sol	0.05	0.041
Rendzines	0.013	0.054
Soil raw minerals	0.036	0.042
Vertisols	0.01	0.01
Little soil advanced to contribution	0.08	0.05

Reported by Cormary and Masson (1964)

3.2 Factor of Vegetation Cover C

This factor is used to express the effect of vegetation cover present in the watersheds. In effect, because of lack of data necessary to calculate the values of the factor C there has recourse to studies conducted on the neighboring areas in our study area.

The spatial distribution map of the vegetation cover factor is obtained directly from the card of occupation of the soil made from satellite images from Google Earth. In fact, the indices of C deductions are chosen in referring to:

- The work of Cormary and Masson (1964) and Masson (1971) in Tunisia.
- The applications of the RUSLE model, especially on the hilly lake Abdessaddok (Heusch 1986).

The values of the factor C obtained are represented in Table 4.

3.3 Factor Practices of Anti-erosive P

This factor reflects the effects of practices that reduce the amount of runoff water and the runoff rate and which reduce to this fact the importance of erosion.

The protection index P used in the USLE model is a report without dimension obtained by comparison of erosion measured on plots where the work is done in the direction of the greatest slope ($P = 1$) and the erosion of plots variously protected and or $P < 1$, all other factors being equal.

The practices anti-erosive the most used at the watershed of El Gouazine are the benches.

The indexing of this factor is derived primarily from experimental results of Masson (1971), Heusch (1986) in Mediterranean area as well as of various compilations (FAO 1993) (Table 5).

Table 4 Factor of the vegetation cover C

Occupation	C (RUSLE)	C (USLE/MUSLE)
Dense forest	0.01	0.1
Degraded forest	0.05	0.1
Annual crop	0.7	0.55
Olive trees	0.104	0.9
Journey	0.55	0.45
Garrigues	0.3	0.36
Mixed unit	0.37	0.72
Vivid gullies	0	0

Reported by Masson (1971), Zante and Collinet (2003)

Table 5 Weighting of the index P according to the slope

Slope (%)	Index P (USLE/MUSLE)	Slope (%)	Index P (RUSLE)
1–2	0.6	0–5	0.1
2–7	0.5	5–15	0.12
7–12	0.6	15–25	0.16
12–24	0.8	25–35	0.18
>24	0.9	>35	0.28

3.4 *Topographic Factor Handset (LS)*

This factor results from the combination of the slope factor S and length L of slope. The action of the angle of the slope on runoff is amplified by the length of the slope, even if the impact of the latter remains limited.

The steep slopes with a fast flow are in general at the origin of major erosion whose importance depends on the geology, of the nature of soil, and the protection by the vegetation cover. The *LS factor* is a function of the length and the angle of the slopes.

3.4.1 Factor of Practices LS According to USLE/MUSLE

Several studies have been carried out for the determination of this factor in our case we are going to work with the equation of Smith et al. (1996) presented below:

$$LS = 1.4 * \left(\frac{A_S}{22.1}\right)^{0.4} * \frac{\sin(\theta * 0.01745)^{1.4}}{0.09}$$

With: θ: the angle of the slope in degrees.

A_s: The specific surface writing under a GIS in the following manner (Moore and Burch 1986):

$$A_s = Flow\,Accumulation * Cell\,Size$$

The surface map (As) is determined by multiplying the card of the runoff accumulation (Flow accumulation) by the pixel size (resolution).

3.4.2 Factor of Practices LS According to RUSLE

In the framework of our study, we used the formula developed by Wischmeier and Smith (1961) which has been used by several authors (Rodríguez et al. 2010; Toumi 2013).

Table 6 Value of 'm' relative to each class of slope

Slope (%)	m
>5	0.5
3–5	0.4
1–3	0.3
<1	0.2

$$LS = \left(flow\,accumulation * \frac{resolution}{22.1}\right)^{m} * (0.065 + 0.045 * S + 0.0065 * S^2)$$

With 'S' is the slope (%) and '*m*' *is* a parameter such as (Table 6).

3.5 Map of Soil Loss

The crossing of the cards of the major factors involved in soil erosion by water allows you to get the card of losses in soil at any point in the watershed (Fig. 2).

3.6 Method of Food and Agriculture Organization (FAO)

The FAO has developed in 1979 a methodology at the national level to both parametric and empirical. This methodology is a generalization of USLE. This formula depends on the following characteristics: the climatic erosivity of rainfall, the nature of the soil, the slope of the land and the occupation of the land.

She made an estimate of annual erosion Es (t/ha/year) according to the following formula:

$$Es = Fm * C1 * C2 * C3$$

With: **Fm**: index of Fournier amended characterizing climatic erosivity of rain;

$$Fm = \frac{\sum_{i=1}^{12} Pi}{P}$$

P_i: being the rain average of the month i (mm),
P: rain annual average (mm).

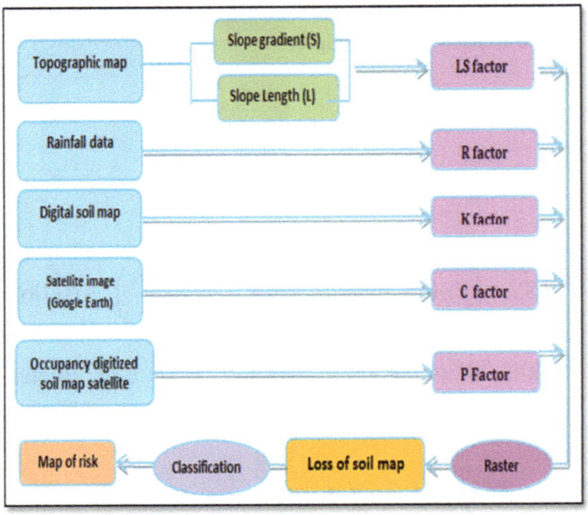

Fig. 2 Flowchart methodological

Table 7 Value of the coefficient C1

Nature of the soil	Coefficient C1
Sandy-loam	0.5–0.6
Silty	0.7–0.8
Silty-clay	0.9–1.1
Clay	1.1–1.2

Source Guide WSC (1995)

- C1: the coefficient of soil texture which depend of its nature and between 0.5 and 1.
- C2: the coefficient which topographic despond of the slope and ranging from 0.5 to 1.5.
- C3: the operating ratio of soils which vary according to the occupation of soils ranging from 0.4 to 1.

C1 is determined from the soil map of the watershed depending on the nature of the ground (Table 7).

C2 is determined from the slopes map of the watershed (Table 8).

C3 is determined from the map of occupation of the soil of the catchment (Table 9).

This work has allowed us to characterize the various parameters of the universal equation of soil loss by consultation of former research and professional studies on the study area. These settings have been distributed on the totality of each watershed. The climatic erosivity of rain is assumed constant throughout the basin for each hilly lake, whereas the other parameters will be presented subsequently in the form of thematic maps.

Table 8 Value of the coefficient C2

Class C2	0–8%	8–30%	>30%
Slope	0.5–0.7	0.8–1.0	1.2–1.5

Source Guide WSC (1995)

Table 9 Value of the coefficient C3

Occupation of ground	Cereal crop	Arboriculture	Journey	Brushland	Forests
C3	0.8–1	0.7	0.6	0.5	0.4

Source Guide WSC (1995)

4 Assessment of Erosion

4.1 The Application of USLE Model

The superposition of the various factors maps involved in water erosion soil to get the card of losses in soil at the catchment. In effect the application of Wischmeier and Smith (1961) formula taking into account the numerical values of five factors given the loss in soil for each point of the catchment. This treatment allows you to divide the territory of the watershed in separate units. Each unit has the values of homogeneous factors of the USLE, with a loss in soil average of all pixels of the unit expressed in (t/ha/year).

Cormary and Masson (1964) on the basis of experiences in the semi-arid Tunisian, proposed a soil loss tolerable of approximately 2.5, 5 and 10 t/ha/year. In fact, on the basis of classification of Masson, we adopted a new classification which better reflects the spatial distribution of soil loss. It has been calculated the erosion on two periods (before management water conservation and soil and after development).

- ***1st scenario: Map of erosion before the management WSC***

The spatial distribution of the land loss of catchment basin El Gouazine before the works of WSC in applying the USLE model is presented by Fig. 3.

The examination of Fig. 3 shows that the sharp erosion of 52 t/ha/year is located at the level of high altitude. While the low erosion (less than 2.5 t/ha/year) is located at the level of low altitude areas and at the level of plains. The total losses of the annual basin are of the order of 53,787 t/year with a rate of average erosion of 2 t/ha/year.

The spatial distribution of different classes of soil loss is presented in Table 10.

Table 10 shows that 76% of the basin is characterized by a loss of ground low (less than 2.5 t/ha/year). While approximately 19% of the basin have a loss of average ground varying from 2.5 to 10 t/ha/year. While only 4% of the basin have a strong soil loss (greater than 10 t/ha/year).

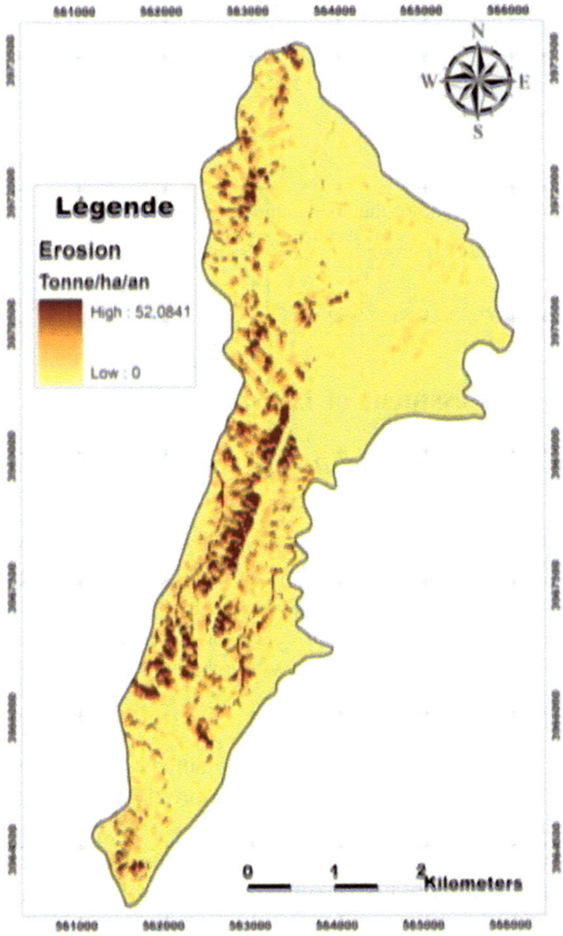

Fig. 3 The spatial distribution of soil losses from the watershed El Gouazine before the WSC management with the USLE model

Table 10 Class of soil loss determined by the USLE model

Class (t/ha/year)	Area (ha)	Surface (%)
0–2.5	1304.31	76.36
2.5–5	176.15	10.31
5–10	143.3	8.38
10–20	56.9	3.33
20–52.08	12	0.69

- ***2nd scenario: Map of erosion after the management WSC***

The crossing of the different maps of factors LS, K, C and P of the RUSLE model with the climatic erosivity of rain obtained by the USLE model allows you to get the card of losses in soil at any point in the watershed after development. The results are illustrated by Fig. 4.

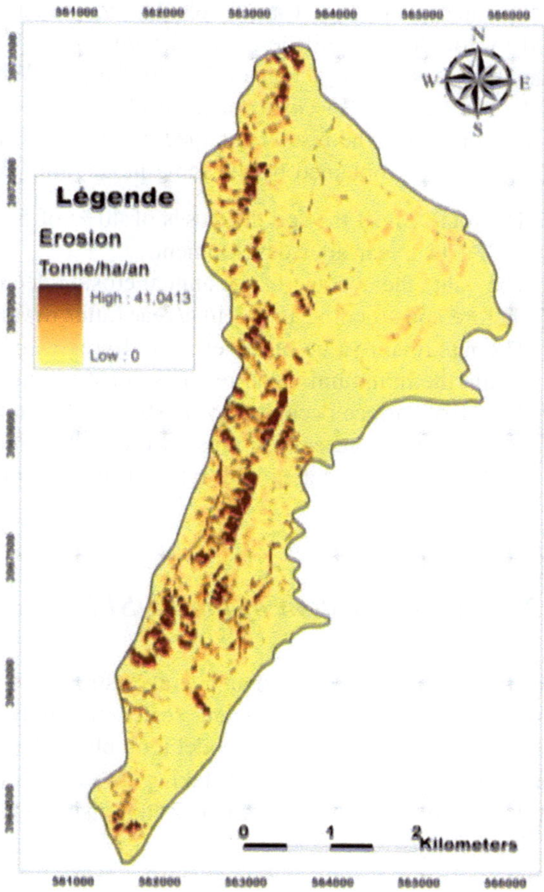

Fig. 4 The spatial distribution of land loss of El Gouazine after the management WSC for the USLE model

Table 11 Class of soil loss determined by the USLE model after the management WSC

Class (t/ha/year)	Area (ha)	Surface (%)
0–2.5	1426.77	90.31
2.5–5	26.52	1.68
5–10	96.65	6.11
10–20	26.52	1.71
20–43.56	2.96	0.19

From Fig. 4 it is found that the average loss by water erosion for all the homogeneous units is approximately 1.5 t/ha/year. The maximum loss and minimum per unit in the watershed are respectively of 44 t/ha/year and 0 t/ha/year. The total losses of the annual basin are of the order of 38,511 t/year. The erosion rate differs from one area to another of the watershed, according to the influence of different factors which control the erosion.

The distribution of the surface on the classes of soil loss adopted is presented in Table 11.

The analysis of Fig. 4 and Table 11 shows that 90% of the basin is characterized by low loss of soil (less than 2.5 t/ha/year). While approximately 8% of the basin have a soil loss average of 2.5–20 t/ha/year. When less than 2% of the surface of the catchment of El Gouazine have a high loss of soil (greater than 20 t/ha/year).

By comparing the results obtained by the USLE model before the management WSC and those obtained by the same model after then, we found that:

- The total loss of the soil decreases of 40%; of 53,787 t/year before planning up to 38,511 t/year after development.
- The value the average of the annual erosion declined by 40%; of 2.04 t/ha/year (before construction) to 1.5 t/ha/year (after development).
- The maximum value of the erosion decreases of 52–44 t/ha/year.
- While the minimum value remains constant equal to 0 t/ha/year.
- The % of surface occupied by an strong erosion decreases of 4–1.9% while the surface which presents an erosion low (<2.5 t/ha/year) increased by 76.4–90% of the total surface area with rates of increase of 15%.

4.2 *The Application of the RUSLE Model*

The RUSLE model provided an average estimate of water erosion at the catchment level of El Gouazine at different scales (monthly, 15 and 30 min) mentioned below. The application of the model took place after modeling of the five factors in thematic maps in raster format, taking into account the numerical values of each card and the multiplication of five factors using a GIS.

4.2.1 Map of Erosion for R (Monthly) of Arldous 1980

- *1st scenario*: *Before the facilities WSC*

The dipped beam maps of factors LS, K, C and P of the RUSLE model with the climatic erosivity of rainfall by the RUSLE model allows you to get the card of losses in soil at any point of the catchment. The results are presented in Fig. 5.

The review of Fig. 5 shows that the high erosion equal to 98.4 t/ha/year is located at the level of highest elevations. While the low erosion (less than 2.5 t/ha/year) is located at the level of low altitude areas and at the level of the plains. The total losses of the annual basin are of the order of 62,731 t/year. The average erosion is of the order of 2.4 t/ha/year.

The spatial distribution of different classes of soil loss adopted is presented in Table 12.

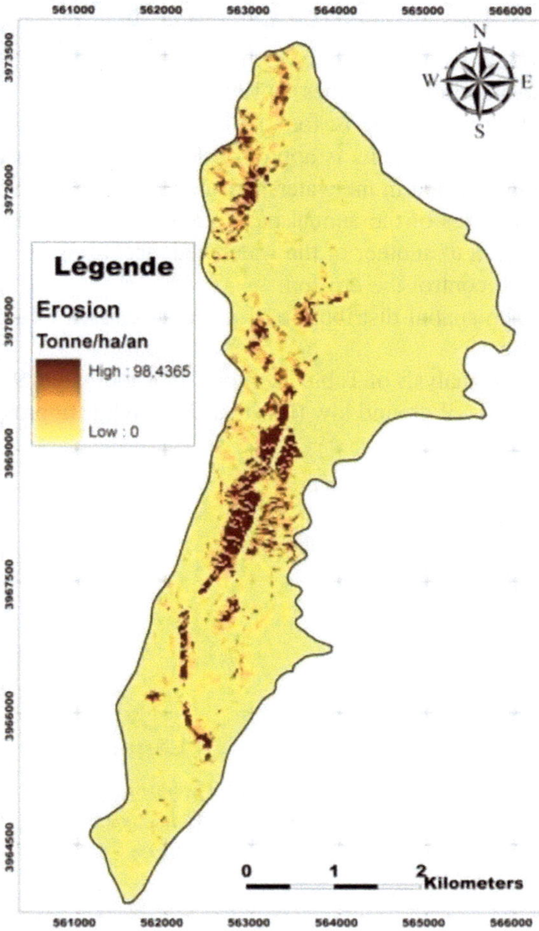

Fig. 5 Map of soil loss determined by the model RUSLE R (monthly) before the facilities WSC of El Gouazine

Table 12 Class of soil loss determined by the RUSLE model (R monthly)

Class (t/ha/year)	Area (ha)	Surface (%)
0–2.5	1391.8	82.2
2.5–5	102.55	6.06
5–10	88.3	5.21
10–20	63.6	3.76
20–100	47.1	2.78

Table 12 shows that more than 82% of the basin is characterized by a loss of ground low (less than 2.5). While approximately 11% of the basin have a soil loss average of 2.5–10 t/ha/year. While less than 7% of the basin have a strong loss of soil (greater than 10 t/ha/year).

- ***2nd scenario: After the management WSC***

Now we will quantify the rate of erosion on the catchment of El Gouazine after implantation of benches in 1996/1997.

From Fig. 6 it is found that the average loss by water erosion for all the homogeneous units is approximately 0.86 t/ha/year. The maximum loss and minimum per unit in the watershed are respectively of 95 t/ha/year and 0 t/ha/year. The total losses of the annual basin are of 22,666 t/year. The erosion rate differs from one area to another of the watershed, according to the influence of different factors which control the erosion.

The spatial distribution of different classes of soil loss adopted is presented in Table 13.

The analysis of Table 13 shows that more than 92% of the basin is characterized by a loss of ground low (less than 2.5). While approximately 6% of the basin have a

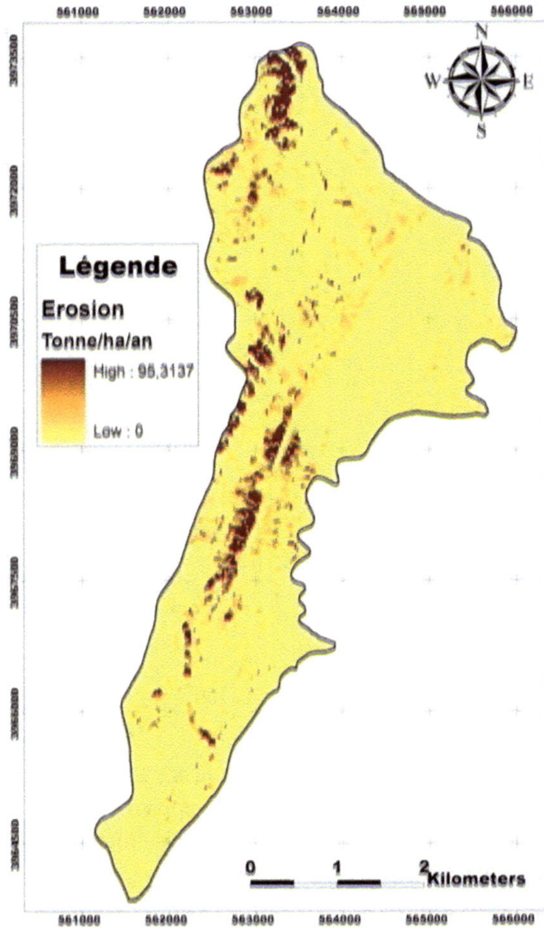

Fig. 6 Map of soil loss determined by the model RUSLE R (monthly) after the management WSC of El Gouazine

Table 13 Class of soil loss determined by the RUSLE model (R monthly)

Class (t/ha/year)	Area (ha)	Surface (%)
0–2.5	1565.7	92.59
2.5–5	60.67	3.59
5–10	42.95	2.54
10–20	18.58	1.1
20–95.31	3.09	0.18

soil loss average of 2.5–20 t/ha/year. While only 1.28% of the basin have a strong loss of soil (greater than 20 t/ha/year) after development.

By comparing the results obtained by the model RUSLE (R monthly) before the WSC and those obtained by the same model after the WSC management, one finds:

- The total loss of the soil decreases of 62,731–22,666 t/year.
- The value the average of the annual erosion decreased from 2.4 t/ha/year (before construction) up to 0.86 t/ha/year (after development), a rate of decline of about 66%; this reflects the effectiveness of these facilities.
- The maximum value of the erosion decreases of 98.43–95.31 t/ha/year.
- While the minimum value remains constant equal to 0 t/ha/year.
- The percentage of surface which contains high erosion decreases of 6.54–1.28%, with a decay rate of 80%.
- The percentage of surface which contains low erosion believes of 82.2–92.52%.

4.2.2 Case of the RUSLE Model with R (15 min)

- *1st scenario*: ***Application of the RUSLE model before the WSC management***

The dipped beam maps of factors LS, K, C and P of the RUSLE model with the climatic erosivity of rainfall by the RUSLE model at no time of 15 min allows you to get the card of losses in soil at any point of the catchment. The following figure illustrates the map of soil loss.

The map of soil loss therefore provides a variety of information concerning the risk of departure of sediment. These information are both quantitative since the results give measures of erosion, expressed in (t/ha/year) for each of the cells of the map, the qualitative result of the allocation of a class of severity of risk and space because of the knowledge of the distribution of the risk of detachment of sediment on the study area. The map of loss in soil has been developed from a classification of erosion rate.

The RUSLE model gives a homogeneous distribution in the polygons more at least uniform.

The spatial distribution of soil losses for the whole watershed El Gouazine is very heterogeneous. In effect, the minimum loss in soils is of the order of 0 t/ha/year and the maximum loss is approximately 333 t/ha/year. The total losses of the annual basin are of the order of 212,273 m^3. The average loss of soils for the whole watershed is of the order of 8 t/ha/year (Fig. 7).

The values obtained, despite their apparent heterogeneity, have a spatial structure defined by the location topo-landscape in which they are drawn.

The parties of the upstream catchment still suffer an erosion more strong: average values are observed for the slopes; the values the least high are located on the intertidal, in the downstream while depressions and in the alluvial valleys. This last element must be qualified: the downstream concavity is unlike the summits, an area of deposition of sediments and not a zone of erosion (Renard et al. 1997) as evidenced by the filling of gullies of catchment (Revel et al. 1989) and the highest thickness of soils in these sectors (Brunet 1957). The risk map eroding should therefore provide in these places of "negative values of erosion" to include the fact that the sediment is deposited. The method used does not allow us to achieve this

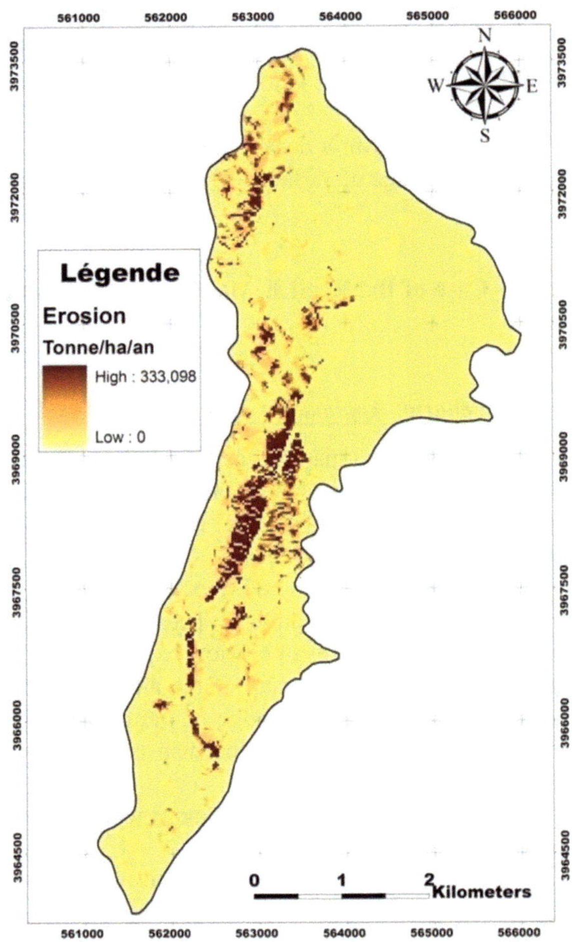

Fig. 7 Map of soil loss determined by the RUSLE model (R15 min) before the WSC management

Table 14 Class of soil loss determined by the RUSLE model (R15 min)

Class (t/ha/year)	Area (ha)	Surface (%)
0–2.5	1169.49	69.1
2.5–5	123.4	7.29
5–10	109.44	6.47
10–20	108.9	6.43
20–333	181.4	10.72

type of distinctions or to quantify the proportion of soil likely to remove in these sectors. However, since they appear on the map as low risk of detachment of sediment, the error introduced does not affect the identification of areas to high risk.

The distribution of the surface on the classes of soil loss adopted is presented in Table 14.

From Table 14 it can be seen more than 69% of the basin is characterized by a low soil loss (less than 2.5). While approximately 13.76% of the basin have a soil loss average of 2.5–10 t/ha/year. While 17.15% of the basin have a strong soil loss (greater than 10 t/ha/year).

- ***2nd scenario: Application of the RUSLE model after the WSC management***

The results obtained are presented in Fig. 8.

The review of Fig. 8 shows that the average loss of soil for the entire catchment area of El Gouazine is 1.46 t/ha/year. The minimum loss in soils is 0 t/ha/year and the maximum loss is approximately 162.4 t/ha/year. The total losses of the annual basin are 38,637.22 t/year. The spatial distribution of different classes of soil loss adopted is presented in Table 15.

Table 15 shows that more than 87% of the basin is characterized by low loss of soil (less than 5). While approximately 9% of the basin have a soil loss average of (5–10 t/ha/year). While less than 4% of the basin have a high soil loss (greater than 10 t/ha/year).

By comparing the results obtained by RUSLE before the WSC management and those obtained by the same model after then, we find:

- The average value of the erosion decreased from 8–1.5 t/ha/year.
- The maximum value of the erosion decreases of 333–162 t/ha/year.
- While the minimum value remains constant equal to 0 t/ha/year.
- The total loss of the soil decreases of more than 81% also (of 212,273 t/ha/year until 38,637 t/ha/year).
- The percentage of surface area occupied by a strong erosion decreases by 81% (from 17.15 to 3.19%), in contrast to the percentage of surface area occupied by an erosion low which increased by more than 26% (from 69.1 to 87.59%).

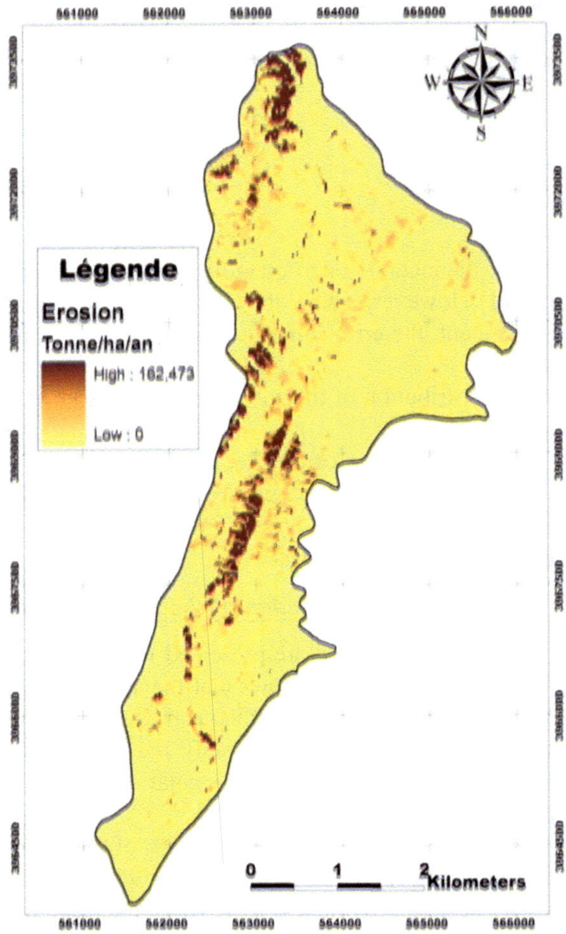

Fig. 8 Map of soil loss determined by the RUSLE model (R15 min) after the WSC management of El Gouazine

Table 15 Class of soil loss determined by the RUSLE model (R15 min)

Class (t/ha/year)	Area (ha)	Surface (%)
0–2.5	1481.18	87.59
2.5–5	96.1	5.68
5–10	59.84	3.54
10–20	39.4	2.33
20–162.47	14.64	0.86

4.2.3 Case of the Model of Wischmeier Revised (RUSLE) with R (30 Min)

- *Application of the RUSLE model R30 min before the WSC management*

The following figure shows the spatial distribution of water erosion throughout the watershed of El Gouazine at no time of 30 min and before the installation of the WSC management.

The review of Fig. 9 shows that the average and the high erosion greater than 20 t/ha/year are located at the level of high altitude of the catchment. While the low erosion (less than 1.5 t/ha/year) is located at the level of areas of low altitude and at the level of the plains.

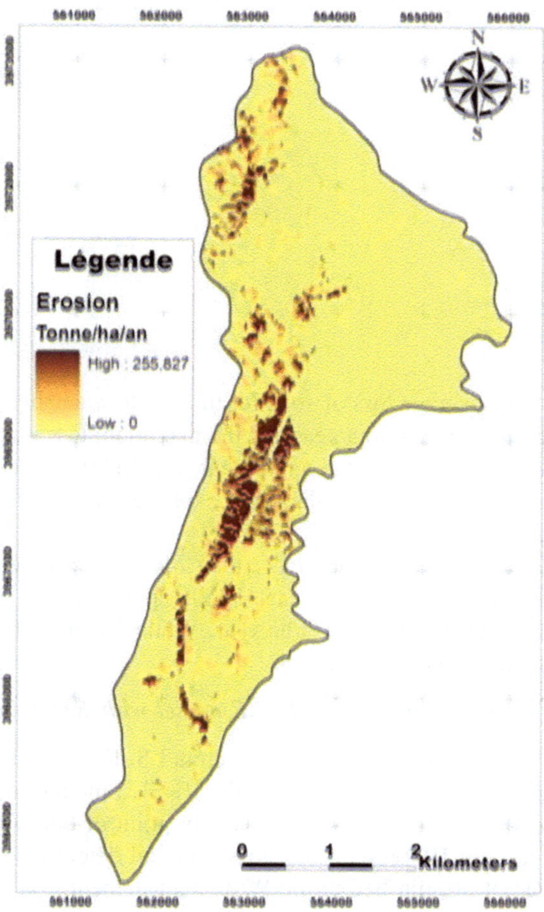

Fig. 9 Map of soil loss determined by the RUSLE model R30 min before the WSC management of El Gouazine

Table 16 Class of soil loss determined by the RUSLE model (R30 min)

Class (t/ha/year)	Area (ha)	Surface (%)
0–2.5	1328.45	74.07
2.5–5	110	6.13
5–10	108.66	6.06
10–20	101.12	5.64
20–255.82	145.25	8.1

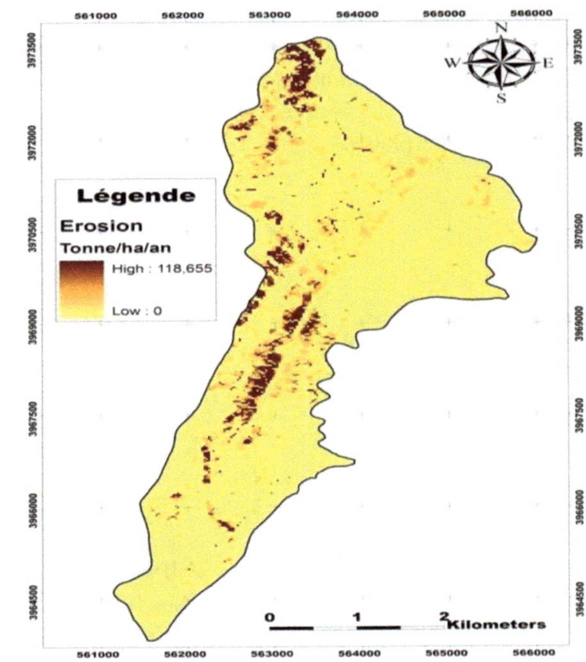

Fig. 10 Map of soil loss determined by the RUSLE model (R30 min) before the WSC management El Gouazine

The average loss of soils for the whole watershed is 6.2 t/ha/year. The minimum loss in soils is 0 t/ha/year and the maximum loss is approximately 255.82 t/ha/year. The total losses of the annual basin is 163,031 t/ha/year.

The distribution of the surface on the classes of soil loss adopted is presented in Table 16.

Table 16 shows that more than 74% of the basin is characterized by low loss of soil (less than 2.5 t/ha/year). While approximately 12.2% of the basin have a soil loss average of 2.5–10 t/ha/year. While 13.74% of the basin have a high soil loss (greater than 15 t/ha/year).

- ***Application of the RUSLE model with R30 min after WSC management***

The dipped beam maps of factors LS, K, C and P of the RUSLE model with the climatic erosivity of acid, by the RUSLE model in no time 30 min, allows you to get the card of losses in soil at any point of our watershed.

Figure 10 presents the losses in soil determined by the RUSLE model at no time of 30 min before the WSC management.

Table 17 Class of soil loss determined by the RUSLE model R(30 min)

Class (t/ha/year)	Area (ha)	Surface (%)
0–2.5	1535.72	90.81
2.5–5	71.42	4.22
5–10	52.51	3.1
10–20	25.08	1.48
20–118.65	6.33	0.37

The review of Fig. 10 shows that the average loss of soils for the whole watershed is about 1.07 t/ha/year. The minimum loss in soils is equal to 0 t/ha/year and the maximum loss is approximately 118.65 t/ha/year. The total losses of the annual basin are of the order of 28,217 t/year. The distribution of the surface on the classes of soil loss adopted is presented in Table 17.

Table 17 shows that more than 90% of the basin of El Gouazine is characterized by a loss of ground low (less than 2.5). While approximately 7.3% of the basin have a soil loss average of 2.5–10 t/ha/year. Then that 1.85% of the basin has a strong soil loss (greater than 10 t/ha/year).

By comparing the results obtained by the model RUSLE (R30 min) before the WSC management and those obtained by the same model after the WSC management, one finds:

- The average value of the erosion decreased from 6.2 to 1.07 t/ha/year.
- The maximum value of the erosion decreases of 256–119 t/ha/year.
- While the minimum value remains constant equal to 0 t/ha/year.
- The total loss of the soil decreases as of 163,031 t/year until 28,216,89 t/year.
- The surface which presents high erosion decreases from 13.7 to 1.9% of the total surface area.
- The low erosion increased to a value 74.07–90.8% of the total surface area.

By comparing the results obtained by the RUSLE model (monthly); RUSLE (R15 min) and those obtained by RUSLE (R30 min), we note for the two periods before and after development that

- ***1st scenario**: **Before the WSC management***

- The application of the RUSLE model shows that the area affected by an erosion tolerated (less than 2.5 t/ha/year) is more cruel to the monthly scale; with a value of 82.2% of the total surface area; that at no time of 30 min (74.07%) and 15 min with a value of 69.1% of the total surface area of the watershed El Gouazine.
- While the area affected by an average erosion (2.5–10 t/ha/year) is more important in the RUSLE model (R15 min) (13.7%) than in the RUSLE model (R30 min) (12.2%) and in the RUSLE model (R (monthly)) (11.3%).
- However that the area affected by an erosion strong (greater than 10 t/ha/year) is less important in the RUSLE model (R monthly) (6.54%) than that in the model RUSLE (R30) (13.7%) and the RUSLE model (R15) (17.2%).

- ***2nd scenario: After the WSC management***

- The area affected by an erosion tolerated (less than 2.5 t/ha/year) is increased for all values calculated in R. In fact, it is more important for the calculation of RUSLE with R monthly with a value of 92.6% of the total surface area that with R at 30 min (90.81%) and not at the time of 15 min with a value of 87.6% of the total surface area of the watershed El Gouazine.
- While the area affected by an average erosion (2.5–10 t/ha/year) decreases for the three methods of calculation. It is more important in the RUSLE model (R15 min) (9.2%) than in the RUSLE model in no time 30 min (7.3%) and at the monthly scale (6.1%).
- Similarly we found that the area affected by an erosion strong (greater than 10 t/ha/year) presents a decrease distinguishable after the WSC management carried out. In effect, it is less important in the RUSLE model(R monthly) (1.3%) than to calculate with the RUSLE model (R30 min) (1.9%) and the RUSLE model (R15 min) (3.19%).

4.2.4 The Application of the Model of Williams (MUSLE)

The MUSLE model (Modified Universal Soil Loss Equation) of Williams and Berndt (1976) allows assessing erosion at the outlet of the sub-basin. It is involving of hydrological parameters and takes into account the sediment from the erosion in ribbon and smile and those from the other major forms of erosion (linear, undermining, solifluction lobes etc.,…). This model is a multiplicative function of four factors of the model of Wischmeier and Smith (1961). The factor of climatic erosivity (R) is substituted by the values of maximum flow rate and the total volume of waters ruisselees.

- ***1st scenario**: **Before the WSC management***

The assessment of water erosion by applying the model MUSLE before the WSC management is presented by the map.

The review of Fig. 11 shows that the average loss of soils for the whole watershed is about 5.5 t/ha/year. The minimum loss in soils is equal 0 t/ha/year and the maximum loss is approximately 140.3 t/ha/year. The total losses of the annual basin are of the order of 144,835 t/year. The distribution of the surface on the classes of soil loss adopted is presented in Table 18.

Table 18 shows that more than 59% of the basin is characterized by a loss of ground tolerated (less than 2.5). While approximately 22.6% of the basin have a soil loss average of 2.5–10 t/ha/year. While 17.16% of the basin have a strong soil loss (greater than 10 t/ha/year).

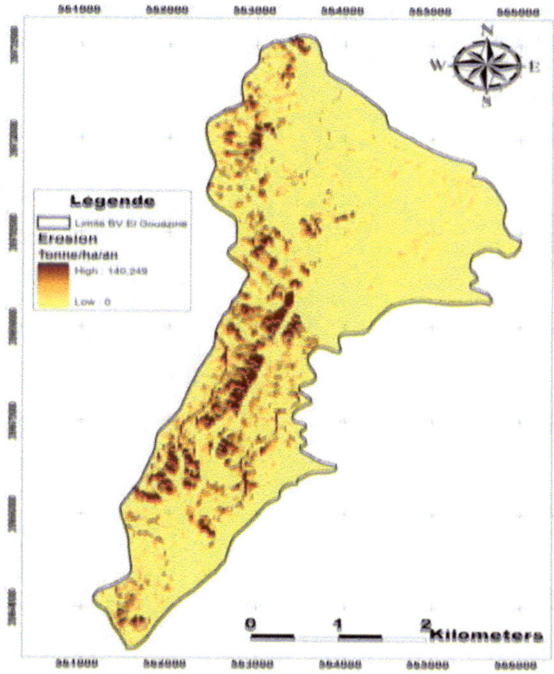

Fig. 11 Map of soil loss determined by the model MUSLE from the USLE model before the WSC management of El Gouazine

Table 18 Class of loss of soil determined by the model MUSLE

Class (t/ha/year)	Area (ha)	Surface (%)
0–2.5	1012.48	59.28
2.5–5	189.34	11.08
5–10	196.81	11.52
10–20	168.27	9.85
20<	124.89	7.31

- ***2nd scenario: After the WSC management***

The map of soil loss after the facilities of conversations of waters and soils made on the catchment of El Gouazine is illustrated by the graph.

From Fig. 12 it is found that the average loss of soil for the entire catchment area is 3.4 t/ha/year. The minimum loss in soils is 0 t/ha/year and the maximum loss is approximately 95.7 t/ha/year. The total losses of the annual basin are of the order of 88,451 t/year. The distribution of the surface on the classes of soil loss adopted is presented in Table 19.

Table 19 shows that more than 66% of the basin is characterized by a loss in soil low (less than 2.5 t/ha/year). While approximately 23% of the basin have a loss in soil average of 2.5–10 t/ha/year. While less than 10% of the basin have a strong loss in soil (greater than 10 t/ha/year).

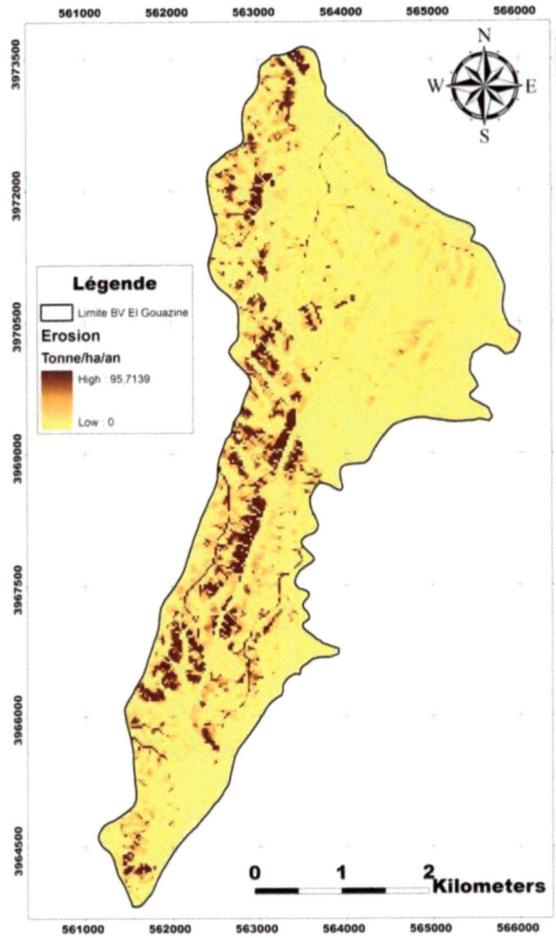

Fig. 12 Map of soil loss determined by the model MUSLE from the USLE model after the WSC management

Table 19 Class of loss in soil determined by the model MUSLE after the WSC management

Class (t/ha/year)	Area (ha)	Surface (%)
0–2.5	1133.08	66.35
2.5–5	227.33	13.31
5–10	171.1	10.01
10–20	110.9	6.5
20<	46.49	2.7

By comparing the results obtained by the model MUSLE before and after the WSC management, one finds:

- The average value of the erosion decreased from 5.5 to 3.36 t/ha/year with a decay rate close to 40%.
- The maximum value of the erosion decreases of 140–95.7 t/ha/year with a rate of decline close to 32%.
- While the minimum value remains constant equal to 0 t/ha/year.
- The total loss of the soil decreases also of 144,835–88,450.53 t/year with a decay rate close to 40%.
- The area affected by an erosion strong decreases of 7.3–2.7% of the watershed El Gouazine with a rate of decline close to 63%.
- The percentage of key surface by an erosion tolerated increased by 59.3–66.4% of the entire surface.

In addition, by comparing the map of loss of soil obtained by the MUSLE models USLE and you find that it has almost the same pace and the same spatial distribution of water erosion on the catchment of El Gouazine.

4.2.5 Loss in Soil by the Formula FAO

An annual estimate of erosion Es (in t/ha/year) is given by the following formula:

$$\mathbf{Es = Fm * C1 * C2 * C3}$$

The application of this equation gives the spatial distribution of erosion as follows.

In applying the formula of the FAO, one obtains a loss in average ground on all the catchment of El Gouazine in the order of 41.3 t/ha/year with a minimum value equal to zero and a maximum value of the order of 207 t/ha/year.

The distribution of the specific erosion on the watershed El Gouazine is illustrated by Table 20.

The application of the formula of the FAO for the catchment of El Gouazine has helped to clarify on the rate of loss in soil and sedimentation (Fig. 13 and Table 20). The results of calculations of losses in soil annual net show that:

- 3.04% of The surface area of the basin is attacked by an erosion low <2.5 t/ha/year.
- 10.6% of The area has a loss in soil medium (10 < Es < 20 t/ha/year).
- In the end, 85.9% of the basin is characterized by a strong erosion (Es > 20 t/ha/year).

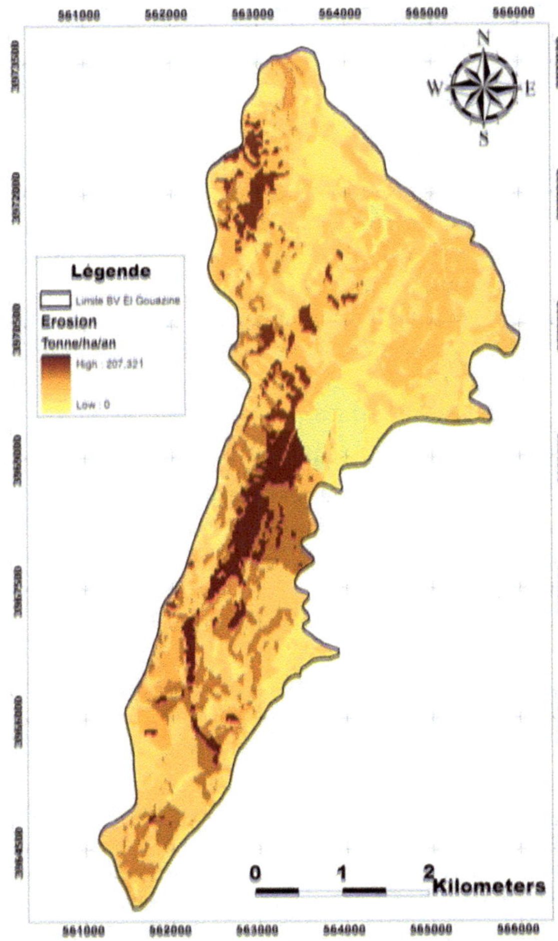

Fig. 13 Map of soil loss determined by the formula FAO

Table 20 Class of the erosion determined by the formula FAO

Class (t/ha/year)	Area (ha)	Surface (%)
0–2.5	52	3.04
2.5–5	–	–
5–10	–	–
10–20	172.2	10.6
20<	1468.6	85.9

5 Comparative Analysis of Results Obtained

In this part we will try to validate the results obtained by the different models (USLE, MUSLE, RUSLE and FAO) comparing them to the values observed during the bathymetric measurements (siltation) at the level of the hilly

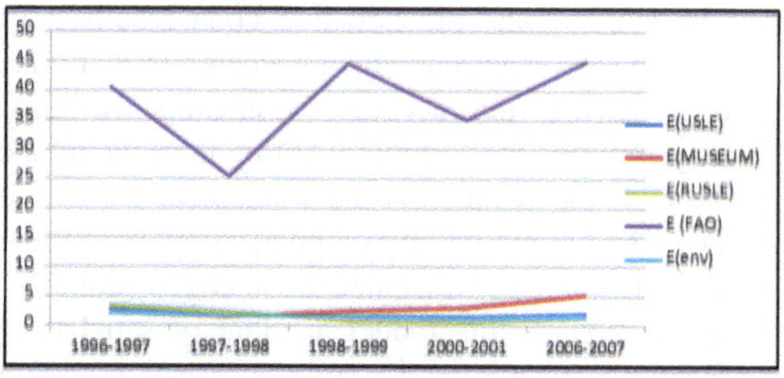

Fig. 14 Variation of the erosion calculated by the 4 models and the erosion measured in the catchment of El Gouazine (t/ha/year)

Table 21 Variation of the quadratic deviation calculated for the 4 models

Erosion	σ
E(USLE)	0.41
E(MUSLE)	1.91
E(RUSLE)	0.75
E(FAO)	37.23

lake El Gouazine (see Fig. 14). In effect, the total erosion in a watershed is the sum of the two basic forms of erosion which are: sheet erosion which can be calculated by the models of erosion and erosion by gullies which cannot be calculated but it can be estimated from the erosion on based on the nature of the soil in the watershed.

To better compare the performance of the models, we proceeded to the calculation of the quadratic deviation, which is the difference between the values calculated by the various models cited previously and measured or observed of the erosion, for each model.

According to Table 21, we note that:

- The values vary from 37.23 for the erosion calculated by the method of the FAO to 0.41 by the USLE model.
- By comparing the various values of the quadratic gap we find that this related to RUSLE and USLE are close and low. In effect, they are more coherent and more close to the erosion measured. We may then conclude that they give a good estimate of the erosion. Whereas, the one corresponding to MUSLE and FAO are more important, the FAO model overestimated the erosion, it gives the values too high of loss in land. It may then be concluded that these two models do not consider well the erosion and it does more suitable to our study area.

In the light of the results obtained and after the fixed objectives for the present study, the following conclusions can be drawn due to the comparative analysis of the results obtained by the different models:

- The USLE model gives a good estimate of the erosion for the watershed El Gouazine by estimating values consistent and valid.
- The MUSLE model has not given good results, the modeling by the MUSLE model overestimated the erosion in the watershed El Gouazine.

- The FAO model did not give good results, he overestimated the erosion for several reason among which we can cite that it is designed to a large scale and it does not take into account the existing facilities in the watershed. In effect, it may improve the formula FAO by introducing a factor which takes into account the existing facilities in the watershed.
- The RUSLE model gives a good estimate of the erosion. In effect, it is the most suitable for our watershed by using R(monthly). The results obtained in the application of this model are that of satisfactory values that can help in the planning of activities for soil conservation and a reduction of siltation of dams. They are providing valuable assistance to policy makers and planners to simulate scenarios of future development of the catchment of El Gouazine and especially plan interventions to combat water erosion.

6 Conclusions and Recommendations

In the news, there are several empirical models which consider the water erosion of the soil at the level of watersheds; these methods vary from simple to more complex. In this context, this work had for objective the mapping and quantification of water erosion on the catchment of El Gouazine by four empirical models which are the universal equation of soil loss (USLE), the revised version of the latter (RUSLE), its amended version (MUSLE) and the equation of the FAO, integrated under a Geographic Information System (GIS). The use of SIG has allowed us to build and to combine these different factors of erosion, in order to obtain a card from synthesis of loss in soil.

In effect, this card of erosion provides information synthetic and systematic on the intensity, the spatial distribution of the phenomenon that will be used to determine the priority intervention zones in order to resolve a large number of problems of crops, soil and watershed management, since the latter is a significant problem at the country level with different degrees of severity.

In addition, the modeling of the erosion by different models, shows that more than half of the surface area of the watershed El Gouazine is affected by an erosion less than or equal to 2.5 t/ha/year, on the other hand approximately 20% of the surface is affected by a loss in soil between 2.5 and 10 t/ha/year. Soil loss of very high value is practically negligible, represents only 4% of the total surface area of the basin. Which proves, in addition, the weakness of the average value of the erosion.

In addition, by comparing the result obtained by the different models before and after the WSC management (1996–1997), we note that there is a remarkable reduction of the loss in land and this is due primarily to the proper functioning of these practices (of the benches essentially) that occupies approximately 43% of the watershed El Gouazine.

Moreover, by comparing the results obtained by each model and the bathymetric measurements we found that the RUSLE model gives results the most adequate for the watershed El Gouazine. In effect, it leads to a good estimate of the erosion.

The lakes are very good sediment traps. They fulfill a role of protection for big dams located downstream. But their rapid clogging goes contrary to an agricultural development. Implanted in fragile environments and low economic activities, they are perceived as an additional resource, rare and vital: the water. To sustain this resource, the management of watersheds to protect these lakes becomes a priority. The nature and the density of these facilities must reconcile the reduction of solid transport without depriving the lake flows into this water by runoff.

In conclusion, this work suggests several prospects of research which can be applied to redesign the same catchment taking into account:

- The study of new scenarios of management.
- The interconnection of groundwater with surface runoff.

References

Arnold JG, Srinivasan R, Williams JR (1998) Wide area soe report modeling and assessment part I: model development. J Am Water Resour Assoc 34(1):73–89

Arnoldus HMJ (1977) Methodology used to determine the maximum potential average annual soil loss due to sheet and rill erosion in Morocco, FAO Soils Bull 34:39–51

Arnoldus HMJ (1980) Methodologies used to determine the maximum potential average soil loss due to sheet and rill erosion in Morocco. FAO Bull 34

Bou Kheir R, Girard M-CL, Khawlie MR, Abadallah C (2001) Soil erosion by water in the Mediterranean circles study and soil management, vol 8(4), pp 231–245

Boussema H-R (1996) Information system for the conservation and management of natural resources. In: International symposium on the role of telecommunications technologies and information on environmental protection, Tunis, Tunisia, 17–19 Apr 1996, pp 112–116

Brown LC, Foster GR (1987) Storm erosivity using idealized intensity distributions. Trans Am Soc Agric Eng 30:379–386

Brunet R (1957) L'érosion accélérée dans le Terrefort toulousain. Revue de Géomorphologie Dynamique, (3-4):33–40

Celik I, Aydin M, Yazici U (1996) A review of the erosion control studies during the republic period in Turkey. Proceedings of 1st International Conference on Land Degradation, p 175–180

Chebbani R, Djilli K, Roose E (1999) Study of the risks of erosion in the catchment of the isser, Algeria. Netw Bull Erosion 19:85–95

Cormary Y, Masson J (1964) Study of water and soil conservation: application to a project type of the formula of soil loss of Wischmeier, 24 pp

De Ploey J, Imeson A, Oldeman L-R (1991) Soil erosion, soil degradation and climatic change. In: Brouwer F-M, Thomas A-J, Chadwick M-J (eds) Land use changes in Europe. Kluwer Academic Publishers, Dordrecht, pp 275–292

FAO (1990) Keep the earth in life: soil erosion, its causes and its remedies. Pedological Bull 50:62

FAO (1993) The state of food and agriculture. Food and Agriculture Organization of the United Nations, Rome

Flanagan DC, Nearing MA (1995) USDA water erosion prediction project: hill slope profile and watershed model documentation. NSERL Report No. 10. USDA-ARS National Soil Erosion Research Laboratory, West Lafayette, IN 47907-1194

Fox KG, Foster GR, Weesies GA, McCool DK, Yoder DC (1997) Predicting soil erosion by water: a guide to conservation planning with the revised universal soil loss equation (RUSLE). Handbook 703, US Department of Agriculture, Washington DC

Heusch B (1986) Fifty years of benches of D. R. S. -C. E. S. In: North Africa: a balance sheet. Cahiers ORSTOM, series pedology, vol XXII(2), pp 153–165

Joftic L, Rosamel Milliman J, Sestini G (1992) Climate change and the Mediterranean. UNEP-E. Arnold Publication, NY

Kassam AH, Velthuizen HT, Mitchell AJB, Fischer GW, Shah MM (1992) AgroEcological Land Resources Assessment for Agricultural Development Planning. A Case Study of Kenya Resources Data Base and Land Productivity Technical Annex 2 Soil Erosion and Productivity. http://www.fao.org

Lewis LA, Verstraeten G, Zhu H (2005) RUSLE applied in a GIS framework: calculating the LS factor and deriving homogeneous patches for estimating soil loss. Int J Geogr Inf Sci 19:809–829

Mansouri T (2001) Spatialized modeling runoff and transport solid of the watersheds of lakes hillside reservoirs of the Tunisian dorsal and Cap Bon. Doctoral thesis in geology at the University of Tunis El Manar, Faculties of Sciences of Tunis, 286 pp

Masson J-M (1971) Erosion by water in the Mediterranean climate. Experimental method of measurement to the scale of the field. Thesis Dr. Engineer, Montpellier, 213 p

Moore I, Burch G (1986) Physical basis of the length-slope factor in the universal soil loss equation. Soil Soc Am J 50:1294–1298

Morgan RPC, Quinton JN, Smith RE, Govers G, Poesen JWA, Auerswald K, Chisci G, Torri D, Styczen ME, Folly AJV (1998) The European soil erosion model (EUROSEM): documentation and user guide. Version 3.6, Cranfield University

Renard KG, Foster GR, Weesies GA, Porter JP (1991) RUSLE: revised universal soil loss equation. J Soil Water Conserv 46:30–33

Renard KG, Freimund JR (1994) Using monthly précipitation data to estimate the R- factor in the revised USLE. J Hydrol 157:287–306

Renard KG, Foster GR, Weesies GA, McCool DK, Yoder DC coordinators (1997) Predicting soil erosion by water: a guide to conservation planning with the revised universal soil loss equation. US Department of Agriculture, Agriculture Handbook 703, 384 pp

Rodríguez JL, García R, Giménez Suárez MC (2010) Comparison of mathematical algorithms for determining the slope angle in GIS environment aplicación de algoritmos matemáticos en la determinación de la inclinación de pendiente en un entorno SIG. Aqua-LAC 2(2):78–82

Roose E (1994) Introduction à la gestion conservatoire de l'eau, de la biomasse et de la fertilité des sols (GCES). Bulletin pédologique de la FAO 70

Revel JC, Coste N, Cavalié J, Costes JL (1989) Premiers résultats expérimentaux sur l'entraînement mécanique des terres par le travail du sol dans le terrefort Toulousain (France). Cahiers de l'ORSTOM, série Pédologie, XXV(1–2):111–118

Shaban A, Khawlie M (1998) Geoenvironmental assessment of riparian areas under extreme climatic events: a case study of representative rivers in Lebanon. In: Mediterranean rivers and riparian zones-processes and management symposium, Zaragosa, Spain, 21 Sept–2 Oct 1998, 25 p

Singh G, Ram B, Chandra S (1981) Soil Loss and Pre-diction Research in India. Bulletin No. T-12/D9, Central Soil and Water Conservation Research Training Institute, Dehra Dun

Smith PH, Holly Fern MT, Lorenz RD, Sromovsky AL, Caldwell JJ, Allison D (1996) Titan's surface, revealed by HST imaging. Icarus 119:336–349. doi:10.1006/icar.1996.0023

Tahiri M, Tabyaoui H, El Hammichi F, Tahiri A, El Hadi H (2014) Assessment and quantification of the erosion and sedimentation from the RUSLE models, MUSLE and deposition integrated in a GIS. Application to the Sous-Bassin of the Oued Sania (Basin of Tahaddart, Gir western north, Morocco). Eur J Sci Res 125:157–178. ISSN 1450-216X/1450-202X

Toumi S (2013) Application of nuclear techniques of remote sensing for the study of water erosion in the catchment area of the oued MINA. Doctoral thesis es-sciences of the higher national school of hydraulics

Williams JR, Berndt HD (1976) Determining the universal soil loss equation's length slop factor for watersheds. In: Proceedings of national soil erosion conference, 25–26 May, pp 217–225

Wischmeier WH, Smith DD (1961) Predicting rainfall erosion losses from cropland east of the Rocky mountains: a guide for selection of practices for soil and water conservation. US Department of Agriculture, Washington, DC

Yu B, Rosewell CJ (1996) A robust estimator of the R-factor for the universal soil loss equation, T ASAE, (39):559–561

Zante P, Collinet J (2003) Mapping of erosion risk on the watershed of the restraint of hilly abdosodok (Tunisian Dorsale)

Hydrological Modeling of Sediment Transport in the Semi-arid Region, Case of Soubella Watershed in Algeria

Mahmoud Hasbaia, André Paquier and Toufik Herizi

Abstract This paper summarizes the major hydrologic studies of the sediment transport in Algeria through a case study of Soubella watershed in Hodna basin (Algeria). Since the 70s, the Algerian Agency of Water Resources (ANRH) has installed a large number of gauging stations, which allowed carrying out many studies about sediment transport. The major contributions aimed principally to model and estimate the suspended sediment transport. Through these studies, the sediment-rating curve is the large discussed model; it is the best significant model for all the basins presented here below. Wadi Soubella watershed (183.5 km^2) is characterized by an aggressive, irregular and violent rainfall. The mean annual precipitation is 288.5 mm (associated with a coefficient of variation Cv = 33% observed during 27 hydrological years 1973–2003). The results show a high specific suspended sediment yields (126 t km^{-2} $year^{-1}$) associated with a higher interannual variability (Cv = 61%). Over the 20 years of this study, total sediment flux is mainly observed during two months: September (45.12%) and October (24.13%). Thus, the early autumn floods are responsible for most of the sediment transport. In Soubella watershed, the sediment-rating curve explains more than 75% of the variance for the whole data (pairs Q-C) at the instantaneous, monthly and seasonally scale and only 70% for the annual scale. A significant regressive relationship (logarithmic) between the sediment-rating curve coefficients (a, b) is obtained at the monthly scale for wadi Soubella and other wadis in Algeria.

Keywords Semiarid region · Algeria · Sediment transport · Hydrologic model · Soubella · Hodna

M. Hasbaia (✉)
Department of Hydraulics, M'Sila University, City Ichebilia, M'sila 28000, Algeria
e-mail: hasbaia_moud@yahoo.fr; mahmoud.hasbaia@univ-msila.dz

A. Paquier
Irstea, UR HHLY, 5 rue de la Doua,
BP32138, 69616 VILLEURBANNE Cedex, France
e-mail: andre.paquier@irstea.fr

T. Herizi
Department of Hydraulics, University of Bejaïa, Bejaïa, Algeria

O. Abdalla et al. (eds.), *Water Resources in Arid Areas: The Way Forward*,
Springer Water, DOI 10.1007/978-3-319-51856-5_14

Notation

A_l Annual liquid yield (Km^3),
A_s Annual sediment yield (Ton),
A_{ss} Specific erosion of soil ($T/km^2/year$)
C Suspended sediment concentration (kg/m^3),
C_e Runoff coefficient (%)
C_T Coefficient of torrential
C_V Coefficient of variation = (tarithmetic mean/standard deviation)
E Runoff (mm)
K_{ER} The rock erodibility
I_l Lithologic index (%)
I_p Energy index of runoff
N_d Number of data
P Average annual rainfall (mm)
Q_l Water flow discharge (m^3/s)
Q_s Sediment discharge (kg/s),
S_b Basin area (km^2)
α Coefficient depending on soil permeability

1 Introduction and State of the Art

This paper aims to summarize and discuss the various hydrologic studies and approaches of the sediment transport in the semiarid region of Algeria through a case study of Soubella watershed.

Algeria is one of the typical semi-arid regions. Its climate is marked by the dominance of intense events with low frequency; this latter feature is clearly revealed by the high spatial and temporal variability of the suspended sediment transport. For example, one flood at Upper Tafna catchment (Northwest of Algeria) caused by a severe storm in July 1989 generated 98% of the annual suspended load (Megnounif et al. 2003). The first dams built in Algeria (until 1900) were quickly filled with sediment. The rate of silting of dams in Algeria in 2010 is estimated at 24%, corresponding to a volume of about 1040 million m^3 (Benblidia et al. 2001).

These examples show the major challenge of this phenomenon, not only for Algeria, but also for the semi-arid region, especially in the Arab Maghreb (Table 1). In Morocco, almost 10% of the dams are filled by sediments; in Tunisia, this proportion is even higher about 25% (Benblidia et al. 2001).

The first measurements of sediment transport in Algeria began in 1946 at Traille gauging station that controls the upper basin of the Isser wadi (Demmak 1982). Using data from a few existing stations, several studies contributed to the estimation of sediment transport in Algerian wadis: (Tixeront 1960; Capolini 1967a, b; Capolini et al. 1969).

Table 1 Specific erosion in the Arab Maghreb (from Probest and Amiotte-Suchet 1992)

Source	Ass (T/km^2/year)	As (10^6 T/year)
Fournier (1960)	60–600	–
Strakhov (1967)	10–50	–
Heusch and Milliès-Lacroix (1971)	265–2569	–
Milliman and Meade (1983)	100	100
Dedkov and Mozzherin (1984)	100–250	–
Walling (1974)	1000–5000	–
Walling and Webb (1981)	>500	–
Snoussi (1988) (For Morocco only)	750	–
Probst and Amiotte-Suchet (1992)	504	100

During the 70s, the National Agency of Water Resources (ANRH) installed a large number of gauging stations, which allowed carrying out a lot of studies about sediment transport (Demmak 1982; Demmak et al. 1991; Terfous et al. 2001; Megnounif et al. 2003; Benkhaled and Remini 2003; Bouanani 2004; Bouteldja 2005; Ghenim et al. 2008; Achite and Ouillon 2007, etc.). These studies showed a high suspended load in the Algerian wadis (Table 2), which is among the highest ones in the world (Probst and Amiotte-Suchet 1992).

The Algerian gauging stations measure the flow discharge ($m^3\ s^{-1}$) and the suspended sediment concentrations (g/l). The flow discharges are directly determined from the rating curves through the water elevation that is measured by means of ladder or float. For each elevation measurement, water samples are taken close to the bank of river to estimate the suspended sediment concentration.

The water samples are filtrated through 145 μm filter paper. The collected mass of sediment is weighted after drying at 105 °C for 24 h. The estimated suspended concentration is considered as the average over the cross section, and the sediment discharge is the product of this concentration by the flow discharge. The number of samples is adapted to the hydrological regime, during low flow with little variation; only one sampling is performed every day. However, during the flood period, the sampling rate increases until reaching a sampling frequency every 15 or 30 min at the flood peak. Bedload is not measured in Algeria; it is estimated using a percentage of suspended sediment load; (Serrat 1999) noted that bedload sediment volume was less than 1% of the suspended load in an intermittent river of Mediterranean type, but for many Algerian wadis, this percentage is ranging from 10% (Khanchoul et al. 2012) to 30% (Achite and Meddi 2004). The lack of data prevents us to discuss, confirm, reject or compare such estimates.

Many studies have been devoted to improve the estimation, the forecast and the spatio-temporal variation of suspended sediment transport in Algerian wadis (Terfous et al. 2001; Megnounif et al. 2003; Benkhaled and Remini 2003;

Table 2 Assessment of the sediment transport in Algeria

Reference	Region	Period	S_b (km^2)	N_d	P (mm)	A_{ss} ($T/km^2/year$)	E (mm)
Megnounif et al. (2003)	The upper-Tafna NW of Algeria	1988–1993	256	1257	345–527	24–4288	43–305
Terfous et al. (2001)	Mouilah wadi NW of Algeria	1977–1993	2650	4034	M 300.90	126.40	18.40
Achite and Meddi (2004)	Haddad wadi NW of Algeria	1973–1995	470	1878	200–379	287	—
Achite and Meddi (2005)	ElAbtal wadi	1973–1995	4126	–	–	–	–
	Sidi A.E.K. Djilali	1973–1995	470	–	–	–	–
	Ain Hamra	1973–1995	2480	–	–	–	–
	Kef Mahboula	1973–1995	680	–	–	–	–
	Takhmaret	1973–1995	1553	–	–	–	–
Ghenim et al. (2008)	Sebdou wadi NW of Algeria	1985–1998	256	–	188–597	107–5876	53–278
Achite and Ouillon (2007)	Abd wadi NW of Algeria	1973–1995	2480	1432	174–303	136	–
Boudjadja et al. (2003)	Allalah wadi	—	295	–	–	2701	–
	Damous wadi	—	577	–	–	2879	–
	Es Sebt wadi	—	112	–	–	2950	–
	Messelmoun wadi	—	218	–	-	3029	–
	Elhachem wadi	—	217	–	–	2905	–
Khanchoul and Jansson (2008)	Mellah wadi NE of Algeria	1972–1997	550	1700	707	373	97
Cherif et al. (2009)	Mekkera wadi NW of Algeria	1950–2001	4102	7042	350–450	111	–
Khanchoul et al. (2012)	Cherf wadi	1975–1994	1710	801	290	347	11

(continued)

Table 2 (continued)

Reference	Region	Period	S_b (km^2)	N_d	P (mm)	A_{ss} ($T/km^2/year$)	E (mm)
	NE of Algeria						
Khanchoul et al. (2012)	Kebir wadi NE Algeria	1976–2008	681	2703	693–775	895	–
Megnounif et al. (2013)	Sebdou wadi NW of Algeria	1973–2004	256	2112	278–538	343	–
Elahcene et al. (2013)	Bellah wadi Central of Algeria[2]	1974–2007	55	3179	519	610	–
Louamri et al. (2013)	Bouhamdane wadi NE of Algeria	1969–2010	1105	2210	589	257	81

Bounani 2004; Ghenim et al. 2008; Achite and Ouillon 2007; Hasbaia et al. 2012; Elahcene et al. 2013; Ghernaout and Remini 2014). The sediment-rating curve approach is widely discussed and checked in these studies.

All these latter studies are based on a hydrologic approach. However, a few studies based on the hydraulic approach exist (Bessenasse et al. 2003; Hasbaia and Benayada 2003; Hasbaia 2011; Benayada and Hasbaia 2013).

2 The Study Area

The Hodna basin with a drainage area of 26,000 km^2 is the fifth basin of Algeria. It is an interior basin, located at 150 km in the south of the Mediterranean coast (Golf of Bejaïa) (Fig. 1). Soubella wadi watershed (183.48 km^2) is located at northeast of Hodna basin. It is limited by Sommam basin at the north and by Chott El Hodna at the south. The main physiographic features are reported in (Table 3).

The Soubella watershed has greater floristic vegetation that extends over about 60% of the surface basin. On the other hand, arable land accounts for 24% and the remaining part of the basin area are bare lands. The Soubella wadi and its tributary El Hammam wadi have a perennial surface flow that decreases from upstream to downstream. Some sources originate in the mountains at the base of slopes from the calcodolomitic cliffs but their flow is very low. The climate of the watershed is semiarid, characterized by high temperature and low rainfall associated with a high spatiotemporal variability. The average temperatures are between 14.20 and 17.60 °C.

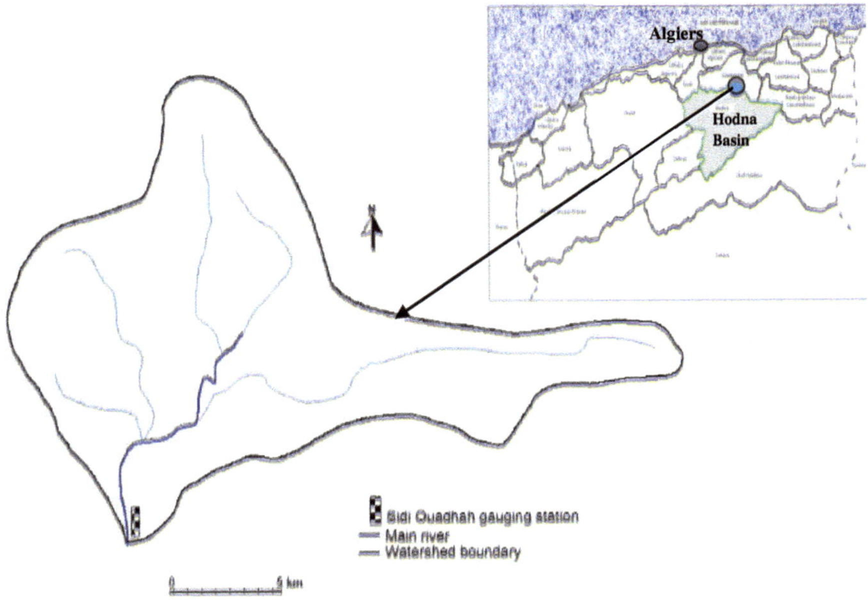

Fig. 1 Soubella wadi watershed location

Table 3 Main characteristics of Soubella wadi watershed

Parameters	Notation	Unit	Value
Area	A	km^2	183.48
Perimeter watershed	P	km	76.18
Maximum altitude	H_{max}	m	1886
Average altitude	H_{avrg}	m	1318.20
Minimum altitude	H_{min}	m	720
Circularity ratio	F_c	–	1.57
Average slope	I	%	18.30
Global slope index	I_g	%	1.96
Length of the main wadi	L_p	km	34.45
Length of equivalent rectangle	L	km	32.50
Width of equivalent rectangle	L	km	5.64
Drainage density	D_d	km/km^2	0.86
Time of concentration	T_c	h	5.41

The warmest months are June, July and August (average temperature between 22.10 and 25.40 °C). The coldest months are December, January and February (average temperature between 5.10 and 6.30 °C). The annual average relative humidity is 61% but humidity can exceed 76%. Sunshine is about 8.10 h/day. The annual rainfall is highly variable, ranging from 130 to 470 mm/year. The mean annual precipitation is about 288.50 mm/year with an interannual coefficient of variation C_V = 33% for 30 hydrological years (1973–2003).

The mean annual discharge at Sidi Ouadah gauging station is 0.135 m^3/s, with a coefficient of variation of 60% recorded during 26 hydrologic years (from 1973/74 to 1998/99). The average annual evapotranspiration is relatively important for the Soubella wadi (about 342 mm) but however, the water balance is classified beneficiary.

3 Hydrologic Data

The study is based on instantaneous measurements of discharge and concentrations of suspended sediment carried out by the ANRH agency. The flows of Soubella wadi are controlled at the gauging station of Sidi Ouadah installed at the watershed outlet located at X = 721.50 km, Y = 268.75 km (Lambert coordinated), and Z = m (Fig. 1). The available data consist of the instantaneous water discharges Q ($m^3\ s^{-1}$) and the suspended sediment concentrations C (kg m^{-3}), measured over several time intervals. These data (with a number of 903 pairs of C and Q measurements) covers 27 years of measurements in various time intervals from March 1973 until Mai 2000. The limited number of data is due to the sampling rate as depending on the hydrological regime (explained in the introduction). In our database, the flow discharge varied from 0.015 to 90.00 $m^3 s^{-1}$ and the suspended sediment concentration varied from 0.10 to 182.00 kg m^{-3} .The two parameters are associated with a high variability, the coefficients of variation are about 225 and 118% respectively.

4 Suspended Sediment Yields

In Algeria, the suspended sediment yield A_s through a section of wadi and during a time interval ($t_{i+1} - t_i$) is calculated by formula (1), the water yield A_l being calculated by a similar formula:

$$A_s = \frac{[Qs_{i+1} + Qs_i]}{2}(t_{i+1} - t_i) \quad (1)$$

Q_{si} is the suspended sediments discharge at time t_i.

Soubella wadi transports annually about 2 million m^3 of water and 44,103 tons of sediment, corresponding to a specific degradation of soil of about 239 T/km^2/year. These average values are associated with a significant variability. During the study period of 21 years, the interannual coefficient of variation C_v is about 157 and 135% for water and sediment yields; respectively. During the hydrologic year 1986–1987, wadi Soubella has transported 14.50 million m^3 of water and about 151,103 tons of sediment. The distribution of the sediment and water loads between seasons (Table 4) shows that the major part of sediment and liquid loads occurs in autumn: about 73 and 67%; respectively. The same trend is observed (with variable rates) for other basins in Algeria (Touïbia et al. 2001; Megnounif et al. 2003; Achite and Meddi 2005; Ghenim et al. 2008; Achite and Ouillon 2007). In the east of Algeria, Khanchoul and Jansson (2008) concluded that 61 and 23% of the total sediment yield of Mellah wadi are observed in winter and spring season respectively.

At the monthly scale, the variability is even higher. The coefficient of variation is about 147 and 161% for the water and sediment yields respectively. During the study period, more than 40% of the total liquid and sediment flux occurs in September due to the violent nature of autumn's floods and the poor condition of the soil (dry and cracked) because of the high temperatures during summer.

Table 4 Distribution of liquid and sediment loads by season (wadi Soubella)

Saison	As (%)	Al (%)
Autumn	72.07	66.92
Winter	2.20	6.96
Spring	15.38	21.00
Summer	10.35	5.11

5 Hydrologic Sediment Transport Modeling

Many contributions to predict sediment transport have been proposed in Algeria, using the hydrological parameters of basins (Table 5).

In order to compare the results of these models, we have used them to estimate the annual sediment yield of Soubella watershed. Through the study period (27 years), the obtained results showed that the best model is the sediment-rating curve with a mean relative error about 34.4%. The others models overestimate the sediment yields of more than 80% associated with a high variability. For this reason, we have studied sediment-rating curve in detail.

The sediment-rating curve is a power function between the suspended sediment concentrations and the flow discharge ($C = aQ^b$ or $Q_s = a\,Q^{b+1}$) (Campbell and Bauder 1940; Crawford 1991; Walling 1974, 1978; Asselman 2000). The second expression is the most used in the hydrologic literature because of its higher correlation coefficient. Leopold and Maddock (1953) showed that b varies between 2 and 3, Bogardi (1974) suggests that a and b reflect the characteristics of the watershed, however Vanoni (1977) considers that a is a factor of soil erodibility.

The sediment rating curve is the widely used model to predict the suspended sediment yield in Algeria. It was tested in several watersheds in Algeria (Terfous and al. 2001; Megnounif et al. 2003; Benkhaled and Remini 2003; Achite and Meddi 2004; Bounani 2004; Bouteldja 2005; Ghenim et al. 2008; Achite and Ouillon 2007, Cherif et al. 2009; Khanchoul et al. 2012; Elahcene et al. 2013; Ghernaout and Remini 2014). In all these studies, the sediment rating curve explained more than 60% of the variance of the suspended sediment transport. The conclusion does not vary much from one time scale to another one; the higher explained variances are noted at the annual scale (Table 6).

The constants a and b depend on the region and the time scale. Rating curves obtained by least squares regression on logarithmic transformed data tend to underestimate the sediment transport rates by about 10% to more than 50% (Ferguson 1986). The best way stands in keeping the curve in its power function (Asselman 2000) unlike this statement, at the daily scale, it overestimates the sediment discharge by 25%. Achite and Ouillon (2007) and Khanchoul and Jansson (2008) concluded that when correction of the bias due to the logarithm transformation was applied (as proposed by Miller (1984)), the sediment yield underestimate decreases from 16 to 10% and, when the data are subdivided into threshold discharge classes, season groups and rising/falling stages, the corresponding rating curve underestimates the sediment yields with a low margin of error (from 0.80 to 3%).

For Wahrane wadi, Benkhaled and Remini (2003) tried to express the constants a and b by the annual runoff, the flow discharge and by the coefficient of hydraulicity. The obtained results are typical of this wadi and cannot be generalized. According Asselman (2000), a and b are not completely independent; they can be connected by a logarithmic relationship for one or more streams. Achite and Ouillon (2007) used a similar relationship to classify floods as aggressive or low.

Table 5 Models of sediment transport yield in Algeria

References	Model	Observations
Tixeront (1960)	$A_s = 354 \,.\, E^{0.15}$ For Tunisian watershed $A_s = 92 \,.\, E^{0.21}$ for the east region of Algeria $A_s = 200 \,.\, E$ for the central region of Algeria	Based on the data collected in 32 Algerian and 9 Tunisian wadis (from 90 to 22,300 km^2) over a period of between 2 and 22 years
Sogreah (1983)	$A_s = \alpha \,.\, E^{0.15}$	Based on the data collected in 30 Algerian basins (from 100 to 300 km^2), with annual rainfall between 300 and 1000 mm
Demmak (1982)	$A_s = 26.62\, I_l + 5.07\, I_p + 9.77\, C_T - 593.56$	A study based on multiple regression applied to 30 Algerian watershed (100–3000 km^2)
Probst and Amiotte-Suchet (1992)	$Ln(A_s) = 4.79 + (0.054)K_{ER} + (0.004)E - (5.6x10^{-5})S_b$	Using the data of 130 basins in the Maghreb, the correlation coefficient of this model equals 0.71
Achite and Meddi (2005)	$A_s = a\, C_e^b$; a and b are constants	This model explains more 60% of the variance of the sediment transport for the five sub-basins (470–2480 km^2) of Mina watershed at the northwest of Algeria
Terfous et al. (2001), Megnounif et al. (2003), Benkhaled and Remini (2003), Bouanani (2004), Achite and Meddi (2004), Ghenim et al. (2008), Achite and Ouillon (2007), Cherif et al. (2009), Khanchoul et al. (2012), Elahcene et al. (2013), Ghernaout and Remini (2014)	$Q_s = a\, C_l^b$	This model named power law (or sediment-rating curve) is validated for all these studies

Table 6 Sediment rating curve for many Algerian wadis (in bold, results from the authors)

References	Wadi	Area (km^2)	years	P (mm)	N_d	*Sediment rating curve model:* $Q_l = a\, Q_s^b$							
						Annual scale		Seasonal scale		Monthly scale		All data	
						a	*b*	*a*	*b*	*a*	*b*	*a*	*b*
Terfous et al. (2001)	Mouilah	2650.0	16	301		0.41	1.70	0.28–0.55	1.5–1.82	–	–	–	–
Benkhaled and Remini (2003)	Wahrane	270.5	14	304–550	3988	1.46–6.43	1.32–1.95	0.45–2.96	1.22–1.53	0.34–3.68	1.13–1.80	–	–
Achite and Meddi (2004)	Haddad	470.0	22	200–379	1878	2.04–58.62	1.23–2.17	10.64–16.89	1.40–1.59	5.79–35.47	1.35–1.74	14.42	1.45
Ghenim et al. (2008)	Sebdou	256.0	13	188–597	1821		–	–	–	–	–	0.26	1.64
Khanchoul and Jansson (2008)	Mellah	550.0	23	707	1700	–	–	0.02–1.57	1.22–1.71	–	–	0.14	1.72
Cherif et al. (2009)	Mekerra	681.0	51	693–775	2703	5.72	1.31	2.75–12.74	1.18–1.35	–	–	–	–
Khanchoul et al. (2012)	Kebir	651.0	24	693–775	2703	–	–	0.60–1.57	1.24–1.44	–	–	0.32	1.47
Elahcene et al. (2013)	Bellah	55.0	32	519	3179			3.43–17.11	1.21–1.74			4.62	1.38
Ghernaout and Remini (2014)	El-Abtal	4900.0	16	278	10,293			4.79–12.08	1.27–1.42	2.63–16.85	1.06–1.64	8.64	1.32
	AEK Djilali		19		3411			26.70–33.24	1.21–1.26	19.66–66.16	1.09–1.47	27.65	1.25
This study	**Soubella**	**183.5**	**26**	**288**	**903**	**7.06–22.71**	**1.07–2.09**	**6.70–16.12**	**1.18–1.36**	**5.58–14.78**	**1.06–1.48**	**11.54**	**1.29**

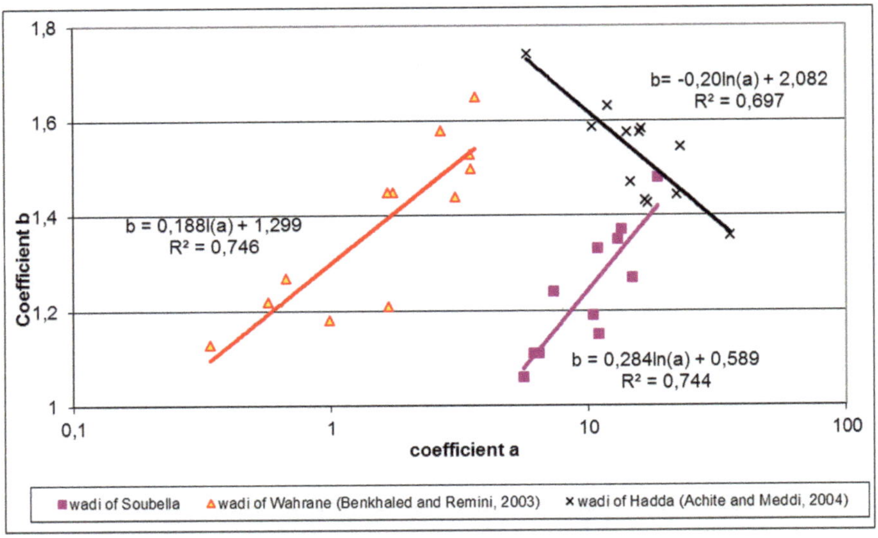

Fig. 2 Correlation between slope/intercept values of sediment-rating curves obtained at monthly scale for Soubella wadi and two others Algerian wadis

In this study, we tried to link a and b. The only significant relationship between a and b was found as a logarithmic expression similar to that obtained by Asselman (2000), Achite and Ouillon (2007) but only for the monthly scale. This logarithmic model is applied to other Algerian wadis (Fig. 2).

This law explained more than 69% of the slope/intercept of the sediment-rating curve variance, at this scale. It is generally believed that the steepness of the rating curve is related to the availability of suspended sediment in a certain area (Asselman 2000). Similar relation is obtained for 12 (a, b) pairs corresponding to 12 selected floods of wadi Soubella (Fig. 3).

The variance explained is high relative to that of the 138 wadi Abd floods, studied by Achite and Ouillon (2007) because the 12 floods are in one single group while the 138 floods of Wadi Abd are classified into 3 classes (average, aggressive and weaker than average floods). The relationship between a and b coefficients is of great interest, especially for poorly gauged basins. It allows deducing or extrapolating the curve in case of lack of data. The variance of sediment transport at wadi Soubella is well explained for all the time scales (interannual, seasonal, monthly, and all data), the explained variance being between 74 and 98%. A few low values (between 60 and 74%) are observed at the interannual scale because of the scatter of the liquid and sediment discharges in the same year from one season to another season and because of the variability of the wet and the dry periods.

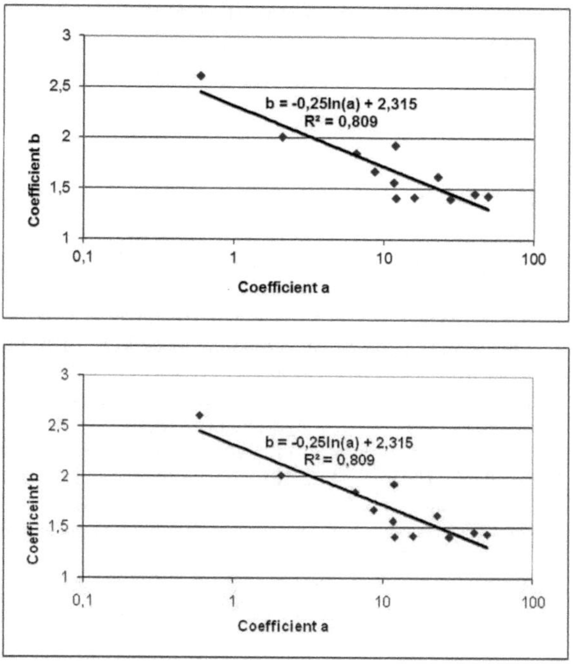

Fig. 3 Correlation between slope/intercept values of sediment-rating curves for 12 floods of Soubella wadi

6 Conclusions

The suspended sediment transport is a challenge for the semi-arid regions, not only because of the high quantities of sediment yield, but also because of the spatio-temporal variability of the phenomenon. Through several studies of many watersheds in Algeria, the sediment transport varies quantitatively and qualitatively from one region to another. Soubella watershed is a sub-basin of Hodna basin. With an area of 183.50 km^2, the average sediment wash-down is 239 T/km^2/year associated with a high variability (interannual coefficient of variation of 135%). At the seasonal scale, more than 70% of the sediment yield occurs in autumn for wadi Soubella; but, for most of the Algerian basins, the suspended sediment transport is intensive in the early autumn and late spring. However, at the monthly scale, the same trend is noticed for wadi Soubella and most of the wadis of Algeria. The explanation of sediment transport using the hydrologic parameters of Algerian basins is not always easy; the few contributions in this field are very limited and cannot be generalized. However, the sediment rating curve was checked for a lot of Algerian watersheds. For Soubella wadi, it explained more than 74% of the variance of the suspended sediment transport at interannual, seasonal and monthly scale. The slope/intercept pairs of the sediment rating curve are not completely independent. For Soubella wadi, we showed that the coefficients (a, b) can be connected by a logarithmic relation at the monthly scale for 12 selected floods. This logarithmic law is very useful in case of lack or deficiency of data, often encountered in the improperly gauged wadis.

References

Achite M, Meddi M (2004) Estimation du transport solide dans le bassin versant de l'oued Haddad (Nord-Ouest algérien). Sécheresse 15(4):367–373

Achite M, Meddi M (2005) Variabilité spatio-temporelle des apports liquide et solide en zone semi-aride. Cas du bassin versant de l'oued Mina (nord-ouest algérien). Revue des sciences de l'eau, 18 (n° spécial): 37–56

Achite M, Ouillon S (2007) Suspended sediment transport in semiarid watershed, Wadi Abd, Algeria (1973–1995). J Hydrol 343:187–202

Asselman N (2000) Fitting and interpretation of sediment-rating curves. J Hydrol 234:228–248

Benayada L, Hasbaia M (2013) Comparisons between unsteady sediment-transport modeling. J Cent South Univ 20(2):536–540. doi:10.1007/s11771-013-1516-9

Benblidia M, Salemn A, Demmak A (2001) Extraction des sédiments dans les retenues. La houille blanche 6(7):76–78

Benkhaled A, Remini B (2003) Analyse de la relation de puissance: débit solide - débit liquide à l'échelle du bassin versant de l'Oued Wahrane (Algérie). Revue des sciences de l'eau 16 (3):333–356

Bessenasse M, Kettab A, Paquier A, Ramez P, Galea G (2003) Simulation numérique de la sédimentation dans les retenues de barrage, cas de la retenue de Zardezas Algérie. Revue des sciences de l'eau 16(2003):103–122

Bogardi J (1974) Sediment transport in alluvial stream. International courses in hydrology, Budapest, Academiai Kiado Press, p 812

Bouanani A (2004) Hydrologie, Transport solide et modélisation Etude de quelques sous basins de la Tafna (NW-Algérie), Phd thesis at Abou Bekr Belkaid University of Tlemcen-Algeria, p 249

Boudjadja A, Messahel M, Pauc H (2003) Ressources hydriques en Algérie du Nord. Revue des sciences de l'eau 16(3):285–304

Bouteldja N (2005) Contribution à la modélisation de l'érosion hydrique dans le bassin versant du Hodna sous bassins versants du Ksob et de Soubella (Algérie), Phd thesis at la Provence Aix-Marseille University, Marseille-France

Campbell FB, Bauder H (1940) A rating-curve method for determining silt-discharge of stream. EOS Trans. Am. Geophys Union 21:603–607

Capolini J (1967a) Etude géomorphologiques des conditions d'envasement du port de Ghazouet, Etude SES 1013/DH2, Alger, (8 plans), p 16

Capolini J (1967b) Bassin versant de l'oued Fodda -Carte de sensibilité à l'érosion, Etude SES Alger, (carte 1/50.000), p 6

Capolini J, Piens S, Ramana R (1969) Bassin versant du Chleff: Oued Rhiou, Sly, Fodda, Deurdeur, Zeddine, Ebda: prévision des apports des crues et des débits solides, Etude SES 13/12/DH2, Alger. p. 41

Cherif E, Errih M, Cherif HM (2009) Modélisation statistique du transport solide du bassin versant de l'Oued Mekerra (Algérie) en zone semi-aride méditerranéenne. Hydrol Sci J 52(2):338–348

Crawford CG (1991) Estimation of suspended—sediment-rating curves and mean suspended sediment load. J Hydrol 129:331–348

Dedkhov AP, Mozzherin DI (1984) Erosiya I stok nanosov na zemle, Izdatelstvo Kasagskogo Universsiteta

Demmak A (1982) Contribution à l'étude de l'érosion et de transport solide en Algérie septentrionale. Phd thesis at Paris University Paris-France. p 323

Demmak A, Ouaar M, Guedjtal A (1991) Quantification de l'érosion à l'exutoire de micro-bassins en zone semi-aride. Utilisation de l'eau des petits bassins versants en zone aride, Ed. AUPELF-UREF, Jhon Libbey Eurotext-Paris, pp 179–188

Elahcene O, Terfous A, Remini B, Ghenaim G, Poulet JB (2013) Etude de la dynamique sédimentaire dans le bassin versant de l'Oued Bellah (Algérie). Hydrol Sci J 58(1):224–236. doi:10.1080/02626667.2012.742530

Ferguson RI (1986) River loads underestimated by rating curves. Water Resour Res 22(1):74–76

Fournier F (1960) Climat et érosion. Presse Universitaire de France Paris

Ghenim A, Seddini A, Terfous A (2008) Temporal variation of the specific sediment yield of the Wadi Mouilah basin (northwest Algeria). Hydrolog Sci J 53(2):448–456

Ghernaout R, Remini B (2014) Impact of suspended sediment load on the silting of SMBA reservoir (Algeria). Environ Earth Sci 72:915–929. doi:10.1007/s12665-014-3125-9

Hasbaia M (2011) Critical study of sediment transport and its impact on the natural stream, PHD thesis at, MOHAMED BOUDIAF University of sciences and Technology of Oran-Algeria. p 149

Hasbaia M, Benayada L (2003) Numerical modelling of bed load in natural channel. In: The 2nd international conference on wady hydrology, 1–4 July 2003, Amman, Jordan

Hasbaia M, Hedjazi A, Benayada L (2012) Variabilité de l'érosion hydrique dans le bassin du Hodna: cas du sous-bassin versant de l'oued Elham, Revue Marocaine des Sciences Agronomique et Vétérinaire 2012(1):28–32

Heusch B, Millieres-Lacroix A (1971) Une méthode pour estimer l'écoulement et l'érosion dans un bassin. Application au Maghreb. Mines et Géologie (Rabat) 33:21–39

Khanchoul K, Jansson MB (2008) Sediment-rating curve developed on stage and seasonal means in discharge classes for the Mellah wadi Ageria. J Compil Swed Soc Anthropol Geogr 227–236

Khanchoul K, Boukhrissa Z, Acidi A, Altschulb R (2012) Estimation of suspended sediment transport in the Kebir drainage basin, Algeria. Quater Int 262(2012):25–31

Leopold LB, Maddock TG (1953) The hydraulic geometry of stream channels and some physiographic implications. Washington D.C. US Geological survey. Professional paper, p 252

Megnounif A, Terfous A, Bouanani A (2003) Production et transport des matières solides en suspension dans le basin versant de la Haute-Tafna (Nord-Ouest Algérie). Revue des sciences de l'eau 16(3):369–380

Megnounif A, Terfous A, Ouillon S (2013) A graphical method to study suspended sediment dynamics during flood events in the Wadi Sebdou, NW Algeria (1973–2004). J Hydrol 497:24–36

Miller DM (1984) Reducing transformation bias in curve fitting. Am Stat 38(2):124–126

Milliman JD, Meadde RH (1983) World wide delivery of river sediment to the ocean. J Geol 91: 1–21

Probst JL, Amiotte-Suchet PA (1992) Fluvial suspended sediment transport and mechanical erosion in the Maghreb (North Africa). Hydrol Sci J 37(6):621–637

Serrat P (1999) Dynamique sédimentaire actuelle d'un système fluvial méditerranéen: l'Agly (France). Comptes Rendus de l'Académie des Sciences, IIA 329:189–196

Snoussi M (1988) Nature estimation et comparaison des flux de matière issus des bassins versants de l'Adour (France), du Sebou, de l'Oum-Er-Rbia et du Sous (Maroc). Impact du climat sur les apports fluviatiles à l'océan. Mémoire de l'institut de Géologie du Bassin de l'Aquitaine No. 22 Bordeaux. France

Sogreah S (1983) Erosion et transport solide au Maghreb. Analyse bibliographique. Rap. Proj. RAB/80/011/PNUD Alger, p 64

Strakhov NMN (1967) Principles of Lithogenesis, vol 1. Oliver and Boyd, Edindburgh, UK

Terfous A, Megnounif A, Bouanani A (2001) Etude de transport solide en suspension dans l'oued Mouilah (Nord-Ouest Algérien). Revue des sciences de l'eau 14(2):173–185

Tixeront J (1960) Débit solide des cours d'eau en Algérie et en Tunisie. IAHS Publication 53:26–42

Touïbia B, Aidaoui A, Gomer D, Achite M (2001) Quantification et variabilité temporelles de l'écoulement solide en zone semi-aride, de l'Algérie du nord. J Sci Hydrol 46(1):41–53

Vanoni VA (1977) Erosion rate from sediment sources, in engineering sedimentation, Manuel ASCE New York, pp 472–480
Walling DE (1974) Suspended sediment and solute yields from a small catchment prior to urbanization. In: Gregory KJ, Walling DE (eds) Fluvial processes in instrumented watersheds, Institute of British geographers special publication, vol 6, pp 169–192
Walling DE (1978) Suspended sediment and solute response characteristics of the river Exe, Devon, England. In: Davidson-Arnott R, Nickling W (eds) Research in fluvial systems. Geoabstracts Norwich, pp 169–197
Walling DE, Webb DW (1981) The reliability of suspended sediment load data. Erosion and sediment transport measurement. In: Proceedings of the florence symposium florence, vol 133. IAHS Publications, pp 177–194

A New Innovative Tool to Measure Soil Erosion

G.N. Zaimes, V. Iakovoglou, I. Kosmadakis, K. Ioannou, P. Koutalakis, G. Ranis, T. Laopoulos and P. Tsardaklis

Abstract Soil erosion is a one of the most serious environmental problems worldwide. This is the result of unsustainable anthropogenic activities such as agriculture, deforestation and urbanization. In addition, the extreme events caused by climate change are expected to accelerate soil erosion. The Mediterranean region, due to the sparse vegetation, topography, frequent wildfires and being inhabited for thousands of years has higher soil erosion rates compared to the rest of Europe. These facts indicate that new tools need to be developed and utilized in order to be able to mitigate effectively and efficiently soil erosion in the Mediterranean. The goal of this study was to develop a new monitoring system of sensors that measures erosion accurately but also measures other important environmental variables that influence erosion. The erosion measurements were based on ultrasonic technology that allows accuracy of up to 1 mm. The system also measures precipitation, soil and air temperature and soil moisture. These measurements can be continuous something very important to understand soil erosion mechanisms. Two systems with sensors were installed in September 2015, in the Kallirahi torrent watershed of Thasos Island in Greece. The specific watershed had a wildfire in August of 2013, and the impact is still evident on the landscape. One of the system of sensors was installed in an area with tree cover and gradual slopes while the other area had no overstory and steep slopes. Evaluating the preliminary results show that the measured variables by the sensor provides very accurate data.

G.N. Zaimes (✉) · V. Iakovoglou · K. Ioannou · P. Koutalakis
Laboratory of Management and Control of Mountainous Waters (Lab of MCMW),
Department of Forestry and Natural Environment Management,
Eastern Macedonia and Thrace Institute of Technology (EMaTTech),
1st km Drama-Mikrohoriou, 66100 Drama, Greece
e-mail: zaimesgeorge@gmail.com

I. Kosmadakis · T. Laopoulos · P. Tsardaklis
Electronics Lab., Physics Department, Aristotle University of Thessaloniki,
52124 Thessaloniki, Greece

G. Ranis
Forest Service of Thasos Island, 64004 Thasos, Greece

O. Abdalla et al. (eds.), *Water Resources in Arid Areas: The Way Forward*,
Springer Water, DOI 10.1007/978-3-319-51856-5_15

Keywords Automated measurements • Continuous measurements • Precipitation • Soil moisture • Soil temperature • Air temperature

1 Introduction

Soil erosion is considered as one of the prime threats worldwide because of the many negative implications it has on humans, society and the environment (Yang et al. 2003). While soil erosion is a natural phenomenon, anthropogenic activities such as agriculture, deforestation and urbanization have accelerated erosion rates (Montgomery 2007). The direct link it has with humans, has led to the exponential increase of the publications on erosion during the last decades (Dekker et al. 2005). In addition, soil erosion prevention has been brought to the forefront and is the reason for the numerous soil conservation efforts worldwide in order to achieve sustainability.

The most obvious problem caused from soil erosion is the loss of the top soil that typically is the most fertile and productive soil (Lal and Stewart 1990; Pimentel et al. 1995). The continuation of loss of the fertile soil will and has had major impacts on agricultural production (substantial decrease) and food cost (substantial increase due to use of more fertilizers). This a serious problem if we incorporate the fact that world's population continues to grow exponentially. In worst cases, agricultural lands are abandoned completely since they become unproductive because of erosion (Young 1998). Soil quality also decreases (by decreasing soil depth, water availability and soil organic matter along with increasing nutrient loss) significantly in natural ecosystems by erosion and can cause major environmental changes to these ecosystems (Pimentel 2006).

The eroded sediment along with the fertilizers and pesticides they carry, typically end up in water bodies. Increased sedimentation and concentrations of the other pollutants alter and degrade the water quality causing serious environmental problems (Bilotta et al. 2012; Dodds 2006; Glendell et al. 2014). In addition, the increased sediment loads along with the increased runoff rates (decreased water holding capacity in eroded soils) have led to the increase of muddy flooding phenomena (Boardman et al. 2006).

Soil erosion has been so severe in some cases, that it has led to land degradation and desertification. This should be no surprise since approximately 75 billion tons of soil are eroded yearly (Pimentel and Kounang 1998). In agricultural areas the soil erosion rates can range from 13 to 40 tons/ha/year (Pimentel and Kounang 1998). Pimentel and Kounang (1998) estimated that soil loss is 13–40 times faster than the soil renewal rate (soil forms slowly). This clearly indicates that soil sustainability will be very difficult to be achieved and significant, effective and efficient conservation efforts need to be implemented.

The new climatic regimes that are being established because of global warming are also expected to impact soil erosion. The new conditions will and have changed rainfall amounts and intensities, number of days of precipitation, ratio of rain to snow, plant biomass production, plant residue decomposition rates, soil microbial activity, evapotranspiration rates and could lead to shifts in land uses (Nearing et al. 2004). These many changes in all these environmental processes make it difficult to exactly predict the impacts of climate change on soil erosion. Still the expected increase in extreme events and particularly the increased rainfall intensity and extended drought periods compared to past conditions (Giupponi and Shechter 2003) should lead to higher surface runoff, higher sediment transport capacity and increased surface soil erosion.

In Europe, the Mediterranean region seems to be the most susceptible to soil erosion, land degradation and desertification (Cerdan et al. 2011). This is due to the natural characteristics of the region, specifically the semi-arid climate, sparse vegetation and steep topography (Davidson 1991; Zaimes et al. 2012). Another major factor is the frequent wildfires in the region that are a natural part of the ecosystem (Shakesby 2011). Burnt areas, especially after the first year of the wildfire, have very high erosion rates, if appropriate conservation measures are not implemented (Neary et al. 2008). Finally, the region has been inhabited by humans for thousands of years that have heavily impacted it (Pearce 1996; Tal 2010). Human inhabitation impacts are very evident in the many unsustainable agricultural lands of the region and the few remaining patches of natural ecosystems. These facts along with potential climate change impacts clearly indicate the need for new tools to mitigate effectively and efficiently soil erosion in the Mediterranean.

To have the most successful conservation efforts, very accurate measurements of erosion are required. Most common soil erosion measurements (e.g. erosion pins, transects) have a accuracy of up to 1 cm (Lawler 1993). In addition, these measurements cannot be taken very frequently. Typically these measurements are taken once every year, or every season although they can also be taken more frequently (every month). More accurate measurements (up to 1 mm) along with continuous measurements would help better understand when erosional processes occur (Lawler 2005). In addition, in order to better understand the mechanism of soil erosion important parameters that influence it should be measure simultaneously. Understanding the mechanisms of erosion can help develop and implement the best suitable conservation practices in different regions.

This study focuses on soil erosion in Greece. In Greece, 26.5% of the country's total land area ($\sim$3.5 million ha) experiences severe soil erosion problems (Mitsios et al. 1995). Despite this fact minimal systematic and holistic efforts have been done to reduce erosion. Institutions in Greece have not used their funds to prepare, design and manage soil conservation measures in the long-term (Barbayiannis et al. 2011). An important first step was the compilation of a soil erosion risk map for Greece (Morvan et al. 2008). In addition, another positive fact is that the number of publications and the research conducted on soil erosion have increased over the years (Koutalakis et al. 2015). Still a strategic management plan for the country is needed due to impending climate change impacts (Koutalakis et al. 2015).

The objective of this study was to develop a system of sensors that measures continuously and very accurately surface soil erosion. This is possible due to the technological evolutions of the last decades that have not been utilized yet as soil erosion assessment tools. Specifically, the new sensor will utilize ultrasonic technology that allows to detect ground level changes continuously with an accuracy of up to 1 mm. The system will also measure other important variables that impact soil erosion directly. These are precipitation, soil and air temperature and soil moisture. Recognizing the exact time an erosional event occured and under what conditions, will provide new insight in soil erosion and help find what the best conservation practices that need to be implemented should be.

2 Study Area

The study area is the island of Thasos in northern Greece (Fig. 1). It was selected because it has a wide range of precipitation, topography (flat areas and steep slopes), different types of vegetation and land-uses, despite occupying a small area, providing an ideal natural laboratory to study surface soil erosion under different conditions. The island is almost circular with a perimeter of approximately 102 km and occupies an area of approximately 378 km^2 (Mallios et al. 2009). The terrain of the inner island is mountainous; the highest peak is Ypsarion with an elevation of 1203 m. The climate is characterized as Mediterranean with an average annual temperature of 15.8 °C and an average annual precipitation of 770 mm (Koutalakis et al. 2014). The island has also

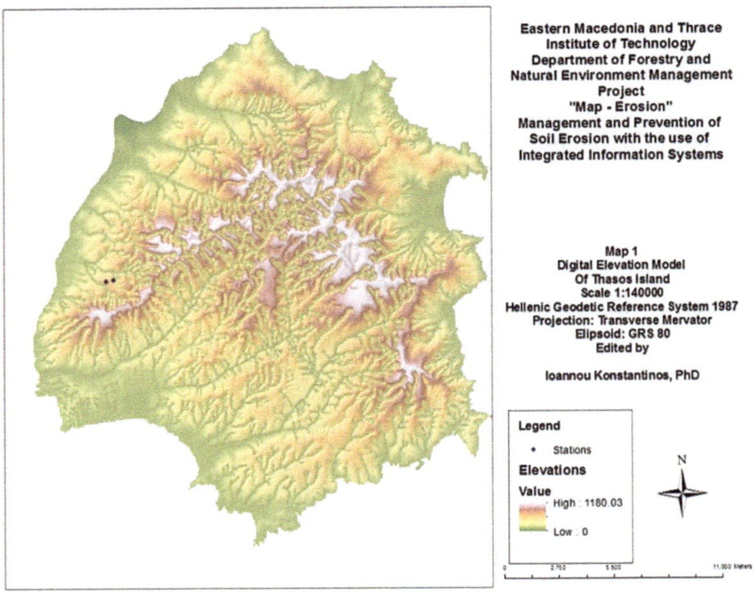

Fig. 1 The study area was Thasos Island. This figure depicts the digital elevation model (DEM) that was developed

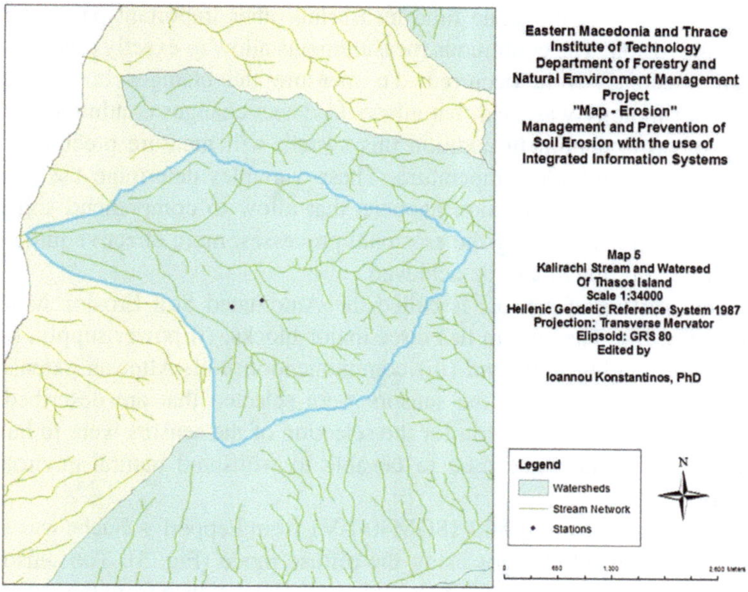

Fig. 2 The two systems of sensors (*circular dots*) were placed in the Kallirahi torrent watershed. The specific watershed was selected because it had experience a wildfire in August 2013

suffered frequent wildfire events during the last 3 decades that burned close to 75% of its wooded areas.

The island has 18 major torrents. Torrents have typically ephemeral or intermittent flow (Emmanouloudis et al. 2011). Compared to streams they have more irregular stream flow (periods of no flow and periods of flash floods), steeper channel slopes and higher sediment transport capacity. The specific watershed selected was Kallirahi torrent that had a wildfire in August of 2013 and its impact is still evident on the landscape (Fig. 2). Watersheds after wildfires tend to have high susceptibility to surface soil erosion.

3 Methods and Materials

The most commonly used erosion measuring techniques do not measure continuously but at specific time intervals (e.g. monthly, seasonally, yearly), although some semi-continuously or continuously measuring techniques have been developed (Lawler 1993, 2005). This is a serious drawback since soil erosion does not happen gradually. Instead it is typically an episodic event that will occur when the proper environmental conditions are established. As a consequence the past measuring techniques do not allow to comprehend erosional processes, completely. Through this study a new system of sensors was developed that measures erosion continuously while at the same time measuring other important parameters from surface soil erosion. The continuous measurements allow to exactly pinpoint in time when an erosional event occurred. To measure the changes in ground level, ultrasonic technology was used that allows to detect changes continuously with an accuracy of up to 1 mm. In addition this system will measure precipitation, soil moisture and air and soil temperature. These variables determine the conditions under which surface soil erosion happens that allow to comprehend soil erosion processes. By comprehending the erosional processes, more effective mitigation of the soil erosion problem can be achieved.

The new system of sensors is called the Automated Soil Erosion Monitoring System (ASEMS). The system has three major blocks: (i) power supply, (ii) measurement and logging units and (iii) communication unit. After an extensive and careful search in the market the sensors were selected that are described in the following paragraph. The criteria for the selection of the sensors were to have high accuracy while at the same time to be able to withstand natural environmental conditions.

The ultrasonic sensor UB800-18GM40-I-V1 from Pepperl + Fuchs was selected to detect ground level changes using the diffuse mode (Fig. 3). The sensor emits ultrasonic pulses and receives the reflected ones. The distance from the sensor to the reflector (ground) is determined by measuring the propagation time.

To measure soil moisture the ML3 ThetaProbe from Delta-T Devices was selected. When the sensor is powered, it creates an electromagnetic wave (similar to FM radio), that is applied into the soil. The sensor is able to detect the influence of soil permittivity to the applied field, resulting in a sufficient measure of soil moisture content (1% volume accuracy). The same sensor is equipped with a 10 KΩ NTC thermistor that can measure soil temperature with an accuracy of 0.5 ° C. Rainfall is recorded by the ADCON's RG1 modern rain gauge (a.k.a. udometer) with an orifice of 200 cm^2. The sensor employs a double tipping bucket assembly

Fig. 3 The UB800 ultrasonic sensor was selected and used

measurement unit. Each tipping action triggers a pulse that equals to 0.2 mm rainfall. Finally the air temperature is measured by a 10 KΩ NTC thermistor from Eliwell. The measuring range is from −40 to 110 °C.

The GP1 data logger from Delta-T Devices Ltd. was selected to perform the measurement procedure and data storage. It is a low power/high accuracy, special purpose data logger, equipped with a secure non-volatile flash memory. Its memory capacity (>500,000 readings) is sufficient enough to store data for several years of continuous recording, depending on the selected measurement time interval.

4 Development, Calibration and Installation

Once the equipment were purchased the testing for the optimal operation was conducted in the laboratory. Based on the laboratory testing a specific placement of the sensors' and wiring to the GP1 data logger was designed for optimal operation (Fig. 4). The ultrasonic sensor was connected to ***CH1*** differential analogue voltage channel, while the soil moisture sensor was connected to ***CH2*** differential analogue voltage channel. Air temperature and soil temperature sensors were connected to ***Temp3*** and ***Temp4*** temperature channels, respectively. The rain gauge was connected to the slow event channel ***Event6***, instead of fast event channel ***Event5***, in order to conserve battery life.

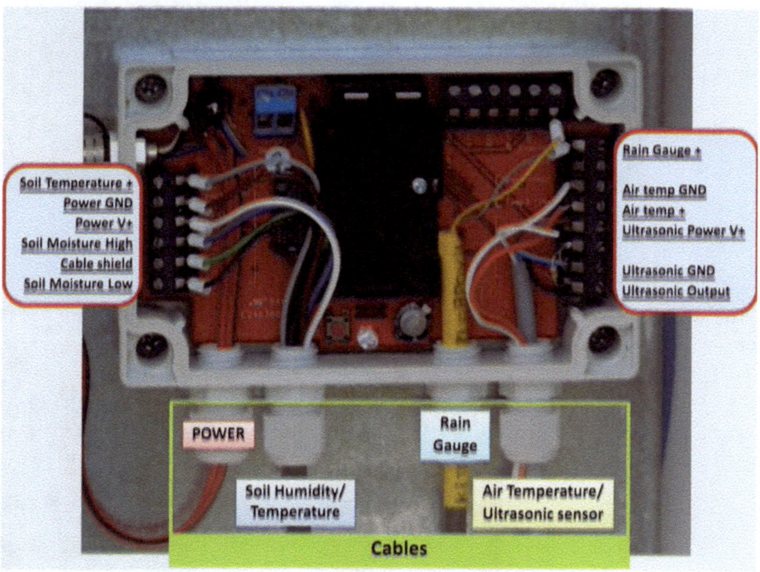

Fig. 4 Sensors' wiring to data logger

Afterwards the ASEMS was place in the field near the Department of Forestry and Natural Environment Management in Drama, Greece. This was also necessary to test how the system works in actual field conditions. After two months, the ASEMS was calibrated and ready to be placed in the study sites on Thasos Island.

The installation procedure was divided into three steps: (i) mast installation, (ii) electronics and solar panel fitting and (iii) sensors' installation. The installation procedure begins with the assembly of the mast (Fig. 5). The assembled mast is positioned near the area of interest and is supported using three equidistant wire ropes as shown in Figs. 5 and 6. When the mast is stabilized, the electronics'

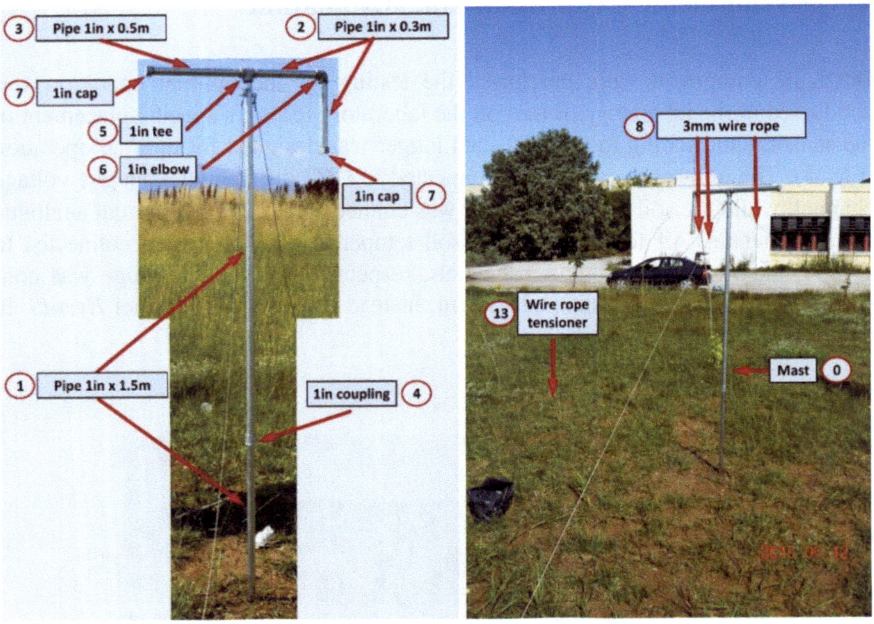

Fig. 5 The mast's assembly (*left picture*) and stabilization (*right picture*)

Fig. 6 The wire rope tensioner

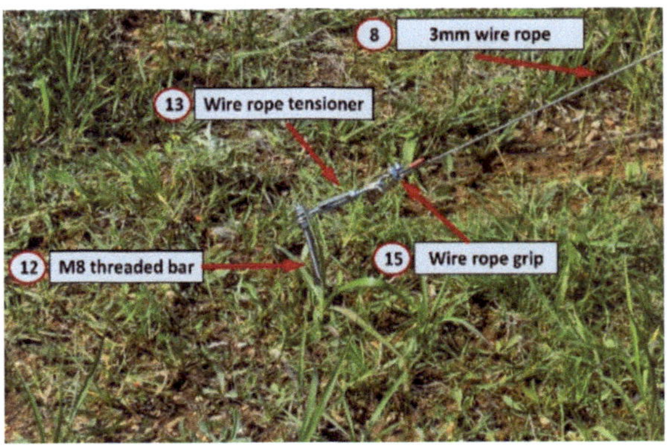

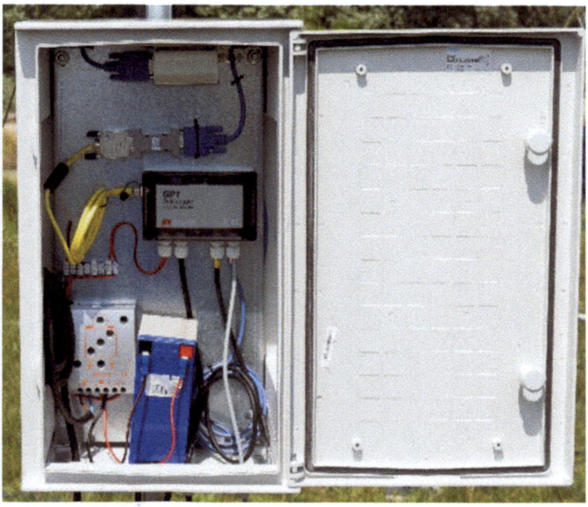

Fig. 7 The electronics' polyester cabinet

cabinet (Fig. 7), the solar panel and some of the sensors can be installed. The polyester cabinet contains all the electronics needed (data logger, modem, battery, etc.).

The polyester cabinet and the solar panel are placed on the mast using the appropriate flat bars and pipe clamps (Figs. 8 and 9). The solar panel is installed facing south and the polyester cabinet is placed in a way that wire ropes do not prevent its door from opening.

The ultrasonic sensor is placed in parallel with the ground level. It is installed in such a way that the expected soil erosion would not affect its stability (Fig. 10). The soil moisture sensor was installed near the ultrasonic sensor, according to manufacturer's instructions. The rain gauge is mounted on the mast using the appropriate hose clamps. Its orifice should be placed parallel to the horizon, for accurate measurements of rainfall (Fig. 11).

After the calibration, two ASEMS were placed in Kallirahi watershed on Thasos Island. This was done in early September of 2015. After extensively surveying the watershed for the most appropriate locations, one ASEMS was installed in an area with tree cover and overall gentle slopes. The second ASEMS was placed in an area with no overstory and with steeper slopes. This way the system would be tested in two very different environments in regards to its effectiveness and functionality. The one site is expected to have high erosion rates (no overstory and steep slopes) while on the other site the erosion rates are expected to be substantially smaller. This would provide the range of potential erosion rates experienced in Greece and in general in the Mediterranean. Finally, to check the proper functionality of the ultrasonic sensor around each of them, 10 erosion pins were placed and are measured, approximately every month. Erosion pins are of the most common methods used by researchers to measure erosion worldwide (Lawler 1993).

Fig. 8 The electronics' cabinet mounting

Fig. 9 The solar panel fitting

Fig. 10 The ultrasonic and soil moisture and temperature sensor installation

Fig. 11 Rain gauge mounting

5 Preliminary Results and Conclusions

Preliminary results show that the ASEMS can provide very accurate data in regards to soil erosion (Fig. 12). Its testing, initially in the laboratory, indicated it could measure the parameters of interest for conditions that exists in the Greek natural environment. Similarly, the field measurements also provided very satisfactory results. Two ASEMS have been placed in the field in September 2015. During this period no significant precipitation events have occured and as a results no major erosional events have been recorded. This was also shown by the measurements of

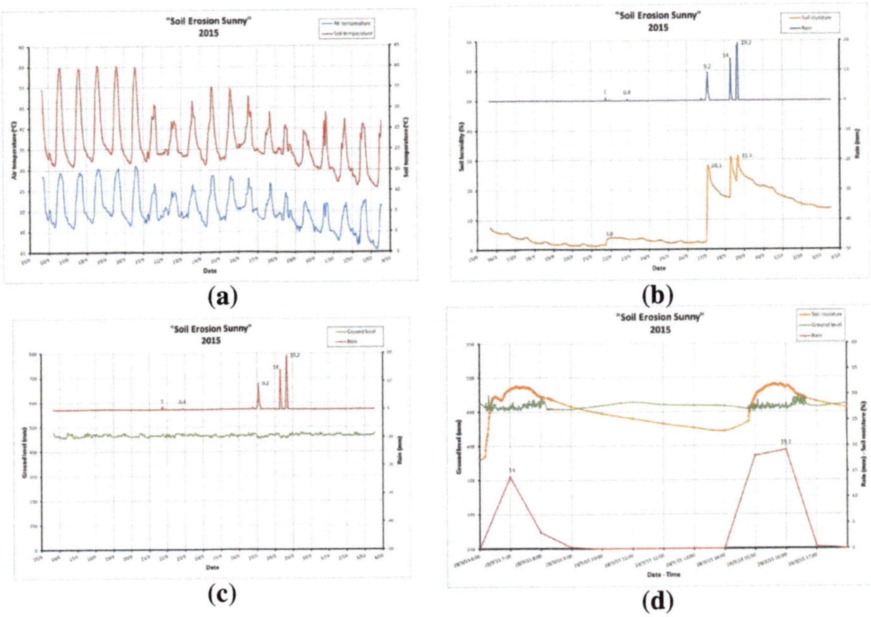

Fig. 12 Graphs from collected data from the station with an overstory tree cover and a shallow slope. **a** Air (*blue line*) and soil (*red line*) temperature (different scales are used to show the temperatures clearly). **b** Rainfall (*blue line*) and soil moisture (*orange line*). **c** Rainfall (*red line*) and ground level (*green line*). **d** Dual rate recording of soil moisture (*orange line*) and ground level (*green line*) triggered by a rainfall event (*red line*)

the erosion pins that surround the ultrasonic sensors. The data collected by one of the ASEMS and the combination of data of the parameters measured are shown in Fig. 12.

In addition, the proposed system offers a very important feature, regarding the logging frequency; called dual rate recording. The data logger is able to change (shorten) the measurement time interval for specific quantities (e.g. soil moisture, soil erosion), when triggered by a predefined parameter (e.g. rainfall) (Fig. 12d). This allows for a more detailed monitoring of the factors correlated to soil erosion. This way the recording could be increased during the erosional event while not having more frequent measurements than needed during the rest of the period. This is a very important feature, since as stated previously, erosion is an episodic event that will happen in a short period of time.

Overall ASEMS has shown great promise. It will measure erosion accurately (up to 1 mm accuracy) while also providing insights on what environmental condition (rainfall, moisture and temperature) cause soil erosion. It should have wide adaptability in the Mediterranean and other regions that face serious erosion problems.

References

Barbayiannis N, Panayotopoulos K, Psaltopoulos D, Skuras D (2011) The influence of policy on soil conservation: a case study from Greece. Land Degrad Dev 22:47–57

Bilotta GS, Burnside NG, Cheek L, Dunbar MJ, Grove MK, Harrison C et al (2012) Developing environment specific water quality guidelines for suspended particulate matter. Water Res 46:2324–2332

Boardman J, Verstraeten G, Bielders C (2006) Muddy floods. In: Boardman J, Poesen J (eds) Soil erosion in Europe. Wiley, Chichester, pp 743–755

Cerdan O, Desprats J-F, Fouché J, Le Bissonnais Y, Cheviron B, Simonneaux V, Raclot D, Mouillot F (2011) Impact of global changes on soil vulnerability in the Mediterranean Basin. In: ASABE—international symposium on erosion and landscape evolution, pp 495–503

Davidson AD (1991) Soil erosion in the Mediterranean Basin. Geography 76(1):71–73

Dekker LW, Oostindie K, Ritsema CJ (2005) Exponential increase of publications related to soil water repellency. Aust J Soil Res 43(3):403–441

Dodds WK (2006) Eutrophication and trophic state in rivers and streams. Limnol Oceanogr 51:671–680

Emmanouloudis D, García Rodríguez JL, Zaimes GN, Giménez Suárez MC, Filippidis E (2011) Euro-Mediterranean torrents: case studies on tools that can improve their management. In: Richards KE (ed) Mountain ecosystems: dynamics, management and conservation. Nova Science Publishers, New York, pp 1–44

Giupponi C, Shechter M (eds) (2003) Climate change in the Mediterranean: socio-economic perspectives of impacts, vulnerability and adaptation. Edward Elgar Publications, Glos

Glendell M, Extence C, Chadd R, Brazier RE (2014) Testing the pressure-specific invertebrate index (PSI) as a tool for determining ecologically relevant targets for reducing sedimentation in streams. Freshwater Biol 59(2):353–367

Koutalakis P, Vlachopoulou V, Zaimes GN, Ioannou K, Iakovoglou V (2014) Assessing soil erosion risk using SWAT and GIS for Thassos Island, Greece. In: Proceedings of the 10th international conference of the Hellenic geographical society, Thessaloniki, Greece

Koutalakis P, Zaimes GN, Iakovoglou Ioannou K (2015) Reviewing Soil Erosion in Greece, World Academy of Science, Engineering and Technology. Int J Environ Chem Ecol Geol Geophys Eng 9(8):816–821

Lal R, Stewart BA (1990) Soil degradation. Springer, New York

Lawler DM (1993) The measurement of river bank erosion and lateral channel change: A review. Earth Surf Proc Land 18:777–821

Lawler DM (2005) Defining the moment of erosion: the principle of thermal consonance timing. Earth Surf Proc Land 30:1597–1615

Mallios Z, Arapaki S, Frantzis A, Katsifarakis KL (2009) Management of water resources of Thassos island in the frame of sustainable development. In: Proceeding of common conferences 11th of EYE and 7th of EEDYP, pp 27–34 (in Greek with an English abstract)

Mitsios J, Pashalidis C, Panagias K (1995) Soil erosion—mitigation techniques to soil erosion. Zymel Editions, Athens

Montgomery DR (2007) Soil erosion and agriculture sustainability. P Natl Acad Sci U S A 104:13268–13272

Morvan X, Saby NPA, Arrouays D, Le Bas C, Jones RJA, Verheijen FGA et al (2008) Soil monitoring in Europe: a review of existing systems and requirements for harmonization. Sci Total Environ 391:1–12

Nearing MA, Pruski FF, O'Neal MR (2004) Expected climate change impacts on soil erosion rates: a review. J Soil Water Conserv 59(1):43–50

Neary DG, Ryan KC, DeBano LF (eds) (2008) Wildland fire in ecosystems: effects of fire on soil and water, USDA. Forest service, general technical report RMRS-GTR-42, pp 250

Pearce F (1996) Deserts on our doorstep. New Sci 151(2037):12–13

Pimentel D (2006) Soil erosion: a food and environmental threat. Environ Develop Sustain 8: 119–137

Pimentel D, Kounang N (1998) Ecology of soil erosion in ecosystems. Ecosystems 1:416–426

Pimentel D, Harvey C, Resosudarmo P, Sinclair K, Kurz D, McNair M et al (1995) Environmental and economic costs of soil erosion and conservation benefits. Science 267:1117–1123

Shakesby RA (2011) Post-wildfire soil erosion in the Mediterranean: review and future research directions. Earth-Sci Rev 105:71–100

Tal A (2010) Desertification. In: Uekötter F (ed) The turning points of environmental history. University of Pittsburgh Press, Pittsburgh, pp 146–161

Yang D, Kanae S, Oki T, Koike T, Musiake K (2003) Global potential soil erosion with reference to land use and climate changes. Hydrol Process 17:2913–2928

Young A (1998) Land resources: now and for the future. Cambridge University Press, Cambridge

Zaimes GN, Emmanouloudis D, Iakovoglou V (2012) Estimating soil erosion in Natura 2000 areas located on three semi-arid Mediterranean islands. J Environ Biol 33:277–282

Part IV
Salinity and Desalination

Heavy Minerals and Granulometric Studies in Coastal Sediments from Keelakarai to Periyapattinam, Gulf of Mannar, East Coast of India

Neelavannan Kannaiyan, M. Suresh Gandhi, R. Elango, G. Sujita and S.M. Hussain

Abstract In this study, granulometric and heavy minerals analysis were carried out in coastal sediment samples from Keelakarai to Periyapattinam, Gulf of Mannar, east coast of India to examine the feasibility of energy conditions that had prevailed during the deposition and its respective heavy mineral assemblages. Out of 40 samples, the backshore area are fine grained, whereas the foreshore area is enriched with mixed grains ranging from coarse to medium grained in size in the alternate samples. Most of the samples were classified under moderately sorted and moderately well sorted in the backshore area, wherein at the foreshore it is poorly to moderately sorted in nature. Most of the sample falls under leptokurtic curve while few other locations fell under mesokurtic. Fine skewed nature is observed in the foreshore samples. Most of the samples are unimodal and few are bimodal. The unimodality at the location reflects the lack of sediment deposition by the rivers. The bimodality is ascribable due to the discharge of fine sediments from small rivers like Gondar and Vaippar. The presence of fine sediments is the reflection of depositional environment which is corroborated with the strong wave divergence that prevails in this region. The coarse grained nature in the foreshore might be due to the prevailing high-energy environment, where the wave is orthogonally converging. Individual heavy mineral weight percentage is analyzed and it varies from 5.5 to 31.55%. The lowest values were noticed in 9 samples, whereas the highest values were observed in the stations 5, 7, and 15 which receives more than 15% of

N. Kannaiyan (✉) · M. Suresh Gandhi · S.M. Hussain
Department of Geology, University of Madras,
Guindy Campus, Chennai 600025, Tamil Nadu, India
e-mail: k.neelavannan@gmail.com

M. Suresh Gandhi
e-mail: msureshgandhi@gmail.com

R. Elango
National Geophysical Research Institute, Hyderabad 500007
Andhra Pradesh, India

G. Sujita
Department of Geology, Anna University, Chennai 600025, Tamil Nadu, India

O. Abdalla et al. (eds.), *Water Resources in Arid Areas: The Way Forward*,
Springer Water, DOI 10.1007/978-3-319-51856-5_16

heavies. Based on the Isodynamic separation studies it is observed that the magnetite and ilmenite were found more at 0.3 A. Further, from the textural analysis, it is clear that the area from Keelakarai to Periyapattinam is generally experiencing the depositional environment as evident from positive skewness with the exception of three locations which are undergoing erosion as evident from negative skewness. This study shows that in the field of placer deposit investigations, it is significant to correlate the techniques like granulometric, Frantz Isodynamic separator and bromoform separation.

Keywords Heavy minerals · Grain size · Coastal sediments · Hydrodynamics · Gulf of mannar

1 Introduction

The Indian Coast of 6500 km stretch is noticeable by the accumulation of different categories and grades of placer deposits. Heavy mineral deposits are mostly concentrated in the states of Tamil Nadu, Kerala, Orrisa and Andhra Pradesh. The beach placer sand deposits of Manavalakurichi in Tamil Nadu, Chavara in Kerala and Chhatrapur in Orrisa are under active exploitation by Indian Rare Earth Ltd. (IREL), Govt. of India (Mohapatra 2015). The beach placer minerals like garnet, ilmenite, zircon, rutile, and monazite which are progressively used in conservative and high-tech uses are in abundant demand in the worldwide market, as there is no sustainable substitute for them available. Heavy mineral studies are useful in understanding the provenance and maturity of the sediments, and energy conditions that prevailed during the deposition and the subsequent environment confined in the depositional basin. The heavy mineral distribution has also been used to deduce the additional increment of heavy mineral from offshore, from Keelakarai to Periyapattinam along the southeastern coast of India. A part from detailed Granulometric and occurrence of beach placers studies carried out by Angusamy and Rajamanickam (2006), this work has been done to understand the magnetic and non-magnetic minerals of beach sands and assemblage of heavy minerals. This paper aims to highlight granulometric-assemblage of heavy mineral in backshore and foreshore and their significance for understanding the provenance of sediments between Keelakarai to Periyapattinam. This study will help us to select beneficiation strategy of the coastal sands.

2 Study Area

The present study area is a part of the east coast of India and the southern part of Taminadu, Gulf of Mannar. The study area extending from Keelakarai to Periyapattinam (20 Km) are shown (Fig. 1). The Government of India declared

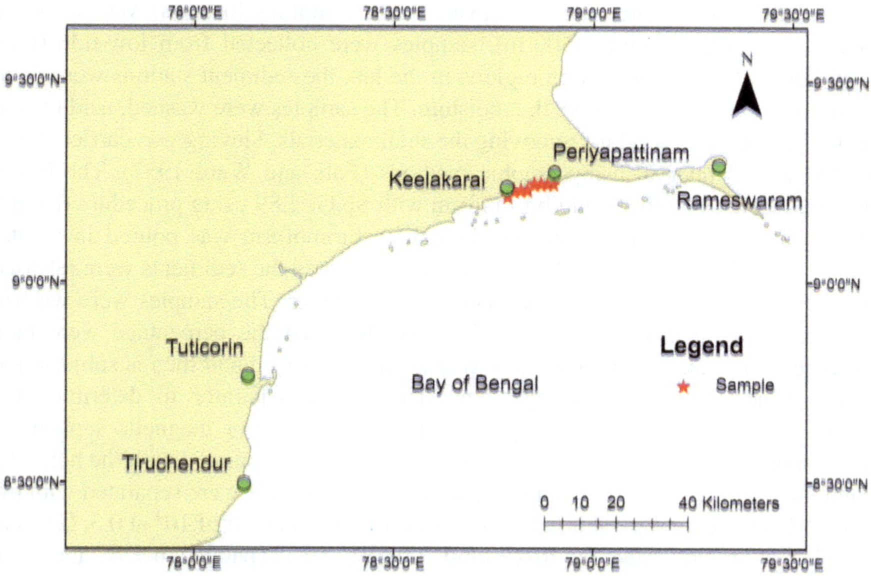

Fig. 1 Study area location map

Gulf of Mannar as one of the main bio-reserve forest, it consists of 21 coral islands in-between Rameswaram and Tuticorin. The Gulf of Mannar coastal zone receives rainfall from North-east monsoon. It is bound in the northeast by Ramehswaram Island east by Bay of Bengal and south by Kanyakumari. The bedrock is located along the coastal areas as isolated patches and pockets with greater intervals between one and the other formation. They are coming under southern part of South Indian granulitic terrain, which covers part of Madurai Block (MB) and the Kerala Khondalite Belt (KKB) and it is mainly composed of charnockites and gneisses. The beach terrace and majority of the bedrock which occurs along the south-eastern coast of Tamilnadu were formed during the Pleistocene and Holocene to present (Sahayam 2010). In this study area, bedrock is predominantly occurring in the inter-tidal regions and predominantly segregated by corals, shell fragments and other marine organisms. The drainage pattern of the area is mainly controlled By Vaigai, Gondar, Vaippar and Vembar river. The study area and sampling locations are shown in Fig. 1.

3 Methodology

A total of 40 sediment samples between Keelakarai anad Periyapattinam beach during the month of January, 2013. The samples were collected with a hand auger from the low-tide, high-tide and berm regions to a depth of 25 cm, maintaining an interval of approximately 1 km between each sampling location. Wherever the beach was wide enough (>500 m), samples were collected from low-tide (LT), high-tide (HT) zones and berm regions In the lab, the sediment stations were dried in an oven at 60 °C to remove the moisture. The samples were washed, acid treated with 10% HCl and dried for removing the shell materials. Sieving was carried out in ASTM at ½ϕ interval using graphic methods (Folk and Ward 1957). The heavy minerals were separated using bromoform with Sp.Gr 2.89 using procedure (Milner 1962) each fraction of the sample along with bromoform was poured in a filter paper. After the heavy and light mineral were separated the sediments were released by opening the cork of the funnel onto the filter paper. The samples were washed with methanol and kept for air drying, weighed and the percentage were then tabulated. The samples are prepared as magnetite-free feed and then is subjected to heavy mineral separation using Frantz Isodynamic separator to determine the behaviour of each mineral during magnetic separation. For magnetic separation, first magnetite and other strong magnetic minerals were separated with the help of a hand magnet, then later remaining magnetic heavy minerals were separated with the help of Isodynamic separator using side slope of 25° and a tilt of 10° at 0.3, 0.5, 0.8, and 1.2 A as suggested by Rosenblum (1958). The separated heavy and light minerals were weighed and their percentage was calculated by using the binocular microscope as referred by Sreenivasa (2014).

4 Results and Discussion

4.1 Granulometricstudies

4.1.1 Mean (Mz)

The backshore phi mean value ranges from 1.71 to 2.57ϕ with the average mean of 2.38ϕ are shown value range in Fig. 2, and Table 1. The majority of the sediment (99%) samples are categorised under fine sand and the rest in medium sand,

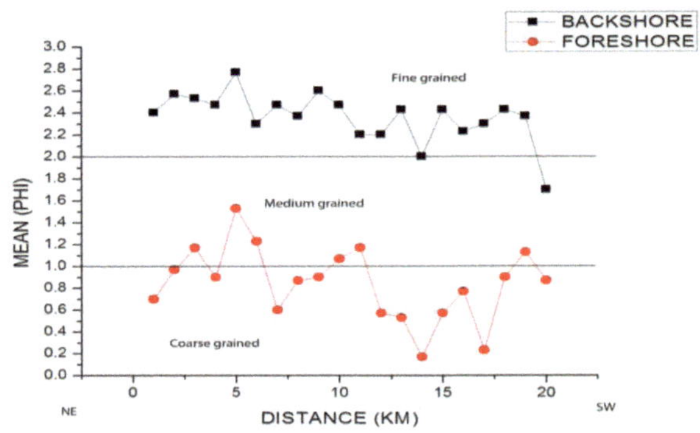

Fig. 2 Mean variation in backshore and foreshore samples

Table 1 Statistical analysis of backshore samples

Stn. No	Mean (Mzϕ)	Std. Dev	Skewness (Ski)	Kurtosis (KG)	Remarks
1	2.4	0.79	0.08	1.33	Fine grained, moderately sorted, near symmetrical, leptokurtic
2	2.57	0.68	0.06	1.28	Fine grained, moderately well sorted, near symmetrical, leptokurtic
3	2.53	0.73	0.02	1.28	Fine grained, moderately well sorted, near symmetrical, leptokurtic
4	2.47	0.77	−0.07	1.18	Fine grained, moderately sorted, near symmetrical, leptokurtic
5	2.77	0.68	0.07	1.13	Fine grained, moderately well sorted, near symmetrical, mesokurtic
6	2.34	0.93	−0.19	1.69	Fine grained, moderately sorted, coarse skewed, very leptokurtic
7	2.47	0.83	−0.05	1.11	Fine grained, moderately sorted, near symmetrical, leptokurtic
8	2.37	0.74	−0.02	1.28	Fine grained, moderately well sorted, near symmetrical, leptokurtic
9	2.6	0.69	0.12	1.41	Fine grained, moderately well sorted, coarse skewed, leptokurtic
10	2.47	0.54	−0.08	1.23	Fine grained, moderately well sorted, near symmetrical, leptokurtic
11	2.25	0.75	−0.12	0.93	Fine grained, moderately sorted, coarse skewed, mesokurtic
12	2.27	0.77	−0.15	1.07	Fine grained, moderately sorted, near symmetrical, mesokurtic
13	2.43	0.72	−0.03	1.64	Fine grained, moderately well sorted, near symmetrical, very leptokurtic
14	2.34	0.93	−0.29	1.01	Fine grained, moderately sorted, coarse skewed, mesokurtic
15	2.43	0.73	−0.09	1.14	Fine grained, moderately sorted, near symmetrical, leptokurtic
16	2.23	0.82	−0.13	1.04	Fine grained, moderately sorted, near symmetrical, mesokurtic
17	2.31	0.71	−0.14	1.14	Fine grained, moderately well sorted, coarse skewed, leptokurtic
18	2.43	0.56	0.06	1.29	Fine grained, moderately well sorted, near symmetrical, leptokurtic
19	2.37	0.65	−0.11	1.46	Fine grained, moderately well sorted, coarse skewed, leptokurtic
20	1.71	1.02	−0.05	0.91	Medium grained, poorly sorted, near symmetrical, mesokurtic

the mean size indicates that the fine sands were deposited at a moderately low energy conditions and the medium sand were deposited at moderate energy condition.

The foreshore phi mean value ranges from 0.7 to 1.53ϕ with average mean of 0.84ϕ. The majority of the sediments (70%) samples are categorised under coarse sand and the rest in medium sand, values are shown in Fig. 2, and Table 1. The mean size indicates that the coarse sand sediments were deposited at a high energy conditions, and the medium sand sediments were deposited at a moderate energy conditions as suggested by Singarasubramanian et al. (2006).

4.1.2 Standard Deviation (σ1)

This measures the sorting of sediments in nature, and indicates the fluctuations in the velocity conditions of the depositing agents (Sahu 1964). For the backshore sediments, standard deviation values ranges between 0.54 and 1.02Φ with an average value of 0.75Φ the values ranges as shown in Fig. 3, and Table 1. The sorting of sediment value ranges from moderately well sorted to moderately sorted in nature. About 50% of the samples fall under moderately well-sorted, while rest of the samples are represented as well sorted. The foreshore values of the sediments range between 0.71 and 1.45Φ with an average value of 1.01Φ values are shown in Fig. 3, and Table 2. The sorting of sediments range from moderately sorted to poorly sorted. About 60% of the samples fall under moderately sorted, while rest of the samples are represented by poorly sorted. Sediments discharged from the river mixes with offshore sediments and with the sediments eroded from a source rock. This type of sorting of sediments may be due to the impact of intermixing and influx of the sediments from both the sea as well as the river (Angusamy and Rajamanickam 2006; Venkatraman et al. 2011).

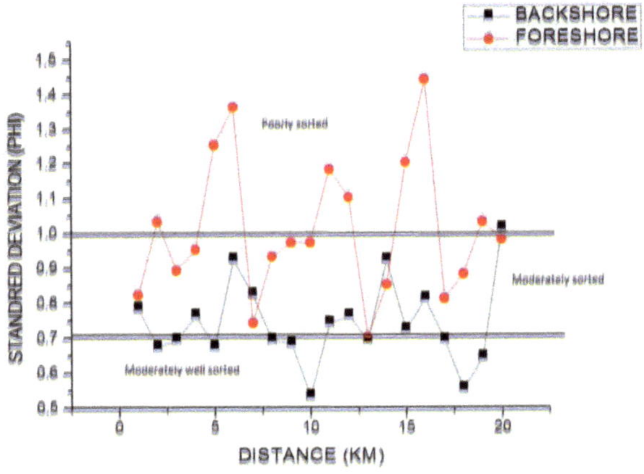

Fig. 3 Standard deviation variation in backshore and foreshore samples

Table 2 Statistical analysis of foreshore samples

Stn. No	Mean	Std. Dev	Skewness	Kurtosis	Remarks
1	0.7	0.83	0.17	1.54	Coarse grained, moderately sorted, fine skewed, very leptokurtic
2	0.97	1.04	0.25	1.17	Coarse grained, poorly sorted, fine skewed, leptokurtic
3	1.17	0.91	0.19	0.95	Medium grained, moderately sorted, fine skewed, mesokurtic
4	0.93	0.96	−0.44	0.79	Coarse grained, moderately sorted, strongly coarse skewed, platykurtic
5	1.53	1.26	−0.09	0.82	Medium grained, poorly sorted, near symmetrical, platykurtic
6	1.23	1.37	0.29	1.13	Medium grained, poorly sorted, fine skewed, leptokurtic
7	0.64	0.75	0.12	1.43	Coarse grained, moderately sorted, fine skewed, leptokurtic
8	0.87	0.94	0.24	1.16	Coarse grained, moderately sorted, fine skewed, leptokurtic
9	0.93	0.98	0.15	0.97	Coarse grained, moderately sorted, fine skewed, mesokurtic
10	1.07	0.98	0.08	0.94	Medium grained, moderately sorted, near symmetrical, mesokurtic
11	1.17	1.19	0.28	0.94	Medium grained, poorly sorted, fine skewed, mesokurtic
12	0.57	1.11	0.14	1.17	Coarse grained, poorly sorted, fine skewed, leptokurtic
13	0.53	0.71	0.13	1.76	Coarse grained, moderately sorted, fine skewed, very leptokurtic
14	0.17	0.86	0.03	1.19	Coarse grained, moderately sorted, near symmetrical, leptokurtic
15	0.57	1.21	0.12	1.26	Coarse grained, poorly sorted, fine skewed, leptokurtic
16	0.77	1.45	0.07	0.77	Coarse grained, poorly sorted, near symmetrical, platykurtic
17	0.23	0.82	−0.06	1.04	Coarse grained, moderately sorted, near symmetrical, mesokurtic
18	0.93	0.89	0.17	1.06	Coarse grained, moderately sorted, fine skewed, mesokurtic
19	1.13	1.04	0.02	0.93	Medium grained, poorly sorted, near symmetrical, mesokurtic
20	0.87	0.99	0.19	1.01	Coarse grained, moderately sorted, fine skewed, mesokurtic

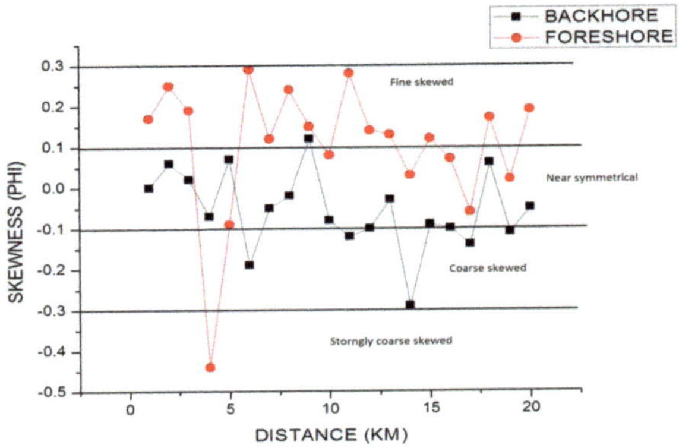

Fig. 4 Skewness variation in backshore and foreshore samples

4.1.3 Skewness (Ski)

This measures the asymmetry of a frequency distribution. The backshore skewness values range between −0.19 and 1.12 with an average value of −0.05 are shown in Fig. 4, and Table 1. The symmetry of the samples varies from near symmetrical skewed to coarse skewed. The foreshore skewness values range between −0.44 and 0.29 with an average value of 0.10 are shown in Fig. 4, and Table 2. The skewness ranges from fine to near symmetrical sediment character. The fine skewed and near symmetrical skewed sediments generally mark the beginning of deposition of fine substance or elimination of coarser portion (Friedman 1961; Duane 1964). The majority of fine skewed nature of sediments indicates that there had been more of riverine contribution. The (50%) positive skewness of sediments indicate, the unidirectional transport or the deposition of sediments in sheltered, low energy environment condition (Brambati 1969).

4.1.4 Kurtosis (KG)

The backshore sediments, the kurtosis values range between 0.91 and 1.69 with an average value of 1.22 are shown in Fig. 5, and Table 1. Followed by the foreshore sediments, kurtosis values range between 0.77 and 1.76 with an average value of 1.10 are shown in Fig. 5, and Table 2. The majority (70%) of the samples fall under leptokurtic to the mesokurtic in nature. Friedman (1962) suggested that extreme high or low values of kurtosis involve that part of the sediment achieved its sorting elsewhere in a high energy environment. The dominance of leptokurtic character of sediments replicate maturity of the sand and variation in the sorting values that are mostly due to continuous addition of finer and coarser materials in changing proportions (Rao et al. 2001).

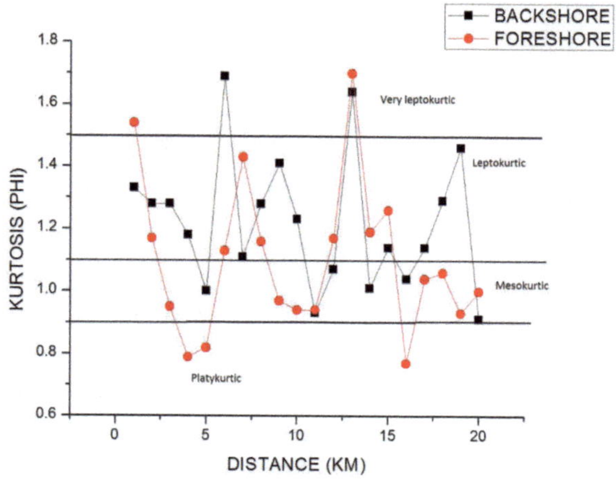

Fig. 5 Kurtosis variation in backshore and foreshore samples

5 Heavy Mineral Studies

5.1 Separation Using Bromoform

A significant amount of the overall bulk heavy mineral fraction was obtained by heavy liquid separation using bromoform as referred by Raslan (2009). From the samples analyzed, the highest heavy mineral concentration was found to be 31.55% in 5th sample, and the lowest heavy minerals concentration was found at 10, 11, 17, 18 and 19th sample. The heavy mineral contents in percentages are given in Fig. 6. Although relative abundance of the heavy mineral component showed general variation along the beach, the highest concentration was present in the 5th sample may be due to the riverine input.

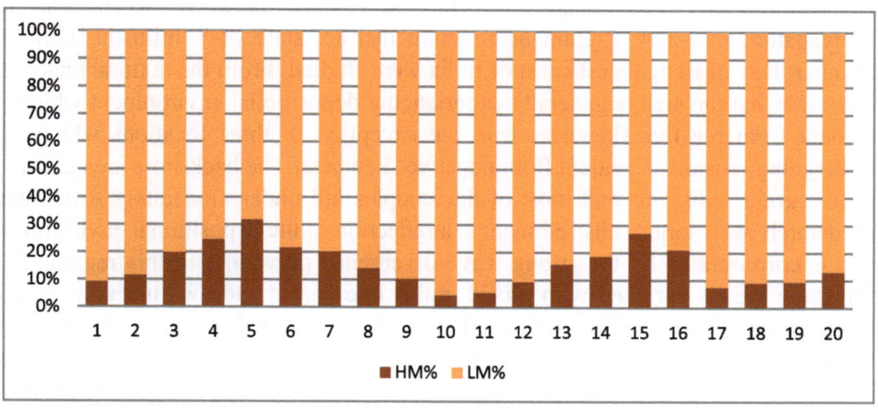

Fig. 6 Percentage of heavy and light mineral separated using bromoform

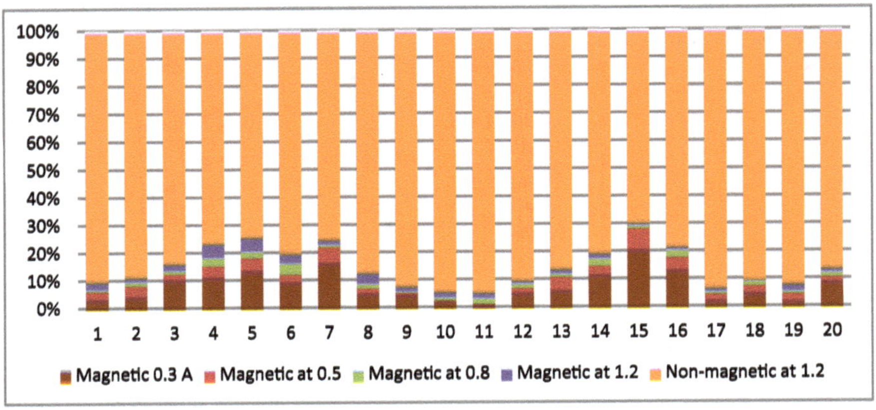

Fig. 7 Minerals separated by Frantz isodynamic separator

6 Frantz Isodynamic Separator

The setting of the separator during this study was (25°) sideslope and (10°) tilt slope. The investigated heavy mineral concentrates, as the magnetic fraction in the range of 0.3, 0.5, 0.8, and 1.2 A of the Frantz Isodynamic separator. Magnetic separation of magnetic minerals and non-magnetic minerals fraction at different current intensities are presented in Fig. 7. From the overall studies, our findings also corroborated with Loveson and Rajamanickam (1988), Angusamy and Rajamanickam (2000).

7 Conclusion

Heavy minerals assemblage in grab sediments between Keelakarai to Periyapattinam a part of Ramanathapuram district, state of Tamilnadu, Gulf of Mannar, Bay of Bengal east coast of India were studied. From the textural analysis, it is clear that the area is generally experiencing depositional environment which is evident from positive skewness with the exception of three locations which are undergoing erosion as evident from negative skewness. The backshore area consists of fine grain sediments and most of the samples fall under moderately sorted and moderately well sorted, this sediment is a reflection of the depositional environment this is corroborated by the strong wave divergence prevailing in this region. The foreshore area is enriched with mixed grains like coarse and medium grain in the alternate sample and it is poorly sorted to moderately sorted in nature which is due to the prevailing high-energy environment, where the wave orthogonally converge. Most of the sample comes under leptokurtic expect in few locations where it is mesokurtic in nature, suggesting that extreme high or low values of kurtosis involve the part of the sediments achieving its sorting elsewhere in a high energy environment. Most of the samples are unimodal and few are bimodal. The unimodality at this location reflects the lack of sediment deposition from the rivers. The bimodality is related to the discharge of fine sediments from smaller rivers like Gundar and Vaippar. Heavy mineral of each sample is analysed and it various from 5.5 to 31.55% average of which has more than 15% of heavies. The higher heavies found in this region may be due to the configuration of the coastline and the slope angle which may favor the enrichment of placer in this region. This study shows the distribution of the heavy minerals in the depositional environments using a techniques such as granulometric analysis and heavy mineral analysis.

Acknowledgements Authors are thankful to Professor and Head, Department of Geology, University of Madras for grant permission to carry out this work and providing lab facilities.

References

Angusamy N, Rajamanickam GV (2000) Distribution of heavy minerals along the beach from Mandapam to Kanyakumari, Tamilnadu. J Geol Soc India 56:199–211

Angusamy N, Rajamanickam GV (2006) Depositional environment of sediments along the southern coast of Tamil Nadu, India. Oceanologia 48:87–102

Brambati A (1969) Stratigraphy and sedimentation of Siwaliks of North Eastern India. In: Proceedings of international seminar in intermontane basins: geology and resources. Chiang Mai, Thailand, pp 427–439

Duane DB (1964) Significance of skewness in recent sediments, Western Pamlico Sound. North Carolina. J Sediment Petrol 34:864–874

Folk RL, Ward MC (1957) Brazos river bars: a study in the significance of grain size parameters. J Sediment Petrol 27:3–27

Friedman GM (1961) Distiction between Dune, Beach and river sands from textural characteristics. J Sediment Petrol 31:514–529

Friedman GM (1962) On sorting, sorting coefficients and the log normality of the grain size distribution of sandstones. J Geol 70(6):737–755

Loveson VJ, Rajamanickam GV (1988) Progradation as evidenced around a submerged ancient port, Periapatnam, Tamil Nadu, India. Ind J Landscape Syst Ecol Stud 12:94–98

Milner HB (1962) Sedimentary petrography, vol 1. George Allen and Union Ltd., pp 643–715

Mohamed Fahmy Raslan MF (2009) Mineralogical and minerallurgical characteristics of Samarsakite-Y, Columbite and Zircon from stream sediments of the Ras Baroud area, Central Eastern Desert Egypt. In: The scientific papers of the Institute of mining of the Wroclaw University of technology, Wroclaw, Poland, No. 128. Mining and Geology XII, pp 179–195

Mohapatra S, Behera P, Das SK (2015) Heavy mineral potentiality and alteration studies for ilmenite in Astaranga beach sands, district Puri, Odisha, India. J Geosci Environ Prot 3:31–37

Prabhakara Rao A, Anilkumar V, Yugandhar Rao A, Ravi GS, Krishnan S (2001) Grain size parameters in the interpretation of depositional environments of coastal sediments between Bendi Creek and Vamsadhara river, East Coast, India. J Ind Assoc Sedimentol 20(1):106–116

Rosenblum S (1958) Magnetic susceptibility of mineral in Frantz isodynamic magnetic separator. Am Minerologist 43:170–173

Sahayam Dajkumar (2010) Distribution of arsenic and mercury in subtropical coastal beachrock, Gulf of Mannar, India. J Earth Syst Sci 119:129–135

Sahu BK (1964) Depositional mechanism from the size analysis of clastic sediments. J Sediment Petrol 34:73–83

Singarasubramanian SR, Mukesh MV, Manoharan K, Murugan S, Bakkiaraj D, John Peter A (2006) Sediment characteristics of the M-9 tsunami event between Rameswaram and Thoothukudi, Gulf of Mannar, Southeast Coast of India. Sci Tsunami Hazards 25:160–172

Sreenivasa A, Jayasheela HM, Bejugam P, Gujar AR (2014) Heavy mineral studies of beach sands of Vagathor, North Goa, India. Int J Mod Eng Res (IJMER) 4:70–78

Venkatraman S, Ramkumar T, Anitha Mary I, Ramesh GR (2011) Variations in texture of beach sediments in the vicinity of the Tirumalairajanar river mouth of India. Int J Sedim Res 26:460–470

Aquatic Mollusks: Occurrences, Identification and Their Use as Bioindicators of Environmental Conditions (Salinity, Trace Elements and Pollution Parameters) in Jordan

Ikhlas Alhejoj, Klaus Bandel and Elias Salameh

Abstract Mollusks living in surface water in Jordan can be used as indicators of water quality. They are predominantly represented by gastropods and only locally by bivalves. Within this framework, gastropod species of the genera *Theodoxus*, *Melanopsis*, *Melanoides*, *Physa*, *Bulinus*, *Galba*, *Hydrobia*, *Ovatella*, *Falsipyrgula*, *Prosostenia*, *Pseudamnicola* proved to be useful. Most commonly encountered are *Theodoxus*, *Melanopsis Melanoides* and *Physa*, and their presence alone presents good information about the quality of the water body they are living in. The results of the study indicate that *Theodoxus* and *Melanopsis* represent indicators of pure and clean fresh water while *Physa* is an indicator of polluted water. Of very special interest is the fauna in King Abdullah Canal influenced by the fauna of Lake Tiberias, represented by *Falsipyrgula*, and the bivalves of the two species of *Unionidae* and of *Corbicula*. Species of *Falsipyrgula*. *Pseudamnicola*, *Melanoides*, and *Prosostenia* tolerate increasing water salinity which when reaching sea water salinity *Ovatella* becomes present. Locally *Pseudamnicola* documents influence of mineral water which can tolerate high Sr^{2+} and Br^{-} concentrations. Other members of Gastropods, especially *Planorbis*, *Gyraulus* and *Ancylus* have almost disappeared from running or standing fresh water in Jordan because of increasing pollution. The bivalves, *Unio*, *Pisidium*, *Sphaerium*, and *Corbicula* are found confined to freshwater environment. With exception of King Abdullah Canal where they used to be useful indicators for the quality of its water, they are rare in Jordanian waters. The study concludes that aquatic mollusks, mainly gastropods, can easily be

I. Alhejoj (✉) · E. Salameh
Department of Geology, The University of Jordan, 11942 Amman, Jordan
e-mail: ekl_hjouj@yahoo.com; i.alhejoj@ju.edu.jo

K. Bandel
Geologisch-Paläontologische Institut und Museum Hamburg,
Universität Hamburg, 20146 Hamburg, Germany

O. Abdalla et al. (eds.), *Water Resources in Arid Areas: The Way Forward*,
Springer Water, DOI 10.1007/978-3-319-51856-5_17

used as biological indicators of environmental quality of surface water bodies. It is a fast, easy and trustful way of observing changes taking place in the environment, especially salinity, added chemical and pollution parameters.

Keywords Biological indicators · Aquatic · Mollusks · Salinity · Pollution · Environmental assessment

1 Introduction

In semi-arid countries, such as Jordan, unpredictable seasonal rainfall and scarcity of water due to increasing demand as a result of population growth represent a great problem and challenge for scientists, planners and policy-makers. Surface water resources in Jordan have been, during the last few decades, negatively impacted by development through the addition of urban, industrial, and sewage wastes to the environment. In this study, the qualities of surface water bodies from chosen locations were studied and found reflected in the types of mollusks species which are living and surviving in them. The prevailing conditions make it imperative to protect Jordan's surface water resources in order to keep these aquatic systems in healthy and productive conditions. This must also apply to other countries with similar environmental conditions.

Aquatic mollusks in Jordan are mainly represented by gastropods and rarely bivalves with about 20 species. They were found on rocks, logs, or dead leaves in springs, river, creeks, canals and lakes with different environmental qualities in fresh and brackish, running and standing, clean and polluted water bodies. Seven species of non-pulmonates plus a number of varieties of *Melanopsis*, in addition to four pulmonates were studied in the present work. Species of gastropods fauna including *Theodoxus*, *Melanopsis*, *Melanoides*, *Physa*, *Bulinus*, *Galba*, *Hydrobia*, *Ovatella*, *Falsipyrgula* and *Pseudamnicola* were identified and studied, while *Planorbis*, *Gyraulus* and *Ancylus* have become so rare in Jordan beyond being of use for the present study. Bivalves were documented with species of the genera *Unio*, *Pisidium*, *Sphaerium*, and *Corbicula*. The idea to use mollusks as biological indicators is based on the presence and absence of distinct species of mollusks and can be utilized to characterize the quality of the environment in which they live. These animals are restricted in living a specific habitat and provide information about the quality of the water where they live. The advantage of this technique is that organisms give information about the environmental quality without the need of classical time-consuming analysis and are thus rapid biomonitoring techniques. Internationally several authors worked on the use of mollusks as biological indicators to determine environment quality status. Pollution of aquatic bodies are widely applied all over the world (Hilsenhoff 1988; Rosenberg and Resh 1993;

Mandaville 2002). Bandel and Salameh (1981) and Alhejoj et al. (2014) studied the indictor values of different organisms in aquatic bodies of Jordan and they noted that these organisms can be considered as an available tool for water quality assessment. Bandel (2000, 2001) studied the mollusk in Jordan and Schütt (1983) determined and described those that had been collected 1978 from Jordan.

This work tries to identify and discuss the occurrence of aquatic mollusks in Jordan: surface water bodies and their value as biological indicators for the quality of their environments by correlating the occurrences of mollusks in springs, creeks, rivers and ponds with the quality of the water in which they live and to put into relation the abundance of certain species with various parameters of water quality, thus relate chemical and physical parameters with biological ones. Also it will highlight the impact of salinity, traces metal, contamination and other stresses on the existence and distribution of mollusks.

2 Materials and Methods

In this study, the occurrences of aquatic mollusks are investigated to identify the environmental quality by correlating their abundances with various physical and chemical parameters of water: temperature (°C), acidity (pH), and electric conductivity (EC). All these parameters were measured directly in the field while biochemical oxygen demand (BOD5), chemical oxygen demand (COD) and ionic composition (e.g., Ca^{2+}, Mg^{2+}, HCO_3^-, Cl^-, NO_3^-, PO_4^3, F^-, SO_4^{2-}, Sr^{2+}, Br^-, etc.) were determined in the laboratory. Samples of mollusks were collected from 19 sites along 12 localities, in the area extending from the Yarmouk River in the north through Zarqa River, Karama reservoir-area, Wadi Shueib, Wadi Hibsan, Wadi Shita, King Abdullah Canal (KAC), and Wadi Atun to Wadi Mujib in the south (Fig. 1) during the late winter to early spring time, from December 2010 to November 2014. This field work was carried out without advanced equipments due to the shallow water of streams and rivers in Jordan and the relatively large size of mollusks. Collecting devices simply consisted of a mesh-screen where mollusks were washed from pebbles, sand or vegetation. The collected organisms were isolated and determined according to their place in the taxonomic system usually down to species level. The freshly collected organisms were documented in the laboratory of the University of Jordan by using an S6 D Leica stereomicroscope with magnification of 6.3×–40× with connected digital camera. The micro structural and morphological features of the mollusc shells were investigated under a scanning electron microscopy (EVO50-Zeiss), in the University of Hamburg.

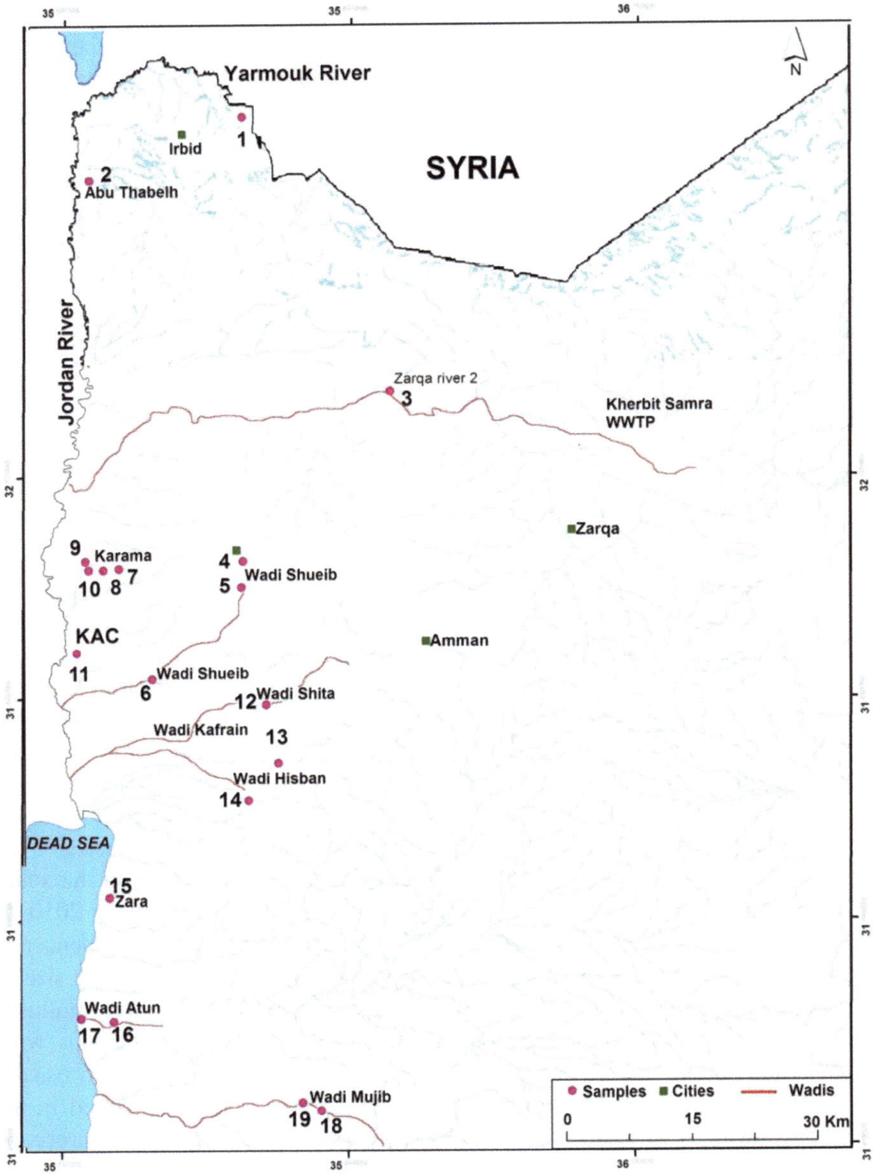

Fig. 1 Location map of the studied sites which cover wadis pouring into the Jordan Valley and the Dead Sea, including the Yarmouk River, Zarqa Rive, Karama reservoir, Wadi Shueib, Wadi Hibsan, Wadi Shita, King Abdullah Canal (KAC), Wadi Atun, and Wadi Mujib

3 Overview About the Mollusks of Jordan

Living individuals of the common Gastropoda belonging to the genera *Theodoxus*, *Pseudamnicola*, Prosostenia, *Hydrobia*, *Falsipyrgula*, *Melanopsis*, *Melanoides*, *Ovatella*, *Galba*, *Bulimus*, and *Physa* were encountered during this study, while *Planorbis*, *Valvata* and *Gyraulus* were noted before (Schütt 1983).

Theodoxus jordani (Sowerby, 1832) occurs in springs and creeks in Jordan along wadis pouring into Dead Sea area (Bandel 2001, Figs. 32–35). The adult shell (teleoconch) consists of about 2.5 whorls and often shells color changing after hatching from the egg capsule from white to black. The black variety is the most common in Jordan, but there are also white ones in caves, or ornamented individuals as in King Abdullah Canal. The aperture (shell mouth) can be closed tightly with a calcareous operculum and this also protects efficiently from crab attacks. The operculum is calcified and provided with a ridge and a peg representing a solid structure on the inner side of the inner lip.

Everywhere within a population of *Theodoxus* distinct and egg cases are quite conspicuous. Often the presence of *Theodoxus* in a stream can be recognized by the presence of the egg capsules attached to hard substrates, stones as well as shells of other gastropods (Bandel 1982, 2001).

Food for intra-capsular development to miniature juvenile stems from nurse eggs which are provided within the egg capsule. The young within such a capsule may survive pollution events that may occur in a creek and which kill all adult specimens. Thus, sometimes, as was observed in Wadi Hisban, only tiny juveniles are found, that had hatched after such a pollution event (Fig. 2a). Egg capsules have cupola-like shape with particles included in the convex upper lid with the flat lower one attached to the substrate (Fig. 2b). When the youngs have developed to the stage of hatching the upper part of the capsule detaches from the lower part and the youngs crawl off. Since development is by nurse eggs, many eggs are originally

Fig. 2 *Theodoxus* from Wadi Hisban creek (2012) **a** under scanning electronic microscope and **b** in the field with their egg capsules

contained in the capsule and of these most serve as food for the few youngs during development. When hatching the shell is solidly calcified and the aperture can be tightly sealed with a calcareous operculum.

Well-fed adults produce many egg capsules and from each of them hatch 3–5 young which begin their feeding on algal crust from the begin of their benthic life. The food of *Theodoxus* consists of algal covers on hard substrates and the snails are very efficient in scraping the rock surfaces. Their many radula teeth used by this process are well equipped for scraping since they are continuously replaced by new ones and some of them in each row are composed in part of iron-oxide and thus harder than the others. *Theodoxus* can be so efficient that *Melanopsis* that normally occurs alongside is found only at the margins of the creek, where softer plant material and sediment is found, while on stones only *Theodoxus* is feeding. On the concrete walls of the KAC (Site 11) in the portions with very clean water *Theodoxus* is predominant, and further downstream in less clean water *Melanopsis* takes its place and is indicative of some small degrees of pollution.

The family Thiaridae is represented by *Melanoides* Olivier, 1804 with *Melanoides tuberculata* Müller, 1774. The elongate shell is up to 5 cm long with high and slender spire and up to 15 whorls. Often even juvenile shells are decollated. Transverse and spiral ribs are generally present, commonly forming tubercles with each other. *Melanoides tuberculata* in Jordan was found to live in pools in Azraq (now extinct here) and it here reached its largest individual sizes. It lived in Karama reservoir lake when water was still less salty- and it can be found in pools next to Al Hasa River- and probably anywhere in Jordan in places with muddy ground covered by clean fresh to brackish water (Fig. 3). M*elanoides* live in the creek that flows through the village Hemma into the Yarmouk, occurs along the Jordan River to the Dead Sea on muddy grounds as well as in similar environment along the lower Mujib River. Its reproduction is again unique since animals usually are parthenogenetic, that is usually only with females present. The youngs grow

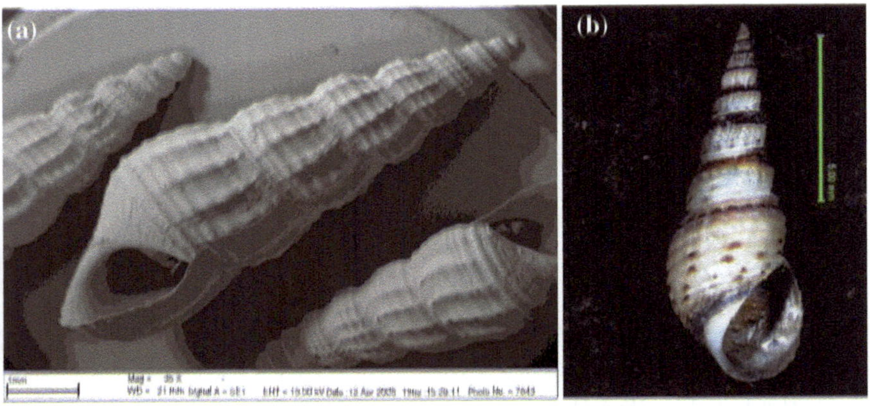

Fig. 3 **a** *Melanoides tuberculata* from KAC (2012) under scanning electron microscopy, **b** *Melanoides tuberculata* from Karama Lake (2011)

within the brood pouch of their mother until the shell becomes relatively large with 0.3–0.4 mm and 5 whorls completed. The initial 1.5–1.7 whorls show an irregularly wrinkled shell, a sculptural pattern, which form due to retarded calcification during early ontogeny (Bandel and Kowalke 1997). But after hatching the shell becomes solid. Its success in rapidly settling all available fresh water substrates in ponds, lakes and channels is connected to its parthenogenetic mode of reproduction and males occurring only as exception. With only one individual a whole population can be started.

The youngs are brooded in a pouch behind the head and this shelter is left only after four of five whorls of the shell completed and well calcified. The youngs released to the environment are thus well protected from attacks and they rapidly reach the stage for reproduction even with no other members of the species nearby. Thus a single individual can start a whole new population of many individuals. The tiny juveniles can easily be transported on the feet or on feathers of water birds and can thus be spread easily over long distances.

Bandel (2000) distinguished among the species of *Melanopsis* Férussac, 1807 quite a number of varieties with *Melanopsis buccinoides* and *Melanopsis costata* perhaps forming the origin of their evolution. *Melanopsis* is an early inhabitant of the Levants with relatives living around the Mediterranean Sea. The solid 1–3 cm high shell is conical, whorls of the spire are rounded and smooth or ornamented by axial and/or spiral ribs. The aperture is egg-shaped with anterior channel and smooth inner lip which is usually thickened by a posterior callus pad. The young snail hatches from its egg with about 1.5 whorls of the shell and very small size. Coloration is dark brown to light reddish brown and arranged in bands. Any clean fresh running water in Jordan would have the well visible *Melanopsis* living in it and, in case it is missing, water is either salty or polluted (Fig. 4). When a new locality with clean running water appears, the first *Melanopsis* will be present soon, carried here by birds. This is due to their reproduction by mode of single eggs, each of which surrounded by a mucus layer and delivered onto the ground. Wading birds

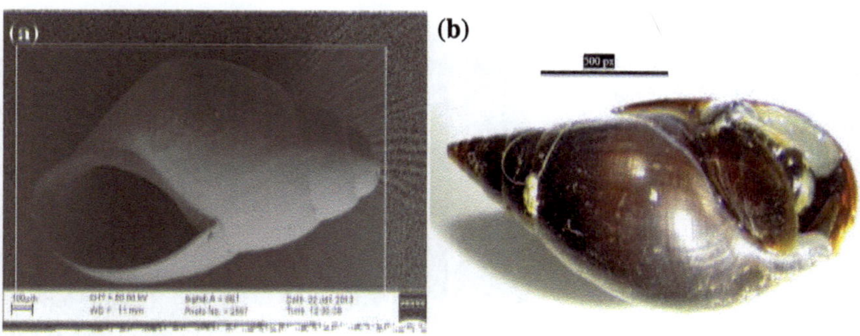

Fig. 4 Juvenile shell of *Melanopsis buccinoidea* from Wadi Hisban (site 13), **a** under scanning electron microscope, the scale bar is 100 μm, **b** *Melanopsis buccinoidea* in the field at the same site

will have them sticking on their legs when they fly off and transport them to all places with running water which can provide them with their preferred food consisting of small arthropods.

Hydrobia Hartmann, 1821 is found close to or in the same places as *Hydrobia acuta* or *Semisalsa contempta* which occurs together with *Pseudamnicola* in brackish water creeks near the Dead Sea and occasionally in different springs with more or less pure fresh water (Fig. 5). It differs from *Pseudamnicola* by having a larger and higher shell with more whorls when fully grown. The shell is about 3.3 mm high and 2 mm wide, consists of about 5 whorls, ovoid elongated, smooth with slightly convex whorls, weakly excavated sutures and has no umbilicus. A rather indistinct ornament of spiral grooves and ridges may be seen, but may also be less evident. The spire is conical and higher than the last whorl. The aperture can be closed with an operculum consisting of few whorls with eccentric nucleus. The embryonic shell consists of a little more than one whorl and the young leaves the egg case crawling on its foot. Numerous small hemispherical egg capsules are attached to hard substrates in the water. The gastropod feeds from growths present on stones much of which consists of diatoms. Only from a creek at the Dead Sea eastern shores *Semisalsa contempta*, was found living in brackish water. *Hydrobia* occurs, as well, living in brackish water and in small artificial ponds collecting spring water.

Pseudamnicola Paulucci, 1878 has a circum-Mediterranean distribution and species in the genus are difficult to distinguish by shell features alone and their anatomy needs to be known as well. It was not only found in the spring at the NE end of the Dead Sea, but also in the spring near Amman-Jerash road (mineral water, old roman ruin) and Zarqa Ma'in springs. The shell of *Pseudamnicola* is around 2 mm high and has only about 4 whorls with the last having more than two thirds of shell height and a thickened margin of the aperture that strengthens the fully grown shell. The small egg capsules are produced in large numbers as was observed in Wadi Atun (Site 16) where greenish cyanobacteria cover and camouflage them. The hatching young has a little more than one whorl of the shell.

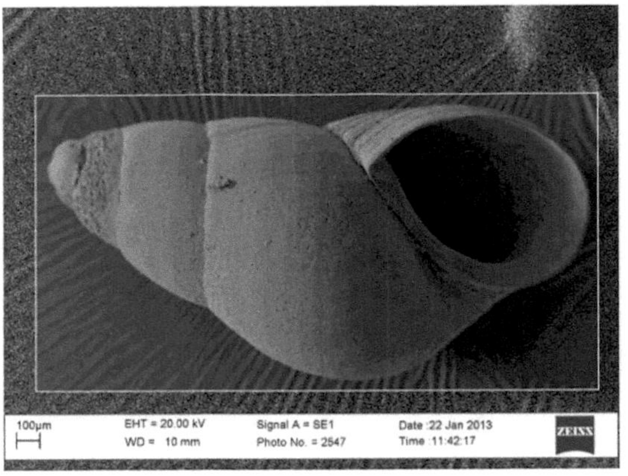

Fig. 5 *Hydrobia acuta* from a creek at the NE edge of the Dead Sea

Falsipyrgula Radoman, 1973 with *Falsipyrgula barroisi* (Dautzenberg 1894) has a shell 2.5–3 mm high and around 1.3 mm wide. The embryonic whorl is only 0.3 mm wide and consists of approximately one whorl. The adult shell consists of 4 whorls. Some morphs have a keel and others are smooth, some are slender, others short. *Falsipyrgula barroisi* lives also in KAC in northern Jordan (Fig. 6a), which receives part of its water from Lake Tiberias.

Ovatella is a member of the Ellobiidae H. and A. Adams in Pfeiffer 1854 which are usually marine pulmonates living in marine and brackish habitats, inhabiting the upper and supralittoral zones of the mangroves of the tropical regions and salt marshes of temporal regions, as well as rocky coasts. *Ovatella myosotis* (Draparnaud 1805) is a common Mediterranean and east Atlantic species that lives in salt marshes. The shell may consist of up to eight whorls, is up to 7.5 mm high, the ornament consists of flattened and wide spiral grooves, the aperture is simple and the shell is covered by brown periostracum that has short spines.

Ovatella myosotis was found living in a small creek below a brackish water spring next to the road to Madaba (Fig. 6b) at its branching off from the main road along the Dead Sea (Schütt 1983, Fig. 9). *Ovatella* lives on the wet zone when salinity reaches the degree for *Salicornia* plants survival. *Hydrobia*, *Pseudamnicola* as well as *Melanoides* are found here under water on hard substrates. *Ovatella* resorbs the inner walls of its shell deposit spawn on wet mud, commonly below surface. The operculum is cast several days after hatching from the egg mass as crawling young. Its occurrence indicates a salinity of the environment that is near marine. It may reach the isolated spots of its occurrence by young or spawn stuck to the feet of wading birds.

Basommatophora are shell bearing air breathing fresh water gastropods without operculum and with one pair of tentacles on their head which bear eyes near their base. Their species in Jordan leave their sticky egg masses as miniature adult. Physidae and Lymnaeidae have no gill but transport a bubble in their pallial cavity. They can thus also utilize badly oxygenated water by taking their oxygen with them. The bubble is actively sucked into the cavity and it is also expelled by

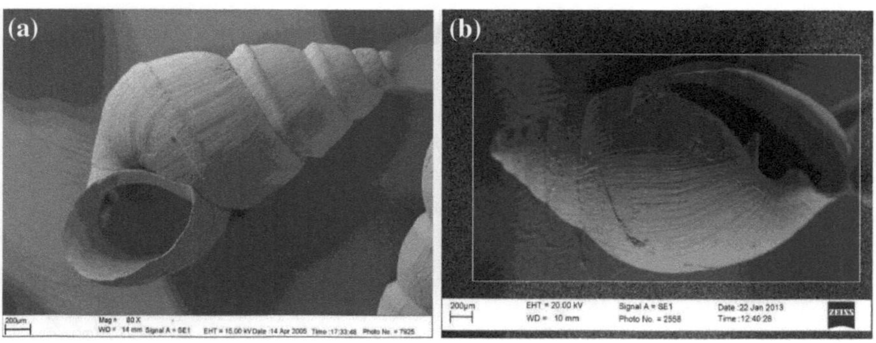

Fig. 6 **a** *Falsipyrgula barroisi* lives also in King Abdullah Canal, **b** *Ovatella* gastropod from the salt creek near the NE end of Dead Sea

muscular contraction of the walls of the lung. Oxygen bubbles can also be collected from plants or cyanobacterial growths, where they are continuously produced when sun light is available. *Lymnaea (Galba) truncatula* (Müller 1774) has a small conical shell with pointed spire and deep sutures between whorls. The shell is up to 10 mm high and 5 mm wide. Individuals were found near Hisban spring in small canals leading to fields (Fig. 7) and near the spring of Wadi Shita.

Bulinus Müller, 1781 has a sinistral shell resembling that of *Physa*. *Bulinus truncatus* (Audouin 1827) and has rounded whorls with the aperture wide and higher than the spire. The living animals appear grey while the similar *Physa* is brown. When crawling, the mantle is not expanded across shell margin, as is the case in *Physa*. *Bulinus truncatus* can be a transmitter of *Schistosoma*, occurs only rarely in Jordan, but may develop large populations rapidly as was found in the pool formed by the ancient water collecting basin of the Romans at Jerash (Schütt 1983, Fig. 12). *Bulinus* was also discovered in springs and puddles formed next to them close to Zarqa River 2013 and here together with *Melanopsis* (Fig. 8). *Bulinus* survived in pools with mixed water of springs and Zarqa River which were not

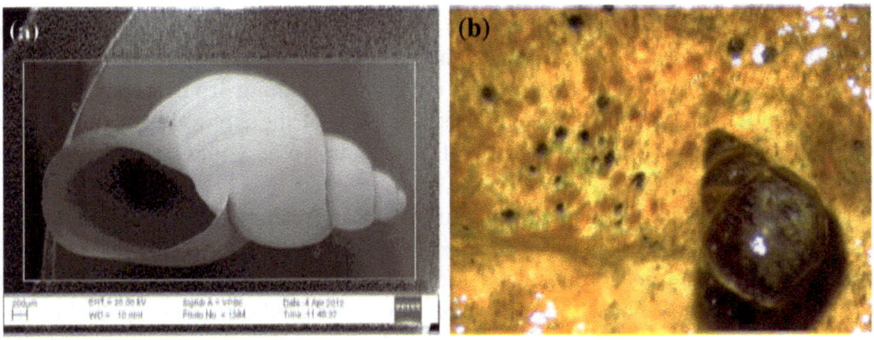

Fig. 7 **a** *Galba truncatula* from Wadi Hisban (2012) documented by SCAN and, **b** as in the field with their eggs

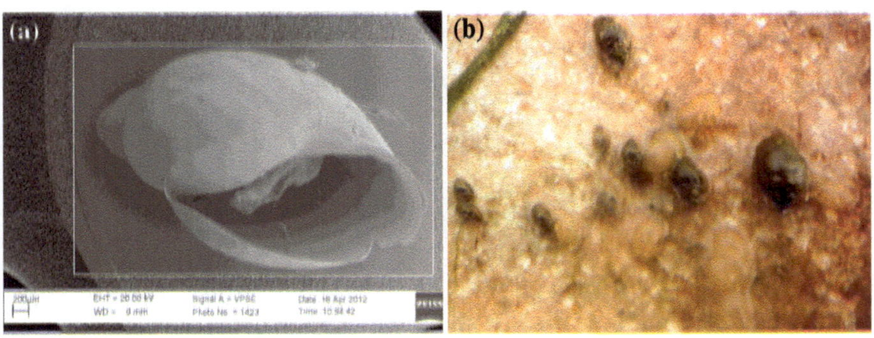

Fig. 8 **a** *Bulinus* gastropoda from Zarqa River, **b** *Bulinus* with gelatinous spawn in the upper Zarqa Valley, March (2013)

entered by *Melanopsis* living in a clear small creek feed from a spring, and both do not enter Zarqa River proper, but disappear before contact to the actual river water, which is obviously not healthy to either one.

Physa Draparnaud, 1801 in Jordan is *Physa acuta* Draparnaud, 1805 and has become one of the most common gastropods living in water. The shell resembles that of *Bulinus* but has a more pointed apex and a shinier smooth surface. Living individuals appear reddish brown with the soft body of the snail seen through the semi-transparent shell. *Physa acuta* is present in weakly polluted creeks, and also in clean rivers such as the Mujib, Shueib creek above and below the treatment plant and even in Zarqa River shortly below its spring at the treatment plant of Amman as in spring 2013. *Physa acuta* has developed into a character species to moderately polluted water in Jordan but does not survive an increase in salinity of the water. On the living *Physa*, finger like appendages of the mantle margin extend onto the outer shell, which is not so in case of *Bulinus*.

Bivalves are represented by species of the genera *Unio*, *Pisidium*, *Sphaerium*, and *Corbicula*, and are of relevance in Jordan only to KAC, but here they may be present in large numbers, especially in its less polluted portion that may reach Deir Alla. This part is very important for Jordan since quite a lot of its water is pumped to Amman to be added to the drinking water sources.

The genus *Unio* is characterized by characteristic mode of development of its off-springs which include a parasitic phase of its juveniles as parasites in fishes (Bandel 1988). The eggs are held in large brood pouches which are connected to the gills. Females of species of the Unionidae have their gills transformed into large pockets in which several hundred-thousand eggs are brooded until they have grown their shells, and they are ready to hatch. When the youngs have evolved to be ready for hatching and have reached the so-called Glochidia-Stage they have a bivalved shell that is mineralized. Their foot has a large gland that produces an elastic ribbon of byssus. Both mineralized valves are held apart by byssus and the valves are connected to each other by strong shell muscles. The Glochidia are expelled into the water by the mother-clam often when it senses the contact to a fish with the tissue of its siphon. When that contact between bivalve siphon and fish has been established the larvae are blown from the clam into the water next to the fish. Larvae at this stage have secreted a long sticky (byssus) thread and the two valves of their shell gape that is the shell is wide open. When the fish sucks in water through the mouth during breathing it washes it along the slits of the gill and the Glochidia larva come in contact with the tissue of the gill. To continue their life they attach by the sticky byssus thread. The larva also has sensory cilia which result in a rapid contraction of both shells when coming in contact with the tissue of the fish, penetrating it with its thorns. This irritation of tissue of the gill results in the reaction by which a blister forms that includes the bivalve larva. This blister will be filled with body liquid of the fish as long as it exists. This inclusion is what the bivalve needs to successfully continue its life as parasite and all larvae which do not succeed in getting hooked to fish tissue will die. The larva feeds on the blood of the fish and the fish transports the larvae of the two species of *Unio* from the canal to all other places in Jordan to which canal water has contact. No individuals outside the canal were observed.

Unio is a bivalve with many species of fresh water in the different parts of the world, with two species living in Jordan. Due to their adaptation as part time parasite on fish the bivalve can move to all areas in a river system which can be reached by some fish living in it.

When the bivalve has reached a stage of development inside of the cyst ready for free life it breaks free and falls from the gill of the fish to the ground to bury in the bottom substrate and lead a hidden way of life for quite some time. The body liquid of the fish is the source of food during that time that could last for weeks or even months and can bridge over problematic times. From there on and until adult stage is reached, often several years later, *Unio* behaves like many bivalves, is half buried in the ground and filters phytoplankton from the water of the river. The juvenile *Unio* hides for an unknown time in the bottom and grows until it may appear close to the bottom surface. A fertile bivalve can live for many years, exact time in Jordan is unknown, but at least 5–10 years.

From getting free and dropping to the bottom of the canal *Unio* continues life as a normal bivalve by pumping water through its gills, catching organisms of the phytoplankton with it and channeling that food to the mouth with aid of ciliary activity along the branches of the gill.

Unio terminalis Bourguignat, 1852 lives in KAC as well as in Lake Tiberias in large populations and was also found in the sediments of the Jordan River before the existence of Lake Lisan (Fig. 9a). The second species here is *Unio semirugatus* Lamarck, 1819 with the shell of rhomboidal shape. *Unio semirugatus* may be the same as *Potomida littoralis* (Lamarck 1801) representing a Circum-Mediterranean species (Fig. 9b). The shell structure consists of an outer prismatic layer with the large prisms composed of fine needles arranged as bundles. The transition to the thick inner layer of nacre is well developed, and the nacre is in a brick-layer mode. It is more regular in the shell than in pearls.

Genus *Corbicula* has a shape of the shell that differs strongly from that of *Unio*, and its shell composition is different. While the shell of *Unio* is composed predominantly of aragonitic material with nacre structure, that of *Corbicula* consists of aragonite in the crossed lamellar structure. *Corbicula* has become introduced to

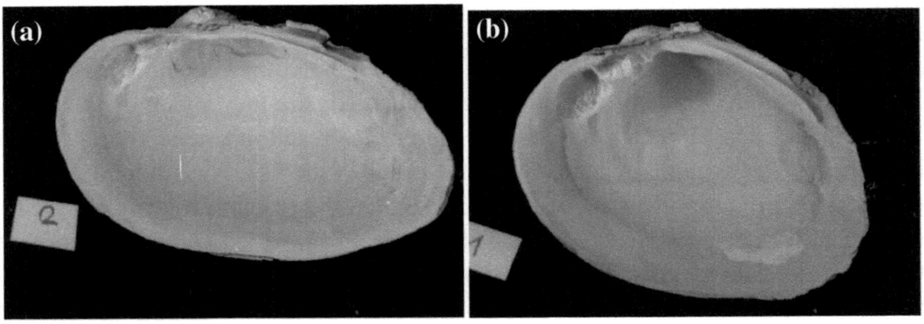

Fig. 9 **a** *Unio terminalis* and, **b** *Unio semirugatus* from KAC

Fig. 10 *Corbicula* from King Abdullah canal

Jordan by man in recent times, originally coming from south-east Asia. *Corbicula fluminalis* (Müller 1774) is very common in KAC (Fig. 10), but was not encountered elsewhere in the country.

Corbicula fluminalis was noted to have sexes separate, but hermaphroditic specimen occur, which change from male to female during their life time. The juveniles are crawling when their shell is only 0.2 mm in size and they hatch from a maternal brood pouch and need approximately one year of growth to be sexually mature and measure 1 cm and afterwards may live for 10 years. During reproduction they can release several hundred off-springs a day and as a whole about 8000 in a year. The larval stage is completed within the brood pouch of the female. It would need to be experimentally evaluated, whether the hatching young can still swim as veliger larva or crawls right away, in case of swimming that would not be more than a few hours. The bivalve is very successful in the Canal, but was not found outside of it.

Genus *Pisidium* occurs with very small individuals among which the females brood the young in a pouch and releases them when they are quite far developed (Bandel 1988). This brood pouch can hold only a few individuals with very high probability to survive when they hatch, due to their relatively large size. They then reach maturity quite rapidly, so that a settlement of a body of water with *Pisidium* can occur in a short time, once a female has come to the water, quite possibly attached to water birds. *Pisidium casertanum* (Poli 1795) was collected from the spring of Wadi Hisban, and in a spring near the bridge across the Zarqa (Schütt 1983). *Pisidium* sp. was described by Nelson (1973) to occur in Azraq Oasis. Quite a number of individuals were found living in King Abdullah Canal next to Abu Habil Mosq Shar Habili together with *Hydrobia*, different types of *Melanopsis* and with *Falsipyrgula*. *Pisidium* can move as rapidly as a gastropod and canals with clean water would be perfect migrating paths for them.

4 Mollusks as Indicators of Environmental Conditions in Jordan

The reasons to use mollusks as bioindicators to examine the quality of water is because of their abundance, diversity and existence in almost every place with water in Jordan from clean to relatively polluted water, running to still (surface) water, and fresh to brackish water. Furthermore, they are easily caught and can usually be detected with the naked eye (macro in size). To collect them it is best to use a screen, and it is not easy to damage them due to their hard shell. Almost all reported on species are fully aquatic and spend most of their life in water, rarely on the moist sand or mud near the water. Regarding the identification of the species, they are well studied in Jordan and they reflect water condition since they are sensitive to changes in wide array of environmental factors and have a relatively rapid response to environmental changes. Jordanian aquatic environments usually represent springs, creeks, and small rivers less commonly ponds and lakes. The amounts of water available in these resources over the year often change drastically in regard to volume and quality of the water and due to seasonal water availability. Mollusks are here good and easily available biological indicators. The requirements for their life are well known, and they can rapidly be determined in the field. Their presence in a stream, spring or pond is a good indicator to environmental factors and stress such as salinity, enrichment of certain trace elements and contamination due to pollution by waste water or oil a documented below.

4.1 Mollusks and Habitat

A clear relation exists between the occurrences and richness of the fauna of mollusks and the environment. This leads to the possibility of use of the presence of mollusks in a specific habitat as marker species to characterize their environment. The type of food they feed on, their need for a specific shelter and the composition and shape of their shell limit their occurrence to specific environments. It is worth mentioning here that water quality degradation and fluctuation of mollusks species in a specific body of water depends not only on human activities, but also on natural stresses. Competition among species, predation by crabs, fish and birds and seasonal variations of water flow and sediment transport such as increased turbidity during floods are of importance.

Competition for food was quite evident in case of *Melanopsis* and *Theodoxus* in Wadi Hisban, site 14 (Fig. 11) and locally on the concrete walls of KAC, site 11. Both gastropods compete for food mainly consisting of the dense growth of algae on the well illuminated hard substrate. *Theodoxus* has partly mineralized teeth of its radula and is thus better equipped to scrape off algae from stony surface, thus

Fig. 11 High occurrences of *Theodoxus* while *Melanopsis* gastropods disappeared in Wadi Hisban spring (site 14)

leaving unsufficient food for *Melanopsis*. When the water becomes more polluted than it is better for *Theodoxus* and is still tolerated by *Melanopsis* and when substrate on which the snails search for food is softer, *Melanopsis* becomes more common or be exclusively present. Thus the presence of numerous individuals of *Theodoxus* does not indicate that environmental conditions are bad for *Melanopsis*, but just that *Theodoxus* is better able to extract food growing on hard surfaces.

The effect of predation can be observed in regard to the presence of the calcareous operculum of *Theodoxus* that can seal the aperture tightly and protects the snail from attacks of crabs while the organic operculum of *Melanopsis* can not prevent the crab *Potamon* to feed on. The effect was observed very clearly in Wadi Mujib where juvenile *Melanopsis* often fall victim to the crab while *Theodoxus* appears to escape attacks.

The chemical composition of the water of each location is listed in Tables 1, 2 and 3. The distribution and diversity of animals and plants living in the water is affected by the ionic content of the water. For instance: calcareous water is an important environment for mollusks fauna because they need the calcium ions to construct their shells. Chemical analyses of the water documented that the highest Sr and Br concentration within Jordanian springs and stream are found in the little stream of Wadi Atun with around 37.42 ppm of Sr^{2+} and Br^{-} concentration of 10.41 mg/L. *Pseudamnicola* occurs here in great numbers, breeding, surviving and tolerating mineral water.

Agriculture related high input of phosphorus and nitrogen cause the eutrophication (nutrient enrichment) problems of Wadi Mujib (site 18) and King Talal dam on Zarqa River. The massive production of organic material due to the growth of algae and cyanobacteria plants has negative impacts on such mollusks with the need for clean and oxygen-rich water. *Theodoxus* and *Melanopsis* do not occur here because they can't live and survive under poor water conditions with low dissolved oxygen, while *Physa* tolerats such condition and can exploit its abundance of food.

From field observations, *Theodoxus* and *Melanopsis* are usually found together, in springs, creeks, ponds and streams with good water quality.

Table 1 Chemical and physical characteristics of surface water samples collected from selected sites of the study area

Sample No.	Location	EC (μS/cm)	Temp (°C)	pH	COD (mg/L)	BOD (mg/L)
1	Yarmouk River	945.69	21.02	8.07	8.94	0
2	Abu Thabelh spring	1920	36.60	6.90	15.80	0.00
3	Zarqa River	2300	17.30	7.90	72.00	9.00
4	W. Shueib (Jadoor spring)	835	18.50	7.63	21.10	20.00
5	W. Shueib	685	18.60	7.47	26.00	5.00
6	W. Shueib	775	18.00	8.45	60.00	35.00
7	Karama (stream)	15630	21.90	8.00	41.20	36.00
8	Karama (Small pond)	16120	21.80	8.00	27.80	15.00
9	Karama (large lake)	20500	22.20	8.20	20.70	10.00
10	Karama Dam	26100	22.30	8.20	35.70	25.00
11	King Abdullah Canal	447.00	20.35	7.93	16.79	15
12	W. Shita (main spring)	630	17.30	7.49	8.30	0.00
13	W. Hisban (Soneh spring)	805	18.5	6.92	10.80	5.00
14	W. Hisban	743	19.2	7.04	1.40	0.00
15	Zara spring	1702	50.30	6.00	4.80	0.02
16	W. Atun upstream	2040	18.00	6.72	5.00	35.00
17	W. Atun downstream	1896	17.5	6.59	4.10	28.00
18	W. Al Mujib Lake	925	17.30	7.71	35.20	5.00
19	W. Al Mujib downstream	1395	17.30	7.90	15.90	0.00

Water in which a high production of plant material takes place witnesses rapid bacterial decay and thus lower oxygen content, such water is populated by *Physa* that can collect bubbles of oxygen from the surface of cynobacterial growths. The snail can also store a bubble of air in its lung when it exploits water that is depleted of oxygen and carry it for breathing in such water. But *Physa* is also tolerant to much chemical water pollution and thus ideally equipped to deal with eutrophication and oxygen depletion.

Soft sediment and standing water is preferred by *Melanoides* while *Melanopsis* avoids muddy ground and standing water. Hydrobia genus occurs only sporadically in Jordanian water and it apparently, prefers flowing water with a lot of vegetation in it.

Physa tolerates pollution much more than does *Bulinus* which also has a left-coiled shell and prefers pond water and slow flowing creeks. *Galba* occurs in running and almost quite water with much vegetation, is rather rare in Jordanian creeks and thus not useful as bioindicator. Both pulmonate gastropods can carry a bubble of air in the mantle cavity as aid for breathing and also as means of locomotion. They reach plants and bottom substrate on their search for food

Table 2 Chemical analyses of the studied surface water (for locations refer to Table 1, concentrations in meq/l, PO_4^{3-} in mg/l)

Sample No.	Ca^{2+}	Mg^{2+}	Na^+	K^+	HCO_3^-	Cl^-	SO_4^{2-}	NO_3^-	PO_4^{3-}
1	3.38	2.56	3.38	0.17	3.87	3.30	1.91	0.19	0.10
2	7.2	4.7	8.72	0.46	7.13	9.34	3.31	0.08	0.20
3	6.18	5.69	11.06	0.84	8.30	11.98	1.55	1.41	0.35
4	4.63	1.66	2.13	0.37	4.01	1.19	1.20	1.40	1.03
5	3.95	1.53	0.55	0.10	4.90	0.77	0.32	0.31	0.06
6	3.99	1.48	1.98	0.20	2.59	1.2	1.46	0.78	1.00
7	37.40	56.36	60.66	5.19	4.69	58.32	87.65	5.02	0.12
8	40.01	58.2	63.11	5.51	5.91	58.45	89.05	5.88	0.17
9	41.93	67.94	96.36	6.23	3.10	83.80	111.07	6.48	0.19
10	54.88	86.07	120.62	8.34	3.15	110.0	153.9	0.003	0.19
11	2.90	2.60	3.23	0.16	3.24	3.63	1.46	0.12	0.11
12	4.43	1.06	0.90	0.05	3.98	0.82	0.65	0.78	0.04
13	4.53	1.75	1.19	0.10	4.09	1.32	1.06	0.71	0.22
14	4.32	2.12	1.01	0.09	4.71	0.89	0.72	0.68	0.18
15	5.03	1.55	8.79	0.31	3.26	10.19	2.10	0.02	0.05
16	5.92	2.92	10.6	0.69	3.80	7.45	8.451	0.83	0.17
17	5.26	2.75	10.31	0.66	3.40	6.59	8.35	0.70	0.03
18	3.73	2.35	2.97	0.25	1.57	2.86	4.39	0.15	0.31
19	5.18	3.83	4.78	0.19	1.57	4.76	7.26	0.11	0.01

keeping attached to the ground by their foot. When the water becomes anaerobic they detach the foot from the ground and the bubble of air in their lung carry them to the surface. This method of avoiding water-free oxygen is not available for species which breathe with a gill as is the case in *Melanopsis* and *Theodoxus*.

Valvata, *Planorbis*, *Gyraulus* and *Ancylus* species occur occasionally in clean vegetated spring environments but are rare in Jordan, therefore none of them is a useful biological-indicators.

The bivalves *Unio* and *Corbicula* need continuously flowing water from which they extract phytoplankton. They have only been encountered in KAC, and on the mud next to them *Falsipyrgula* is common. That small gastropod entered here from Lake Tiberias. It feeds on plankton cells that have settled to the ground of the canal. Here also the bivalves Pisidium and Sphaerium occur and they also live in strongly vegetated irrigation canals in the Jordan Valley. They occur only rarely in Jordanian waters and cannot be used as biological indicators. *Prosostomia* (*Semisalsa*) and *Pseudamnicola* prefer running water while *Ovatella* may move on the wet mud next to it. They were found together with *Melanoides* in a salty water creek near the Dead Sea that becomes increasingly more salty on its way to the Dead Sea.

Table 3 Trace elements concentrations in mg/l in the water samples collected from the study area (for locations refer to Table 1)

No.	Fe^{2+} ppm	Pb^{2+} ppm	Zn^{2+} ppm	Mo^{2+} ppm	Cu^{+} ppm	Mn^{2+} ppm	Sr^{2+} ppm	Ba^{2+} ppm	Br^{-} mg/l	F^{-} mg/l
1	0.021	<0.01	<0.01	<0.01	<0.01	<0.01	0.320	<0.01	0.210	0.112
2	0.019	<0.01	0.02	<0.01	<0.01	0.029	6.283	<0.01	1.947	0.035
3	0.014	<0.01	<0.01	<0.01	<0.01	<0.01	3.014	<0.01	0.606	0.440
4	0.021	<0.01	<0.01	<0.01	<0.01	0.109	0.503	<0.01	0.293	0.240
5	0.017	<0.01	0.030	<0.01	<0.01	<0.01	0.590	<0.01	0.205	0.244
6	0.022	<0.01	0.043	<0.01	<0.01	<0.01	1.085	<0.01	0.448	0.190
7	0.017	<0.01	<0.01	<0.01	<0.01	0.1	32.82	<0.01	6.421	0.975
8	0.016	<0.01	<0.01	<0.01	<0.01	0.101	31.62	<0.01	6.795	1.254
9	0.015	<0.01	<0.01	<0.01	<0.01	0.102	6.36	<0.01	0.864	0.417
10	0.021	<0.01	<0.01	<0.01	<0.01	0.108	50.21	<0.01	15.720	0.967
11	0.016	<0.01	<0.01	<0.01	<0.01	<0.01	0.453	<0.01	0.021	0.420
12	0.016	<0.01	<0.01	<0.01	<0.01	<0.01	0.430	<0.01	0.275	0.703
13	0.021	<0.01	<0.01	<0.01	<0.01	<0.01	0.350	<0.01	0.286	0.519
14	0.022	<0.01	<0.01	<0.01	<0.01	0.110	2.100	<0.01	0.128	0.409
15	<0.01	<0.01	<0.01	<0.01	0.192	5.36	<0.01	0.699	0.509	0.409
16	0.024	<0.01	0.020	<0.01	<0.01	0.025	37.42	<0.01	10.400	0.893
17	0.023	<0.01	0.020	<0.01	<0.01	0.025	37.390	<0.01	10.410	0.892
18	0.021	<0.01	<0.01	<0.01	<0.01	0.044	11.360	<0.01	1.506	0.691
19	0.020	<0.01	<0.01	<0.01	<0.01	<0.01	1.990	<0.01	0.355	0.635

4.2 *Mollusks and Salinity*

Salinity refers to the salt content in the environment coming from natural and human activities. High salt contents commonly occur in semiarid and arid region where evaporation exceeds precipitation. Like many other countries with arid climate, Jordan's surface and ground water face salinity problems. Salameh (1996) referred the major factors contributing to rapid degradation of the quality of fresh water to salinization and pollution. According to Salameh (1996, 2001) the main reasons for quality degradation of the waters in Jordan are treated and untreated wastewater, rock weathering (e.g., dissolution of evaporates from rock formation such as Lisan Formation), industrial activities, irrigation return flow, high salt content in precipitation water and arid climate with insufficient precipitation and high evaporation. Quite a number of aquatic organisms live in brackish environments and may even be confined to them. The salt level present in aquatic systems was measured here by electronic conductivity unit (EC) in micro Siemens per cm (μS/cm) in the field, using electronic probe (WTW equipment).

The salinity of the different sites was measured to range from 567 to 26,100 μS/cm during late winter to early spring season. The highest EC value in the water was found close to the Dead Sea in regions such as Karama Dam area (site 10). In

Jordan freshwater bodies have usually less than 2000 μS/cm EC values. Some slightly salty water ranges from around 2000 to 3000 μS/cm and brackish water could range from 4000 to 10,000 μS/cm, and saline water have conductivities of more than 10,000 μS/cm.

Salinity fluctuations strongly affect the distribution and abundance of mollusks. Each species within the mollusks fauna has its own optimum or preferred salinity level and each one has a different requirement regarding the salinity of the water it lives in. The relative upper limit tolerance to salinity in the here treated aquatic mollusks was observed and is discussed in this study (Fig. 12).

Accordingly, *Theodoxus* and *Melanopsis* are very vulnerable to salinity and are restricted to freshwater with little tolerance. Their abundance was highest in cool running freshwater creeks and springs with a salinity range of 685–1395 μS/cm and they were completely absent from slightly salty water, brackish and saline water. But *Melanopsis* very rarely occur in quite clean warm water which is found alone without *Theodoxus* in springs south of Wadi Yarmouk.

Melanoides tolerates higher salinity than is the case in any of the varieties of *Melanopsis*. A similar water quality may be also accepted by *Hydrobia* but the species of this genus occurs only sporadically in Jordanian water and it thus, is not very useful as indicator species.

Melanoides noted in the Karama Dam, tolerates salinity ranges of about 23,600 μS/cm as found during the winter season 2010. In early spring 2011, the decline of rainfall resulted in a decreasing amount of water entering the reservoir and the salinity went up to 26,100 μS/cm. The former huge population of *Melanoides* disappeared and non have lived there since.

Our field observation shows that *Prosostenia* (*Semisalsa*) and *Pseudamnicola* prefer creeks with brackish water with high salt contents which reach to 23,500 μS/cm. *Ovatella* is a salt loving gastropod and in creeks with the salinity reaching that of the sea. *Ovatella* can be present near the Dead Sea.

Physa, Galba, Bulinus, Pisidium, Sphaerium, Falsipyrgula, Unio and Corbicula are particularly intolerant to salinity and prefer freshwater environment. In case of

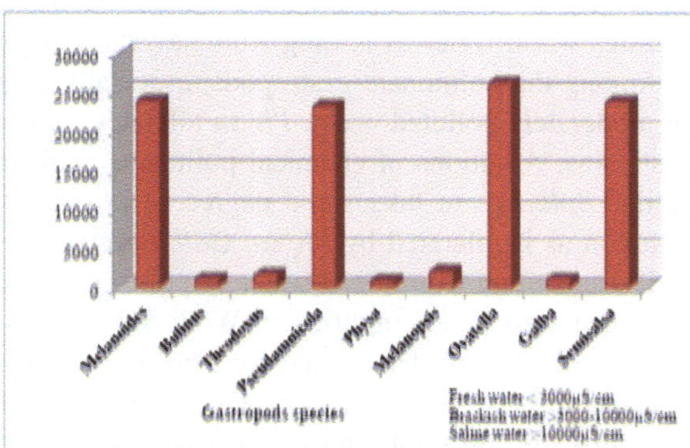

Fig. 12 Relative occurrences of gastropod species in fresh, brackish, and saline water in the study area

Physa that is intolerant to raised salinity and is of special interest, since it can live in relatively strongly polluted water and is the first mollusk that was found in Zarqa River downstream of the waste water treatment plant of Amman.

4.3 Mollusac and Pollution

Several types of pollutants affect water bodies in Jordan, including: sediment, organic wastes, excess nutrients and toxic substances, which result from human activities (anthropogenic) or from natural stresses (Salameh 1996). Impacts of contamination and sediment load (turbidity) are known to result in significant changes to the richness and diversity of aquatic mollusks. But according to Salameh and Bannayan (1993) and Salameh (1996), the major pollution sources in Jordan are inefficient wastewater treatment plants, cesspools, industrial wastewater, irrigation return flows, and solid waste disposal.

Theodoxus in Jordan is clearly restricted to clean water and therefore in creeks it only occurs near spring site, since most water courses in the downstream areas are polluted to a degree more than *Theodoxus* can tolerate. Its tolerance is less than that of *Melanopsis* which occurs even downstream connected to some pollution. But streams as that of the Zarqa have neither *Theodoxus* nor *Melanopsis* surviving, while *Physa* may occur sporadically in that river as found below the treatment plant and near the Jordan Valley. Below Mujib dam Mujib River has turned to such a low degree of pollution that *Theodoxus* reappeared, after being absent from the streams which feed the reservoir of the lake due to pollution. The water leaving the lake is settled by *Melanopsis*, and further downriver the water of the stream improved so much, that *Theodoxus* also appeared again.

The begin of KAC at the Yarmouk River holds *Theodoxus*, and the species can spread to the south in that Canal periodically reaching the area near Deir Alla. The Jordan River is free of this snail (and all other snails as well), but also relatively clean looking water such as that of lower Wadi Al Hasa has no *Theodoxus*. *Melanopsis* is less delicate regarding water purity than *Theodoxus*. It reacts very sensitively to chemical pollution, for example by sewage or traces of oil. It also tolerates warm water as issuing from several springs in Jordan.

Among the pulmonate gastropods, especially *Physa* is a good indicator species for moderately polluted water. *Physa* tolerates some degree of house sewage added to streams. Additions of chemical pollutants as for example some oil kills it. The similar *Bulinus* can take some degree of pollutants but not up to the degree of pollution as is tolerated by *Physa*. *Galba* prefers clean water.

The transition from clean water to polluted water within the course of a single creek can well be exemplified in Wadi Shueib. Just below Es Salt some springs provide the clean water needed for *Theodoxus*, and *Melanopsis* (Fig. 13a). When that water mixes with water of the creek coming from town the snails disappear, but *Melanopsis* is still found in isolated localities with spring water entering the creek. All snails leave the water after the outlet of Es Salt treatment plant effluent joins the

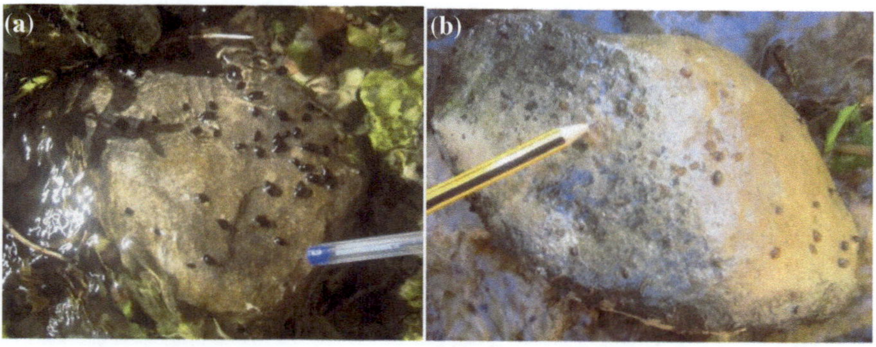

Fig. 13 In Wadi Shueib stream, aquatic mollusks fauna document a shift in community composition related to human activity, **a** *Theodoxus* and *Melanopsis* gastropods are present in clean water of Jadoor spring (site 4), **b** *Physa* associated with organic pollution-tolerant organisms are found in the creek downstream Salt sewage treatment plant (site 6)

creek (site 6). The quality of this creek, downstream of the treatment plant deteriorates (Oxygen degradation, high amount of organic sludge). The elevated phosphorus concentration results in strong growth of bacteria until first *Physa* is noted, initially found best by the presence of their gelatoinue eggs (Fig. 13b), Here also members of the fauna that can deal with organic pollution such as the sewage worm (*Tubifex*) and blood worms (red larvae of Chironomidae) are common. A similar assemblage of organic pollution-tolerant organisms are present in the pond close to Wadi Muijb dam (site 18) and a few kilometers downstream Amman treatment plant, also connected to the first appearance of *Physa* on rocks.

In the case of Wadi Mujib below the dam (site 18) a change to much better water quality from that of the lake and also from some of its tributary streams was observed. Here initially nutrient tolerant species (e.g., *chironomids*, *Tubifex* and *Physa*) are associated to a high nutrient production by strong growth of algae and phytoplankton (Fig. 14a). Here also the content of dissolved oxygen is relatively low. But still close to the dam *Melanopsis* appears, fish and frogs are common and within a short distance of about 2 km downriver the biological communities change completely documenting a rapid improvement in water quality also due to the turbulent flow of the river across many rapids in the stream bed. The river bed is stony and air mixes with the river water, and nutrient tolerant organisms such as Tubifex and red chironomid larvae are replaced by the pollution intolerant organisms such as *Theodoxus*, and *Melanopsis*, site 19 (Fig. 14b). In general, the river is cleaned by a self-purification process, but there are still many nutrients in the water which support the growth of a rich phytoplankton and the mud that may settle in quiet puddles on the side turns black just below the surface, and the lower sides of stones are black. The high quality water containing many planktonic organisms here even enables the growth of fresh water sponges covering rocks in the river rapids.

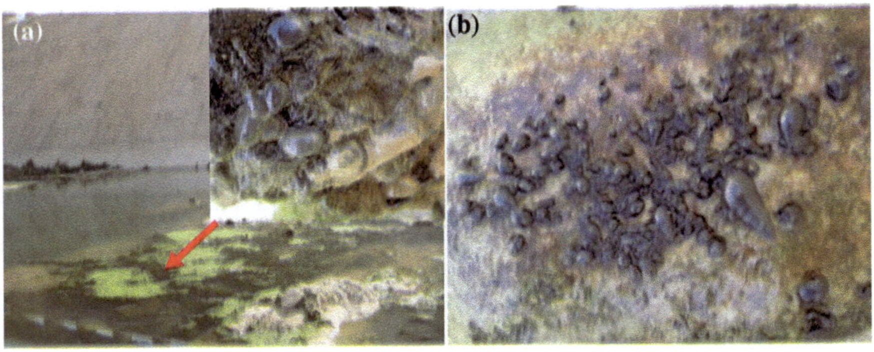

Fig. 14 The occurrences of bioindicators along Wadi Mujib (site 18: lake downstream the dam; site 19: creek in Wadi Mujib about 2 km downstream of Wadi Mujib dam), **a** *Physa acuta* eggs from Mujib Lake affected by eutropification (site 18). **b** Large number of *theodoxus* and *Melanopsis* in Wadi Mujib water downstream of the dam (site 19)

In Zarqa River case, degradation of water quality is mostly a result of wastewater treatment outflow and junk in the River bed. The Zarqa River downstream of Khirbet es Samra treatment plant is affected by a rich autotrophic production of organic material, predominantly of bacterial origin. After a short distance of flow gastropod *Physa* is found associated with frog, Tubifex, red larvae of chironomids, and other river insect larvae. Later fish appear as a characteristic organism to this type of water in Jordan. In a general way, composition of life within the river does not change much all the way down to the Jordan Valley but gastropod *Physa* continues to represent the character organism. Aquatic mollusks have varied tolerance levels to pollution, but generally are considered intolerant organisms and they cannot survive in strongly polluted water.

5 Conclusion

The present study concludes that mollusk organisms especially the gastropod group is a very important biological indicator and can be utilized as a rapid and very useful marker of the chemistry and grade of pollution of the environment. Gastropods are considered as one of the best candidates for use as bioindicators of water qualities.

Among gastropods, *Theodoxus* stands for water of high quality, *Melanopsis* is indicative of water with slight pollution, *Melanoides* as an indicator of quiet water and tolerance to some increased salinity, and *Physa* as a good indicator for moderately to quite strongly polluted water. The presence of *Pseudamnicola* usually documents influence of mineral water influence and *Ovatella* stands for not polluted but salty water that approaches the salinity of the sea water. *Melanoides* in contrast to *Melanopsis* prefers muddy ground and standing water. It may tolerate an increase in salinity, but no longer that of Karama reservoir. Here it had lived in huge number a few years ago, but now has disappeared due to raised content of salt in the lake. *Melanoides* tolerate higher salinity than is the case in any of the varieties of *Melanopsis*. A similar water quality may be also accepted by *Hydrobia* but the species of this genus occurs only sporadically in Jordanian water and it thus, is not very useful as indicator species. It apparently prefers flowing water with a lot of vegetation in it.

Special conditions are required in case of bivalves, with *Unio* and *Corbicula* which prefer continuously flowing water that contains much phytoplankton. Together with *Falsipyrgula* they reflect the influence of Lake Tiberias. The bivalves *Pisidium* and *Sphaerium* occur only rarely in Jordanian waters and are of little use for pollution assessment added to that their small size. Regarding the water salinity, gastropods Ovatella *Prosostenia* and *Pseudamnicola* prefer creeks with brackish water and each of them have different tolerance degree to salinity. *Bulinus*, *Galba* and *Physa* have different tolerance to pollution with *Physa* the most important of these with high tolerance to pollution but low tolerance to raised salinity. In general, *Physa* gastropods could tolerate organic pollution and can be considered as a good indicator of eutrophic water bodies. *Valvata*, *Planorbis*, *Gyraulus* and *Ancylus* have been noted to occur in clean water in Jordan but they are rare and thus of no use for water quality assessment.

References

Alhejoj I, Bandel K, Salameh E (2014) Macrofaunal and floral species in Jordan and their use as environmental bio-indicators. University of Jordan

Bandel K, Kowalke T (1997) Systematic value of the larval shell of fossil and modern Vanikoridae, Pickworthiidae and thegenus Fossarus (Caenogastropoda, Mollusca). Berliner Geowiss Abh E25:31–29, Berlin

Bandel K (2000) Speciation among the Melanopsidae (Caenogastropoda). Special emphasis to the Melanopsidae of the Pannonian Lake at Pontian time (Late Miocene) and the Pleistocene and Recent of Jordan. Mitteilungen aus dem Geologisch-Paläontologischen Institut der Universität Hamburg 84:131–208

Bandel (2001) The history of Theodoxus and Neritina connected with description and systematic evaluation of related Neritimorpha (Gastropoda). Mitteilungen aus dem Geologisch-Paläontologischen Institut der Universität Hamburg 85:65–164

Bandel K, Salameh E (1981) Hydrochemical und hydrobiological research of the pollution of the waters of the Amman Zerka area (Jordan). Schriftenreihe Deutsche Gesell. Techn. Zusammenarbeit GTZ 94:1–60 (Eschborn)

Bandel K (1982) Morphologie und Bildung der frühontogenetischen Gehäuse bei conchiferen Mollusken. Facies 7:1–198

Bandel K (1988) Stages in the ontogeny and a model oft the evolution of bivalves (Mollusca). Paläontologische Zeitschrift 62:217–254

Hilsenhoff WL (1988) Rapid field assessment of organic pollution with a family-level biotic index. J N Am Benthol Soc 7(1):65–68

Mandaville SM (2002) Benthic Macroinvertebrates in Freshwaters—Taxa Tolerance Values, Metrics, and Protocols. Soil & Water Conservation Society of Metro Halifax

Nelson B (1973) Azraq: desert oasis. Cox & Wyman Ltd., London, 436 p

Salameh E (1996) Water Quality Degradation in Jordan (impacts on environment, economy
Salameh E (2001) Water shortage and environmental degradation. In: Living with water scarcity, water resources in Jordan Badia region the way forwarded. Publication of Al al-Bayt University, Mafraq, Jordan. pp 71–87
Salameh E, Bannayan H (1993) Water resources of Jordan; present status and future potentials. FES, RSCN, Amman
Schütt H (1983) Die Molluskenfauna der Süßgewässer im Einzugsgebiet des Orontes unter Berücksichtigung benachbarter Flußsysteme.- Archic fuer Molluskenkunde, 113: 1–91. Frankfurt a.M

Techno-economical Comparison of MED and RO Seawater Desalination in a Large Power and Water Cogeneration Plant in Iran

Babak Golkar, Ramin Haghighi Khoshkhoo and Aliasghar Poursarvandi

Abstract Seawater desalination is one of the major solutions to overcome water crisis in MENA region. In this paper, two main scenarios have been studied for desalting seawater from the source of the Persian Gulf. First scenario is based on two 170 MW gas turbines, one 160 MW steam turbine and 143,000 m^3/day MED thermal desalination, while the second plant is based on two 170 MW gas turbines, one 95 MW steam turbine and 143,000 m^3/day SWRO membrane desalination. Technical and economical modeling of each scenario has been done and their levelized water costs have been compared. Results show that levelized water cost in cogeneration plant with membrane desalination is lower than cogeneration plant with thermal desalination in the range of 0 (Energy conversion agreement) to 30 \$cent/m^3 for fuel price. For Persian Gulf FOB gas price of 8.6 \$cent/m^3, MED water cost is 1.16 \$/m^3 and SWRO water cost is 0.99 \$/m^3. Sensitivity analysis has been done for a range of availability in power and water production, years of operation, and capital cost margin.

Keywords Desalination · MED · RO · Cogeneration of water and power · Water cost

Nomenclature

Latin symbols

A	Availability
C	Cost
CC	Capital cost
D	Days of a year

B. Golkar · R.H. Khoshkhoo (✉)
Mechanical and Energy Department of Abbaspour College of Engineering,
Shahid Beheshti University, Tehran, Iran
e-mail: R_haghighi@sbu.ac.ir

A. Poursarvandi
Research and Development Department of Monenco
Iran Consulting Engineers, Tehran, Iran

O. Abdalla et al. (eds.), *Water Resources in Arid Areas: The Way Forward*,
Springer Water, DOI 10.1007/978-3-319-51856-5_18

E	Net electrical power (kW)
e^*	Unit capital cost of power block (\$/kW)
F	Fuel consumption (m^3/h)
H	Hours of a day
i	Interest rate
L	Ratio of loan to total capital cost
N	Year
n	nth year
O&M	Annual O&M cost
P	Price
T	Tax related to profit
w^*	Unit capital cost of water block (\$/m^3/day)

Greek symbols

Φ Annual operation and maintenance cost (\$/m^3)

Subscripts

c	Related to construction period
f	Related to fuel
l	Related to loan
o	Related to operation period
p	Related to power block
real	Related to capital cost affected by loan and construction period
w	Related to water block

1 Introduction

Desalination is the process of reducing total dissolved solids (TDS) in water. Desalinated water is mainly used in domestic and industrial applications. Desalination and wastewater recycling are two rainfall-independent water resources (Fischetti 2007). Most of the efforts in recent decades are related to decrease the cost of desalination. Desalination has been mostly used in arid areas such as MENA region and Australia as one of the tools to combat water crisis. According to the International Desalination Association (IDA), desalination capacity was 66.5 million m^3/d in 2011, (Henthorne 2012) and 78.4 million m^3/d in 2013 (meed.com 2012). The traditional desalination process is heating up the saline water at a pressure lower than atmospheric pressure which is called vacuum distillation. Because of lower boiling temperature, industrial waste heat can be used as the heat source. Other commercial desalination method is using semi-permeable membranes to remove salts by adding pressure to feed water which is reverse osmosis technology (Fritzmann et al. 2007). Typically Reverse Osmosis (RO) technology uses less energy than thermal technologies. Cost of desalinated water depends on both energy price and type of technology (Thiel 2015). In power and water cogeneration plant, required energy for desalination process is supplied by power block. Excess produced electricity will be fed into the grid. Different technologies can be used in power block and in water block. Dual purpose plants are energy efficient and can be a feasible option in MENA region (Hamed 2005; Misra and Kupitz 2004). The water cost trend for desalination process and re-use process are shown in Fig. 1. As can be seen, desalinated water cost is decreased to about 0.5 US$/m^3, but it is still slightly higher than re-use water cost. In Fig. 1, area A is a region that the price concepts are not applicable and areas B and C are regions that over exploit the natural resources and regions with abundant natural resources respectively (Reddy and Ghaffour 2007).

In 2008 Wittholz et al. derived a levelized costs or unit production costs (UPC) of water for desalination technology according to the capacity of water production (Fig. 2). As depicted in Fig. 2, thermal technologies are most expensive than membrane-based ones. It is also evident that the UPC is much lower for brackish water (BW) when compared with sea water (SW). This is well understood that lower total dissolved solids (TDS) will result in lower produced water cost (Wittholz et al. 2008).

Fig. 1 Trends in desalination and conventional water costs

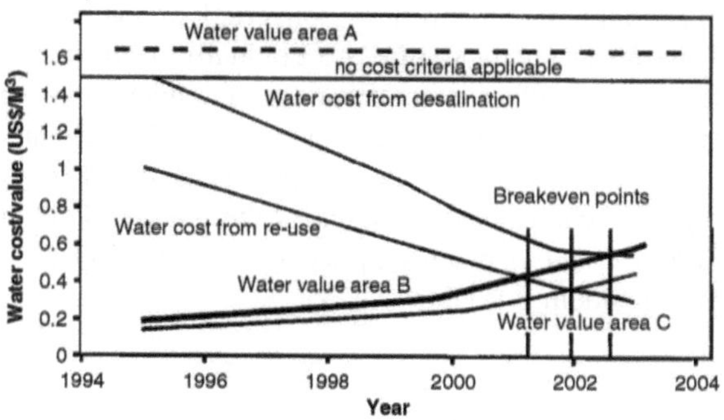

Fig. 2 Trends in water costs for various desalination technologies

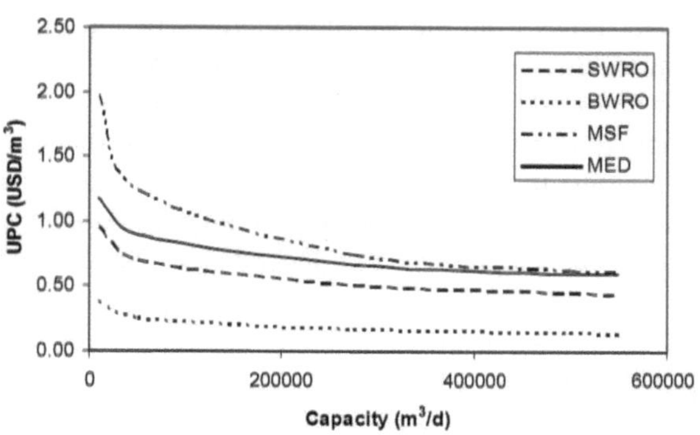

Because of the increase of power demand and aging of older power stations and according to the water crisis in Iran, both power and water generation plants are needed to meet the demand in both products. The principal goal of the present work is to compare the levelized cost of water in a cogeneration of water and power plant between two desalination technologies of MED[1] and RO[2] cogenerating with the standard combined cycle power plant of *MAPNA* (*Niam* Pattern, consisting two MGT70 gas turbines and one steam turbine).

2 Case Study of Cogeneration System

2.1 *Water Block*

2.1.1 MED Technology

MED systems are consisting of a number of effects (evaporators) that are connected together. Steam, which can be produced using waste heat or using stand-alone boiler, is the driver of MED evaporating sea water in the first effect. Part of sea water is sprayed on the tubes in each effect and the steam flows inside the tubes. By transferring heat, steam is condensed and leaves the effect as the distillated water. A fraction of sea water is converted to pure steam which is fed into the next evaporator and the rest is converted to the more concentrated water or brine. Pressure and temperature are gradually decreased through the effects. The vapor leaving last effect is condensed in a condenser using sea water. Figure 3 shows a simplified schematic of a MED process (Marie Zak 2012).

According to the latest experiences of MED desalination plants in MENA region, technical specifications of thermal desalination technology according to *Ras Laffan C* MED plant supplied by Sidem in Qatar are given by Table 1 (Veolia Water Sidem 2011).

2.1.2 RO Technology

Water flows from the less concentrated region to the more concentrated region in nature. But in reverse osmosis process, water is driven by pumps from the more concentrated region to the less concentrated region. There have been many improvements in RO technology in recent years and it usually uses less energy comparing to the thermal technologies. Nano-composite, nano-tube membranes and chlorine resistant membranes are three examples of research and development successful outputs in RO technology. Nano-composite, nano-tube membranes

[1]Multi Effect Distillation.

[2]Reverse Osmosis.

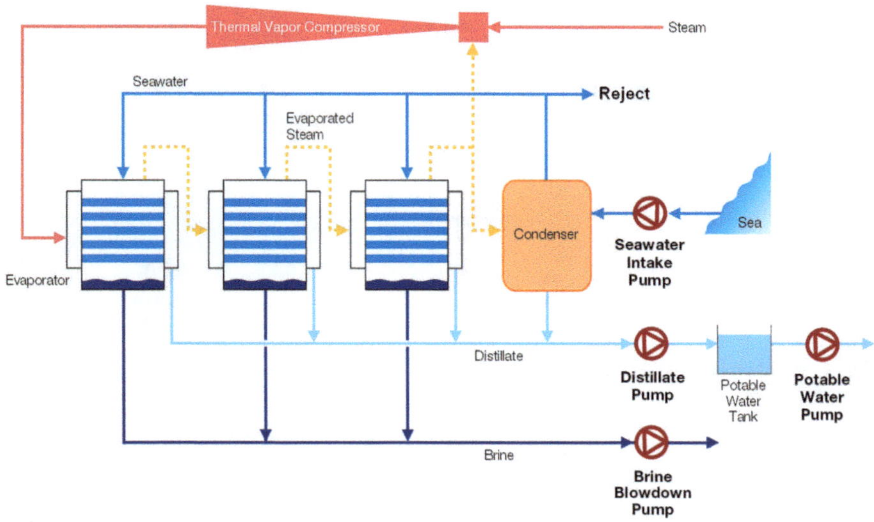

Fig. 3 MED desalination schematic

Table 1 Technical specifications of MED desalination

Parameter	Value
Motive steam pressure (bar A)	3.2 (saturated)
Number of effects	8
GOR	11.1
Capacity (each unit) (m^3/day)	28,600

reduce the energy usage of desalination plant and chlorine resistant membranes reduce the fouling in plant (Fritzmann et al. 2007). A schematic block diagram for a sea water reverse osmosis (SWRO) plant has been given in Fig. 4. Since the concentrate leaves the membrane module with a high pressure, an Energy recovery device (EDR) is usullay used to reover its energy based on a hydro-turbine or pressure exchanger.

2.2 *Power Block*

The power block configurations are based on MGT70 gas turbines. MGT70 is made by TUGA which is one of the subsidiary companies of MAPNA group. Two power configurations have been modeled. First scenario is consisted of two MGT70 gas turbines, two HRSG with SF and one Condensing steam turbine with once through cooling system coupled with a SWRO desalination plant. Second scenario is consisted of two MGT70 gas turbines, two HRSG with SF and one back pressure steam turbine coupled with a MED desalination plant. Technical specifications of power block are given in Table 2 (MAPNA Co. 2004a, b).

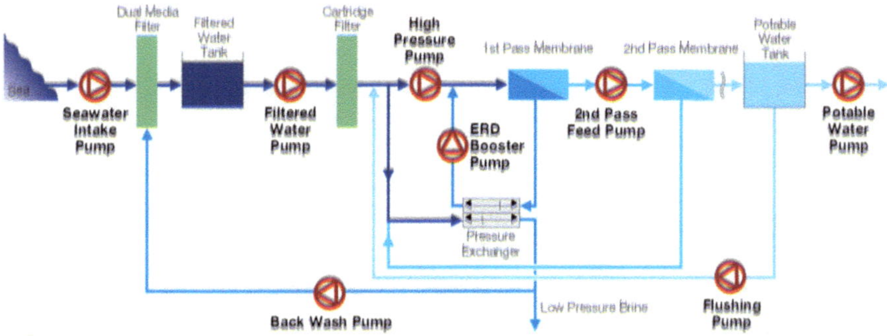

Fig. 4 A schematic SWRO desalination plant

Table 2 Technical specifications of power block

Parameter			Value
Gas turbine nominal power (MW)			170
Steam turbine nominal power		Condensing type	160
		Back-pressure type	95
HRSG	High pressure section	Pressure (bar A)	90
		Temperature (°C)	520
		Steam flow (ton/hr)	241.2
	Low pressure section	Pressure (bar A)	8.5
		Temperature (°C)	230
		Steam flow (ton/hr)	32.5
	Temperature at SF downstream (°C)		592.6
	Stack temperature (°C)		114.2

2.3 *Site Conditions*

Site conditions are assumed to be same as Qeshm Island. Dry bulb temperature, relative humidity, seawater temperature and seawater TDS are given in Table 3 (MAPNA Co. 2011).

Depending on seawater properties and desalination technology details, specific electricity consumption has different values. Persian Gulf seawater properties are taken as the case study and are supposed to be constant for both thermal and membrane technology, and therefore, the average specific electrical energy consumption for MED and RO desalination processes (considering intake pumps, additive dozing systems, pretreatment and post treatment) are considered as Table 4 (Corrado Sommariva 2010).

Table 3 Site conditions

Parameter	Value
Dry bulb temperature (°C)	26.5
Relative humidity (%)	69.8
Seawater temperature (°C)	35
Seawater TDS (ppm)	42,000

Table 4 Specific electricity consumption in MED and RO plant

Parameter	Value
MED specific electrical energy consumption (kWh/m^3)	1.5
RO (with ERD) specific electrical energy consumption (kWh/m^3)	4.5

2.4 *Cogeneration Configurations*

Two cogeneration systems have been studied in this research, as mentioned before. These two scenarios are:

2 × MGT70 + 2 × HRSG + 1 × CST + RO (there after named RO + CST)
2 × MGT70 + 2 × HRSG + 1 × BPST + MED (there after named MED + BPST)

3 Technical Simulation

Thermoflex module of Thermoflow-V.20 software was used for the thermodynamic simulation of combined power and water plant. Technical specifications of each plant in two configurations are given in Table 5. Simulated schematic for two mentioned configuration systems are shown in Figs. 5 and 6.

Table 5 Technical specification for two scenarios

Specifications		Value	
		RO + CST	MED + BPST
Total desalinated water capacity (m^3/day)		143,000	143,000 (5 × 28,600)
Total installed power (MW)		495	435
Net actual power (MW)		424.7	384.2
Auxiliary power usage (MW)	Water block	26.8	9
	Power block	12.7	11
Fuel consumption (m^3/h)		95,717	95,717
Net electrical efficiency (%)		47.70	43.14
Net cogeneration efficiency (%)		NA	83.59

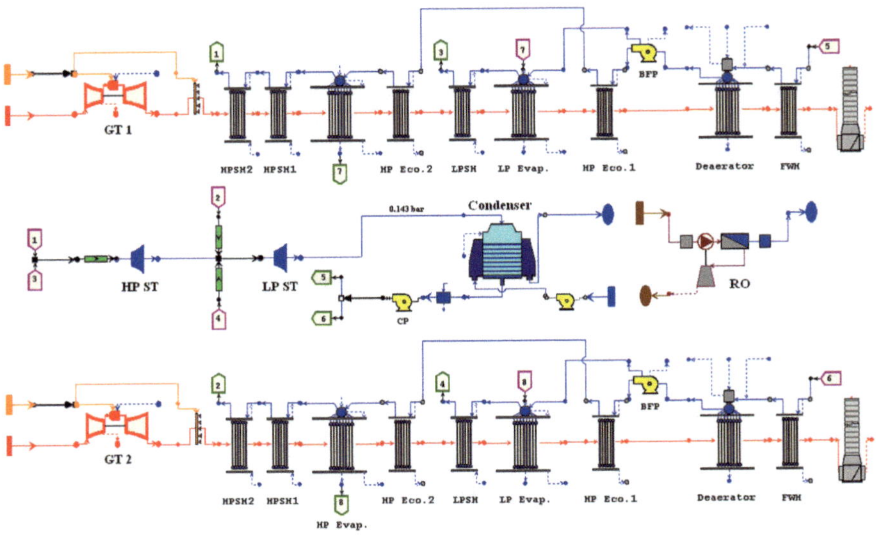

Fig. 5 RO + CST cogeneration of water and power schematic

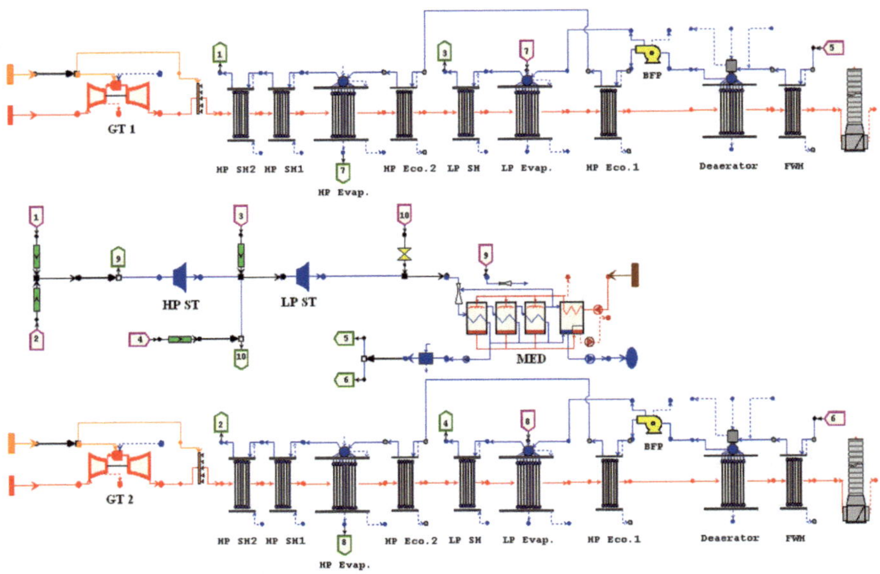

Fig. 6 MED + BPST cogeneration of water and power schematic

4 Economic Equations

4.1 Costs

4.1.1 Water Block Capital Cost

Unit capital cost of desalination plants is set to 1300 $\$/(m^3/day)$ for MED and 900 $\$/(m^3/day)$ for RO desalination plant considering recent desalination plants installed in Persian Gulf region (Tri-Tech, Itochu and Sasakura, Doosan, Veolia Sidem, Aquatech, Aqualyng, Hitachi and Hyflux Companies). The capital cost of water block is calculated by Eq. (1).

$$CC_w = W \times w^* \quad (1)$$

where W and w^* are the water capacity and the specific water capital cost, respectively.

4.1.2 Power Block Capital Cost

Unit cost of combined cycle power plant utilizing MGT70 gas turbine is considered 765 $/kW (Combined cycle with condenser steam turbine), in which, for MED + BPST scenario the condenser cost was subtracted from total capital cost. The capital cost of power block is calculated by Eq. (2).

$$CC_p = E \times e^* \quad (2)$$

where E and e^* are the power capacity and specific power capital cost, respectively.

4.1.3 Running Costs

Running costs (cost which are repeating annually) are consisted of water block operation and maintenance cost, power block operating and maintenance (O&M) cost and fuel cost.

Annual O&M cost for water block per cubic meter of desalinated water (Φ) is given in Table 6 (Torzewski 2015). For the power block annual O&M cost, it is set to 2% of power block capital cost.

Annual fuel price is calculated by Eq. (3).

$$C_f = F \times A \times P_f \times D \times H \quad (3)$$

where F represents the fuel consumption per hour, A is the availability of fuel consumption in year, P_f is the fuel price, D is the number of days in year, and H is the number of hours in day.

Table 6 O&M cost of desalination per m³ of produced water

Technology	Φ ($/m³)
RO	0.23
MED	0.17

4.2 Revenues

Revenues of Cogeneration of Water and Power (CWP) plant is consist of water revenue and power revenue according to Eqs. (4) and (5)

$$R_p = E \times A \times C_p \times D \times H \tag{4}$$

$$R_w = W \times A \times C_w \times D \tag{5}$$

where C_p and C_w are the units cost of electricity ($/kWh) and unit cost of water ($/m^3), respectively.

4.3 Levelized Cost Calculation Procedure

Considering loan for capital secure, the real capital cost of water and power block is calculated by Eqs. (6) and (7).

$$CC_{real_w} = (CC_w) \times \left[(1 - L\%) + \frac{L\%}{N_L} \sum_{N_C}^{N_L+N_C} \left(\frac{1+i_L}{1+i}\right)^n\right] \tag{6}$$

$$CC_{real_p} = (CC_P) \times \left[(1 - L\%) + \frac{L\%}{N_L} \sum_{N_C}^{N_L+N_C} \left(\frac{1+i_L}{1+i}\right)^n\right] \tag{7}$$

where L represents the loan percentage for the capital cost fund, N_L represents the loan payback time, N_C represents the construction time, i_L represents the loan payback interest rate and i represents the interest rate.

For calculating the levelized cost of water, firstly, levelized cost of electricity is calculated considering a combined cycle power plant with condensing steam turbine without any desalination plant according to Eq. (8). Then, using the calculated electricity levelized cost, the water levelized cost for both of RO + CST and MED + BPST scenarios is calculated by Eqs. (9) and (10).

$$CC_{real_p(\text{CST with no desalination})} = (1 - T) \sum_{n=N_C}^{N_L+N_C} \frac{(R_P - O\&M_P - C_f)_{(\text{CST with no desalination})}}{(1+i)^n} \tag{8}$$

Table 7 Economic assumptions

Parameter	Value
Discount rate (%)	12
Construction period (years)	1
Operation period (years)	20
Ratio of loan to total capital cost (%)	75
Loan payback discount rate (%)	6
Loan payback period (Years)	5
Availability[a] (%)	90
Tax to profit (%)	25
Fuel price ($\$/m^3_{fuel}$)	0[b]–0.3

[a]Annual operation time fraction
[b]Fuel price is set to zero in Energy Conversion Agreement condition (ECA)

$$CC_{real_w} = (1 - T) \sum_{n=N_C}^{N_L+N_C} \frac{(R_w - O\&M_w)}{(1+i)^n} \tag{9}$$

$$CC_{real_{w+p}} = (1 - T) \sum_{n=N_C}^{N_L+N_C} \frac{(R_P + R_w - O\&M_P - O\&M_w - C_f)}{(1+i)^n} \tag{10}$$

where T represents the tax associated with net profit.

4.4 Economic Assumptions

Economic assumptions of the study are given by Table 7.

5 Results and Discussion

5.1 Sensitivity Analysis of Power and Water Cost Against Fuel Price

Applying Eqs. (3)–(10), and considering technical specifications of Table 5, levelized cost of electricity and water manipulating fuel price can be calculated which are illustrated in Figs. 7 and 8. .

As it is shown in Fig. 7, levelized cost of electricity for fuel price range of 0–30 cent/m^3_{fuel} varies between 2.4 and 8.8 $cent/kWh. It is obvious that higher electrical efficiencies will result in lower slip of electricity cost with higher stability against fuel price. Based on technical simulation and economical relationships,

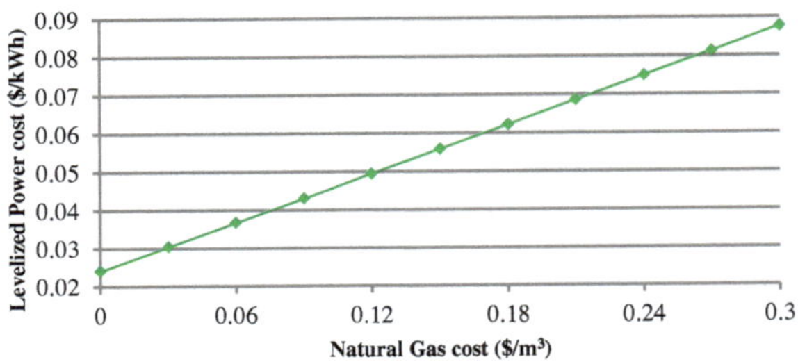

Fig. 7 Levelized cost of electricity against fuel price

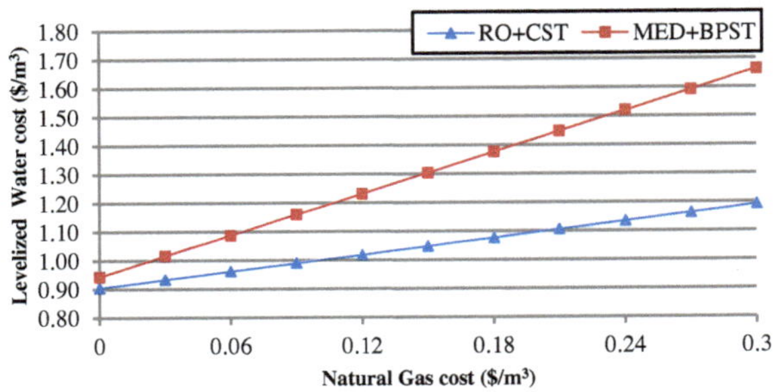

Fig. 8 Levelized cost of water against fuel price

Fig. 8 is showing that levelized cost of water for RO + CST scenario is lower than MED + BPST scenario. Also RO + CST scenario has higher stability against fuel price. Because of higher electrical efficiency for RO + CST scenario compared with MED + BPST, external costs of increase in fuel price have higher effect on MED + BPST scenario.

5.2 Sensitivity Analysis of Power and Water Cost Economic Parameters

Sensitivity analysis of water cost against fuel price in different water block availabilities considering other parameters to be constant is according to Fig. 9.

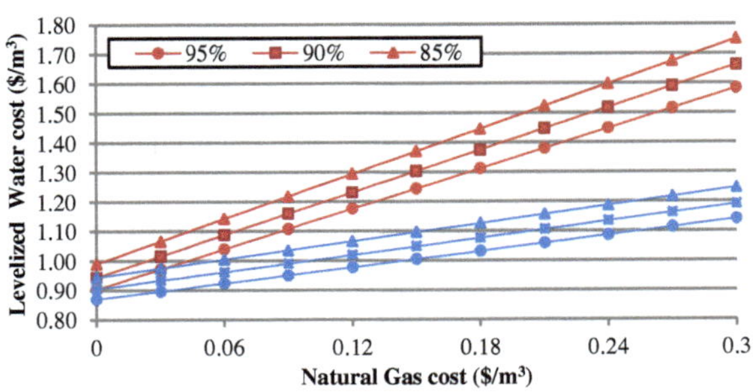

Fig. 9 Levelized cost of water against fuel price in different water block availabilities

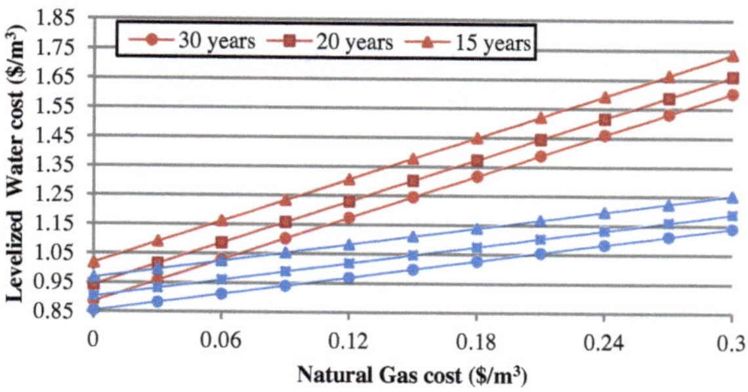

Fig. 10 Levelized cost of water against fuel price in different operation periods

Availability of water block is criterion for maintenance and reparation time i.e. availability of 90% means the water block facilities need about 36 days in a year for maintenance, and in this period, no potable water is generated. The importance of availability in higher fuel prices is much more than low fuel prices. Also the MED + BPST scenario has higher sensibility to availability than RO + CST scenario; the average increase in water cost of MED + BPST is 5% by reducing availability to 85% from 90%, but the average increase in water cost of RO + CST is 4.4% considering such decrement in availability. This result is showing that the minimum price of water should be set equal to cost of water at which availability is minimum, considering other parameters to be constant. Sensitivity analysis of water cost against fuel price in different operation periods considering other parameters to be constant is according to Fig. 10.

Operation period is a criterion of equipments quality and valid operation. As it is shown in Fig. 10, by decreasing operation period for both water and power block by 5 years, levelized cost of water in two scenarios of RO + CST and MED + BPST are increasing by 6.2 and 5.95% in average. In equal operation period, again levelized water cost of RO + CST scenario is lower than MED + BPST scenario. Because of relatively high interest rate, initial years of operation have more importance than latest years of operation. With this effect, decreasing 5 years of operation affects the levelized water cost more than increase of 10 years of operation. Considering no degradation in water production, if the operation period of RO + CST considered 5 years lower than MED + BPST at 20 years of operation, the levelized water cost of MED + BPST scenario is lower than RO + CST scenario for fuel price below 1.7 $cent/m^3. Also if anyway the operation period of MED + BPST scenario could be kept up to 30 years, the levelized cost of water in this scenario is lower than RO + CST for fuel price below 6 $cent/m^3. For 15 years of operation, the levelized cost of water in thermal plant is higher at any price of natural gas. Levelized cost of water against capital cost multiplier at fuel price of 8.6 $cent/m^3 (Persian Gulf FOB[3]—infomine.com September 30th 2015) has been given in Fig. 11. It is shown that for equal ranges of ±20% change in capital cost of water block, the cost of water for RO + CST

[3]Free On Board (Global price).

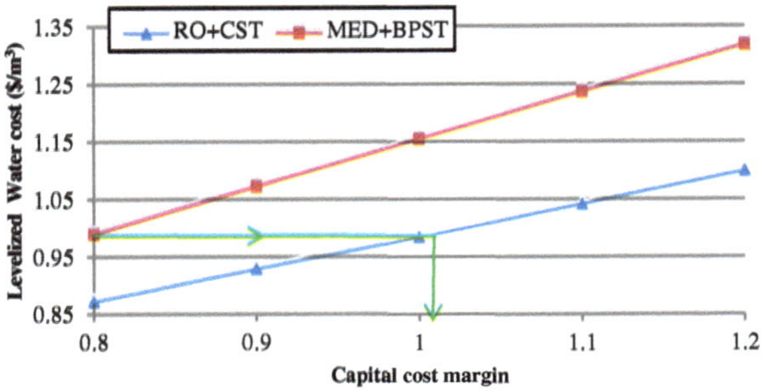

Fig. 11 Levelized cost of water against desalination capital cost multiplier in fuel price of 8.6 \$cent/m^3

scenario is remaining bellow MED + BPST scenario. In a specified capital cost, it has been show that by 20% decrease in MED capital cost (1040 \$/m^3/day) and 1% increase in RO capital cost (909 \$/m^3/day), levelized cost of water production for both RO + CST and MED + BPST is equal in fuel price of 8.6 \$cent/m^3.

6 Conclusions

In this paper, the levelized water cost of desalination in a large scale water production capacity of 143,000 was compared for both MED and RO plants. The specification of the MED plant was taken same as Ras Laffan C plant in Qatar. The study was done on the basis of cogeneration of water and power plant near the south coast of Iran, considering Persian Gulf seawater properties. Power block in both scenarios was based on *MAPNA Niam* pattern combined cycle with *MGT70* gas turbine and in membrane desalination the steam turbine is condensing type and in thermal desalination is back-pressure type. The results of technical and economical studies showed that maximum price of water for RO + CST scenario is nearly 1.2 \$/m^3 while this value for MED + BPST with 38% increase is about 1.66 \$/m^3. Considering range of ±20% equal variation in desalination capital cost, the water cost of RO + CST will remain under MED + BPST but the cost of water for two scenarios are equal for 20% decrease in capital cost for MED and +1% increase in capital cost for RO.

Acknowledgements We acknowledge the support of the Iran Power Plant Projects Management Company (MAPNA Group)—Investment Project Division for financial and technical support

References

Corrado Sommariva ILF Consulting Engineers (2010) Desalination and advanced water treatment. Balaban Desalination Publication

Desalination industry enjoys growth spurt as scarcity starts to bite. https://www.globalwaterintel.com/desalination-industry-enjoys-growth-spurt-scarcity-starts-bite/

Fischetti M (2007) Fresh from the sea. Sci Am. http://www.scientificamerican.com/article/fresh-from-the-sea/
Fritzmann C, Löwenberg J, Melin T, Wintgens T (2007) State of the art of reverse osmosis desalination. Desalination 216:1–76. doi:0011-9164/07/$
Hamed OA (2005) Overview of hybrid desalination systems-current status and future prospects. Desalination 186:207–214. doi:0011-9164/07/$
Henthorne L (2012) The current state of desalination. International Desalination Association
Marie Zak G (2012) Thermal desalination: structural optimization and integration in clean power and water. Dissertation, Massachusetts Institute of Technology
MAPNA Co. (2004a) Heat recovery steam generator heat balance diagram. In: Project: 22 combined cycle power units
MAPNA Co. (2004b) MGT70 specification. Mapna Turbine Engineering and Manufacturing Co.
MAPNA Co. (2011) Qeshm power and water cogeneration. In: Process flow diagram with heat and mass balance
Misra BM, Kupitz J (2004) The role of nuclear desalination in meeting the potable water needs in water scarce areas in the next decades. Desalination 166:1–9. doi:0011-9164/07/$
Reddy KV, Ghaffour N (2007) Overview of the cost of desalinated water and costing methodologies. Desalination 205:340–353. doi:0011-9164/07/$
Thiel GP (2015) Salty solutions. Phys Today
Torzewski A (2015) Dr. Rudolf Baten, MAPNA group desalination workshop. Fichtner GmbH & Co. KG
Veolia Water Sidem (2011) General Presentation—December
Wittholz MK, O'Neill BK, Colby CB, Lewis D (2008) Estimating the cost of desalination plants using a cost database. Desalination 229:10–20. doi:0011-9164/07/$
www.infomine.com/investment/metal-prices/natural-gas/1-year/. 1 Sept 2015

Modeling Dispersion of Brine Discharges from Multiple Desalination Outfalls

Anton Purnama

Abstract Discharging brine effluents through long sea outfalls are an economic disposal strategy for coastal seawater desalination plants. The interactions of two or more brine discharge plumes are expected as many desalination outfalls often tend to be closely clustered together along open coastlines. A far field mathematical model using a two-dimensional advection diffusion equation in a highly simplified flat seabed is presented to study the dispersion and merging of brine discharge plumes in shallow coastal waters. The analytical solutions are illustrated graphically by plotting contours of concentration to replicate the overlapping plumes following discharges from multiple desalination outfalls. To assess the potential environmental impacts, the radius and the concentration at the end of the allocated mixing zone around the outfall are formulated. The compounded concentration at the edge of the regulatory mixing zone is then used as a measure for assessing the effectiveness of multiport diffusers over the single (port) outfall discharge. It is found that the modern engineering practice which installs a multiport diffuser at the end of the outfall pipe does minimize the impacts.

Keywords Environmental impact assessment · Flat seabed · Mathematical model · Multiport diffuser · Regulatory mixing zone

1 Introduction

In hot and arid climate countries such as Oman and the rest of the Arabian Gulf, Middle East and North Africa countries, seawater desalination is the reliable solution to meet the growing demands for water due to population growth and economic and social development and to reduce the reliance on depleting groundwater resources (Ahmad and Baddour 2014; Roberts et al. 2010; Voutchkov

A. Purnama (✉)
Department of Mathematics and Statistics, Sultan Qaboos University, Muscat, Sultanate of Oman
e-mail: antonp@squ.edu.om

O. Abdalla et al. (eds.), *Water Resources in Arid Areas: The Way Forward*, Springer Water, DOI 10.1007/978-3-319-51856-5_19

2011). To manage the future electricity and water demands in the Sultanate, Oman Power and Water Procurement (OPWP) was established in 2005 to procure the production of electricity and desalinated water and to ensure the adequacy of generation of resources for new power and desalination capacities. According to the 7-year statement for the period 2015 to 2021 (OPWP 2015), OPWP forecasted that the projected total demand for the desalinated water will increase at an average rate of 7% per year from 948,000 m^3/d in 2014. Additional desalinated water requirements will be made in 2015 by expansion at Barka I plant with a capacity of 57,000 m^3/d, and at Al-Ghubrah plant with a capacity of 191,000 m^3/d. Further expansion in 2016 will be made at Sur plant with a capacity of 48,000 m^3/d. For future requirements, a new plant will be built by 2017 at Quriyat with a capacity of 200,000 m^3/d. By 2018, a new Barka III plant with a capacity of 281,000 m^3/d and a new Sohar II plant with a capacity of 250,000 m^3/d will be built. Another expansion will be made by 2019 at Sur with a capacity of 55,000 m^3/d and at Duqm with a capacity of 60,000 m^3/d. OPWP also welcomes the opportunity for combining power generation with water desalination so as to benefit from economies of co-location and co-procurement. Power generation and seawater desalination plants extract large volumes of seawater and discharge back (unwanted) hot wastewater via marine outfall systems into the sea, containing a high salt concentration (hypersaline brine), together with occasional discharges of corrosion products, toxic antifoulants and antiscalants used (Lattemann and Hopner 2008; Roberts et al. 2010; Voutchkov 2011).

An outfall is a long pipeline that discharges brine effluents from coastal desalination plants to the ocean (Bleninger and Jirka 2008; Purnama et al. 2011). Immediately after release from the outfall, vigorous and rapid dilution of concentrate brine is governed by the effluent buoyancy, momentum of the discharge and its interaction with the receiving sea currents. At the end of this near field mixing stage, the established steady discharge brine plume then continues to drift away with the currents. There are several approaches adopted to mitigate the potential environmental effects of brine discharge. To avoid direct impacts from high salinity in the near field, especially for the co-location power and seawater desalination plant, brine effluents are diluted with seawater used for cooling in the power plant (Purnama et al. 2011). An engineering solution to achieve high dilution is to install a properly designed diffuser system at the end of the outfall pipe (Bleninger and Jirka 2008). Further, by locating the outfall in a favorable oceanographic site, high dilution of the discharge brine plume can be enhanced into the far field.

If continuous discharge of brine effluents can not be avoided, then it should be done as optimally as possible to ensure minimal environmental impact. For a shorter outfall, it is observed that the discharge plumes are spreading and heading towards the coastline (Al-Barwani and Purnama 2008; Purnama and Al-Barwani 2006). Coastal regions and beaches are important to Oman for fisheries, local recreation and tourism and for conservation areas. Oman has many but scattered and remote coastal communities typically living in small fishing villages. The growing tourism industry and fisheries industry are among the most important and profitable industries in Oman. For a long sea outfall, one factor affecting the dispersion of brine effluents in the far field is the seabed depth profiles (Al-Barwani and Purnama 2005; Kay 1987; Purnama and Al-Barwani 2004; Purnama et al. 2004), which typically range between a sloping sandy beach and a mountainous coast with rock sea cliffs, where water depth gets very deep within a short distance from the coastline, and thus variations in water depth become insignificant. Therefore, we introduce a highly simplified model of flat seabed with a constant water depth, $h = h_0$ (Fig. 1), where h_0 is an arbitrary reference water depth and the beach is assumed to be a continuation of the rock sea cliffs.

In the far field modelling, the beach is considered to be straight and the sea wide, and the brine discharges plume is assumed to be vertically well-mixed over the water depth (Chin 1985; Kay 1987). The coastal (drift) current is assumed to be steady with a speed U_0 and remains in the x-direction parallel to the beach (positive to right of the discharge site) at all times. For simplicity, other complexities such as tidal motions, density and temperature are also ignored. The dispersion processes are represented by eddy diffusivities D_0, and diffusion in the x-direction is neglected, as the plumes in steady currents become very elongated in the x-direction (Al-Barwani and Purnama 2008; Purnama and Al-Barwani 2006).

For multiple desalination outfalls, we represent the first (single) long outfall discharging brine continuously at a constant rate Q_0 as a single point source at $(x_0 = 0, y_0 = \alpha h_0)$; the second outfall discharging at a rate Q_1 as another point source at $(x_1 = -\ell h_0, y_1 = (\alpha + h)h_0)$; the third outfall discharging at a rate Q_2 as a point source at $(x_2 = -2\ell h_0, y_2 = (\alpha + 2h)h_0)$; and so on, where $hh_0 > 0$ is the outfall's (offshore) and $\ell h_0 > 0$ the (along the shore) separation distances. As illustrated in Fig. 1, the point sources at (x_k, y_k), where $x_k = -k\ell h_0$, $y_k = (\alpha + kh)h_0$, represent the kth long sea outfall discharging brine effluent with a constant rate Q_{k+1}.

The long-term environmental impact assessments of brine discharge from coastal desalination plants have become increasingly important, both as a result of public concern and scientific awareness and because of the increasing scale of seawater desalination activity (Lattemann and Hopner 2008; Jenkins et al. 2012; Roberts

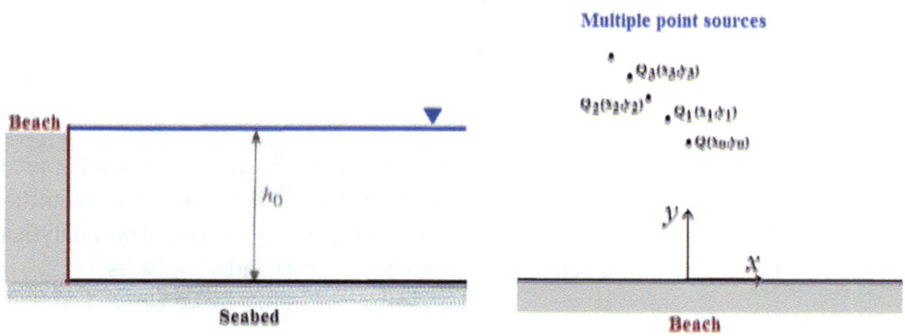

Fig. 1 Cross-section view of a flat seabed (*left*); and plan view of multiple point sources in a flat seabed (*right*)

et al. 2010). Desalination brine outfall discharge is currently regulated by imposing a concentration limit value at the allocated mixing zone (Ahmad and Baddour 2014; Jenkins et al. 2012). For multiple brine outfalls operated closely together along open coastlines, where the cumulative impacts from neighbouring outfalls are strongly inter-dependent and of a capacity limit of the receiving waters (Jirka et al. 2004), how relevant and adequate are the current regulations? The use of mathematical models has been a key strategy for the basis of sound engineering design and for assessing the potential impacts (Bleninger and Jirka 2008; Chin 1985; Jenkins et al. 2012). In terms of the practical applicability, it is well recognized that the present far field mathematical model can be applied as a tool to perform preliminary worst-case assessments. If this easy-to-use assessment indicates no far-field impacts at all, no further action is needed and the use of more sophisticated and time-consuming three-dimensional hydrodynamic and water quality modeling can be avoided.

The aim of this paper is to investigate if the standard mixing zone regulations can still be used for assessing the long-term impacts from the discharge of multiple brine outfalls. Using the simple far field model, the radius and the concentration at the end of the mixing zone are formulated analytically. The results can be used to answer questions such as, if a new seawater desalination plant is to be built on a coastline where an existing plant is operated, how can one calculate the compounded concentration at the end of the mixing zone? And in particular, will a scenario ever arise in which the existing plant is no longer meeting the regulatory requirements?

2 Single Outfall Discharges

First, we assume that the brine effluents are being discharged with a constant rate Q_0 from the single point source at $(0, \alpha h_0)$, where the value of α is sufficiently large so that the brine plumes do not feel the presence of the vertical beach at $y = 0$. The two-dimensional advection-diffusion equation for the plume concentration $c(x, y)$ is given by (Kay 1987; Purnama et al. 2004)

$$h_0 U_0 \frac{\partial c}{\partial x} - h_0 D_0 \frac{\partial^2 c}{\partial y^2} = Q_0 \delta(x) \delta(y - \alpha h_0), \tag{1}$$

where δ is the Dirac delta function used to represent the position of a point source. For graphical representation of the solution, we use dimensionless quantities $c(x, y) = C(X, Y) Q_0 / h_0^2 U_0$, $\lambda = U_0 h_0 / D_0$, $x = X h_0$ and $y = Y h_0$. The analytical solution of Eq. (1), in its dimensionless form for $X \geq 0$ and $Y \geq 0$, is

$$C(X,Y)=\sqrt{\frac{\lambda}{4\pi X}}\exp\left\{-\frac{\lambda(Y-\alpha)^2}{4X}\right\}. \qquad (2)$$

The model parameter $\lambda = U_0 h_0 / D_0$ represents the brine plume elongation in the x-direction, which reflects the uncertainty in sea conditions. Figure 2 shows the graphs of λ for some relevant measured values of U_0 and D_0 for a shallow water of depth $h_0 = 6$ m. The larger values of λ are mostly due to a stronger current U_0 with less longitudinal dispersivity D_0 (Kay 1987).

The model application is illustrated graphically by plotting contours of the concentration given by solution (2), and the value of $\lambda = 0.1$ is used in all plots, unless stated otherwise. Contours of Eq. (2) are plotted in Fig. 3 for two values of $\alpha = 60$ and $\alpha = 90$ (which corresponds, for a shallow water depth of 6 m, to a desalination outfall length of 360 and 540 m respectively). It is shown that the concentration value of 0.004 following discharges from the single outfall at $\alpha = 60$ will eventually reach the vertical beach $y = 0$ at about 1200 m downstream. However, for discharges from the long outfall at $\alpha = 90$, the concentration value of 0.004 is spreading parallel to the beach, as though it does not feel the presence of the beach. Therefore, the value of $\alpha \geq 90$ will be used in the subsequent plots to represent a long sea outfall where significant parts of the concentration discharge plumes are not reaching the vertical beach at $y = 0$.

The potential environmental impacts of an ocean outfall discharge are usually regulated using the concept of a regulatory mixing zone (RMZ), *an allocated impact zone* or a region of non-compliance around the outfall where water quality criteria can be exceeded (USEPA 1991). Thus, water quality requirements are specified at the edge of the mixing zone rather than by the end of the outfall pipe (Jirka et al. 2004; Purnama et al. 2011). The standard regulatory guidelines for assessing the potential impact of brine discharges into the sea consists of two key

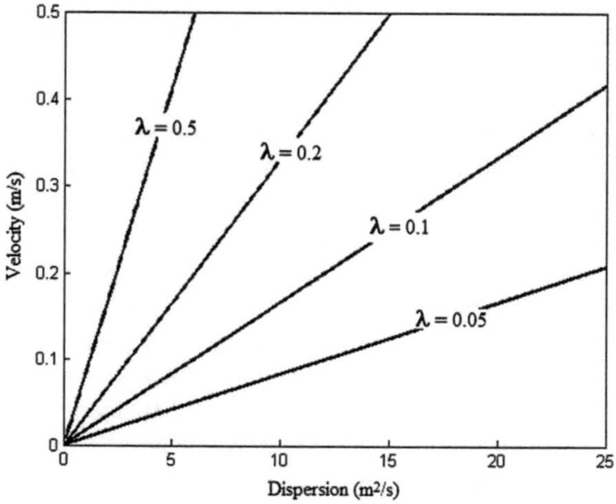

Fig. 2 The model parameter λ for a shallow water depth of 6 m

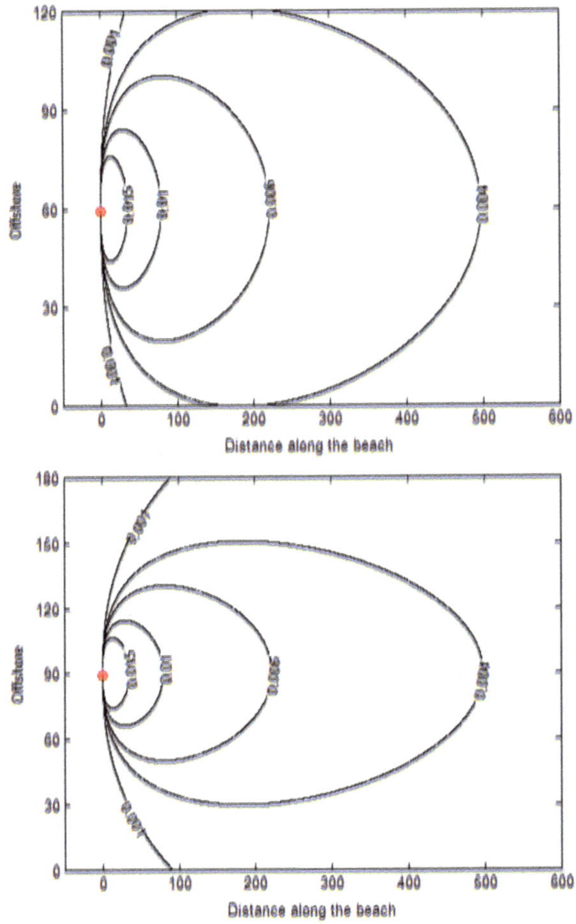

Fig. 3 Contours of concentration for a single point source: (*top*) $\alpha = 60$ and (*bottom*) $\alpha = 90$

elements: a (salinity) concentration limit and a point of compliance expressed as a distance from the outfall discharge (Ahmad and Baddour 2014; Jenkins et al. 2012). In the Sultanate of Oman, the excess salinity should not be more than 2 ppt above ambient, and the point of compliance for the concentration limit is the boundary of the circular mixing zone of radius 300 m (Sultanate of Oman 2005).

However, for modeling purposes the concentration limits can also be expressed as a minimum dilution (Ahmad and Baddour 2014), which is defined as the ratio of the concentration at the end of the outfall pipe to that of the maximum allowable concentration at the end of the RMZ (Jirka et al. 2004). The acceptable excess salinity at the mixing zone boundary normally should not exceed 5%, which is equivalent to a minimum dilution not less than 20:1. The radius R of the RMZ varies from 50 to 500 m (Ahmad and Baddour 2014; Jenkins et al. 2012).

The maximum concentration value at the edge of the RMZ of radius R from the single point source at $(0, \alpha)$ is $C_* = 0.5/R\sqrt{\pi}$ which occurs at $X = R = 1/\lambda$ and

$Y = \alpha$. To investigate the effects of uncertainty in sea conditions, the minimum dilution $1/C_*$ at the end of the RMZ is given in Table 1. Both the minimum dilution value and radius R increase as the model parameter value $\lambda = 0.1$ reduces to $\lambda = 0.02$. Thus, the minimum dilution and radius values are smaller for favourable oceanographic mixing condition (i.e. stronger ambient current), and a smaller concentration value and a larger circular mixing zone is required for a calm sea condition.

3 Two Outfalls Discharges

We consider two long brine outfalls located ℓh_0 distance apart as illustrated in Fig. 4 (Purnama and Shao 2015), where the single long outfall of length $\alpha \geq 90$ is represented as a point source at the position $(0, \alpha h_0)$ discharging brine with rate Q_0. Similarly, for $h \geq 0$ we represent the (new) second outfall of length $(\alpha + h)h_0$ as another point source at $(-\ell h_0, (\alpha + h)h_0)$ discharging with different constant rate Q_1. If these two outfalls are operated by one large plant discharging, then a

Table 1 The RMZ for a shallow water depth of 6 m

λ	$R h_0$ (m)	$1/C_*$	C_* (%)
0.1	60	35:1	2.9
0.09	67	39:1	2.6
0.08	75	44:1	2.3
0.07	86	51:1	2.0
0.06	100	59:1	1.7
0.05	120	71:1	1.4
0.04	150	89:1	1.1
0.03	200	118:1	0.8
0.02	300	177:1	0.6

Fig. 4 Plan view of the two point sources in a flat seabed

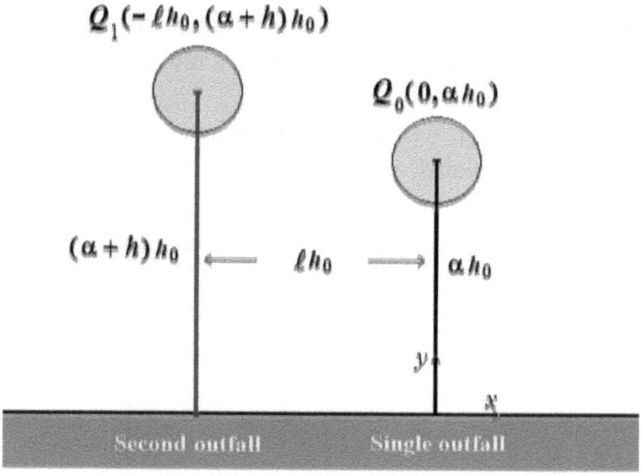

combined total of discharge rate is defined as $Q = Q_0 + Q_1$. For two independent plants discharging, we will assume that $Q = Q_0$, which is the value of the original discharge rate of the single outfall discussed in the previous section.

By applying a linear superposition, the (dimensionless) two-dimensional advection-diffusion equation for the plume concentration $C(X, Y)$ from the two point sources is given by

$$\frac{\partial C}{\partial X} - \frac{1}{\lambda}\frac{\partial^2 C}{\partial Y^2} = q_0\,\delta(X)\delta(Y-\alpha) + q_1\,\delta(X+\ell)\delta(Y-(\alpha+h)), \tag{3}$$

where $q_1 = Q_1/Q$ and $q_0 = Q_0/Q$. The solution of Eq. (3) is given by

$$C(X,Y) = \sqrt{\frac{\lambda}{4\pi X}}q_0\exp\left\{-\frac{\lambda(Y-\alpha)^2}{4X}\right\} + \sqrt{\frac{\lambda}{4\pi(X+\ell)}}q_1\exp\left\{-\frac{\lambda(Y-(\alpha+h))^2}{4(X+\ell)}\right\}. \tag{4}$$

The interaction of two plumes is also governed by the outfall lengths α and h, ℓ the outfall's separation distance between them, and the individual discharge rates q_1 and q_0.

Figure 5 shows a typical merging plume following discharges from two point sources with a separation distance $\ell = 30$ (which, for a shallow water depth of 6 m, corresponds to a distance of 180 m), and discharging at equal rate $q_0 = q_1 = 1$, and it is clear that for $X \geq 0$, the merging plume is spreading as one.

Substituting $Y = \alpha$ and $X = R = 1/\lambda$ in Eq. (4), we obtain the compounded concentration at the end of the RMZ as

$$C_{\mathrm{m}} = q_0 C_* \left[1 + \frac{q_1}{q_0}\frac{1}{\sqrt{1+\ell/R}}\exp\left\{-\frac{(h/R)^2}{4(1+\ell/R)}\right\}\right]. \tag{5}$$

For illustration, the concentration at the end of the RMZ is shown in Fig. 6 when $h/R = 2$. The value of compounded concentration C_m/q_0C_* decreases as the separation distance ℓ/R increases. That is, the contribution of the second outfall is smaller if its length $\alpha + h$ is longer, and its discharge rate q_1 is smaller, than that of the single outfall.

For the case where the two brine outfalls are operated independently by two desalination plants, the separation distance would be extremely large (of the order of 10 km) and the two brine outfalls discharge plumes are well separated. Thus, each (single) outfall will produce a maximum concentration value $C_* = 0.5/R\sqrt{\pi}$ at the edge of the RMZ.

Next, if the separation distance between the two outfalls ℓh_0 is of the order of kilometers, then from Eq. (5), the compounded concentration C_{m} at the edge of the RMZ can be approximated by

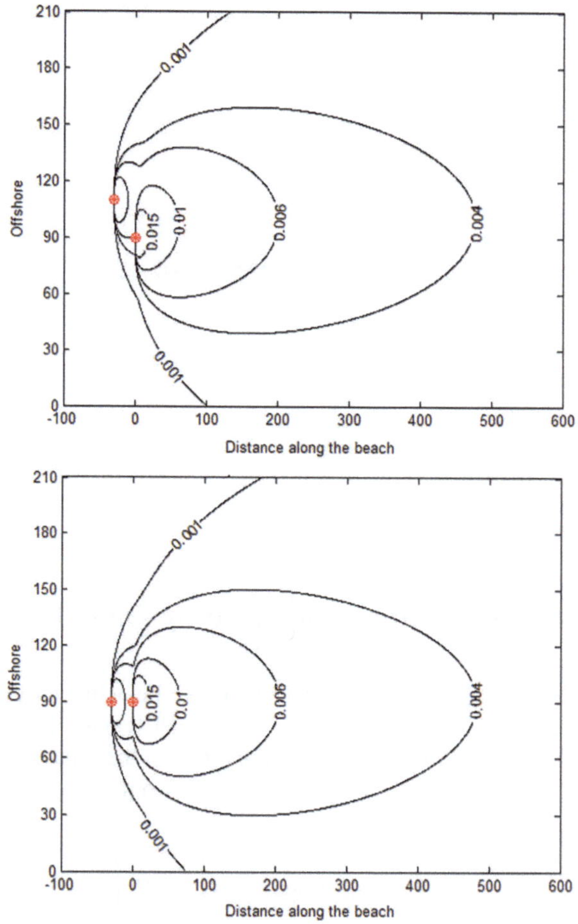

Fig. 5 Contours of concentration for two point sources with $\ell = 30$ and $\alpha = 90$: (*top*) $h = 20$ and (*bottom*) $h = 0$

$$C_{\mathrm{m}} \approx q_0\, C_* \left(1 + \frac{q_1}{q_0}\frac{1}{\sqrt{1+\ell/R}}\right).$$

It is easy to see, for example, if $\ell/R \geq 333$ (which, for a shallow water depth of 6 m, corresponds to a distance of 2 km), then $C_{\mathrm{m}} \approx q_0 C_*[1 + 0.055(q_1/q_0)]$. If both outfalls are discharging at an equal rate $q_0 = q_1$, then the compounded concentration is about 5.5% higher than $q_0 C_*$. However, if $q_1 = 0.5q_0$, the concentration is about 2.8% higher than $q_0 C_*$.

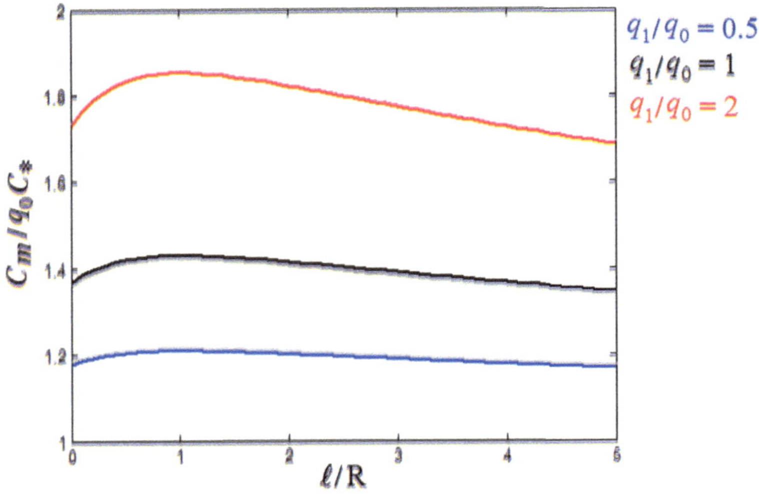

Fig. 6 Compounded concentration at the end of the RMZ for two point sources

4 Two Outfalls Operated by One Plant

Modifying the rates of discharge can be achieved through a relatively less expensive method than extending the outfall lengths. Thus, if the two brine outfalls are operated by one desalination plant, then the total discharges can be shared between them, i.e. $q_0 + q_1 = 1$. For $1 + \ell/R = 0.5(h/R)^2$, Eq. (5) has a maximum value

$$C_{\max} = C_* \left[1 - q_1 \left(1 - \sqrt{\frac{2}{e}} \frac{1}{h/R} \right) \right].$$

As shown in Fig. 7, for $q_1 > 0$ and $h/R \geq 0.86$, then $C_{\max} \leq C_*$. That is, the maximum concentration value at the end of the RMZ is smaller than that of the single outfall C_*. This result agrees with the finding that the total brine effluent load can be allocated optimally between two outfalls to minimize the potential environmental impact (Purnama and Shao 2015; Smith and Purnama 1999). However, economically it is cheaper to install a two-port diffuser at the end of the (single) outfall pipe than build a new sea outfall.

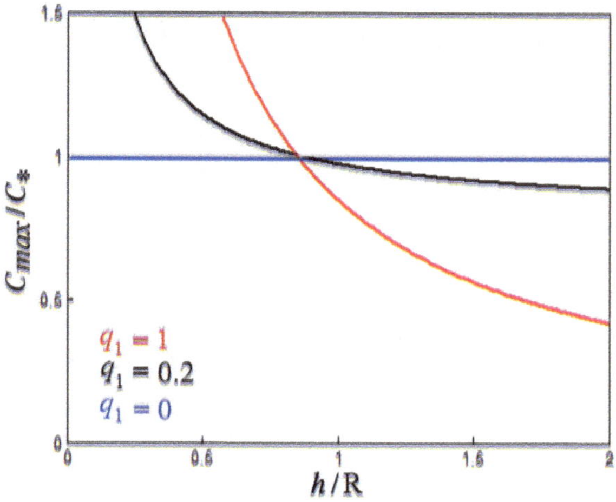

Fig. 7 Maximum concentration value at the end of the RMZ for two point sources when $q_0 + q_1 = 1$

5 Single Outfall with a Two-Port Diffuser

For the case of a two-port diffuser installed at the end of the outfall pipe, the separation distance between two outfall's ports would be short and is about $\ell/R \leq 0.01$. For example, when $\lambda = 0.1$ and for a shallow water depth of 6 m, this gives ℓh_0 to be less than 0.6 m. The compounded concentration value at the end of the RMZ as given in Eq. (5) can be reduced to

$$C_{\mathrm{m}} \approx C_* \left[q_0 + q_1 \exp\left\{ -\left(\frac{h}{2R}\right)^2 \right\} \right]$$
$$\approx C_* \left[1 - q_1 \left(\frac{h}{2R}\right)^2 \left\{ 1 - \frac{1}{2}\left(\frac{h}{2R}\right)^2 + \cdots \right\} \right],$$

since $q_0 + q_1 = 1$. Again, it is easy to verify that the value of the concentration is smaller than C_*. Figure 8 shows that as the discharge rate q_1 increases and the port's offshore separation distance h/R gets longer (in the range of 1.5–6 m for a shallow water depth of 6 m), the compounded concentration value gets smaller than C_*. This result demonstrates that installing a two-port diffuser at the end of the outfall pipe will improve the mixing and dilution of the brine effluent discharge plume in shallow coastal waters.

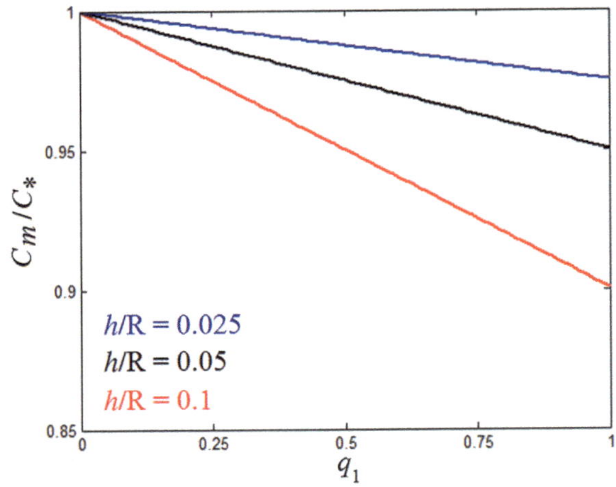

Fig. 8 Compounded concentration at the end of the RMZ for a two-port diffuser

6 Single Outfall with a Multiport Diffuser

The modern engineering practice for brine discharge from desalination plants is to distribute the large volume of brine effluent over a large expanse by installing a multiport diffuser at the end of the outfall pipeline to substantially improve the mixing and dilution of brine discharge plumes in the coastal waters (Bleninger and Jirka 2008; Purnama et al. 2011). A multiport diffuser is a line structure consisting of many closely spaced ports or nozzles designed to release a series of brine effluent plumes into the receiving coastal water. Setting both values of hh_0 and ℓh_0 as small compared to αh_0, the multiple point sources depicted in Fig. 1 represents a design of a multiport diffuser.

For a multiport diffuser with n ports discharging, the total brine discharge load is distributed equally, and so each port discharges at a constant rate of $q_k = 1/n + 1$. The solution for the brine plume concentration $C(X,\ Y)$ is given by

$$C(X, Y) = \sum_{k=0}^{n} q_k \sqrt{\frac{\lambda}{4\,\pi(X + k\ell)}}\,\exp\left\{-\frac{\lambda(Y - (\alpha + kh))^2}{4(X + k\ell)}\right\}.$$

Figure 9 shows a typical merging plume following discharges from four point sources with separation distance $\ell = h = 1$ and each point source discharging at an equal rate of 0.25, and downstream of the last point source (i.e., for $X \geq 0$), the merging plume is spreading as one.

Similarly, on substituting $Y = \alpha$ and $X = R = 1/\lambda$, we obtain the compounded concentration at the end of the RMZ

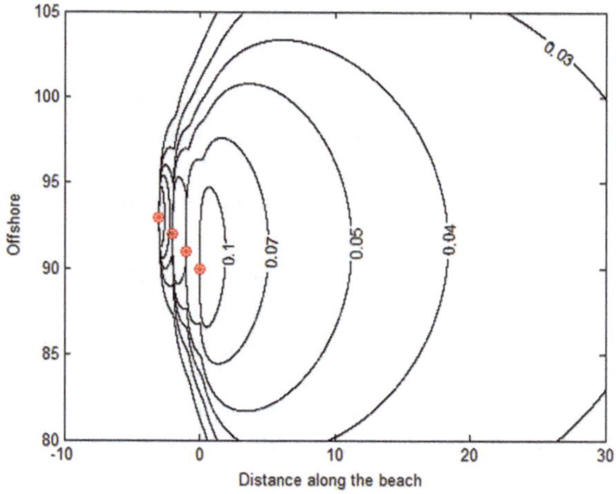

Fig. 9 Contours of concentration for four point sources with $\ell = h = 1$ and $\alpha = 90$

Table 2 The multiplier of C_m/C_* in Eq. (6)

n	$h/R = 0.025$	$h/R = 0.05$	$h/R = 0.1$
4	0.999	0.996	0.985
9	0.996	0.982	0.929
14	0.989	0.958	0.831
19	0.981	0.923	0.691
24	0.969	0.878	0.510

$$C_{\mathrm{m}} = C_* \left[\sum_{k=0}^{n} q_k \, \frac{1}{\sqrt{1 + k\ell/R}} \exp\left\{ -\frac{(kh/R)^2}{4(1 + k\ell/R)} \right\} \right].$$

Again, on assuming $\ell/R \leq 0.01$, the compounded concentration value can be approximated to

$$C_{\mathrm{m}} \approx C_* \sum_{k=0}^{n} \frac{1}{n+1} \left\{ 1 - \left(\frac{kh}{2R} \right)^2 + \cdots \right\} \approx C_* \left\{ 1 - \frac{n(2n+1)}{24} \left(\frac{h}{R} \right)^2 + \cdots \right\}. \tag{6}$$

From Table 2, we noted that as the number of points n and the separation distance h/R increases, the compounded concentration value becomes smaller than C_*. This finding is in agreement with the general fact that in coastal waters, a multiport diffuser improves the mixing of brine effluent plumes substantially (Bleninger and Jirka 2008; Purnama et al. 2011).

7 Concluding Remarks

The simple models proposed can be extended to any type of passive pollutant continuously being discharged into the sea from a point source, such as treated effluents from coastal sewage works, toxic contaminants from industrial installations in coastal areas, and hot water discharged from coastal power stations. However, implementation issues related to the control of discharge rates, reliability and cost effectiveness of the marine outfall are not addressed.

The potential impacts of multiple brine outfalls operated closely together are strongly inter-dependent of the capacity of the coastal waters and compounded from neighbouring outfalls. Using a simple far field model, the radius and the compounded concentration at the end of the RMZ are formulated for multiple point sources. The results show that if two desalination outfalls are independently operated and at a large distance apart, the concentration value at the end of the RMZ can be kept small as long as the new outfall discharges at a rate smaller than that of the single outfall. If the integrated total brine effluent discharge load can be shared between two outfalls, then the maximum concentration value is smaller than that of the single outfall if the new outfall length is longer than the mixing zone. A similar result is also found for the case of an outfall installed with multiport diffusers.

References

Ahmad N, Baddour RE (2014) A review of sources, effects, disposal methods, and regulations of brine into marine environments. Ocean Coast Manag 87:1–7

Al-Barwani HH, Purnama A (2005) The effect of the beach profile on dispersion of brine waste discharges. Int J Manag Syst 21:1–14

Al-Barwani HH, Purnama A (2008) Simulating brine plumes discharged into the seawaters. Desalination 221:608–613

Bleninger T, Jirka GH (2008) Modelling and environmentally sound management of brine discharges from desalination plants. Desalination 221:585–597

Chin DA (1985) Outfall dilution: the role of a far-field model. J Environ Eng 111:473–486

Jenkins SA, Paduan JD, Roberts PJW, Schlenk D, Weis JS (2012) Management of brine discharges to coastal waters recommendations of a science advisory panel. Technical report 694, California Water Resources Board, California, USA

Jirka GH, Bleninger T, Burrows R, Larsen T (2004) Environmental quality standards in the EC-water framework directive: consequences for water pollution control for point sources. European Water Management Online (EWMO) www.ewaonline.de

Kay A (1987) The effect of cross-stream depth variations upon contaminant dispersion in a vertically well-mixed current. Estuar Coast Shelf Sci 24:177–204

Lattemann S, Hopner T (2008) Environmental impact and impact assessment of seawater desalination. Desalination 220:1–15

Oman Power and Water Procurement (2015) OPWP's 7-year Statement (2015-2021), Muscat, Sultanate of Oman. www.omanpwp.com

Purnama A, Al-Barwani HH (2004) Some criteria to minimize the impact of brine discharge into the sea. Desalination 171:167–172

Purnama A, Al-Barwani HH (2006) Spreading of brine waste discharges into the Gulf of Oman. Desalination 195:26–31

Purnama A, Al-Barwani HH, Al-Rawahi MS, Al-Harthi TM (2004) Dispersion models for brine discharges from desalination plants of Oman. Kuwait J Sci Eng 31:1–13

Purnama A, Al-Barwani HH, Bleninger T, Doneker RL (2011) CORMIX simulations of brine discharges from Barka plants, Oman. Desalin Water Treat 32:329–338

Purnama A, Shao DD (2015) Modeling brine discharge dispersion from two adjacent desalination outfalls in coastal waters. Desalination 362:68–73

Roberts DA, Johnston EL, Knott NA (2010) Impacts of desalination plants discharges on the marine environment: a critical review of published studies. Water Res 44:5117–5128

Smith R, Purnama A (1999) Two outfalls in an estuary: optimal wasteload allocation. J Eng Math 35:273–283

Sultanate of Oman (2005) Ministerial Decision No. 159/2005: promulgating the bylaws to discharge liquid waste in the marine environment, Minister's Office, Ministry of Regional Municipalities. Environment and Water Resources, Muscat, Sultanate of Oman

USEPA (1991) Technical support document for water quality-based toxics control. EPA 505-290-001. U.S. Environmental Protection Agency, Office of Water, Washington, DC

Voutchkov N (2011) Overview of seawater concentrate disposal alternatives. Desalination 273:205–219

Assessment of Hydro-chemical Processes Inducing the Groundwater Salinisation in Coastal Regions: Case Study of the Salalah Plain, Sultanate of Oman

Brahim Askri and Razan Ali Al-Shanfari

Abstract Salalah plain aquifer, located in the southeast of Sultanate Oman, has been extensively used for agricultural, industrial and municipal purposes since the early 1970s. Over abstraction to satisfy the growing need of freshwater demand has contributed to the groundwater salinisation by seawater intrusion. The aim of this study is to evaluate the impact of this phenomenon on the groundwater quality in the Salalah plain. The occurrence of seawater intrusion in this plain was analysed using a hydrodynamic and hydrochemical approaches. The first approach was based on the analysis of groundwater table depth and salinity data available during the period from 1984 to 2009. The groundwater piezometric heads recorded in the upstream and intermediate regions of the plain showed seasonal fluctuations due to the monsoon. In the coastal region, the piezometric heads showed small fluctuations, which are indicative of a quasi-static regime. The groundwater salinity showed large variations and generally increased in the flow direction from north to south. Near the shoreline, the groundwater salinity has increased by 8420 μS/cm during the period from 1984 to 1994. In the study area, 11 groundwater samples were collected during October 2015 and analysed for EC, pH and major ions. Results indicate that groundwater chemistry was classified as Na–Cl type in one well, Ca–Mg–Cl type in three wells and Ca–Cl type in seven wells. The first two water types indicate the effect of seawater intrusion and comparatively the Ca–Cl water type highlights the effect of ion exchange reactions on the mineralisation of groundwater. The high concentrations of nitrate recorded especially in the Hafah agricultural area confirm the impact of fertilizers application on groundwater quality.

Keywords Salalah coastal plain · Groundwater salinity · Saturation index

B. Askri (✉) · R. Ali Al-Shanfari
Department of Built and Natural Environment, Caledonian College of Engineering, Seeb, Sultanate of Oman
e-mail: askrib@yahoo.com

O. Abdalla et al. (eds.), *Water Resources in Arid Areas: The Way Forward*, Springer Water, DOI 10.1007/978-3-319-51856-5_20

1 Introduction

Saltwater intrusion is a global environmental issue that degrades groundwater quality by increasing salinisation in many coastal aquifers (Park et al. 2005; Bahar and Reza 2010; Askri 2015). This process occurs due to groundwater overexploitation to meet the growing requirement of freshwater for agricultural, industrial and drinking purposes (Singh 2014; Yüce 2005). In coastal regions, the excessive groundwater abstraction may cause a decline of piezometric heads and induce a modification of natural flow system (Kouzana et al. 2009). The effect of seawater intrusion on groundwater salinisation has been studied extensively during the last two decades using different approaches. Sherif et al. (2011) showed that the recently seawater intrusion has become the principal factor of water salinity in the aquifer of wadi Ham, UAE. Arslan (2013) classified the groundwater in Bafra coastal plain, Turkey, into three groups affected by saltwater intrusion of different levels. Kuzana et al. (2009) indicated that saltwater intrusion from the Mediterranean Sea is the main cause of high groundwater salinity observed in the Korba aquifer, northern Tunisia. In this coastal aquifer, Zghibi et al. (2013) highlighted the occurrence of direct cation exchange associated with saltwater intrusion and dissolution processes linked to cations exchange. Agoubi et al. (2013) studied the causes of groundwater salinisation in the marine Jeffara aquifer, southern Tunisia. They showed that the saltwater mixing with groundwater is the consequence of seawater intrusion.

Coastal groundwater aquifers in Sultanate of Oman are supplying about 99% of freshwater demand (Rajmohan et al. 2009). The rapid growth of population in Salalah city along with economic growth enhanced the groundwater salinisation through over abstraction of groundwater for drinking, municipal, agricultural and industrial purposes. Clarck et al. (1987) indicated significant drops of piezometric heads in the Salalah plain aquifer since the early 1980s. As consequence, evidence of seawater intrusion due to the excessive groundwater abstraction became apparent mainly in the coastal sector (D&MI 1992). This study concerns the impact of seawater intrusion on the groundwater quality in the Salalah plain. The objectives are (i) to analyse the dynamics of seawater intrusion in the shallow groundwater aquifer, (ii) to explore the hydrochemical processes regulating the groundwater mineralization and (iii) to evaluate the degree of suitability of groundwater for irrigation purpose.

2 Materials and Methods

2.1 *Study Area*

Salalah is the capital of Dhofar Governorate in southeastern Sultanate of Oman. It is the second largest city in the country with a population of 134,000 inhabitants and a yearly population growth rate of 2.3% (Oman Census 2003). The study area is a sector of the Salalah coastal plain (Fig. 1). Jabal Al-Qara (mountain) and Arabian Sea represent, respectively, the northern and southern boundaries of the area. Wadis exist in the plain and flow to the sea. Two recharge and flood-protecting dams have been built in 1993 and 2007 on Sahalnawt and Zarsis wadis, respectively. Since 2003, treated wastewater was injected into the groundwater to create a hydraulic water barrier to reduce the amount of saltwater that can penetrate into the freshwater (Shammas 2007).

Most of precipitation in Salalah region results from the monsoon occurring between late June and early September. It is a combination of rain, fog, drizzle and mist. The annual rainfall amounts vary from 45 and 154 mm in the plain and from 230 to 450 mm in Jabal AL-Qara (Shammas 2007). Cyclonic storms occur once every 3–7 years, affect the plain and cause loss of life and extensive damage of infrastructures. Daily temperature ranges from 27 °C in January to 32 °C in June. Humidity varies from 40% in February to 97% in July. The evaporation is very high and ca reach 1700 mm/year.

The Salalah plain is an intensively irrigated area. Fruits such as carica papaya, bananas and coconut palms, grass and vegetables are cultivated mainly in the Hafa agricultural area. Recently, large-scale grass cultivation has been implemented in Sahalnawt and Jarziz farms.

Fig. 1 Location map of the study area *filled circle* represents sampling wells in October 2015, *filled triangle* represents monitoring wells from May 1982 to February 2007. *Line A–B* is a cross section clarified in Fig. 2

The study area is a part of the down-faulting of the Salalah plain of Oman. This plain resulted from the major tectonic activity, which was occurring in Dhofar region during the refting between the Arabian and African plates (Roger and Platel 1987). It has two distinct formations: the Fars Group of the Tertiary age (middle to early Miocene age) and the Mushsayl Formation of the Tertiary age (Early Miocene to Oligocene) (Fig. 2). The Fars group has a maximum saturated thickness of 120 m, while the permeable and saline Mughsayl Formation has a thickness of about 800 m (Shammas and Jacks 2007). The Fars group comprises Nar and Adawnib Formations. The main aquifer in the plain contains carbonate and conglomeratic marine deposits of Adawnib Formation. The aquifer thickness is varying from 60 to 70 m. Wadi alluvium including gravel, gravely clays and calcarenite deposits are overlying the aquifer.

The groundwater is considered confined aquifer in Jabal Al-Qara and unconfined aquifer in the Salalah plain. It is characterised by a storage coefficient of 3%, hydraulic conductivity ranging from of 10 to 1000 m/day and transmissivity varying from 1000 to 200,000 m^2/day (MAF-ICBA 2012). In general, the groundwater is flowing from north to south with piezometric heads at about 10 m above mean sea level near the piedmont of Jabal Al-Qara, declining to about sea level near the coast (Fig. 3). The low hydraulic gradients of this aquifer (about 1 m/km) can be explained by its hydraulic connection to the sea and its high transmissivity. In this case, any depletion caused by the groundwater pumping can be compensated by the inflows from the sea.

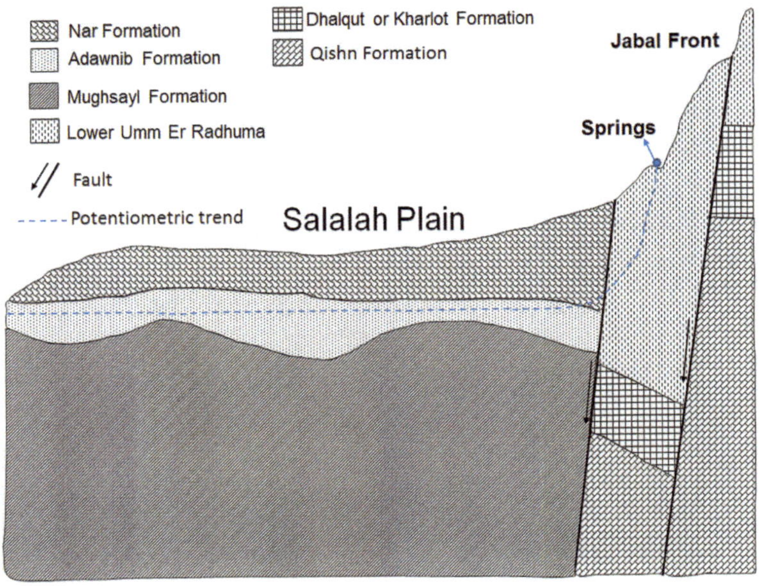

Fig. 2 Hydrogeological cross-section of Salalah plain and adjacent Jabal Al-Qara (after Geo Resources Consultancy 2004)

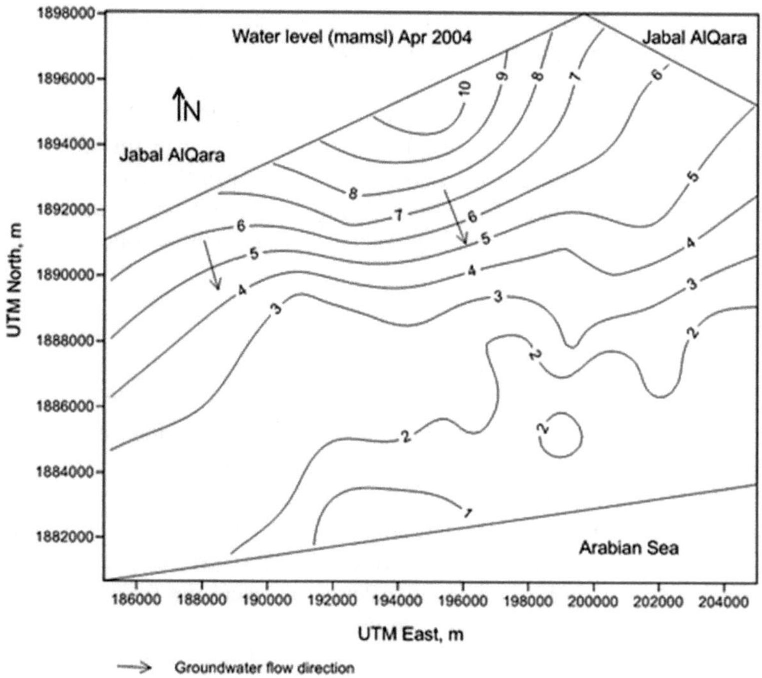

Fig. 3 Contour of groundwater piezometric heads in the study area (Askri et al. 2016)

The domestic, municipal, agricultural and industrial water needs in the Salalah plain are provided by the groundwater. The potential water resources of this aquifer are about 874 Mm^3 of which 42% is considered as fresh groundwater and the remaining 58% as brackish groundwater (MAF-ICBA 2012). The main sources of recharge are the underflow from Jabal Al-Qara to the plain, waters mobilised in Sahalnawt and Zarsis dams, infiltrations of precipitation falling on the plain, infiltrations of floodwater flowing through the wadis and injection of treated wastewater. The groundwater aquifer is exploited by about 1500 wells and 17 boreholes. The total groundwater abstraction is estimated at 69 Mm^3, out of which 46% is used for irrigation (Askri et al. 2016).

2.2 *Sampling and Data Collections*

Historical data of groundwater table depth and electrical conductivity (EC) recorded in the Salalah plain were collected from the Ministry of Regional Municipalities and Water Resources, Sultanate of Oman. The hydraulic heads in the groundwater aquifer were measured as the height of the groundwater table above mean sea level. They were calculated by subtracting the groundwater table depth from the land surface altitudes. These data were used to investigate the dynamic of seawater intrusion in the plain aquifer by means of spatial and temporal variations of the hydraulic heads and salinity of the groundwater aquifer.

In October 2015, eleven groundwater samples were collected in the study area for major anion and cation analyses. The concentrations of major ions were measured using titration method at the Laboratory of Water in Salalah city, Directorate General of Water, Sultanate of Oman. Assuming chloride to be a conservative ion, the fraction of seawater in the mixed water was estimated as follow (Appelo and Postma 2005):

$$f_{sea} = \frac{C_{Cl(sample)} - C_{Cl(freshwater)}}{C_{Cl(seawater)} - C_{Cl(freshwater)}} \quad (1)$$

where $C_{Cl(sample)}$, $C_{Cl(freshwater)}$ and $C_{Cl(seawater)}$ are the chloride concentrations of groundwater sample, freshwater and seawater, respectively.

PHREEQC software was used as a specification program to calculate the saturation indices (SI) (Parkhurst and Appelo 1999). This parameter is showing if water will trend to precipitate (SI > 0) or dissolve (SI < 0) a given mineral (Hamzaoui-Azaza et al. 2011). Specific calculations were run for calcite, aragonite, dolomite and halite minerals In order to understand the effect of saline sources on groundwater quality. The processes inducing groundwater salinisation were identified using the following molar ratios: Na/Cl, Ca/Mg and (Ca + Mg)/HCO_3.

The electrical conductivity (EC) and the sodium adsorption ratio (SAR) were used to assess the degree of groundwater suitability for irrigation use. Indeed, the water availability to the plants decreases with the increase of these two parameters values. The sodium hazard is expressed by the SAR, which is calculated using the following equation:

$$SAR = \frac{\text{Na}}{\sqrt{\frac{(\text{Ca}+\text{Mg})}{2}}} \quad (2)$$

where units are expressed in milli-equivalent per liter.

The assessment of irrigation waters was made using the diagram of US Salinity Laboratory Staff (US Salinity Lab Staff 1954), where SAR is plotted against EC.

3 Results and Discussion

3.1 *Temporal Variations of Piezometric Heads and Salinity of Groundwater*

Data of monthly groundwater table depth in three irrigation wells located in upstream (F-63), intermediate (F-68) and downstream (F-56) regions of the study area were available from May 1984 to February 2007. The piezometric heads of the groundwater show seasonal and long-term fluctuations (Fig. 4). The high piezometric heads observed during monsoon were usually followed by decreasing trend during post monsoon. The small declines are explained by the high transmissivity of the karstic aquifer and its hydraulic connection to the sea. The long-term fluctuations of the groundwater piezometric heads show different patterns with respect to well location. The piezometric heads recorded in the wells F-63 and F-68 declined until June 1998, after which they increased until February 2007. The decreasing trend observed during the first period may be explained by the increasing number of wells implemented in the intermediate and upstream regions

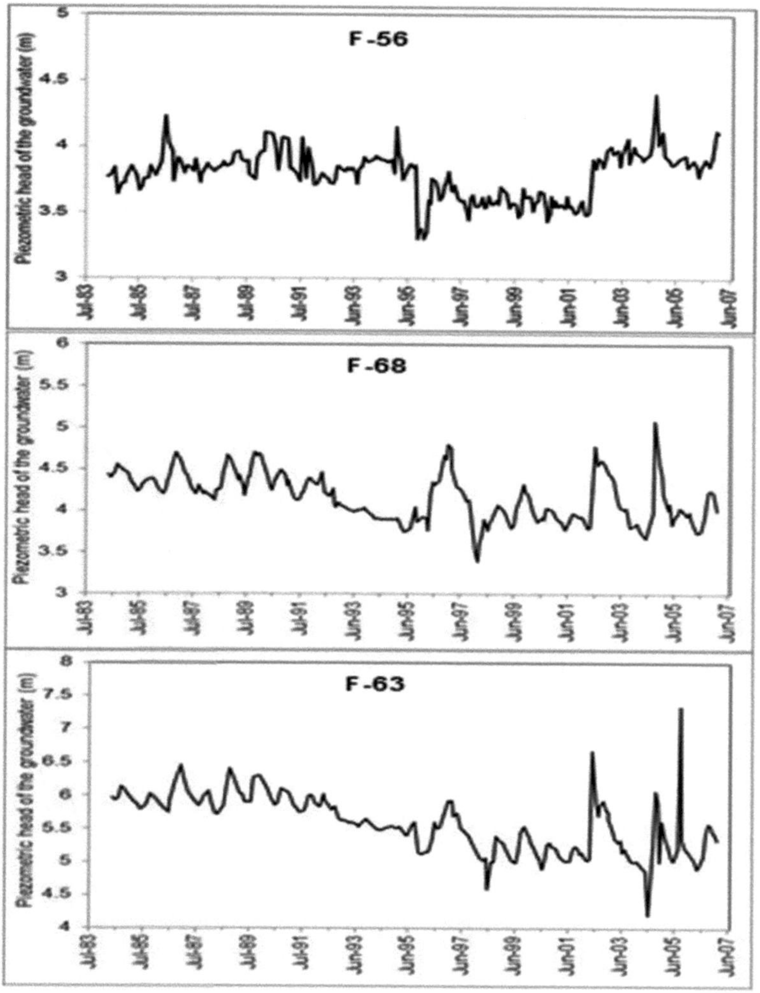

Fig. 4 Plots of monitored groundwater piezometric heads from 1984 to 2007

of the study area. The piezometric head recorded in the well F-56 was approximately constant until June 1995. Then, it decreased continuously until April 2002. In May 2002, this piezometric head increased by 40 cm after which it was maintained approximately constant until August 2006. The groundwater regime in this well seems to be in equilibrium, which means that the water abstraction by pumping was compensated by recharge through the rainwater infiltration and lateral flow from the sea.

Data of monthly groundwater salinity in the wells F56, F-63 and F-68 were available from 1984 to 1994. During this period, the EC increased by 8420, 152 and 83 μS/cm for the wells F-56, F-68 and F-63, respectively (Fig. 5). These results

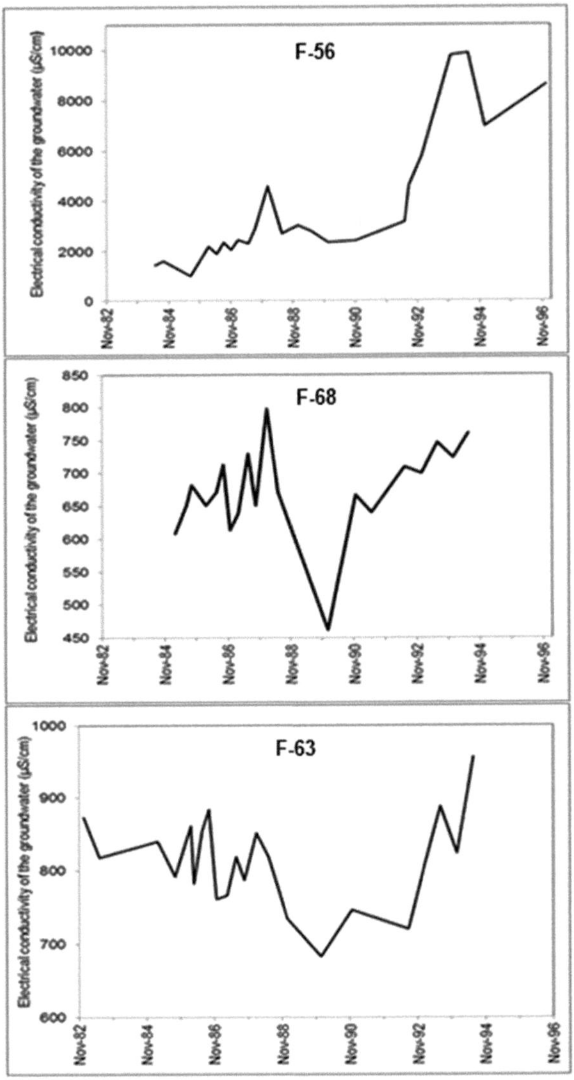

Fig. 5 Plots of monitored groundwater salinity from 1982 to 1994

show that the groundwater salinity has increased in the study area at different rates. This rate was very high in the coastal area probably due to a strong seawater intrusion, which has been favored by overpumping. The groundwater salinity increased slowly in the upstream and intermediate of the study area. The location of these regions far from the sea and from the agricultural areas suggest the mineral dissolution as the main source of the groundwater salinisation.

3.2 *Spatial Variability of Groundwater Salinity*

Figure 6 shows the spatial distributions of the groundwater salinity in the study area in 1998, 2004, 2006 and 2009. The EC of the groundwater showed large variations and generally increased in the flow direction from north to south. Comparing the maps for the years 1998 and 2004, it is evident that the area of fresh waters has increased and the area of saline waters has decreased mainly in eastern corner of the study area. This amelioration of the groundwater quality can be explained by the

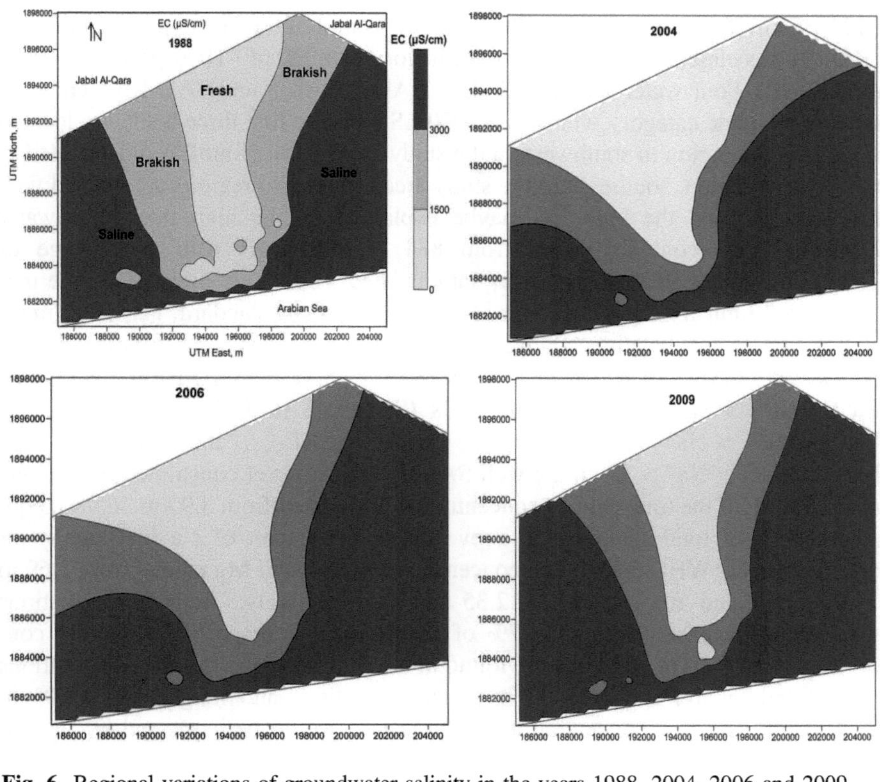

Fig. 6 Regional variations of groundwater salinity in the years 1988, 2004, 2006 and 2009

positive effect of the groundwater injection by the treated wastewater. The comparison between the salinity maps recorded during the years 2004 and 2009 shows an opposite trend as the area of fresh waters has decreased, while the area of saline waters has increased mainly in eastern corner of the study area. The deterioration of the groundwater quality during this period was observed despite the increase in the artificial recharge from the treated water from 2.0 to about 7.0 mm^3 per year during the period from 2003 to 2009 (Shammas 2007). During the last year, only 53 mm of rainfall was recorded in the Salalah plain. Thus, the groundwater salinisation can be explained by the evapo-concentration of some salts, which is the consequence of high rate of evaporation and low quantity of rainfall.

3.3 Major Ion Trends

In October 2015, the pH values in the study area were in the range of 7.1–7.8, showing that the samples are neutral to slightly alkaline (Table 1). In this case, the dissolution of carbonate minerals may be the origin of the groundwater alkalinity. The EC of the waters varied from 1360 to 9310 μS/cm with an average of 3124 μS/cm. Only one sample collected in the center of the Hafah agricultural area (SAL-9) was classified into fresh water category with EC of <1000 μS/cm (Mondal et al. 2010). Four water samples (SAL-3, SAL-4, SAL-5 and SAL-10) were classified into saline category with EC > 3000 μS/cm. The first three wells are located in the Awqad region in southwestern the study area and the fourth well is located in the Dahariz region, southeastern the study area. These two regions are located near the shoreline and the high EC maybe explained by the high rate of seawater intrusion. TDS contents varied from 884 to 6051 mg/l with an average of 2030 mg/l. About 91% of the water samples had TDS value higher than the recommended limit for drinking water as per WHO (2004) standard, i.e., 1000 mg/l.

The order of relative abundance of major cations was (molar basis): $Na^+ > Ca^{2+} > Mg^{2+} > K^+$ in four wells (SAL-4, SAL-5, SAL-10 and SAL-11), $Ca^{2+} > Na^{2+} > Mg^{2+} > K^+$ in four wells (SAL-1, SAL-3, SAL-6 and SAL-8), $Ca^{2+} > Mg^{2+} > Na^+ > K^+$ in two wells (SAL-2 and SAL-7) and $Mg^{2+} > Ca^{2+} > Na^+ > K^+$ in the well SAL-9. The Na level constituted on average about 39.0% of the total cations concentration and varied from 3.92 to 36.98 meq/l. About 55% of total samples fall above the suitable limit of Na (8.7 meq/l) for drinking water (WHO 2004). The concentrations of Ca and Mg ranged from 5.69 to 14.02 meq/l and from 4.94 to 12.35 meq/l, respectively. Their concentrations represented on average 33 and 27% of all the cations, respectively. Higher concentrations of Na, Ca and Mg were found in the well SAL-10. The K concentrations represented on average only 0.5% of all the cations and ranged from 0.05 to 0.44 meq/l. The low concentration of this cation would be explained by its tendency to be fixed by clay minerals.

Table 1 Chemical results of the major ions meq/l

Water type	Well label	EC (µS/cm)	TDS (mg)	pH	Cl	NO_3	SO_4	HCO_3	F	Ca	Mg	Na	K
Fresh	SAL-9	1360	884	7.24	13.99	2.71	2.29	3.33	0.02	5.69	6.42	3.92	0.05
Brakish	SAL-1	1580	1027	7.81	12.69	5.71	3.12	3.62	0.02	8.28	6.17	6.31	0.07
Brakish	SAL-2	1700	1105	7.8	13.76	4.71	3.35	3.97	0.02	8.53	6.58	6.53	0.08
Brakish	SAL-8	2060	1339	7.51	19.80	5.21	2.83	4.00	0.02	9.88	8.39	8.70	0.10
Brakish	SAL-6	2250	1462	7.48	16.92	4.50	2.04	3.48	0.01	8.48	6.83	6.96	0.08
Brakish	SAL-11	2340	1521	7.78	9.53	3.43	3.68	3.92	0.04	6.99	4.94	9.57	0.11
Brakish	SAL-7	2460	1599	7.19	18.22	4.14	4.31	4.30	0.01	11.48	9.22	6.09	0.07
Saline	SAL-3	3190	2073	7.32	18.33	7.21	3.89	4.25	0.04	10.48	7.65	9.22	0.11
Saline	SAL-5	3600	2340	7.17	19.20	2.86	3.08	3.66	0.02	9.48	7.49	13.05	0.15
Saline	SAL-4	4510	2931	7.14	19.80	6.71	3.66	3.97	0.04	11.13	8.81	14.79	0.17
Saline	SAL-10	9310	6051	7.1	25.66	8.57	0.81	6.63	0.06	14.02	12.35	36.98	0.44

The order of anion abundance are $Cl^- > HCO_3^- > SO_4^{2-} > NO_3^-$ in two wells (SAL-11 and SAL-5), $Cl^- > NO_3^- > HCO_3^- > SO_4^{2-}$ in eight wells (SAL-1, SAL-2, SAL-3, SAL-4, SAL-6, SAL-8, SAL-9 and SAL-10), and $Cl^- > SO_4^{2-} > HCO_3^- > NO_3^-$ in one well (SAL-7). The Cl level represented about 58% of all the anions and ranged from 9.53 to 25.66 meq/l. Higher Cl concentration in Dahariz region (SAL-10) suggests seawater intrusion. However, the relatively low concentrations of this anion in the upstream region (SAL-1 and SAL-11) indicate the lateral recharge from Jabal Al-Qara. The suitable limit of chloride (250 mg/l) for drinking water (WHO 2004) has been exceeded in total water samples. The SO_4 and HCO_3 concentrations ranged from 0.81 to 4.31 meq/l and 3.33 to 6.63 meq/l, respectively. Their average concentrations represent 10 and 14% of all the anions, respectively. Total water samples had SO_4 concentration below WHO standard, i.e., 250 mg/l. The increase of SO_4 concentration from the inland to the shoreline can be explained by a possible seawater intrusion, which has a high SO_4 concentration. This interpretation is also justified by the absence of geological formations containing SO_4 in the study area. The HCO_3 can come from the following three sources: dissolution of carbonate minerals, soil CO_2 or bacterial degradation of the organic contamination (Bahar and Reza 2010). The NO_3 concentrations varied from 2.71 to 8.57 meq/l. Its average concentration represent about 17% of all the anions. About 64% of total waters exceeded the drinking water standard for nitrate (50 mg/l). The highest NO_3 concentrations recorded in Hafah agricultural area shows an anthropogenic pollution, which may be introduced by agricultural activities (crops residue, use of nitrogen fertilizers and animal manure).

The seawater fraction in the groundwater varied from 1.3 to 4.3% in the wells SAL-11 and SAL-10, respectively (Fig. 7). The first well is located 9.3 km far from the coastline, while the second one is located 1.2 km far from the coastline.

3.4 *Water Types*

The hydrochemical processes operating in the groundwater aquifer were analysed using the Piper diagram (Fig. 8). Trilinear diagram analysis indicates that all brackish waters are Ca, Mg, Na, Cl or Ca, Na, Mg, Cl which means high groundwater episodes are washing salinised groundwater from Na–Cl to Ca and Mg that coming from carbonate aquifer in the study area.

Three water types were found in the study area. They are in the order of Ca–Cl > mixed Ca–Mg–Cl > Na–Cl. The low mineralised Ca–Cl type freshwater and brackish water was identified in seven wells (SAL-1, SAL-2, SAL-3, SAL-6, SAL-7, SAL-8 and SAL-9) located mainly in northwestern the study area. Cl and Ca were the dominant anion and cation, respectively. This water type was influenced by the reverse ion exchange process. Water types (mixed Ca–Mg–Cl and Na–Cl) are present in four wells (SAL-4, SAL-5, SAL-10 and SAL-11) and indicate the

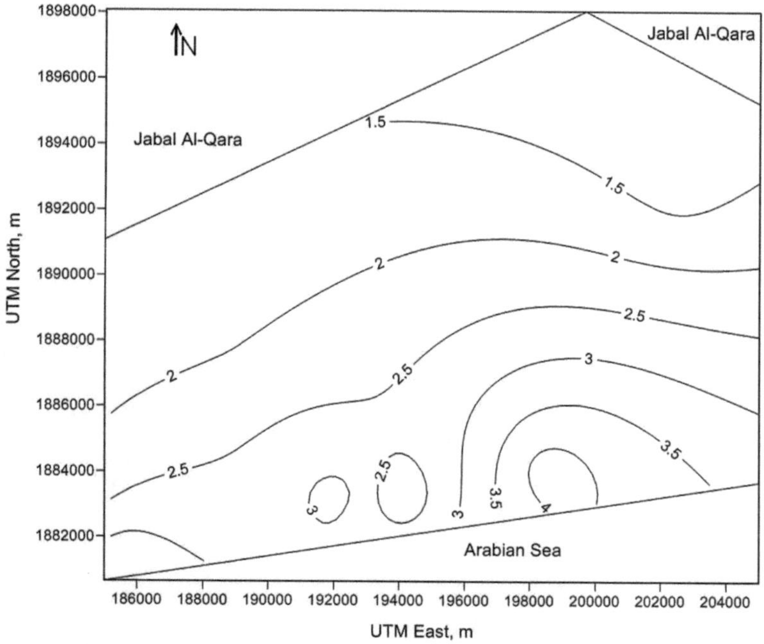

Fig. 7 Contour of seawater fraction in the groundwater

mixing of high salinity water caused from seawater intrusion with existing water. These wells seem to be influenced by a strong seawater intrusion. These results indicate an active saltwater intrusion followed by reverse ion exchange reactions in the study area.

3.5 Effects of Mineral Dissolution

Approximately the total water samples were under-saturated with respect to halite and supersaturated with respect to araganite, calcite and dolomite minerals (Table 2). These results show that these mineral phases are contributing to the groundwater mineralisation in the study area. Thus, the precipitation of calcite, araganite and dolomite minerals, and dissolution of halite are probable. Precipitation of the first three minerals may occur during the cation exchange process (Mtoni et al. 2012).

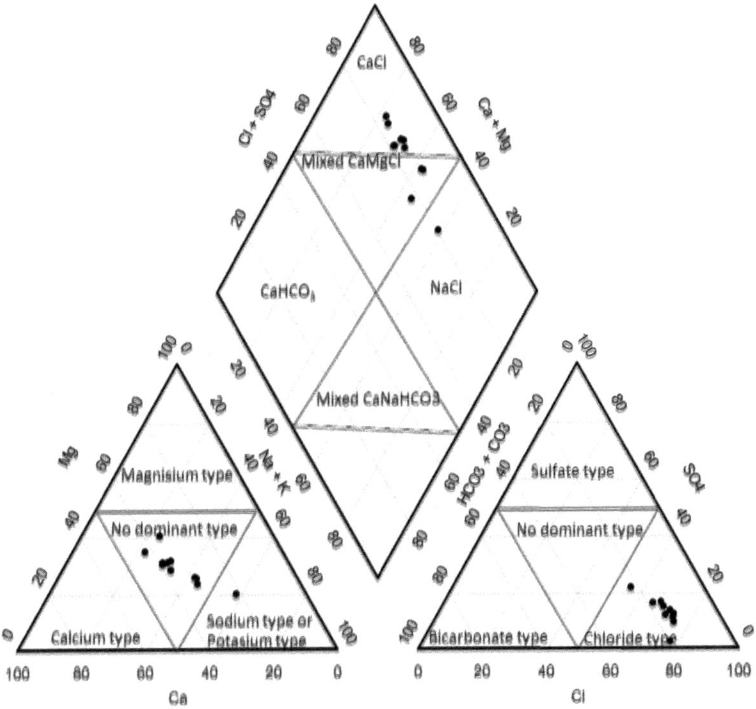

Fig. 8 Piper plot of groundwater hydrochemical data in the study area

Table 2 Molar ratios and saturation indexes for calcite, aragonite, dolomite and halite

Well label	Ca/Mg	Ca/ (HCO_3 + SO_4)	SI of calcite	SI of aragonite	SI of dolomite	SI of halite
SAL-1	170.41	0.11	0.81	0.67	1.63	-5.81
SAL-2	235.57	0.10	0.85	0.7	1.71	-5.76
SAL-3	117.39	0.08	0.046	0.32	0.92	-5.50
SAL4	70.27	0.05	0.27	0.12	0.57	-5.27
SAL-5	93.67	0.06	0.22	0.08	0.47	-5.33
SAL-6	159.17	0.09	0.48	0.34	1.01	-5.65
SAL-7	146.25	0.08	0.38	0.23	0.79	-5.68
SAL-8	188.62	0.07	0.61	0.47	1.28	-5.49
SAL-9	313.17	0.09	0.08	-0.06	0.35	-5.97
SAL-10	51.57	0.02	0.5	0.35	1.08	-4.78
SAL-11	145.21	0.10	0.75	0.61	1.48	-5.75

3.6 Sources of Groundwater Salinisation

Some molar ratios can be used to determine the sources of groundwater salinisation in coastal regions (Leboeuf et al. 2003; Yüce 2005; Mondal et al. 2010; Arslan 2013). In this context, the Ca/Mg and Na/Cl molar ratios reflect the degree of seawater intrusion. This process is indicated by a high Ca/Mg (>1) molar ratio and relatively low Na/Cl molar ratio (<0.86) (Vengosh et al. 1997). Table 2 shows the selected molar ratios of the groundwater in the study area. The Na/Cl ratio was less than 0.86 in 82% of total water samples, while the Ca/Mg ratio was higher than 1 in 91% of the total water samples. These results show that the process of seawater intrusion has a predominant effect on groundwater quality in the study area.

The processes of mineral dissolution-precipitation in the study area were alos analysed using some molar ratios. For example, a Na/Cl molar ratio close to 1 shows that halite dissolution is responsible for sodium, whereas the Na/Cl molar ratio greater than 1 typically shows that the reactions of silicate weathering are the main origin of sodium (Meybeck 1987). The $(Ca + Mg)/HCO_3$ molar ratio can determine the sources of Ca and Mg in groundwater. These two cations originate from the dissolution of carbonate minerals if this ratio is equal to 0.5 (Sami 1992). The release of sodium from silicate weathering reactions may be a source of sodium in two wells situated in the intermediate region of the Salalah plain. Na/Cl molar ratio was less than 1 in 82% of total samples, indicating that the reverse ion exchange process is the main cause of the significant reduction of Na concentration. Potassium or sodium in seawater replaces magnesium or calcium adsorbed onto the surface of clays during the occurrence of the seawater intrusion. This phenomenon caused a relative depletion of sodium in groundwater. The $(Ca + Mg)/HCO_3$ molar ratio was higher than 3.04 in total water samples showing that an excess of alkalinity in the groundwater was balanced by the sodium and potassium. The excess of calcium and magnesium over bicarbonates indicates extra sources of these cations.

3.7 Suitability of Groundwater for Irrigation Use

The SAR values varied in the study area from 1.6 to 10.2 with a mean of 3.6. Three water samples fall into the category C3-S1 (high salinity with low sodium) and one water sample falls into the category C4-S4 (very high salinity with very high sodium). The other water samples fall into the categories C4-S3, C4-S2, S4-S1 and C3-S2 (Fig. 9).

Generally, the water samples show high to very high salinity hazard with medium to low alkali hazard. The most of these samples fall into the category high salinity with low to medium sodium. Thus, the most of the water samples may not be suitable for irrigation under ordinary conditions. They can be used occasionally if soils are permeable, drainage is adequate, irrigation water is applied in excess to provide leaching, and salt tolerant crops are selected.

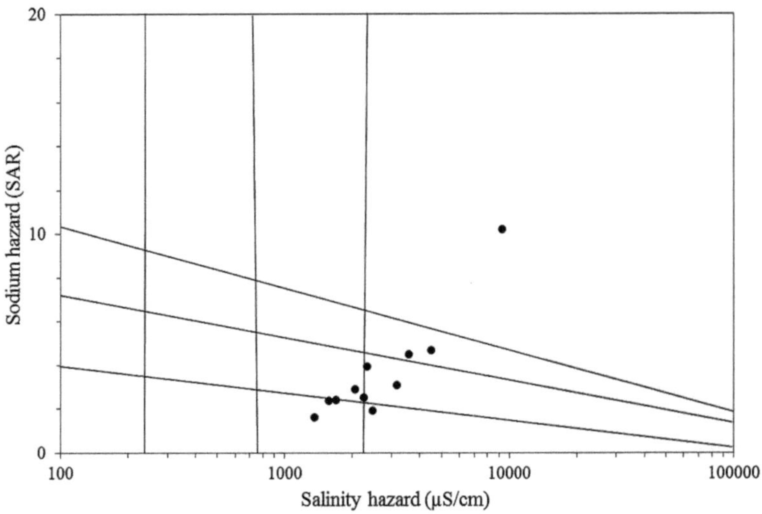

Fig. 9 US Salinity Laboratory Staff diagram plot of the groundwater samples

4 Conclusions

Available data of groundwater table depth and salinity in the Salalah plain aquifer were analysed to understand the dynamics of seawater intrusion in this plain. The piezometric heads recorded in the upstream and intermediate regions had a decreasing trend during the period from 1984 to 1998, which may be explained by the increasing number of wells implemented in these regions. The groundwater regime in the coastal area was in equilibrium in terms of recharge-discharge. However, the rate of seawater intrusion was very high.

In October 2015, the groundwater chemistry in the Salalah plain was strongly influenced by seawater intrusion, reverse ion exchange process and anthropogenic pollutants. Three water types were identified in the study area: Ca–Cl type in seven wells, mixed Ca–Mg–Cl type in three wells and Na–Cl type in one well. The first seven wells were under the influence of the reverse ion exchange, while the last four wells were affected by a strong seawater intrusion. The occurrence of this phenomenon was also demonstrated by Na/Cl and Ca/Mg molar ratios. The high nitrate concentrations recorded in the Hafah agricultural area highlights the impact of agricultural activities on the groundwater quality. The use of groundwater in Salalah plain for irrigation purpose require certain conditions related to the soil texture, implementation of a drainage system and selection of salt tolerant crops.

Acknowledgements The authors are thankful to the Department of Monitoring of Water Resources, Ministry of Regional Municipalities and Water Resources, Sultanate of Oman for providing necessary data. The authors are also thankful to the anonymous reviewers for their helpful comments.

References

Agoubi B, Kharroubi A, Abichou T, Abida H (2013) Hydrochemical and geoelectrical investigation of Marine Jeffara Aquifer, southeastern Tunisia. Appl Water Sci 3:415–429

Appelo CAJ, Postma D (2005) Geochemistry, groundwater and pollution, seconded. Balkema, Rotterdam

Arslan H (2013) Application of multivariate statistical techniques in the assessment of groundwater quality in seawater intrusion area in Bafra Plain, Turkey. Environ Monit Assess 185:2439–2452

Askri B (2015) Hydrochemical processes regulating groundwater quality in the coastal plain of Al Musanaah, Sultanate of Oman. J Afr Earth Sci 106:87–98

Askri B, Abdelkader TA, Razan AS, Rachida B, Khater KF (2016) Isotopic and geochemical identifications of groundwater salinisation processes in Salalah coastal plain, Sultanate of Oman. Chem Erde 76:243–255

Bahar MM, Reza MS (2010) Hydrochemical characteristics and quality assessment of shallow groundwater in a coastal area of Southwest Bangladesh. Environ Earth Sci 61:1065–1073

Clark ID, Fritz P, Quinn OP, Rippon PW, Nash H, Al-Said SB (1987) Modern and fossil groundwater in an arid environment: a look at the hydrogeology of southern Oman. In: Proceedings of use of stable isotopes in water resources development, IAEA-SM-299, Vienna, Austria, pp 167–187

D&MI (1992) Water and wastewater master plan for Salalah: Office of the Minister of State and Governor of Dhofar; Salalah, Sultanate of Oman

Geo Resources Consultancy (2004) Detailed water resources management and planning study for the Salalah region (Part A, B & C), final report prepared for MRMEWR, Sultanate of Oman

Hamzaoui-Azaza F, Ketata M, Bouhlila R, Gueddari M, Riberio L (2011) Hydrogeochemical characteristics and assessment of drinking water quality in Zeuss-Koutine aquifer, southeastern Tunisia. Environ Monit Assess 174:283–298

Kouzana L, Mammou AA, Felfoul M (2009) Seawater intrusion and associated processes: case of the Korba aquifer (Cap-Bon, Tunisia). C R Geosci 341:21–35

Leboeuf PP, Bosh AP, Calvaches ML, Vallejos A, Andreu JM (2003) Strontium, SO4/Cl amd Mg/Ca ratios as tracers for the evolution of seawater in coastal aquifers: the example of Castell de Ferro Aquifer (SE Spain). Academie des Sci Geosci 335(14):1039–1048

MAF-ICBA (2012) Oman salinity strategy, Ministry of Agriculture and Fisheries. Sultanate of Oman. International Center for Biosaline Agriculture, Dubai, UAE

Meybeck M (1987) Global chemical weathering of surficial rocks estimated from river dissolved loads. Am J Sci 287:401–428

Mondal NC, Singh VP, Singh VS, Saxena VK (2010) Determining the interaction between groundwater and saline water through groundwater major ions chemistry. J Hydrol 388:100–111

Mtoni Y, Mjemah IC, Bakundukize C, Camp MV, Martens K, Walraevens K (2012) Saltwater intrusion and nitrate pollution in the coastal aquifer of Dar es Salaam, Tanzania. Environ Earth Sci. doi:10.1007/s12665-012-2197-7

Oman Census (2003) Ministry of National Economic, Sultanate of Oman

Park SC, Yun ST, Chae GT, Yoo IS, Shin KS, Heo CH, Lee SK (2005) Regional hydrochemical study on salinization of coastal aquifers, western coastal area of South Korea. J Hydrol 313:182–194

Parkhurst DL, Appelo CAJ (1999) User's guide to PHREEQC (version 2)—a computer program for speciation, batch-reaction, one-dimensional transport, and inverse geochemical calculations: U.S. geological survey water-resources investigations report 99-4259, 312 pp

Rajmohan N, Al-Futaisi A, Al Touqi S (2009) Geochemical process regulating groundwater quality in a coastal region with complex contamination sources: Barka, Sultanate of Oman. Environ Earth Sci 59:385–398

Roger J, Platel JP (1987) Geological mapping and mineral exploration programme in Southern Dhofar. Final Report. Bureau de Recherches Geologiques et minieres, 87 OMN 091. Ministry of petroleum and minerals, directorate general of minerals

Sami K (1992) Recharge mechanisms and geochemical processes in a semi-arid sedimentary basin, Eastern cape, South Africa. J Hydrol 139:27–48

Shammas MI (2007) Sustainable management of the Salalah coastal aquifer in Oman using an integrated approach. Ph.D. thesis, Department of Land and Water Resources Engineering, Royal Institute of Technology (KTH), Sweden

Shammas MI, Jacks G (2007) Seawater intrusion in the Salalah plain aquifer, Oman. Environ Geol 53:575–587

Sherif M, Mohamed M, Kacimov A, Shetty A (2011) Assessment of groundwater quality in the northeastern coastal area of UAE as precursor for desalination. Desalination 273:436–446

Singh A (2014) Optimization modelling for seawater intrusion management. J Hydrol 508:43–52

USSL (1954) Diagnosis and improvement of saline and alkali soils, handbook, vol 60. USDA, Washington, 147 pp

Vengosh A, Gill J, Reyes A, Thoresberg K (1997) A multi-isotope investigation of the origin of groundwater salinity in Salinas valley, California. American Geophysical Union, San Francisco, California

World Health Organization (2004) Guidelines for drinking water quality, vol. 1 recommendations (3rd). WHO, Geneva

Yüce G (2005) Determination of the recharge area and salinization degree of karst springs in the Lamas Basin (Turkey). Isot Environ Health Stud 41(4):391–404

Zghibi A, Tarhouni L, Zouhri L (2013) Assessment of seawater intrusion and nitrate contamination on the groundwater quality in the Korba coastal plain of Cap-Bon (North-east of Tunisia). J Afr Earth Sci 87:1–12

Part V
Wastewater Treatment and Reuse

Maximum Use of Treated Wastewater in Agriculture

Ahmed Al-Busaidi and Mushtaque Ahmed

Abstract This study contributed to the existing knowledge on urban treated wastewater reused for agriculture in MENA region by identifying means to optimize wastewater reuse by taking into consideration various parameters such as return to farmers, groundwater quality and impacts on soil and groundwater. The aim of the study was to maximize the usage of treated wastewater reuse in conjunction with groundwater or any available water resource by taking into consideration their quality and quantity, in addition to the environmental, economic and agronomic components. The study was conducted at Sultan Qaboos University, Oman with three types of crops (eggplant, radish and okra). All grown crops were irrigated either by A: 50% groundwater and 50% treated wastewater or B: 100% groundwater or C: 75% treated wastewater and 25% groundwater or D: 100% treated wastewater. Physical and chemical properties of the soil were almost similar and did not show significant differences with both waters. Moreover, some chemical properties such as total carbon and some major elements (N, P, K) were significantly increased ($p < 0.05$) when treated wastewater was applied. A significant increases in plant productivity was noticed when plants were irrigated with treated wastewater. Whereas, there were insignificant changes in heavy metals of tested crops between different treatments.

Keywords Treated wastewater · Groundwater · Heavy metals · Yield

A. Al-Busaidi (✉) · M. Ahmed
Department of Soils, Water and Agricultural Engineering,
College of Agricultural & Marine Sciences, Sultan Qaboos University,
P.O. Box 34, Al-Khoud 123, Muscat, Oman
e-mail: ahmed99@squ.edu.om

O. Abdalla et al. (eds.), *Water Resources in Arid Areas: The Way Forward*,
Springer Water, DOI 10.1007/978-3-319-51856-5_21

1 Introduction

Shortage of conventional water resources in Middle East is creating a problem in sustaining agricultural productivities (World Bank 2012). Therefore, there is a need for a good management of existing water supplies including non-conventional water resources such as treated wastewater. Treated wastewater is a good sources of agricultural water that could contribute to the conservation of water resources and improving soil fertility (Xu et al. 2010). Oman is one of the countries that considered treated wastewater application and have a national policy to recycle all treated wastewater (Al-Ajmy 2002).

Applying treated wastewater in agriculture could help in conserving freshwater and reducing fertilizer application. However, it may contain different concentrations of some pathogens or toxic salts such as heavy metals. Moreover, other contaminants like emerging pollutants can be found in treated wastewater (Mohammad and Mazahareh 2003). Treated wastewater is widely used in different countries of the world and many users and researchers have recognized its benefits (Mujeriego and Sala 1991; Levine and Asano 2004; Bedbabis et al. 2010). The continuous use of treated wastewater in irrigation increases the total soluble salts in the soil. The cation exchange capacity values are increased by increasing the period of using treated wastewater for irrigation, especially in the surface layer (0–30 cm). Moreover, Cu, Fe, Pb, Co, Zn and Mn were added to the irrigated lands when treated wastewater was applied (Selem et al. 2000). The accumulation of heavy metals in the edible part of plant could adversely affect human health (Abd-Elfattah et al. 2002).

The rapid development of Oman's urbanization, increase in population and increase in agricultural production has led to high demand for water and urgent need to use treated wastewater as alternative source of freshwater in agriculture. However, treated wastewater could have some limitations such as presence of heavy metals which could accumulate in soil, plant and fruits. High concentrations of heavy metals in plant fruits could affect human health and cause many environmental problems. However, the conjunctive use of treated wastewater and groundwater resources could be employed, helping to safeguard farmer's income and sustain food production. Despite this promising option, more research and education efforts are needed to ensure proper use of treated wastewater for agricultural production. Therefore, the aim of the study was to maximize the application of treated wastewater in conjunction with groundwater by taking into consideration their quality and quantity, in addition to the environmental, economic and agronomic components.

2 Materials and Methods

The study was conducted in open field at the Agricultural Experiment Station, Sultan Qaboos University. Plots of 9 m^2 were made and each plot had a line of maize, sweet corn and okra. Each plot was irrigated either by A water (50% treated

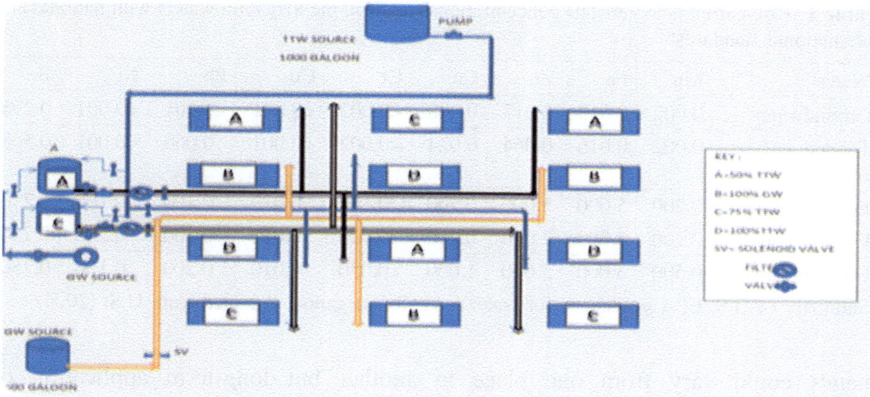

Fig. 1 Field experimental design with different treatments

wastewater and 50% groundwater) or B water (100% groundwater) or C water (25% groundwater and 75% treated wastewater) or D water (100% treated wastewater) as shown in Fig. 1. Plants were daily irrigated based on evapotranspiration. Plant samples were taken at the end of the study, whereas, soil samples were taken before and at the end of the study at depth of 0–30 cm.

Plants growth and yield of each crop irrigated by different waters were monitored. Fruits quality and quantity were assessed. Samples from soil and plants were taken for different physical, chemical and biological analyses. All physicochemical analysis for soil and plants were done in soil and water labs (SQU) following standard methods (Gee and Bauder 1986) and using inductively coupled plasma (ICP) instrument for metals analysis. Soil and plant Nitrogen (N) were analyzed in Rumais Research Laboratory (Ministry of Agriculture and Fisheries).

The study was repeated for three seasons and the data were statically analyzed using analysis of variance with Duncan's Multiple Range Test for the comparison of the mean values (SPSS 1998).

3 Results and Discussion

3.1 Heavy Metals in Irrigation Water

Water is the source of life for all living things and plants cannot survive without it. Table 1 demonstrates heavy metals concentrations in the irrigation waters used in the study. Comparing the used waters with national and international standards, it can be seen that elements concentrations mentioned in Table 1 had lower values than applied standards. However, long term usage of some waters may accumulate some harmful elements in soil and plant tissues if mismanagement occurs. In some studies, it was found that wastewaters could carry appreciable amounts of trace toxic metals (Yadar et al. 2002; Brar et al. 2000). The concentrations of those

Table 1 Comparing heavy metals concentrations (mg/l) in the irrigation waters with national and international standards[a]

Water	Mn	Fe	Zn	Cu	Cr	Cd	Pb	Ni	B
Groundwater	0.002	0.013	0.013	0.008	<0.002	<0.001	<0.001	<0.001	0.295
Treated wastewater	0.002	0.016	0.064	0.024	<0.002	<0.001	0.066	<0.001	0.508
EPA standard	0.200	5.000	5.000	0.500	0.100	0.010	0.100	0.100	0.750
FAO standard	0.200	5.000	2.000	0.200	0.100	0.010	0.500	0.200	0.750
Omani standard	0.500	5.000	5.000	1.000	0.050	0.010	0.200	0.100	0.750

[a]Summary of U.S. EPA guidelines for water reuse for irrigation. Adapted from U.S. (2004)

metals could vary from one place to another but long-term application of wastewaters on agricultural lands often results in accumulation of those metals on the irrigated soils (Rattan et al. 2002). Rattan et al. (2002) found high concentration of Fe, Cu, Co, Pb, Mn, Cr, As, Ni Cd,, and Zn in sewage effluents compared to well water. On the other side, irrigation by sewage effluent increased soil organic matter contents as well as heavy metals compared to the well water.

3.2 *Soil Physicochemical Properties*

Total carbon and organic matter are good indicators for soil fertility. Their values were increased when the land was irrigated by treated wastewater compared to groundwater (Fig. 2). That was expected since treated wastewater is usually rich with different organic and in-organic materials that could improve soil fertility and plant growth.

In addition to organic matter, treated wastewater is full of different nutrients when it compared to surface or ground waters (Fig. 3).

For soil micro-elements concentrations, Table 2 is showing a close values for the measured heavy metals of groundwater and treated wastewater. However,

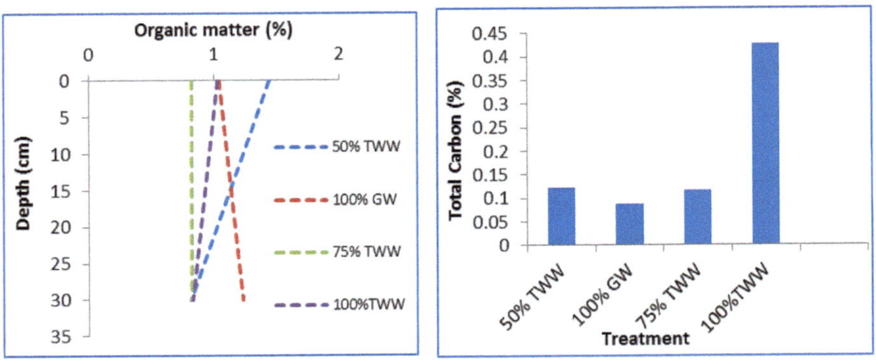

Fig. 2 Soil organic matter and total carbon

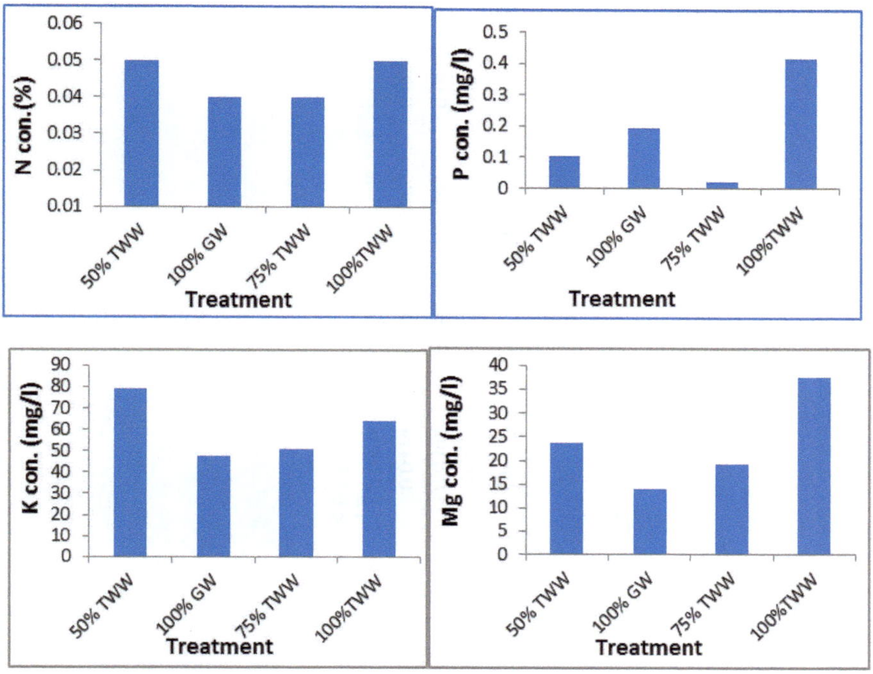

Fig. 3 Soil elements concentrations

significant differences were found between some treatments. Therefore, that increase could be an indicator for long run changes in soil properties and accumulation of some salts due to treated wastewater application if mismanaged was practiced. This expectation was confirmed with the findings of Palaniswami and Sree Ramulu (1994) and Bansal et al. (1992) studies when they applied wastewater for a long period. Rattan et al. (2004) noticed a build-up of Cu, Mn, As, Ni, Cr, Zn, Co, Pb and Fe in the sewage irrigated soils, over the well water-irrigated ones especially with Fe. Moreover, some soils have accumulated more than 70 mg kg^{-1} total Zn that could cause phyto-toxicity problem (Rattan et al. 2004).

3.3 *Crop Physicochemical Analysis*

From Table 3, it can be seen that treated wastewater gave the best yield for all three crops compared to groundwater. Plant growth and productivity were enhanced by different nutrients added from treated wastewater. That was also indicated by Stewart et al. (1990) and Abohassan et al. (1988) where they identified the beneficial effects of treated sewage water on some trees grown in Australia and Saudi Arabia.

Table 2 Mean concentration of heavy metals (mg/l) in soil samples*

Treatment	Mn	Cd	Fe	Zn	B	Ba	Cr	Co	Pb	Ni
50%TWW	0.018a	0.001a	0.330a	0.026a	0.166c	0.118a	0.043a	0.058a	0.196b	0.005a
100%GW	0.018a	0.001a	0.331a	0.001b	0.171b	0.123a	0.039a	0.060a	0.219a	0.011a
75%TWW	0.016a	0.001a	0.334a	0.001b	0.088d	0.087b	0.041a	0.061a	0.220a	0.005a
100%TWW	0.018a	0.001a	0.345a	0.003b	0.309a	0.110a	0.039a	0.062a	0.234a	0.001a

*Means in the column with same letter indicate no difference at Duncan's multiple range test at $p < 0.05$

Table 3 Average weight (kg) of some crops grown in the study

Treatment	Sweet corn	Okra	Maize
50%TWW	0.091	12.500	1.273
100%GW	0.141	11.091	1.193
75%TWW	0.085	10.556	1.160
100%TWW	0.090	13.958	1.593

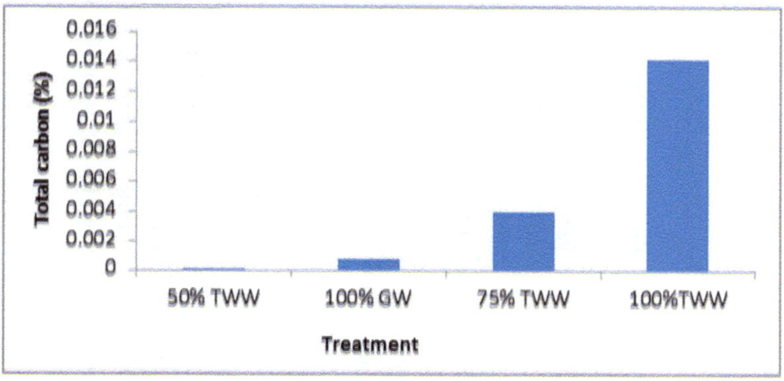

Fig. 4 Percentage of total carbon in maize plant leaves

Maize leaves were the best indicator for carbon content. Therefore, it can be seen from Fig. 4 that treated wastewater got the highest values compared to other treatments. It could be a reflection for what was found in water and soil samples. In similar study done by Abd-Elfattah et al. (2002), they found significant differences between leaves grown in soils irrigated with treated wastewater and leaves grown in soils irrigated with Nile water of both seasons.

Usually there is a direct relationship between salts found in the irrigation water and irrigated land. Whereas, for plants, root selectivity and present of salts in different forms could play a role in elements movement and translocation from soil to plant. Table 4 indicating significant differences ($p < 0.05$) in concentration of different elements in all applied treatments. However, concentration of heavy metals in the edible parts of all grown plants were within the level of the international standards (Table 5). This results was in parallel with Pescod (1992) findings, when he reported a normal concentration of heavy metals in seeds grown in treated wastewater. This could be due to low bio-available of heavy metal in the soil and crop uptake. Therefore, it do not represent a threat to the quality grown crops.

In general, different nutrients and heavy metals are presents in treated wastewater. However, movements and availability of these elements to the plant is highly dependent on their concentration and the environmental conditions surrounding the plant (Izzo et al. 1993). Same results were also found by Mahdi et al. (1998) when they reported that concentration of the elements in different crops is affected by tree

Table 4 Heavy metals concentration (mg/l) in tested crops*

Treatment	Element conc. (mg/l)								
	Mn	Cd	Fe	Zn	B	Cr	Co	Pb	Ni
Okra									
50%TWW	0.173d	0.001a	1.224c	0.357b	0.521c	0.060a	0.069a	0.252b	0.006b
100%GW	0.190c	0.001a	1.365b	0.364b	0.336d	0.071a	0.075a	0.229c	0.013b
75%TWW	0.242b	0.001a	2.372a	0.482a	0.745b	0.083a	0.087a	0.255b	0.127a
100%TWW	0.263a	0.001a	1.177d	0.495a	0.862a	0.057a	0.073a	0.300a	0.014b
Sweet corn									
50%TWW	0.177b	0.001a	1.295b	0.329c	0.073c	0.122b	0.091b	0.222c	0.068b
100%GW	0.204a	0.001a	1.582a	0.613a	0.492a	0.215a	0.100 a	0.191d	0.104a
75%TWW	0.152c	0.001a	1.584a	0.301c	0.122b	0.061c	0.070c	0.240b	0.011c
100%TWW	0.127d	0.001a	0.889c	0.400b	0.062c	0.064c	0.072c	0.444a	0.003d
Maize									
50%TWW	0.457b	0.001a	2.365b	0.256a	0.903a	0.136d	0.064a	0.210c	0.037a
100%GW	0.463a	0.001a	2.362b	0.219c	0.717c	0.146c	0.074a	0.213c	0.047a
75%TWW	0.366d	0.001a	2.279d	0.189d	0.454d	0.151b	0.075a	0.280a	0.040a
100%TWW	0.393c	0.001a	2.483a	0.226b	0.832b	0.181a	0.073a	0.241b	0.052a

*Means in the column in each crop have the same letter indicate no difference according to Duncan's multiple range test at $p < 0.05$

Table 5 Guideline for safe limits of heavy metals

Samples	Standards	Fe	Zn	Cu	Pb	Cd	Mn	Cr
Water (mg L-1)	Indian standard (Awashthi 2000)	NA	5.0	0.05	0.10	0.01	0.1	0.05
	WHO/FAO (2007)	NA	2.0	0.20	5.0	0.01	0.2	0.10
	European union standards (EU 2002)	–	–	–	–	–	–	–
	USEPA (2010)	NA	2.00	1.00	.015	.005	.05	0.10
	Kabata-Pendias (2010)	0.8	NA	NA	NA	NA	NA	NA
Soil (mgkg-1)	Indian standard (Awashthi 2000)	NA	300–600	135–270	250–500	3–6	NA	NA
	WHO/FAO (2007)	–	–	–	–	–	–	–
	European union standards (EU 2002)	NA	300	140	300	3.0	NA	150
	USEPA (2010)	NL	200	50	300	3.0	80	NA
	Kabata-Pendias (2010)	1000	NA	NA	NA	NA	NA	NA
Plant (mgkg-1)	Indian standard (Awashthi 2000)	NL	50.0	30.0	2.5	1.5	NL	NA
	WHO/FAO (2007)	450	60.0	40.0	5.0	0.2	500	5.0
	European union standards (EU 2002)	NL	60	40	0.30	0.20	NL	NA
	USEPA (2010)	–	–	–	–	–	–	–

Source CPCB (2002)

age and species. Longer exposure to treated wastewater did not indicate major effects on fruit minerals, including heavy metals. Sampling over longer period of time is needed to confirm the changes in nutrient composition over time. Therefore, in the present study it can be seen that treated wastewater treatment sometime got the highest values for heavy metals which could be an indication for heavy metal accumulation with long term application if treated wastewater is used without proper management. This prediction could be similar to Abd-Elfattah et al. (2002) findings when they found a significant difference in fruit contents of heavy metals and trace elements (Ni, Cd, Fe, Pb, Cu, Mn, Zn) between fruits produced by treated wastewater compared with Nile water in both seasons. Abd-Elfattah et al. (2002) detected some heavy metals accumulated in the edible part of some crops that could affect human health.

Finally the findings of this study are supported by many researches such as, Omran et al. (1988) when they conclude that long run application of treated wastewater to orange plants did not show any significant problems. Furthermore, no phyto-toxicity problem was observed in all tested crops due to irrigation with treated wastewater (Hamad et al. 1995). This finding was in parallel with El Mardi et al. (1995) study when they found that no symptom of phyto-toxicity problem was observed in date palm leaves and fruits.

4 Conclusions

No significant differences were found in some soil physical and chemical properties due to irrigation by groundwater and treated wastewater. Whereas, the concentrations of some chemical properties such as total carbon and some major elements (N, K, Mg), were significantly increased due to application of treated wastewater as compared to groundwater. Small increase in some heavy metals concentration in soil irrigated with treated wastewater compared to groundwater. However, values of heavy metals for all treatments were within the international standards.

There was a significance increases in plant growth and productivity when plants were irrigated with treated wastewater compared to groundwater. Moreover, there was an increase in heavy metal concentration of plants irrigated with treated wastewater compared to groundwater. However, all measured values were within the international standards.

Finally, it can be concluded that treated wastewater is a good source for different nutrients that can improve soil fertility and plant growth. However, a continuous monitoring is required to ensure its quality.

Acknowledgements The authors would like to thank staff from Sultan Qaboos University and Ministry of Agriculture and Fisheries for their support in collecting and analyzing soil, water and plant samples. Special thanks to USAID (FABRI) for their financial support.

References

Abd-Elfattah A, Shehata SM, Talab AS (2002) Evaluation of irrigation with either raw municipal treated waste or river water on elements up take and yield of lettuce and potato plants. Egypt J Soil Sci 42(4):705–714

Abohassan AA, Kherallah IE, Kandeel SA (1988) Effect of sewage effluent irrigation regimes on wood quality of *Presopi juliflera* grown in Riyadh Region. Arab Gulf J Sci Res 136(l):45–53

Al-Ajmy I (2002) Wastewater in the Sultanate and its effects on environment. In: International conference on wastewater management and its effect on the environment in hot and arid countries, Ministry of Regional Municipalities, Environment and Water Resources, Sultanate of Oman

Bansal RL, Nayyar VK, Takkar PN (1992) Accumulation and bioavailability of Zn, Cu, Mn and Fe in soils polluted with industrial waste water. J Indian Soc Soil Sci 39:795–799

Bedbabis S, Giuseppe F, Bechir BR, Makki B (2010) Effects of irrigation with treated wastewater on olive tree growth, yield and leaf mineral elements at short time. Hortic Sci 126:345–350

Brar MS, Mahli SS, Singh AP, Arora CL, Gill KS (2000) Sewer water irrigation effects on some potentially toxic trace elements in soil and potato plants in northwestern India. Can J Soil Sci 80:465–471

CPCB (Central Pollution Control Board) (2002) Parivesh, Newsletter from CPCB. Available at http://www.cpcb.delhi.nic.in/legislation/ch15dec02a.htms

El Mardi MO, Salama SB, Consolacion EC, Al-Shabibi MS (1995) Effect of treated sewage water on vegetative and reproductive growth of date palm. Commun Soil Sci Plant Anal 26 (11&12):1895–1904

Gee GW, Bauder JW (1986) Particle-size analysis. In: Klute A (ed) Methods of Soil analysis, physical and mineralogical methods, Part 1, 2nd edn. Agronomy series No. 9. American Society of Agronomy and Soil Science Society of America Inc. Publ. Madison, Wisconsin USA, pp 383–404

Hamad I, Nizam AA, Suleiman MS (1995) Biological detection of contamination in vegetables irrigated with brackish and underground water. In: Proceedings of the Arab regional workshop on the use of saline, brackish, and treated sewage effluent water in irrigated agriculture, Muscat, Sultanate of Oman, pp 481–531

Izzo K, Scagnozzi A, Belligno A, Navari-Izzo F (1993) Influence of NaCl treatment or Ca, K, Na interrelations in maize shoots. In: Fragesoand MAC, Beusichem ML (eds) Optimization of plant nutrition. Kluwer Academic Publishers, Dordrecht, pp 577–582

Levine A, Sanot A (2004) Recovering sustainable water from waste water. Environ sci Technol 201A

Mahdi OE, Salama SB, Consolacion EC, Al-Solomi M (1998) Effect of treated sewage water on the concentration of certain nutrient elements in date palm leaves and fruits. Commun Soil Sci Plant Anal 29:5–6

Mohammed MJ, Mazahareh N (2003) Changes in soil fertility parameters in response to irrigation of forage crops with secondary treated waste water. Commun Soil Sci Plant Anal 34:1281–1294

Mujeriego R, Salal (1991) Golf course irrigation with reclaimed waste water. Sci Technol 24 (9):161–172

Omran MS, Waly TM, Abo-Eleinain EM, Elnashar BMB (1988) Effect of sewage irrigation on yield, tree components, and heavy metals accumulation in navel orange trees. Biol Wastes BIWAED 23:17–24

Palaniswami C, Sree Ramulu US (1994) Effects of continuous irrigation with paper factory effluent on soil properties. J Indian Soc Soil Sci 42:139–140

Pescod M (1992) Wastewater treatment and use in agriculture. Bull. FAO #47 (125) (Rome)

Rattan RK, Datta SP, Chandra S, Sahran N (2002) Heavy metals and environmental quality: Indian scenario. Fertile. News 47(11):21–40

Rattan RK, Datta SP, Chhankar PK, Suribabu K, Singh AK (2004) Long term impact of irrigation with sewage effluents on heavy metal contents in soils, crops and groundwater—a case study. Agric Ecosyst Environ 109:310–322

Selem MM, Amir S, Abdel-Aziz SM, Kandil MF, Mansour SF (2000) Effect of irrigation with treated waste water on some chemical characteristics of soils and plants. Egypt J Soil Sci 40 (1/2):49–59

SPSS Inc. (1998) SPSS Base 8.0 for windows user's guide. SPSS Inc., Chicago

Stewart HTL, Hopmans P, Flinn DW (1990) Nutrient accumulation in trees and soil following irrigation with municipal effluent. Aust Environ Pollut 63(2):155–177

U.S Environmental Protection Agency (2004) Guidelines for Waste Water Reuse, Municipal support office of wastewater management. Washington, DC, EPA/625/R-04/108

World Bank (2012) Renewable energy desalination: an emerging solution to close the water gap in the Middle East and North Africa. World Bank, Washington, DC

Xu J, Laosheng W, Andrew CC, Yuan Z (2010) Impact of long-term reclaimed wastewater irrigation on agricultural soils: a preliminary assessment. J Hazard Mater 183:780–786

Yadar RK, Goyal R, Sharma RK, Dubey SK, Minchas RS (2002) Post irrigation impact of domestic sewage effluent on composition of soils, crops and ground water—a case study. J Environ Int 28(6):481–486

Potential of Treated Wastewater Usage for Adaptation to Climate Change: Jordan as a Success Story

Mohammed A. Salahat, Mohammed I. Al-Qinna and Raed A. Badran

Abstract Jordan sustainable development is obstructed by severe water scarcity that induces imbalances and shortages of water supply for various uses especially under high population growth rate, sudden immigrations, and climate change. Reserving water for drinking by treating wastewater treatment plants (WWTPs) effluent and reusing it for non-drinking could be a solution. This paper investigated the capability and contribution of the existing WWTPs' effluent for reuse in agriculture sector as a climate change adaptive measure. The paper provided clear understanding for the current and future climate changes impacts, developed climate change and water policies, current water resources and demands for agriculture sector, and suggested adaptive measures. Further, it emphasized on characterizing the WWTPs and quantification of effluent taking into account the satisfaction to Jordanian standards and guidelines. Major WWTP's effluents are within Jordanian standards; however some WWTP's have concerns to microbial quality that restricts their reuse. Samra WWTP effluent can be used for highly restricted class of cooked vegetables, parks, and playgrounds. The results demonstrated that wastewater reuse can be a crucial part of Jordanian water budget, can solve environmental problems, and can be a feasible adaptive option when managed properly. Further recommendations for WWTP operations, managements, reuse, and monitoring are included.

Keywords Climate change adaptation · Treated wastewater reuse · Agriculture

M.A. Salahat (✉) · M.I. Al-Qinna
Department of Land Management and Environment,
Faculty of Land Management and Environment,
The Hashemite University, Zarqa, Jordan
e-mail: mdsalahat@hu.edu.jo

R.A. Badran
Head of Waste Water Reuse Section,
Water Authority of Jordan (WAJ), Amman, Jordan

O. Abdalla et al. (eds.), *Water Resources in Arid Areas: The Way Forward*,
Springer Water, DOI 10.1007/978-3-319-51856-5_22

1 Introduction

Jordan, like many countries, is overwhelmed by water stress as a result of low precipitation rates, irrigation demands, industrial pollution, and untreated municipal sewage; consequently, Jordan is ranked among the top four "water poor" countries in the world (MoEnv 2014). Fresh water resources in Jordan suffer severe shortage and are under the water poverty line of 1000 m^3/capita/year. On a per capita basis, available water from renewable sources is anticipated to fall from 150 to 90 m^3/capita/year in year 2003 to the year 2025, (MWI 2009). A report issued by the Arab Forum for Environment and Development (AFED) declared that "Arab countries (including Jordan) are in many ways among the most vulnerable in the world to the potential impacts of climate change, in a region which already suffers from aridity, recurrent drought and water scarcity" (AFED 2009).

Jordan is experiencing a high population growth largely due to migrations of refugees and Jordanian returnees from other countries as a result to the political crises throughout the region, rural to urban internal migration, and higher urbanization. Those factors put increased stress on water availability and require a delicate balance to be struck between household and irrigation demands for water (Ammary 2007). The Jordanian government is following a sustainability plan to manage the water scarcity. Aspects of sustainability plan of water management is to address vital issues as enhancing conventional and non-conventional water resources, modeling future water demands according to water use and growth rates, reducing water pollution through management and regulatory plans, and improving technologies for water resources management and reuse of treated waste water.

The stress on water resources are significantly exacerbated by climate change. The Intergovernmental Panel on Climate Change (IPCC) expects that by the 2020s around 0.5 billion people could suffer increased water stress as a result of climate change (IPCC 2001). In Jordan, climate change could result in rainfall reduction and temperatures increase, leading to less water for municipal and agricultural uses (MoEnv 2014). Climate change is limiting Jordan's efforts to achieve its objectives conceived under the United Nations' Millennium Development Goals (MDGs). Efficient planning for climate risks management is the main factor to achieve MDG targets. This poses serious challenges for water resources management. "Adaptation to climate change is therefore a moral, economic and social imperative: action is needed now and water management should be a central element in the adaptation strategy of any country" (UN 2009).

Among the possible water sector adaptations to climate change, several mega projects are either implemented (e.g. Deisi project) or under construction (e.g. the Red Sea–Dead Sea canal). The non-renewable Deisi aquifer water transportation project can supply up to 100 MCM/year upon completion, however this amount is enough only to replenish the anticipated increase in water demand. The Red Sea–Dead Sea canal project faces two main barriers: the very high capital costs, and transboundary political limitations. Non-conventional water resources in Jordan (wastewater reuse and desalinization) offer options for climate change adaptation.

Seawater desalination requires high capital investment costs, and desalination of brackish water shows a good opportunity, but its quantity is low (around 14 MCM/year as estimated in 2013) (MWI 2015). Wastewater reuse has a low marginal cost and is proven to be a highly feasible option in Jordan. The goal of this article is to explore the potential of using treated wastewater in Jordan as a climate change adaptation measure through economical exchange of freshwater in the agriculture sector.

2 Current and Future Climate Change

Global warming and climate change (CC) is happening and is having more recognition among scientist and local, regional, or national entities. Jordan ratified the UNFCCC in 1994, and from that year started to work to fulfill its obligations by assigning the Ministry of Environment (MoEnv) as a focal point. The fulfillment of the national obligation to UNFCCC implies that Jordan should have the human, organizational, institutional and financial resources for developing the required tasks and functions on a permanent basis (NEEDS 2010).

According to the Jordanian first, second, and third national communications report submitted to UNFCCC in 1999, 2009, and 2014, respectively, the estimated historical changes in climatic variables were significantly indicative for increase in temperature and reduction in precipitation (MoEnv 1999, 2009, 2014). The scenario projections suggest warmer and drier climate, warmer summer, drier autumn and winter, with chances of heat waves, more drought and thus a highly stressed water balance, and trends of extreme events. In respect to recent studies, the future projections using multi-model ensembles suggest an "extremely likely" rise in mean temperature between 3.1 and 5.1 °C by the end of year 2100, and a "likely" reduction in precipitation from 15 to 21% in most of the country (MoEnv 2014).

2.1 Impacts of Climate Change on Water Resources

All conducted water studies in Jordan identify water stress as a major barrier for sustainable development; and recognizes climate change as a magnifier. The impact sensitivity of climate change on water sector is high with negative effect. Expected impacts are less groundwater recharge and replenishment, groundwater depletion and salinization, water contamination, drought, soil erosion and desertification, disappearance or reduction in springs, violations and vandalism, social conflicts and economic stresses (Abdulla et al. 2009; Altz-Stamm 2012; MoEnv 2014).

Agricultural activities and production is one of the most climate-sensitive sectors. Thus, under climate change conditions the agriculture production is predicted to reduce significantly. According to Al-Bakri et al. (2010), a 2 °C temperature increase and 20% precipitation reduction will decrease wheat yield by 21%, 35% for barley, and 10% for tuber and root crops.

2.2 *Climate Change Policy and Water*

Jordan has put enormous efforts to cope with climate change. The country has conducted excellent capacity building projects. The country has been successful in the assessment of climate change challenges and the measures needed. On the other hand, implementation of those measures has been in lower pace due to lack of financial resources, technical capacity, and weak linkages with national plans (MoEnv 2007, 2013a).

Jordan is classified as a Non-Annex I party and thus is obligated for few commitments to UNFCC. Some of these commitments are to prepare for climate change adaptation, address climate change considerations in the relevant policies and action plans (MoEnv 2013a).

On the other hand, Jordan has developed policies and strategies, and proposed plans to sustain environmental resources. These policies included measures related to the successful implementation of integrated resources management (IRM); creating enabling environment, defining institutional roles, and establishing management tools. For example, the Ministry of Water and Irrigation (MWI) has developed a comprehensive water strategy entitled "Water for Life" for the period 2008 to 2022 (MWI 2009), and was revised in 2012. The strategy mainly focuses on effective water demand management and supply operations, and institutional reform. Climate change is well addressed in the strategy's vision. (MoEnv 2012, 2013b, c).

3 Water Demand for Agriculture

3.1 *Agricultural Land Area*

Most of Jordan's land is under arid and semiarid climate, making it highly sensitive to water availability and climate change. At the same time, land use is dynamic with continuous changes between different uses. The main cause is the increased pressure on the limited natural resources, particularly water, due to high population growth rate. Land use is dominated by rangeland; non-cultivated areas, while agricultural land forms a small part of the total area (DOS 2015).

According to the department of statistics (DOS) (2013a, b), the total agricultural lands in Jordan is estimated to be about 2.6×10^5 ha of which rainfed lands contributes about 60.4% of the total agricultural lands (i.e. 1.6×10^5 ha), while irrigated lands represent about 39.6% of the total agricultural lands (i.e. 1.0×10^5 ha). Most of the water in irrigated lands are used for trees (4.5×10^4 ha) and for vegetables (4.7×10^4 ha), while irrigated lands allocated for forage crops are only 1.1×10^4 ha. On the other hand, 74% of rainfed lands are planted with forage crops (1.1×10^4 ha), while trees and vegetables represent only 24.5 and 1.5% of total rainfed lands, respectively.

3.2 Water Resources for Agriculture Sector

Jordan extended its conventional irrigation water resources and included non-conventional resources; mainly treated wastewater reuse and desalination. The conventional water resources consist of twelve groundwater basins (renewable and nonrenewable) providing 251 MCM/year and forming 55% of total water used for agriculture, and surface water providing 105 MCM/year making 23% of the agricultural use (Fig. 1). According to (MWI 2015) 53.7% of total water use is directed for irrigation purposes. The major sources for irrigation water for the year 2012 are treated wastewater reuse of 100 MCM/year compromising 22% of agricultural water use, groundwater, and surface water. There is no data available about amount of desalinized water used for agriculture since most of it is at the farm level.

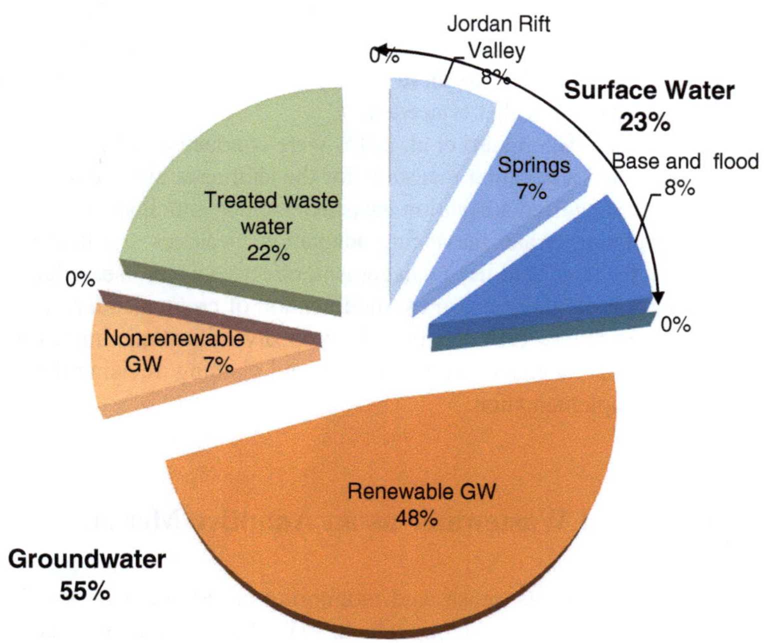

Fig. 1 Agricultural water resources in Jordan for 2012

The Red Sea/Dead Sea canal (RSDS) project is the largest mega project to desalinize sea water. It is planned to desalinize about (80–100) MCM/year at the northern intake location of Aqaba.

Al-Omari et al. (2013) modeled water demand under RSDS canal project until 2050, and they anticipated that Amman and Zarqa domestic demand will be satisfied until the year 2050. Furthermore, for the RSDS scenario they found that, the agricultural demand deficit in the Jordan Valley, the largest and the main agricultural area, will drop to about 85 MCM for the year 2050 as a result of the increased treated wastewater flow to the valley.

3.3 Suggested Adaptive Measures for Agricultural Water Shortage

Climate change will affect agricultural yield and accompanied services (i.e. Food production, food security, people employed in agriculture) directly because of variations in temperature and rainfall that affect production, and indirectly through stresses on water resources. Other impacts of CC may include increases in seasonal temperature variability, and frequency of temperature extremes, (e.g., frost and heat waves), decrease in water availability (drought), shorter winters, shifting in growing seasons, short life of wild flowers and thus honey production, frequency distribution and intensity of pest and disease occurrences, fire incidence, changes in soil quality, and failure to meet chilling requirements (MoEnv 2009). Given the broad potential impacts of CC on water and agriculture, there is a need for both general and specific measures to adapt to these impacts. A safe and permanent source of water is just one adaptation that is needed.

Several studies (i.e. Hammouri et al. 2015) were conducted in Jordan to address the major impacts and adaptation measures for the shortages in water especially in agriculture sector. Suggested adaptation measures include anticipatory, autonomous and planned adaptation; water harvesting adaptation measures and land resources adaptation measures (e.g. land-use management, effective water use, reducing need for water, water conservation measures, modification of crop calendar); adopting a “Conservation Agriculture (CA)” approach; crop diversification; improved water management; better strategies on crop selection and planting, revising the policies; and action plans implementation.

4 Using Treated Wastewater as an Adaptive Measure

Wastewater treatment plants, reuse, and monitoring are each the responsibility of different governmental bodies. Among which, WWTPs operation and maintenance is the responsibility of Water Authority of Jordan (WAJ), Jordan Valley Authority

(JVA) is in charge for maintaining water delivery system and distributing water to the farms. Ministry of Environment (MoEnv) monitors water quality of surface water resources, and Jordan Food and Drug Administration (JFDA) is monitoring fresh fruits and vegetables quality (Al-Momani 2011; Seder and Abdel-Jabbar 2011).

In 2009, a new strategy "Water for Life 2008–2022" was issued based on vision-driven change effort (MWI 2009). The strategy aims to sustainable management of existing water resources through a strategic and integrated approach. The strategy emphasized MWI goals and water management actions. Farmer's capability to react with the strategy is a major part of current status and future goals in regard of using irrigation water. Management of water resources with focusing on sustainability of current and anticipated use, pollution protection for resources, conveying; distributing; applying; and using water with highest efficiency possible, is well stressed out in the strategy.

Adapting to climate change makes the affected societies forge ahead with efforts to cope with a vague future of water supply. One solution is to save more water for domestic use by using treated wastewater in industry or irrigation which presents a win-win situation that provides abundant high-quality, non fresh water to other purposes while reserving drinking water for domestic use.

4.1 Existing Wastewater Treatment Plants

As a result of Jordan's ambitious campaign since the 1980s, 62.4% of the population currently is connected to wastewater plants network (MWI 2015). According to (MWI 2015), there are 31 wastewater treatment plants (WWTPs) (Fig. 2) serving the country with a designed total capacity of 606,305 m^3/day (equivalent to 221.3 MCM/year). The number of WWTPs has almost doubled since 1993 (14 WWTPs with total capacity of 58 MCM/year), indicating government and donors' efforts to utilize the treated wastewater as a nonconventional water resource.

4.1.1 Existing WWTPs Characteristics

The major existing WWTPs in terms of capacity, inlet flow (m^3/day), adopted treatment technology, year of commissioning, design BOD, and the produced liquid sludge (m^3/day) are shown in Table 1. In 2013, national capacity of Jordan's WWTPs was about 323,950 m^3/day (equivalent to 118.2 MCM/year) of raw wastewater. The main adopted treatment technologies are activated sludge, waste stabilization ponds, extended aeration, trickling filters, and oxidation ditches.

The total designed flow for all 31 WWTPs is about 606,305 m^3/year, by which the largest WWTP is the Samra WWTP in Zarqa that is designed to treat 364,000 m^3/day of raw wastewater (60% of the total designed flow of all WWTPs), followed by South Amman WWTP that is designed to hold 52,000 m^3/day (8.6%

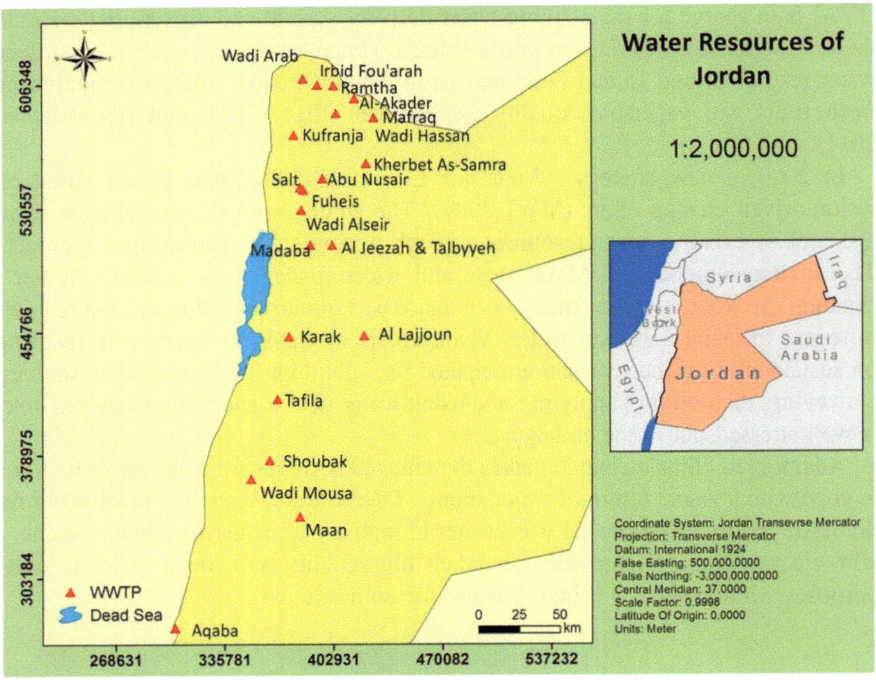

Fig. 2 Spatial location of the existing WWTPs in Jordan

of the total designed flow of all WWTPs). In terms of actual received raw wastewater flow, the Samara WWTP is receiving 71.2% (i.e. 230,606 m^3/day) of the total flow rate that estimated about 323,950 m^3/day (i.e. 121 MCM/year).

The Samra WWTP treats wastewater released from the Zarqa river basin, which is part of the two populated cities of Greater Amman and Zarqa. The Samra WWTP drains its effluent into the King Talal Dam, which provides irrigation water for the Jordan Valley.

In the term of sufficiency of the designed flow as compared to actual inlet flow, all existing WWTPs have been upgraded to contain almost double the inlet flow. Before upgrading, most of WWTPs suffered from over-capacity flow, especially the Samra WWTP. Samra WWTP was the first plant in Jordan established in 1982 based on stabilization pond technology with a 68,000 m^3 daily inflow capacity and had served only 300,000 citizens at that time. However, due to the sharp population growth rate and sudden migrations (due to unrest situation in neighboring countries), the plant now serves about 2.26 million people living in the Greater Amman and Zarqa areas and thus was overloaded especially in 2002 (before the main upgrade) where it was receiving about 186,000 m^3 daily raw wastewater.

The latest upgrade from 2003 to 2008 was implemented on a Build-Operate-Transfer (BOT) basis. The plant is redesigned and upgraded to accommodate an average daily flow of 420,000 m^3 and a peak daily flow of

Table 1 Wastewater treatment plants in Jordan for the year 2013

No.	Plant name	Service governorate	Designed flow (m^3/day)	Inlet flow (m^3/day)	Treatment technology	Year of commissioning	Liquid sludge (m^3/day)	Remarks
1.	Samra	Amman, Zarqa	364,000	230,606	AS	1984 Old 2008 New	3000	Good
2.	Aqaba- Mechanical	Aqaba	12,000	9845.5	EA	2005	232	Good
3.	Aqaba-Natural	Aqaba	9000	6730.6	WSP	1987	150	Good
4.	Madaba	Madaba	7600	5172	AS	2005 New	250	Good
5.	Irbid (Central)	Irbid	11,023	8132	TF and AS	1987	210	Good (will be upgraded soon)
6.	Sult	Balqa	7700	5290.7	EA	1981 Old 1994 Upgrade	130	Good
7.	Jerash	Jerash	9000	3680.8	OD	1983	100	Will be upgraded
8.	Mafraq	Mafraq	6050	2008.8	WSP	1988	47	Will be upgraded
9.	AinBasha (Baqa)	Amman, Balqa	14,900	10,208.6	TF	1987 Old 2000 Upgrade	250	Good
10.	Karak	Karak	5500	1753.4	TF	1988	10	Will be upgraded
11.	Abu-Nuseir	Amman	4000	2570.8	AS R,B,C	1986	60	Good
12.	Tafila	Tafila	7500	1380	TF	1988	8	Will be upgraded
13.	Ramtha	Ramtha	7400	3488.3	AS	(1987) 2004 New	100	Good
14.	Ma'an	Ma'an	5772	3170.8	EA	(1989) 2009 New	100	Good
15.	Kufranja	Ajloun	9000	2763	TF	1989	60	Will be upgraded
16.	Wadi-Essir	Amman	4000	3623.9	Aeration lagoons	1997	86	Good
17.	Wadi Al Arab	Irbid	21,000	10,264	EA	1999	240	Good

(continued)

Table 1 (continued)

No.	Plant name	Service governorate	Designed flow (m^3/day)	Inlet flow (m^3/day)	Treatment technology	Year of commissioning	Liquid sludge (m^3/day)	Remarks
18.	WadiMousa	Ma'an	3400	3028.9	EA	2000	100	Good
19.	Wadi Hassan	Irbid	1600	1131.8	OD	2001	40	Good
20.	Tal-Almantah	Balqa	400	300	TF & AS	2005	7	Good
21.	Al-Ekeder	Mafraq	4000	3907.8	WSP	2005	92	Good
22.	Al-Lajjoun	Karak	1200	853.1	WSP	2005	20	Under upgrading
23.	Fuheis	Amman, Balqa	2400	2221	AS	1997	16	Good
24.	Al-Jiza	Amman	4500	703.9	AS	2008	17	Good
25.	Al-Merad	Jerash	9000	1000	AS	2011 (2010)	24	Good
26.	Shoobak	Ma'an	350	100	WSP	(2010)	2	Good
27.	Al-Mansorah	Ma'an	50	15	WSP	(2010)	0.4	Good
28.	South Amman	Amman	52,000			Under construction		
29.	Mu'tah and Adnaniyyah	Karak	7060			Under construction		
30	Shallaleh	Irbid	13,700			Under construction		
31.	Shouna Shamaliyyah	Irbid	1200			Under construction		

AS Activated Sludge, *WSP* Waste Stab Ponds, *EA* Extended Aeration, *TF* Trickling filter, *OD* Oxidation Ditch
The data was obtained from MWI (2015)

840,000 m^3. Currently, the plant consists of primary sedimentation tanks, aeration tanks, anaerobic sludge digesters, biogas power generators and hydro-power generators (utilizing the elevation change between Amman and Samra), and an odor control system (MWI 2015).

According to MWI (2015), there are some WWTPs that still not functioning yet, namely; South Amman, Mu'tah and Adnaniyyah, Shallaleh, and Shouna Shamaliyyah because they are under construction or commissioning and will be operating soon.

4.1.2 Guidelines for Using Treated Wastewater

Utilizing of treated wastewater for agricultural purposes entails certain restrictions to be developed and applied to ensure public safety and health. Main suggested crops to be grown under treated wastewater include industrial crops and forest trees, parks and playgrounds, cooked vegetables, field crops, or fruit trees. Non-food crops reduce human contact with the treated wastewater, which ease water quality standards and thus less treatment requirements than other uses.

Worldwide, there are many organizations that have developed standards and guidelines for using treated wastewater. The World Health Organization (WHO) guidelines are the most important ones, and they are focused mainly towards developing countries. The goal of the guidelines is to protect humans (citizens, agricultural workers, irrigated fields neighbours, etc.) from contagion. WHO guidelines focus on the microbiological quality and the required treatment technique to obtain that quality. The main technique is the stabilisation ponds method because it has lower cost, simpler, and high efficient in removing the parasites which are considered the most infectious agent in the developing world. Other important guidelines were developed by US Environmental Protection Agency (EPA 1992), the "Guidelines for Water Reuse", and were updated in 2004.

Jordan developed its own regulations in 1982 that were very restrictive and prohibited the reuse of the treated wastewater for agriculture. Revisions were made due to the increased awareness of the opportunities of using treated wastewater and due to the progress in technology. In 1995 MWI issued the first Jordanian standard for wastewater reuse and was updated in 2002 and 2006 for different qualities of water resources (Table 2). The Jordanian standard is based on WHO guidelines with some modifications to meet the local requirements and conditions.

Table 3 shows the actual tests of main effluent criteria for the year 2013 with the values higher than Jordanian standards had been bolded. All the treated wastewater plants have effluent pH lower than the standards (6–9 units), with highest value for Ma'an WWTP. The relatively low pH's permits the use of effluent for irrigation. Total suspended solids (TSS) is a water quality parameter representing all the suspended solids including organic and inorganic material. TSS comprises all particles less than 2 μm, though bacteria and algae can also part of TSS. Suspended particles can gain more heat from solar radiation than water molecules, and transfer it to water molecules by conduction. Warmer water holds dissolved oxygen less than colder water, so dissolved oxygen levels will decrease. In addition, the higher surface water temperature can result in water stratification, or layering, and the two layers do not mix up. Decomposition and respiration often occur in the lower layers, they can become too hypoxic (low dissolved oxygen levels) for organisms to survive. All WWTPs have TSS within limits of B and C categories except for four WWTPs; natural Aqaba (constructed in 1987 with waste stab ponds mechanism), Karak (was constructed in 1988 with trickling filter mechanism; needs upgrade), Al-Ekeder and Al-Lajjoun (was constructed in 2005 waste stab ponds mechanism and reaches their maximum capacity; need extension and upgrade).

Values of total dissolved solids for WWTPs are lower than the standards except for Tal-Almantah plant that exceeded the standards for the three main categories. High TDS concentration in wastewater could cause toxicity through increases in salinity, or could lower the efficiency of biological treatment by resulting in low COD removal efficiencies.

The results in Table 3 show that most of the WWTPs have *Escherichia coli* higher than the Jordanian standards for categories A and B except for Samra, Aqaba—Mechanical, Abu-Nuseir, Ramtha, Ma'an, Wadi-Essir, Wadi Mousa, and Wadi Hassan WWTPs. Thus, the exceeding WWTPs must have higher attention regarding sterilizing and chlorination.

Regarding COD, all WWTPs have values within Jordanian standards for categories B and C except Al-Ekeder, and Al-Lajjoun WWTPs where their COD exceeds the standards for all categories.

4.2 Quantification of Treated Wastewater Reuse

Treated wastewater is critical element in the Jordan's water budget and strategy. The government in association with partners as USAID, GIZ, and others had developed several projects (e.g. Water Reuse Implementation Project and Reuse in Industry, Agriculture and Landscaping (RIAL, etc.) to improve the reuse of treated wastewater benefits and reform associated policies and regulations.

Currently, WAJ is the governmental entity responsible of WWTPs, and of collecting water in areas not served by the retail water utilities, while retail water utilities are responsible of collecting wastewater within their geographic extension. In Jordan, irrigation water management entails efficient water delivery and sustaining water resources, especially in Jordan Valley. Fresh surface water is supplied to treatment plants in trade of treated wastewater used for irrigation.

In 2012, approximately 98.6% of the produced treated wastewater is used for agriculture representing about 22% of total irrigation water resources, while 1.4% of that amount is used for industrial activities. Treated wastewater contributed to about 56% of the total water resources used for irrigation in the North and Middle Jordan Valley and this percentage is increasing on an annual basis because of higher effluents from Samra WWTP and other plants releasing treated water toward the Jordan Valley, such as Wadi Al Arab WWTP, Wadi EsSir, Kufranjah, and Sult.

Table 2 Criteria for treated wastewater reuse in irrigation and their allowable limits

Parameter	Unit	Jordanian Standards 2006[a]		
		A	B	C
BOD	mg/l	30	200	300
COD	mg/l	100	500	500
DO	mg/l	>2	–	–
TSS	mg/l	50	150	150
TDS	mg/l	1500	1500	1500
pH	Unit	6–9	6–9	6–9
Turbidity	NTU	10	–	–
Nitrate	mg/l	30	45	45
Total nitrogen	mg/l	45	70	70
Total PO_4^{-2}	mg/	30	30	30
Escherichia coli	MPN/100 ml	100	1000	Not applicable
Intestinal nematodes	Egg/l	≤ 1	≤ 1	≤ 1

[a]*A* represents cooked vegetables, parks, playgrounds and sides of roads within city limits, *B* represents fruit trees, sides of roads outside city limits, and landscape, and *C* represents field crops, industrial crops and forest trees

There are two main methods to reuse treated wastewater; the first one is the indirect way, e.g. using the water after mixing with fresh water, and direct way by using water without mixing. According to MWI (2015), the total amount treated wastewater produced for the year 2013 is about 121 MCM, where direct use is estimated 27.9 MCM representing 23.1% of the total amount produced. Direct treated wastewater reuse was achieved through contracting with about 215 farmers to irrigate an area of about 14,184 dunums (Table 4). For the year 2013, the Samra WWTP accounts for 47% of the total direct treated wastewater reuse (13.12 MCM) irrigating about 3990 dunums, followed by Aqaba- Mechanical that accounts for 12% (3.347 MCM), then Aqaba-Natural 6.7% (1.867 MCM) irrigating about 1,580 dunums, Madaba 6.5% (1.808 MCM) irrigating about 1,220 dunums, Ramtha 5.7% (1.591 MCM) irrigating about 1,302 dunums, Al-Ekeder 3.8% (1.051MCM) irrigating about 1069 dunums, and Wadi Mousa 3.6% (0.992 MCM) 1069 dunums.

The southern part of Jordan; namely Wadi Mousa hosted one of the earliest pilot projects funded by USAID to reuse treated wastewater in a participating farm, the second stage was to irrigate nearby farms (Seder and Abdel-Jabbar 2011). Another pilot project was in the northern part of the country; namely Wadi Hassan. The effluent was used in University of Irbid campus for lawn irrigation (KfW Development Bank 2006). The general reuse of treated wastewater is directed for irrigating fodder, olives, fruit trees, palm trees, forest, nursery, wind breaks, turf, and landscape plants. The MWI have created a monitoring utility that is responsible for regular monitoring and testing of the treated wastewater quality, as well as the soil and the plant for environmental precautions.

Table 3 Treated wastewater quality referred to the Jordanian standards

Plant name	pH	BOD_5 (mg/L)	COD (mg/L)	TSS (mg/L)	TDS (mg/L)	*E. coli* (MPN/100 ml)	Within/not with Jordanian standards
Samra	7.85	9.91	71.09	17	1109.82	18	Within
Aqaba–Mechanical	7.89	4.65	26.09	5.45	552.09	6	Within
Aqaba–Natural	8.08		**340.55 a**	**221.73 a b c**	**767.82**	**51,293 a b**	**COD, TSS, *E. coli***
Madaba	8.03	8.64	58.14	13.05	1178	**254,722 a b**	***E. coli***
Irbid (Central)	8.13	**34.18 a**	**183.77 a**	**87.64 a**	1064.73	**208,971 a b**	**BOD, COD, TSS, *E. coli***
Sult	7.95		88.77	**55.82 a**	827.73	**14,620 a b**	**TSS, *E. coli***
Jerash	7.86		**412.86 a**	**106.14 a**	1408.36	**277,353 a b**	**COD, TSS, *E. coli***
Mafraq	7.98		**364.18 a**	**125.09 a**	1032.36	**2,628,923 a b**	**COD, TSS *E. coli***
AinBasha (Baqa)	8.09	**49.5**	**109.52 a**	33	1169.34	**1,027,803 a b**	**BOD, COD, *E. coli***
Karak	7.92		**393.95 a**	**190.55 a b c**	963.64	**3,060,615 a b**	**COD, TSS *E. coli***
Abu-Nuseir	7.64	6.98	58.79	8.68	1084.61	5	Within
Tafila	8	**49.55 a**	**214.82 a**	**97.86 a**	796.73	**2,244,119 a b**	**BOD, COD, TSS *E. coli***
Ramtha	8.09		56.41	12.5	1393.45	57	Within
Ma'an	8.39		46.45	11.64	1054.64	12	Within
Kufranja	8.1		**331.18 a**	**96.32 a**	1077.27	**2,150,640 a b**	**COD, TSS, *E. coli***
Wadi-Essir	7.87		**120.09 a**	45.09	864.73	38	**COD**
Wadi Al Arab	8.05	19.98	88	22.73	984.77	**45,121 a b**	***E. coli***
WadiMousa	8.02		47	7.35	835.55	3	Within
Wadi Hassan	7.76		56.73	8.14	1107.82	11	Within
Tal-Almantah	6.87	**35.18 a**	**179.36 a**	**97.77 a**	**1877.18 a b c**	**4985 a b**	**BOD, COD, TSS, TDS, *E. coli***
Al-Ekeder	8.1		**556.09 a b c**	**283.18 a b c**	1241.45	**422,582 a b**	**COD, TSS, *E. coli***

(continued)

Table 3 (continued)

Plant name	pH	BOD_5 (mg/L)	COD (mg/L)	TSS (mg/L)	TDS (mg/L)	*E. coli* (MPN/100 ml)	Within/not with Jordanian standards
Al-Lajjoun	8.17		**547.73 a b c**	**284.09 a b c**	1491.18	**22,658 a b**	**COD, TSS, *E. coli***
Fuheis	7.94	4	**134.77 a**	**100 a**	980.64	**11,388 a b**	**COD, TSS, *E. coli***
Al-Jiza	8	7.82	76.82	14.91	1271.45	**2130 a b**	***E. coli***

BOD5represents the biochemical oxygen demand at 20 °C over 5 days and is a measure of the biodegradable organic matter in the wastewater
(a) represents exceeding Jordanian standards of cooked vegetables, parks, playgrounds and sides of roads within city limits, (b) represents exceeding Jordanian standards of fruit trees, sides of roads outside city limits, and landscape, and (c) represents exceeding Jordanian standards of field crops, industrial crops and forest trees

Taking into account the percent of directly treated wastewater use either inside the WWTP vicinity or through contracts with farmers, some WWTPs are fully committed for direct use as Aqaba-Mechanical, Aqaba-Natural, Madaba, Mafraq, Ramtha, Kufranja, Wadi Mousa, Wadi Hassan, Al-Ekeder, and Al-Jiza (Talbiea). On the other hand, only 15% of the total produced treated wastewater from Samara WWTP is implemented for direct use. Thus indirect reuse of treated wastewater in majority is obtained from Greater Amman. Unfortunately, 77% of the total produced treated wastewater is indirectly used. The treated wastewater from Samra WWTP runs into Zarqa River and feeds King Talal Dam (KTD). Then, it is mixed with collected rainwater to flow thereafter through King Abdulla Canal to be used in irrigating Jordan Valley farms. The diluted mixed wastewater is reused to irrigate 4000 farms forming 100,000 dunums (i.e. 25 times larger than the area irrigated with direct use), mostly using drip irrigation (Vallentin et al. 2009; Ulaimat 2012). Similar situations to Samra plant exist at Sult, Jerash, Ain Basha (Baqa), Abu-Nuseir, Ma'an, Wadi EsSir, and Fuheis ranging from 4 to 46.3% direct use. On the other hand, other WWTPs have no direct use such as Wadi Al Arab, Tal-Almantah, and Al-Lajjoun plants where treated wastewater flows through river and valley to be used totally indirectly.

5 Reuse of Treated Wastewater as a Success Story for Climate Change Adaptation

Counter acting increasing water demand due to population growth, rapid urbanization, and threats of climate change by allocating new resources is, on its own, a monumental challenge. In Jordan, this challenge is further exacerbated by the need to satisfy the competing water demands of various sectors. Water is considered the critical constraint for sustainable economic development of Jordan. Jordan is a small, but growing country with fragile and limited water resources and economic situation where water per capita has dropped to 86 M^3/capita/year: thus ranking Jordan as the water poorest country of the world.

Ensuring an adequate, safe, and secure water supply is economic development hurdle for Jordan. Searching for the most affordable adaptation options does not always yield the safest option. There are high risks of water quality deterioration due to rapid growth, and unplanned land use actions threaten the safety of the remaining unsecured and limited water resources. While treated wastewater reuse has been used for a long time in Jordan, the question remains: Is this option is safe and sustainable? Given the competing demands for freshwater resources among the different uses, can this conflict encourage using wastewater for irrigation? Looking over the history of water demand and supply in Jordan, treated wastewater reuse can help satisfy freshwater demand with low treatment cost.

From environmental standpoint, the tremendous amounts of produced wastewater in urban areas are considered a problem by itself. The associated health

Table 4 Treated wastewater reuse attributes

No.	WWTP name	Amount of outlet (MCM/year)	No. of agreements	Area irrigated (dunum)	Direct Amount of water used (MCM/year)	Crops	Excess water destination	Direct used water (%)
1.	Samra	87.527	34	3990	13.12	Fodder, olives	King Talal dam	15
2.	Aqaba–Mechanical	3.347	1	–	3.347	Industry, turf	–	100
3.	Aqaba–Natural	1.867	4	1580	1.867	Palm, wind breaks, turf	–	100
4.	Madaba	1.808	27	1220.3	1.808	Fodder, olives, nursery	–	100
5.	Irbid (Central)	2.877	None	–	–	–	Jordan river	0.0
6.	Sult	2.464	5	99.54	0.109	Olives, fruit trees	Wadi Shoaib valley	4.4
7.	Jerash (Al-Merad)	0.416	1	27.5	0.03	–	King Talal dam	7.2
8.	Mafraq	0.572	18	660.20	0.572	Fodder	–	100
9.	AinBasha (Baqa)	4.701	15	436.589	0.638	Nursery, polo playground	King Talal dam	13.6
10.	Karak	0.548	8	608.93	0.548	Fodder, forest, olives	–	100
11.	Abu-Nuseir	0.983	1	75	0.18	Al Ordon street (landscape)	Berain valley	18.5
12.	Tafila	0.327	None	–	–	–	GhourFifa	0.0
13.	Ramtha	1.591	22	1302.0	1.591	Fodder	–	100
14.	Ma'an	0.902	9	382	0.418	Fodder	The valley	46.3
15.	Kufranja	0.916	10	811.62	0.916		–	100

(continued)

Table 4 (continued)

No.	WWTP name	Amount of outlet (MCM/year)	No. of agreements	Area irrigated (dunum)	Direct Amount of water used (MCM/year)	Crops	Excess water destination	Direct used water (%)
						Fodder, forest trees		
16.	Wadi ESsir	1.582	1	61.82	0.068	Olives	Kafrain dam	4.3
17.	Wadi Al Arab	4.006	None	–	–	–	Jordan river	0.0
18.	Wadi Mousa	0.992	38	1069	0.992	Fodder, olives	–	100
19.	Wadi Hassan	0.435	1	721	0.435	Olives, fruit trees	–	100
20.	Tal-Almantah	0.129	None	–	–	–	–	0
21.	Al-Ekeder	1.051	17	1068.65	1.051	Olives, fruit trees	–	100
22.	Al-Lajjoun	0.257	None	–	–	–	Lajjoun valley	0.0
23.	Fuheis	0.829	1	30	0.033	Fodder	WadiShoaib valley	4.0
24.	Al-Jiza (Talbiea)	0.146	None	–	0.146	Forest trees, landscape plants	–	100
25.	Al-Merad		2	40	0.044	–	King Talal dam	10.58
26	Shallaleh	0.772	None				Jordan valley	0

problems (e.g. pollution, odor, vectors) and the investments needed in wastewater management infrastructure (mostly notable storage) is another problem. Thus, treated wastewater reuse results in environmental, economic, and public health advantages over effluent disposal.

Using treated wastewater for irrigation supplements available water for agriculture and thus increases the amount of freshwater available for other uses. Also, in comparison with other treatments such as desalinization, wastewater reuse is cheaper. However, in order to ensure environmental and public safety, appropriate water policies, standards, monitoring procedures, qualified regulators, technical expertise, and massive investments in operation and maintenance are required. According to Ghneim (2010), the existence of wastewater networks and treatment plants are important for successful wastewater reuse in agriculture. Though, the more important is the existence of appropriate policies, policy instruments for the implementation of those policies, legislations, institutional frameworks, and regulations.

The Jordanian government, through its entities and divisions, has put great efforts to establish a capable policy for wastewater standards. It has been able to develop legal and institutional framework to manage wastewater through adopting various policies relevant to wastewater, e.g. sanitation, pricing, environment, and public health.

According to current treated wastewater effluent chemistry as compared to standards, wastewater reuse has caused an excellent way to get rid of the wastewater, and thus enhanced the environment and public health. In addition, wastewater reuse has increased agricultural production and enabled farmers to utilize a valuable resource of plant nutrients that exists in the water. WHO (2006) stated that "the safe use of wastewater in agriculture would maximize public health gains and environmental protection".

High population growth lead to higher municipal water demands, though amount of generated wastewater increases. Achieved through proper treatment, wastewater is currently suitable for many uses (e.g. irrigation, industry, or aquifer recharge). Thus, treated wastewater is considered a crucial component of Jordan's renewable water resources. Although public acceptance for the reuse of treated wastewater is still low, the increasing number of successful projects for using wastewater in Jordan as indicated by the increasing number of contracts with farmers is an indication of a promising future. In addition, wastewater can be a dependable water supply resource even in drought waves.

The success of centralized wastewater treatment (e.g. Samra WWTP after being upgraded under BOT management) in either entire or partial use for agriculture production in Jordan Valley, demonstrates the success for treated wastewater to be an adaptive method for climate change to substitute freshwater in irrigated agriculture. On the other hand, water quality is one of the major barriers for wastewater reuse. Historically, some WWTPs in Jordan were receiving water more than their design capacity, which resulted in a poor quality effluent. Despite this, and according to ACWUA (2010), Jordan is one of the most pioneer countries in treated wastewater quality control and reuse safety. The country has developed and applied a safety plan for agricultural production irrigated with a mixture of freshwater and treated wastewater, i.e. a crop monitoring program (CMP) for fruits and vegetables production in Jordan Valley. Yet, this plan is developed for Jordan Valley and needs scale-up to a national level.

Limitations of the treated wastewater standards restrict agricultural use to certain crops and thus will shift the land use towards applicable classes. Illegal actions from farms might be another serious concern, however the governmental agencies and periodically monitoring the use per flow from the source till the final destination to allocate these lands and take the necessary actions.

According to above, the case of use of wastewater treatment in the Jordanian agriculture is a rational option. In fact, Jordan is considered the most advanced country in MENA region in using wastewater for irrigation (other countries in the region had adapted wastewater reuse in their water budget mainly for landscape irrigation and ground water recharge i.e. UAE and Qatar). Wastewater is a crucial part of the Jordan's national water budget rather than a pollutant. In addition, treated wastewater reuse for irrigation has been implemented as a management tool of water demand, allowing reallocating freshwater to the domestic use rather than for irrigation.

6 Conclusions

In Jordan, it is predicted that climate change will result in reduced rainfall, increased temperatures, increased chances of heat waves, more frequent droughts, and more extreme weather events. These changes will adversely affect the all life sectors, especially the existing fragile water systems (e.g. less surface water replenishment, groundwater depletion and salinization, desertification) and the agricultural sector, which the country is heavily dependent upon.

The government of Jordan has taken various initiatives (e.g. policies, strategies, action plans) to adapt and mitigate climate change; nevertheless, implementation has been limited partly due to fund shortage, technical capacity, and fragile linkage among related governmental entities. Among the adaptive actions being made, is reserving water for domestic use by recycling, treating, and using WWTPs effluent for other purposes (e.g. industrial or irrigation) could be a solution. By questioning the capability and trust worthiness of Jordanian wastewater production for reuse as an adaptive measure to climate change, this paper was able to demonstrate that treated wastewater is a crucial renewable water resource and is a major constituent of the water budget, and can solve environmental problems and be considered a feasible adaption option.

Through intensive government and donors efforts, thirty one WWTPs were established with total designed capacity of 221.3 MCM/year. These plants have contributed to satisfy about 22% of total irrigation water demands and thus reserving more drinking water. Approximately 98.6% of the total treated wastewater is utilized for irrigation while 1.4% is used for industrial activities.

The treated wastewater production for the year 2013 is estimated about 221.3 MCM/year, from which direct treated wastewater reuse represent only 23.1% of the total amount produced. Direct treated wastewater reuse was achieved through contracting with about 215 farmers to irrigate an area of about 14,184 dunums. The government has managed to institutionally arrange the existing WWTPs operation, reuse, monitoring and quality control under the responsibilities of different bodies (e.g. MWI, WAJ, JVA, MoEnv, MoA, JFDA, and RSS).

Most wastewater treatment plants have been upgraded to accommodate twice the current inlet flow and thus are operating at full efficiency. In comparison with Jordanian standards for treated wastewater reuse, the main WWTP (Samra) and other plants are found to be within the standards and thus can be used for A-class (e.g. cooked vegetables, parks, playgrounds and sides of roads within city limits) or B-class (e.g. fruit trees, sides of roads outside city limits, and landscape). On the other hand, the majority of WWTPs' wastewater quality has exceeded the allowable thresholds especially concerning microbial quality and thus restricting the reuse to only C-class (e.g. field crops, industrial crops and forest trees) which puts a pressure on the WWTPs operators in disinfecting the water before use.

The public acceptance for the reuse is still low, but the farmers have no other alternative for water supply. Samra WWTP success story represents a strong example of best quality control and safety schemes for reuse. The Samra; largest WWTP, is a clear indicative of trustable sustainable and adaptation success of WWTP reuse.

To ensure the wastewater reuse environmental and health sound sustainability, appropriate water policies, standards and monitor, enabling institutional setting, qualified regulators, scientific skills, and high operation and maintenance expenses are required. Water quality is the main barrier for utilizing wastewater in irrigation. Limitations of the treated wastewater standards restrict the agriculture use to certain crops and thus shift the landuse. Thus, the WWTPs should be upgraded based on quality monitoring results. The national standards and regulations should periodically modify to fulfill the requirements of WHO, FAO, and EPA guidelines. Finally, farmers in handle with the reuse of treated wastewater should be environmentally trained while their farm lands should be intensively monitored for any possible soil-water-plant systems contaminations.

Acknowledgements The authors would like to thank Water Authority of Jordan for providing the data used in accomplishing this work. Thanks are also extended to Dr. Anne Dare, Purdue University for her valued inputs and kind revising of the manuscript.

References

Abdulla FA, Eshtawi T, Assaf H (2009) Assessment of the impact of potential climate change on the water balance of semi-arid watershed. Water Resour Manage 23:2051–2068

ACWUA (2010) Wastewater reuse in Arab countries. Comparative compilation of information and reference list. ACWUA working group on wastewater reuse, March 2010. http://www.ais.unwater.org/ais/pluginfile.php/356/mod_page/content/124/Jordan_Summary-Report-Country Casestudies_final.pdf

AFED (Arab Forum for Environment and Development) (2009) Mostafa KT, Najib WS. Published with technical publications and environment & development magazine, Beirut, Lebanon http://www.afedonline.org
Al-Bakri JT, Suleiman A, Abdulla F, Ayad J (2010) Potential impacts of climate change on the rainfed agriculture of a semi-arid basin in Jordan. Phys Chem Earth 35:125–134
Al-Momani SS (2011) State of the wastewater management in the Arab countries. The Hashemite Kingdom of Jordan. Country report. Arab Water Council (AWC), Expert Consultation. UAE-Dubai 22–24 May 2011. www.arabwatercouncil.org/administrator/.../Jordan-Country-Report.pdf
Al-Omari A, Salman A, Karablieh E (2013) The Red Dead Canal project: an adaptation option to climate change in Jordan. Desalin Water Treat. doi:10.1080/19443994.2013.819168
Altz-Stamm A (2012) Jordan's water resource challenges and the prospects for sustainability. http://www.caee.utexas.edu/prof/maidment/giswr2012/TermPaper/Altz-Stamm.pdf. Accessed 29 Dec 2014
Ammary BY (2007) Wastewater reuse in Jordan: present status and future plans. Desalination 211:164–176
DOS (2013a) Jordan statistical year book. Department of Statistics, Jordan
DOS (2013b) The main structure and land use in agricultural Census. Department of Statistics, Jordan
DOS (2015) Jordan in figures. Department of Statistics, Jordan
Environmental Protection Agency (EPA) (1992) Guidelines for water reuse, EPA/625/R-92/004, September (1992) US Agency for International Development, Washington DC. http://nepis.epa.gov/Exe/ZyPURL.cgi?Dockey=30004JK8.TXT
Ghneim A (2010) Wastewater reuse and management in the Middle East and North Africa: a case study of Jordan. Ph.D. thesis. Tag der wissenschaftlichenAussprache: 18.06.2010, Berlin 2010, D 83. https://opus4.kobv.de/opus4-tuberlin/files/2758/Ghneim_Gesamt.pdf
Hammouri N, AL-Qinna M, Salahat M, Adamowski J, Prasher SH (2015) Community based adaptation options for climate change impacts on water resources: the case of Jordan. J Water Land Dev 26(VII–IX):3–17. doi:10.1515/jwld-2015-0013
Intergovernmental Panel on Climate Change report on Climate Change (2001) Working group II: impacts, adaptation and vulnerability. Cambridge University Press, Cambridge
KfW Development Bank (2006) Jordan: sewage disposal greater Irbid I Ex post evaluation report, 2006. https://www.kfw-entwicklungsbank.de/migration/Entwicklungsbank-Startseite/Development-Finance/Evaluation/Results-and-Publications/PDF-Dokumente-E-K/Jordan_Irbid_2006.pdf
MoEnv (1999) Jordan's first national communication to the United Nations Framework Convention on Climate Change (UNFCCC). Ministry of Environment, Amman, Jordan
MoEnv (2007) National capacity self-assessment for global environmental. Ministry of Environment, Jordan
MoEnv (2009) Jordan's second National Communication to the United Nations Framework Convention on Climate Change (UNFCCC). Ministry of Environment, Amman, Jordan
MoEnv (2012) Assessment of direct and indirect impacts of climate change scenarios (water resources study, vol. 1). Climate change adaptation in the Zarqa River Basin. Ministry of Environment, Jordan
MoEnv (2013a) Policy-oriented National Priority Research topics in climate change, Biodiversity, and combating desertification (2013-2020) with guidelines, procedures, tools and potential funding sources to support their implementation in the Hashemite Kingdome of Jordan. Developing policy-relevant capacity for implementation of the global environmental conventions in Jordan, The CB-2 Project, February 2013, Environmental. Ministry of Environment, Jordan
MoEnv (2013b) Adaptation to climate change in the Zarqa River basin: development of policy options for adaptation to climate change and Integrated water resources management (IWRM). Ministry of Environment, Jordan
MoEnv (2013c) Adaptation to Climate Change in the Zarqa River basin: guiding manual for identifying the nest adaptation options. Ministry of Environment, Jordan

MoEnv (2014) Jordan's third National Communication to the United Nations Framework Convention on Climate Change (UNFCCC). Ministry of Environment, Amman, Jordan

MWI (2009) Water for life: Jordan's water strategy, 2008-2022. Ministry of Water and Irrigation, Jordan

MWI (2015) Jordan water sector: facts and figures 2013. Ministry of Water and Irrigation Publication supported by Deutsche GesellschaftfürInternationaleZusammenarbeit (GIZ) GmbH

NEEDS (2010) National environmental and economic development study for climate change, Jordan national report submitted to the United Nation framework convention on climate change, October 2010

Seder N, Abdel-Jabbar S (2011) Safe use of treated wastewater in agriculture. Arab Countries Water Utilities Association (ACWUA), December 2011. http://www.ais.unwater.org/ais/pluginfile.php/356/mod_page/content/124/Jordan_-_Case_Study(new).pdf

Ulaimat AA (2012) Wastewater reuse management in Jordan. Solution submitted to the solutions platform at the 6th world water forum in Marseille, retrieved on March 10, 2012, http://www.solutionsforwater.org/solutions/wastewater-reuse-management-in-jordan-applications-solutions

UN (2009) Guidance on water and adaptation to climate change. United Nations Economic Commission for Europe: convention on the protection and use of transboundary watercourses and international lakes. New York, Geneva. ISBN: 978-92-1-117010-8

Vallentin A, Schlick J, Klingel F, Bracken P, Werner C (2009) Use of treated wastewater in agriculture Jordan Valley, Jordan: case study of sustainable sanitation projects. Sustainable Sanitation Alliance (SuSanA). http://www.susana.org/_resources/documents/default/2-78-en-susana-cs-jordan-treated-wastewater-reuse-2009.pdf

WHO (2006) WHO Guidelines for the safe use of wastewater, excreta and greywater. Vol I: policy and regulatory aspects. World Health Organization

Effect of Operational Changes in Wastewater Treatment Plants on Biochemical Oxygen Demand and Total Suspended Solid Removal

Mustafa Bob

Abstract The research presented in this paper investigates the effect of a significant operational change occurred in a wastewater treatment plant in Madinah city, Saudi Arabia on the effluent water quality. The operational change involved the cancellation of the primary settling tanks of the plant, and as a result, the raw wastewater received at the plant directly entered the aeration tanks. The effect of this change on treated wastewater quality as determined by Biological Oxygen Demand (BOD) and total suspended solids (TSS) concentration was evaluated. These parameters were measured for one month before and one month after the cancelation of the primary settling tanks and the average values were calculated. Results showed that the average removal percentage of BOD and TSS in the last month before the operational change was implemented was 89 and 88%, respectively. The average removal percentage for the same parameters in the first month immediately after implementing the operational changes was 91 and 92% respectively, which shows that there is no significant difference in the removal efficiency for these parameters after the operational change. These results indicate that the treated wastewater quality was not affected by the operational change implemented at the plant. Results of this research provide useful information for the authorities responsible for wastewater treatment as well as for the scientific community.

Keywords Madinah city · Wastewater treatment · Primary settling tank · BOD · TSS

M. Bob (✉)
Civil Engineering Department, College of Engineering,
Taibah University, Madinah City, Saudi Arabia
e-mail: mustafambob@gmail.com

O. Abdalla et al. (eds.), *Water Resources in Arid Areas: The Way Forward*,
Springer Water, DOI 10.1007/978-3-319-51856-5_23

1 Introduction

Proper treatment of wastewater in urban areas is of a critical importance and is considered an important element in the development of countries since the disposal of untreated wastewater can have serious impacts on human health and the environment. In addition, treated wastewater that meets reuse standards can contribute to minimizing water scarcity problems. Ways of wastewater treatment range from very simple, cheap and less efficient processes to advanced, costly and efficient processes (Ab Halim et al. 2015). There are many factors that govern the selection of a suitable process among these wastewater treatment processes. These factors include the circumstances at local area such as the climate, economy, social attributes, availability of standards and monitoring actions, land availability, effluent discharge options and effluent reuse applications (Ahmad et al. 2008). Examples of widely-used wastewater treatment technologies include activated sludge processes, sequencing batch reactors and constructed wetlands (Kalbar et al. 2012).

In biological treatment of wastewater, the degradation of organic compounds by bacteria is a key step (Chiban et al. 2013), but it usually results in a big amount of sludge that needs handling and disposal. For the removal of heavy metals and other undesirable cations and anions from wastewater, membrane techniques (e.g. nanofiltration), oxidation, coagulation/flocculation, ion-exchange and adsorption are all employed (Igwe and Abia 2003; Ozturk and Bektas 2004; Babel and Kurniawan 2003; Zhao and Sengupta 1998). Although these processes are effective, they have some disadvantages which include, for example, incomplete removal of metals, high energy requirements, and generation of waste products that require careful handling and disposal (Soudani et al. 2011a). In recent years, environmentally friendly materials have been employed in wastewater treatment. In particular, the use of some plant materials in the adsorption process has been investigated in laboratory experiments (Soudani et al. 2011a, b; Chiban et al. 2011) and good results were obtained. For example, Soudani et al. (2011a) investigated the use of a Mediterranean plant in removing some pollutants from raw domestic and industrial wastewater. Specifically, they conducted laboratory batch adsorption experiments using dried *Carpobrotus edulis* as a new cheap adsorbent. Their results showed that the removal percentage of phosphate and nitrate ions from municipal wastewaters by dried *C. edulis* was found to be −96 and −97%, respectively. Removal rate for the heavy metals tested ranged from 94% (cadmium) to 99% (lead). These results indicate efficient performance but in the case of industrial wastewater treatment, proper disposal of the dried plants is required as suggested by the authors.

Many countries around the world suffer from fresh water shortage and are trying to find economical and efficient ways to improve treated wastewater quality to close the gap in water supply. Saudi Arabia is one of these countries. The country is characterized by an arid environment and is known as the largest country in the world without rivers and lakes (Bob et al. 2014; Al-Harbi et al. 2009). The country occupies about 70% of the Arabian Peninsula and had a water withdrawal of about

20 billion m^3 in 2010 (Kajenthira et al. 2012). About 90% of the water demand in the country is met from groundwater (Abderrahman and Al-Harazin 2008). Since wastewater is usually the only non-conventional resource that will increase as population increases, and as groundwater resources in Saudi Arabia are increasingly depleted, it is clear that an effective wastewater treatment and reuse is urgently needed.

The wastewater treatment system in Saudi Arabia is still at early stages as the main treatment method is the conventional activated sludge system. Nevertheless, conventional activated sludge is still in use in many other countries. These systems, generally require large surface areas for treatment and biomass separation. Minimizing the cost of operation while producing high quality treated water is the ultimate goal of all water treatment plants, not only in Saudi Arabia but all over the world. In tertiary wastewater treatment plants, the main processes are usually the primary settling, the aeration in the activated sludge tanks and the filtration and disinfection process. In an attempt to minimize the cost of wastewater treatment, the authorities at Madinah city Wastewater Treatment Plant implemented a significant operational change by cancelling the primary settling tanks at the plant, and redirecting the raw wastewater to aeration tanks. This paper investigates the effect of this step on treated wastewater quality. Effluent water quality as determined by Biological Oxygen Demand (BOD) and Total Suspended Solids (TSS) concentration was monitored for a month after the implementation of the change and compared to the effluent quality before the change, and the results are presented here. Results of this research provide useful information to the authorities responsible for wastewater treatment as well as for the scientific community.

2 Description of the Wastewater Treatment Plant

Madinah city Wastewater Treatment Plant is located in the northern side of the city and it is the only wastewater treatment plant in the city. The city is located in the western part of the Saudi Arabia at about 600 m above mean sea level (Matsah and Hossain 1993) and is characterized by an arid climate. The plant currently treats about 200,000 m^3 of raw wastewater per day and there are plans to extend the plant to treat 400,000 m^3/day in the near future. The plant employs a tertiary treatment system that includes primary treatment in primary clarifiers, secondary treatment (activated sludge) as well as filtration and disinfection using chlorine. A total of about 70 tons of sludge is produced per day at the plant. Figure 1 shows a simple schematic of the treatment processes employed at the plant.

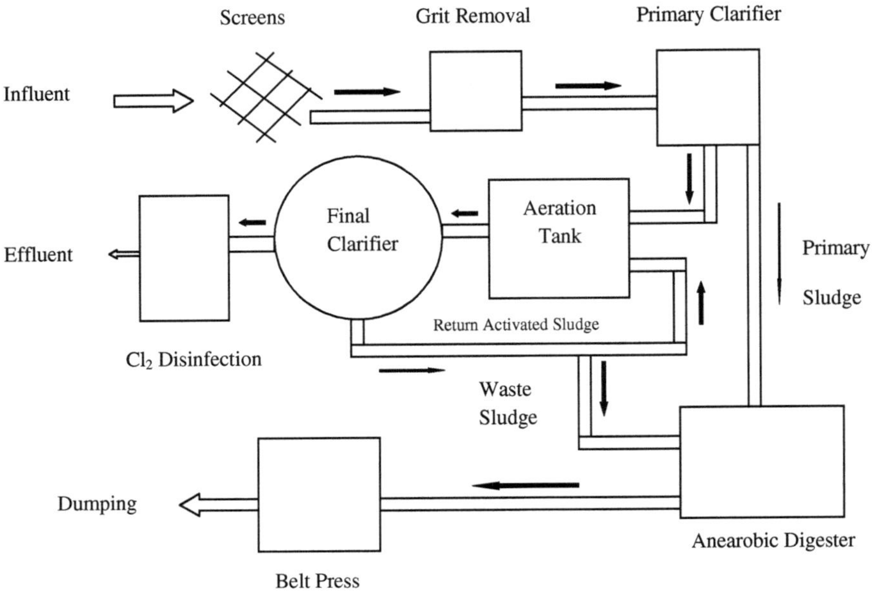

Fig. 1 Schematic representation of the treatment steps employed at the Madinah city wastewater treatment plant

3 Materials and Methods

BOD and TSS concentration are measured on a daily basis in raw and treated wastewater composite samples that are collected daily by laboratory technician using an automatic composite sampler. In addition, samples are collected from the aeration tanks and analyzed for mixed liquor suspended solids (MLSS) and mixed liquor volatile suspended solids concentration (MLVSS). The sampling device is designed to automatically collect samples every one hour throughout the whole day resulting in a 5–6 L sample every day. This sampling technique ensures that the samples collected represent and account for all the variations in the influent and effluent wastewater characteristics throughout the day. BOD and TSS concentration are measured using the standard method for the examination of water and wastewater.

4 Results and Discussion

Table 1 shows the BOD and TSS concentration in the raw and treated wastewater samples collected for a period of approximately one month before the implementation of the operational change in the treatment plant. The statistical parameters for the concentration data are also presented in Table 1. As can be seen in Table 1,

Table 1 BOD and TSS concentration in raw and treated wastewater before cancellation of the primary clarifiers, along with the statistical parameters for the concentration data. All concentration are in mg/L

Sample	$BOD_{influent}$	$BOD_{effluent}$	$TSS_{influent}$	$TSS_{effluent}$
Day 1	139	–	132	15.00
Day 2	103	12.00	126	13.00
Day 3	115	11.00	124	11.00
Day 4	95	9.00	132	14.00
Day 5	109	12.00	118	16.00
Day 6	113	11.00	124	15.00
Day 7	107	12.00	126	16.00
Day 8	115	11.00	118	12.00
Day 9	97	11.00	132	14.00
Day 10	119	12.00	124	12.00
Day 11	107	11.00	136	18.00
Day 12	127	12.00	134	16.00
Day 13	97	11.00	122	15.00
Day 14	113	12.00	124	16.00
Day 15	119	12.00	122	14.00
Day 16	101	12.00	158	13.00
Day 17	94	11.00	144	13.00
Day 18	113	12.00	138	15.00
Day 19	121	13.00	136	15.00
Day 20	117	12.00	132	16.00
Day 21	157	NA	136	12.00
Day 22	121	12.00	144	16.00
Day 23	117	12.00	132	16.00
Day 24	103	NA	128	14.00
Day 25	101	NA	138	NA
Day 26	104	12.00	132	15.00
Day 27	NA	NA	136	15.00
Day 28	NA	NA	142	NA
Day 29	NA	NA	144	32.00
Day 30	NA	15.00	134	30.00
Day 31	NA	15.00	NA	21.00
Average	112	11.88	132	15.86
Standard deviation	13.99	1.23	8.88	4.64
COV (%)	12	10.3	6.7	29.3

NA Not available

the average BOD concentration of the raw water before implementing the operational change was 112 mg/L and that of the treated water was 11.88 mg/L. The average TSS concentration of the raw water in the 30 days period was 132 mg/L, while the average effluent concentration for the same parameter was 15.86 mg/L. Effluent BOD concentration (11.88 mg/L) and effluent TSS concentration (15.86 mg/L) indicate that treated water quality is generally acceptable based on these two parameters. Based on the influent and effluent concentration, the plant achieved 89% BOD removal and 88% TSS removal before the removal of the primary clarifiers.

It is well understood that primary clarifiers are employed in wastewater treatment plants as one of the initial steps, mainly to get rid of all settling substances by means of gravity (Tchobanoglous and Burton 1991). A significant amount of sludge usually accumulate at the bottom of these clarifiers, and this sludge is usually pumped and sent for treatment as can be seen in Fig. 1. At the Madinah city Wastewater Treatment Plant, primary clarifiers were removed from the treatment steps as mentioned above to save some of the high cost associated with primary sludge handling and treatment.

The absence of primary clarifiers is common in many wastewater treatment plants around the world. However, it is critical to evaluate the effect of this change (i.e. removal of primary clarifiers) to make sure that treated water quality is not affected at this particular plant since the absence of primary clarifiers was not included in the original design.

Table 2 shows data for BOD and TSS concentration in the raw and treated wastewater after the removal of the primary clarifiers. As can be seen in Table 2, the average BOD concentration in the raw and treated wastewater in the 30 days period of samples collection was 134.43 and 13 mg/L, respectively. As for the TSS, the average concentration for the raw and treated wastewater was 203.37 and 16.3 mg/L, respectively. These data indicate that the plant achieved 91% BOD removal and 92% TSS removal after the cancellation of the primary clarifiers.

Comparing the plant performance before and after the operational change shows that there is no significant difference in the plant performance before and after the change based on BOD and TSS removal percentages. This is conceivable since the main purpose of the primary clarifiers, as mentioned earlier, is to get rid of settable substances, and since the wastewater passes, immediately after the aeration process, to the final clarifier (see Fig. 1) where settable substances can still be removed efficiently. The MLSS and The MLVSS concentration in the aeration tank before and after the operational change is shown in Fig. 2. As can be seen in Fig. 2, there was no significant difference in these parameters before and after the change. These parameters are used to calculate another important parameter, namely the food to microorganism ratio (F/M) which is used as indicative for the activated sludge process performance. Using the values in Fig. 2, together with the daily flow rate of raw wastewater the F/M before and after the operational change was calculated as 0.12 and 0.21, respectively. Slightly higher F/M was achieved at the plant after the operational change which indicates that more food was available to the microorganism, in which case the bacteria remains active and dispersed. This may have contributed to the foaming problem that is noticed in the aeration tank following the operational change. Foaming is, however, a widespread problem in biological wastewater treatment plants all over the world. It usually results from the activity of

Table 2 BOD and TSS concentration in raw and treated wastewater after cancellation of primary clarifiers, along with the statisical parameters for the concentration data. All concentration are in mg/L

Sample	$BOD_{influent}$	$BOD_{effluent}$	$TSS_{influent}$	$TSS_{effluent}$
Day 1	188	14.00	264	21.00
Day 2	156	11.00	252	13.00
Day 3	94	12.00	228	18.00
Day 4	109	10.00	248	16.00
Day 5	127	13.00	268	16.00
Day 6	151	12.00	248	14.00
Day 7	143	10.00	256	16.00
Day 8	119	11.00	284	14.00
Day 9	128	12.00	267	18.00
Day 10	185	9.00	232	14.00
Day 11	166	11.00	216	13.00
Day 12	143	12.00	232	12.00
Day 13	137	10.00	244	11.00
Day 14	149	13.00	232	12.00
Day 15	151	11.00	272	16.00
Day 16	163	14.00	264	18.00
Day 17	139	12.00	256	18.00
Day 18	115	13.00	124	15.00
Day 19	156	18.00	127	21.00
Day 20	132	17.00	107	27.00
Day 21	127	18.00	128	21.00
Day 22	105	15.00	136	16.00
Day 23	110	18.00	164	17.00
Day 24	128	13.00	148	15.00
Day 25	110	14.00	132	16.00
Day 26	108	16.00	128	16.00
Day 27	131	10.00	148	14.00
Day 28	108	16.00	152	16.00
Day 29	128	16.00	188	17.00
Day 30	127	15.00	156	18.00
Average	134.43	13.2	203.36	16.3
Standard deviation	22.92	2.58	57.37	3.20
COV (%)	0.17	0.19	0.28	0.19

certain bacteria along with the gas bubbling in the system and the presence of some hydrophobic surfaces (Hug 2006). Panel (a) of Fig. 3 clearly shows the formation of foam on the top of some of the aeration tanks in the plant. In addition, a photo for one of the primary tanks, which appears empty as a result of removing primary

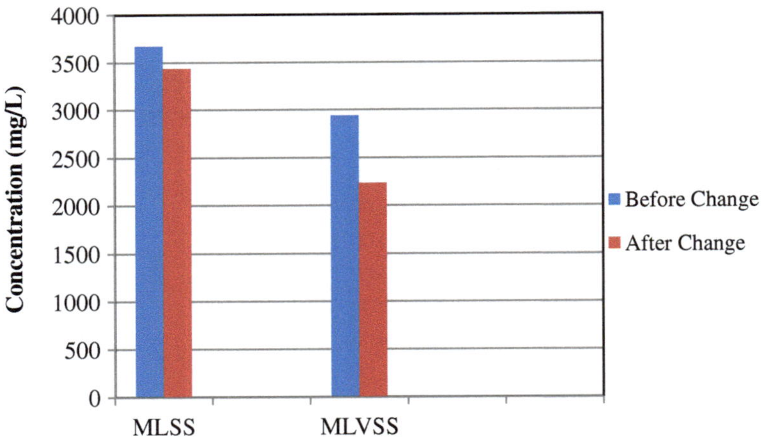

Fig. 2 MLSS and MLVSS concentration in the aeration tank before and after the implementation of the operation change

settling from the treatment steps, and a photo for one of the final sedimentation tanks are shown in panel (b) and (c) of Fig. 3, respectively. Prevention of the foam and hence eliminating the problems associated with it can be done through the addition of some polyelectrolytes to promote sedimentation in the system or by implementation of measures that increase the plant tolerance to foaming [e.g. reducing the activity of the bacterial by adding small amounts of chlorine as demonstrated by Hug (2006)].

Nevertheless, this operational change resulted in saving in primary sludge handling and also, following this operational change a significant improvement in the air quality (i.e. decrease in the bad odor) was noticed at the plant. It is understood that primary clarifiers can contribute significantly to odor problems in wastewater treatment plant since they hold the raw wastewater for while, and as such by removing these clarifiers an improvement in the air quality at the plant is expected, as was the case for this particular treatment plant.

Green technologies and processes are desirable in wastewater treatment, as well as other systems. The use of environmentally friendly products in wastewater treatment as discussed in the introduction section which has shown good results as demonstrated by Soudani et al. (2011a) is a step towards green technology. In general, any action or a process that results in reducing waste and pollution by changing patterns of production and consumption (commonly referred to as source reduction) is considered a green technology (Green Technology 2010). The removing of the primary settling tanks as discussed above achieved a reduction in the energy and waste at the plant while achieving the same treatment goals and as such it can be considered as one step towards a green treatment plant. While it is true that other wastewater treatment processes can achieve similar or even better performance than this treatment plant as reported in literature, the performance of this treatment plant should mainly be compared to its performance before

Fig. 3 **Panel a** A photo of some of the aeration tanks at the plant. Foam formation is clearly visible in the tanks. **Panel b** A photo of one of the empty primary settling tanks after removal from operation. **Panel c** A photo of one of the final sedimentation tanks

implementing this significant operational change as the focus of this study is to evaluate the effect of this change.

5 Conclusion

The effect of removing the primary clarifies from the treatment steps in Madinah city Wastewater Treatment Plant is evaluated based on BOD and TSS concentration in raw and treated wastewater. The removal percentage of BOD at this treatment plant before and after implementing the operational change is found to be 89 and 91%, respectively, indicating no significant difference. As for the TSS, the removal percentage is found to be 92% after the implementation of the operational change as compared to 88% before the change. A significant saving is the cost of handling the primary sludge is expected at this plant as a result of this change. In addition, this operational change also resulted in a significant improvement in the air quality in the plant as reflected in the absence of the strong odor that prevailed before implementing this change. Formation of foams is observed on the top of the aeration tanks following the implementation of this change. Future research is needed to conduct air quality measurements at the plant to quantitatively assess the improvement in the air quality in and around the plant and to investigate the foaming problem in details.

Acknowledgements The authors thank the authorities at the Madinah Wastewater Treatment Plant for providing important data and information and for useful discussion. This research was partially funded by the National Plan for Science, Technology and Innovation (MAARIFAH)—King Abdel Aziz City for Science and Technology—the Kingdom of Saudi Arabia, award number (10WAT104705).

References

Abderrahman W, Al-Harazin I (2008) Assessment of climate changes on water resources in the Kingdom of Saudi Arabia. In: Proceedings of the GCC environment and sustainable development symposium, Dhahran, Saudi Arabia, 28–30 Jan

Ab Halim M, Nor Anuar A, Azmi S, Jamal Abdul, Abdul Wahab N, Ujang Z, Shraim A, Bob M (2015) Aerobic sludge granulation at high temperatures for domestic wastewater treatment. Bioresour Technol 185:445–449

Ahmad M, Bajahlan A, Hammad W (2008) Industrial effluent quality, pollution monitoring and environmental management. J Environ Monit Assess 147:297–306

Al-Harbi O, Hussain G, Lafouza O (2009) Irrigation water quality evaluation of Al-Mendasah groundwater and drainage water, Al Medainah Al Monwarah Region, Saudi Arabia. Int J Soil Sci 4:123–141

Babel S, Kurniawan T (2003) Low-cost adsorbents for heavy metals uptake from contaminated water, a review. J Hazard Mater 97:219–243

Bob M, Abd Rahman N, Taher S, Elamin A (2014) Multi-objective assessment of groundwater quality in Madinah City, Saudi Arabia. J Water Qual Expo Health. doi:10.1007/s12403-014-0112-z

Chiban M, Soudani A, Sinan F, Persin M (2011) Wastewater treatment by batch adsorption method onto micro-particles of dried Withania frutescens plant as a new adsorbent. J Environ Manage 95:S61–S65

Chiban M, Soudani A, Zerbet M, Sinan F (2013) Wastewater treatment processes. In: Valdez CJ, Maradona EM (eds) Handbook of wastewater treatment: biological methods, technology and environmental impact. Nova science publishers, Inc., New York. ISBN: 978-1-62257-591-6

Green Technology Organization (2010). http://www.green-technology.org/what.htm

Hug T (2006) Characterization and controlling of foam and scum in activated sludge systems. Ph. D. Dissertation. Available at http://dx.doi.org/10.3929/ethz-a-005180592

Igwe J, Abia A (2003) Maize cob and husk as adsorbents for the removal of heavy metals from waste water. J Phys Sci 2:83–92

Kajenthira A, Siddiqi A, Anadon L (2012) A new case for promoting wastewater reuse in Saudi Arabia: bringing energy into the water equation. J Environ Manage 102:184–192

Kalbar P, Karmakar S, Asolekar S (2012) Assessment of wastewater treatment technologies: life cycle approach. Water Environ J:1747–6585

Matsah M, Hossain D (1993) Ground conditions in Al-Madinah Al-Munawarah. J King Abdel Aziz Univ Earth Sci 6:47–77

Ozturk N, Bektas T (2004) Nitrate removal from aqueous solution by adsorption onto various materials. J Hazard Mater 112:155

Soudani A, Chiban M, Zerbet M, Sinan F (2011a) Use of Mediterranean plant as a natural adsorbent for municipal wastewater treatment. J Environ Chem Ecotoxicol 3:199–205

Soudani A, Chiban M, Zerbet M, Sinan F (2011b) Bensergao wastewater treatment by adsorption on dried Launea arborescens plant as an environmentally friendly material. Phys Chem News 58:17–19

Tchobanoglous G, Burton F (1991) Wastewater engineering: treatment, disposal and reuse. McGraw Hill, New York

Zhao D, Sengupta A (1998) Ultimate removal of phosphate from wastewater using a new class of polymeric ion exchangers. J Water Res 32:1613–1625

Treatment of Industrial Wastewater and Its Reuse for Sustainable Agriculture Practices—A Green Concept

Singanan Malairajan

Abstract Fresh water is very much essential for the survival of human beings. In recent years, water pollution issues have been the focus of increasing global concern. In developing countries, industrial sectors consuming large amount of the available fresh water for their productions. At the same time, it releases equal volume of wastewater into the environment and causes ecosystem damages. In the concept of environmental and economic sustainability, a proper wastewater management and water reuse system can help to a greater extent in development of national economy. Most of the currently used water treatment technologies are very much costly and not eco-friendly in nature. In this context, a new search for cheap and low cost wastewater treatment technology is essential. Recently, a new biocarbon technology in water and wastewater purification process is introduced in pilot scale project. The biocarbon is generated using a novel medicinal plant *Tridax procumbens* (Asteraceae). The characteristics of the biocarbon is unique and having good potential for the removal of inorganic and organic components in water and wastewater system. As a model trial, a saline water sample containing TDS of 3000 mg/L was introduced in the reactor system, after equilibrium time of 3 h; the concentration of outlet water was 425 mg/L. The removal of color from an industrial wastewater was performed with the initial concentration of 100 mg/L with optimum biocarbon dose of 2.5 g/100 mL. An excellent result (98.5%) of color removal is achieved in 150 min. These results suggest that, the treated waste water can be reused for sustainable agricultural practices.Please check and confirm if the author's name and initial are correct.Corrected

Keywords Desalination · Biocarbon · Water reuse · Green technology

---

S. Malairajan (✉)
Water and Food Chemistry Research Laboratory,
PG and Research Department of Chemistry,
Presidency College (Autonomous), Chennai 600005, Tamil Nadu, India
e-mail: msinganan@yahoo.com

O. Abdalla et al. (eds.), *Water Resources in Arid Areas: The Way Forward*,
Springer Water, DOI 10.1007/978-3-319-51856-5_24

1 Introduction

Fresh water is a valuable resource for sustainable ecosystem. Demands on surface and ground water resources for household, commercial, industrial, and agricultural purposes are seriously increasing. In many parts of the world, there is already a widespread scarcity, gradual destruction and increased pollution of freshwater resources. Treating and reusing the wastewater for landscape development as well as certain non-potable applications are smart ways of reducing pressure on high-quality water supplies.

Surface and ground water contamination with respect to heavy metals by industrial and domestic effluents has increased in recent years due to their dangerous effect on aquatic flora and fauna even in relatively low concentrations. Heavy metals such as As, Hg, Pb, Cd, Cr, Ni, Zn, Cu and Fe are largely found in industrial wastewater. Heavy metals are non-biodegradable and their presence in receiving water bodies causes bioaccumulation in living organisms, which leads to several health problems in animals, plants and human beings (Bernard et al. 2013; Oghenerobor et al. 2014). All the water bodies including rivers, canals and the drainage canals are being used by industries, industrial estates and municipalities of towns and cities for the discharge of effluents from industrial, construction, transportation and other activities. Therefore, it is expected that surface and ground water may get contaminated widely (Naseem 2012).

Greater environmental awareness among public in recent years and the demand for cleaner environment has compelled the treatment of industrial and domestic effluent by the concerned establishments. As such there has been a great deal of research into finding cost-effective methods for the removal of contaminants from wastewater. In recent years considerable attention has been dedicated to the study of removal of heavy metal ions from solution by adsorption/biosorption. Natural materials or wastes from agricultural operations may have potential to be used as low cost adsorbents, as they represent unused resources and widely available and are environmentally friendly.

Many physico-chemical methods have been proposed for the removal heavy metals from industrial effluents. Adsorption is one of the effective purification and separation technique used in industry especially in water and wastewater treatments (Guixia et al. 2011).

Various cheaper materials, including soil, silica, activated carbon, green coconut shell and others, have been used to remove different pollutants from industrial effluents for their safe disposal into the biosphere (Mahmoodi et al. 2011; Kumar and Meikap 2014). Biosorption is a relatively new process that has proven very promising in the removal of contaminants from aqueous effluents using low-cost adsorbents derived from agricultural materials (Nomanbhay and Palanisamy 2005).

In this study an attempt was made to use, the leaves of *Tridax procumbens* (Asteraceae) for the preparation of biocarbon for the potential removal of color and dissolved solids from an industrial wastewater. The treated industrial wastewater is applied in pilot scale for cultivation of *Setaria glauca* a fodder grass for livestock applications and *Solanum melongena* for small scale irrigation. The selected medicinal herb is found in agricultural fields in Tamil Nadu, India throughout the seasons. The present work investigates the potential use of biocarbon of *T. procumbens* (Asteraceae) for biosorption of color as well as dissolved solids present in the grey water system.

2 Materials and Methods

2.1 Preparation of Biocarbon

T. procumbens (Asteraceae) is an important medicinal plant widely distributed in agricultural fields. The plant leaves were collected and air dried for 48 h. The dried leaves were grounded in ball mills and the screened homogeneous powder was used for the preparation of biocarbon. It is prepared by treating the leaves powder with the concentrated sulphuric acid (Sp. gr.1.84) in a weight ratio of 1:1.8 (biomaterial: acid). The resulting black product is kept in an air-free oven, maintained at 160 ± 5 °C for 6 h followed by washing with double distilled water until free of excess acid, then dried at 105 ± 5 °C. The particle size of biocarbon screened between 100 and 120 μm was used (Singanan 2015). This material is used as adsorbent in the treatment of industrial wastewater. The processing of biocarbon preparation is presented in the Fig. 1.

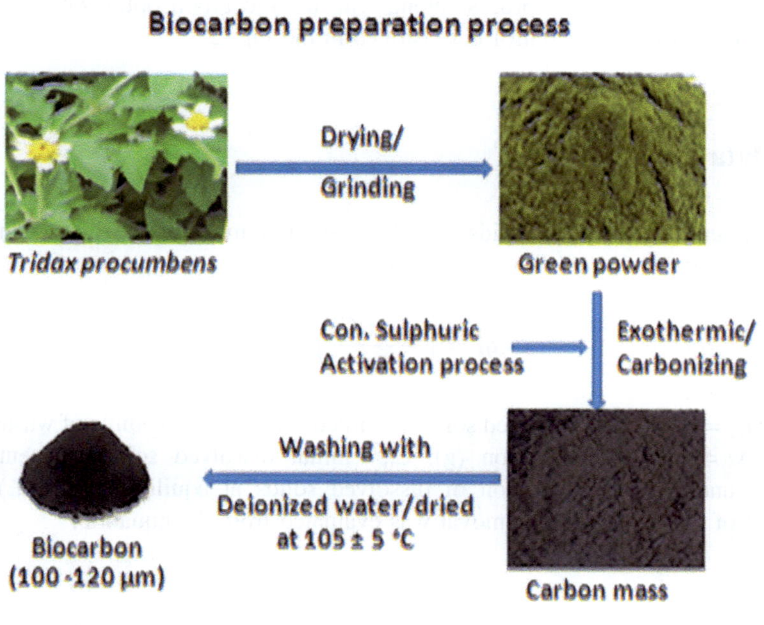

Fig. 1 Processing of preparation of biocarbon

2.2 Collection and Analysis of Grey Water

The grey water sample is collected from a combined effluent treatment plant (CETP) from an industrial area in northern district of Tamil Nadu, India. The grey water is passed through the specific screening system for the removal any dirt materials. A dirt free grey water sample is collected and stored in clean polythene containers for further experimental purposes. The important physico—chemical characteristics such as pH, EC, alkalinity, SS, TDS, EC, DO, BOD, COD were performed by using standard methods outlined in APHA (1985) manual.

2.3 Treatment of Grey Water by Biosorption Process

The biosorption process is carried out at room temperature of 28 ± 2 °C. For the evaluation of treatment capacity of and optimization of biocarbon dose, 100 mL of grey water sample containing TDS of 3000 mg/L is used in a series of eight Erlenmeyer flask of 250 mL capacity with the addition of 0.5, 1.0, 1.5, 2.0, 2.5, 3.0, 3.5 and 4.0 g biocarbon dose and equilibrated using Remi orbital shaker system for the contact time of 150 min at 300 rpm. For the removal of color, 100 mL of grey water sample is taken in a 250 mL Erlenmeyer flask, 2.5 g/100 mL of biocarbon is added and equilibrated at the same speed and aliquots of samples was withdrawn for each 30 min and the color is estimated 620 nm. The experiments were carried out in triplicates and the average values obtained and used for further applications. The grey water treatment process along with the biocarbon optimization process using saline water as a control is illustrated in the Fig. 2.

3 Data Analysis

The amount of dissolved solids adsorbed per unit mass of the biosorbent was evaluated by using the following equation:

$$q_e = \left(\frac{C_o - C_e}{w}\right) \times V \tag{1}$$

where q_e = amount of dissolved solids adsorbed (mg/g), V = volume of wastewater (mL), w = mass of biocarbon (g), C_o = initial dissolved solids concentration (mg/L) and C_e = concentration of dissolved solids at equilibrium (mg/L). The percent of dissolved solids removal was evaluated from the equation:

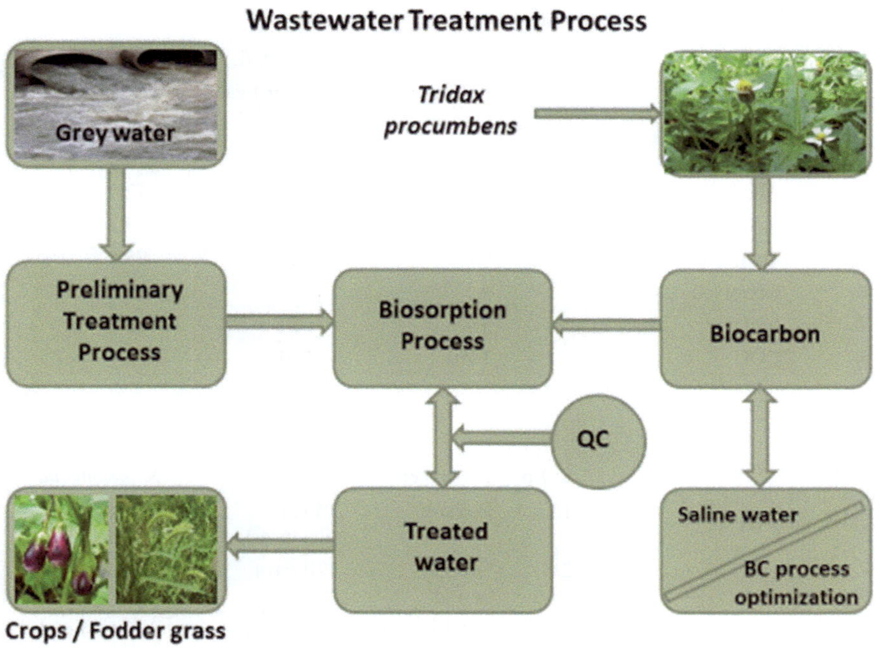

Fig. 2 Treatment process of grey water

$$\% \text{ Removal} = \left(\frac{C_o - C_e}{C_o}\right) \times 100 \tag{2}$$

The analytical data were analysed and standard deviations of the statistical tests were carried out using programme of analysis of variance (ANOVA) by using SPSS program.

4 Results

The findings on the essential characteristics of grey water quality before and after treatment was presented in Table 1. It is well known that, total dissolved solids, DO, BOD and COD are very important in the treatment of a wastewater. In Table 2, after the viable use of treated wastewater, the productive results of the crop yield and the growth characteristics of *S. glauca* and *S. melongena* was presented. The characteristics of biocarbon, mainly morphological features (SEM) of the adsorbent before and after biosorption process are illustrated in Fig. 3a, b. For the purpose of quantification of biocarbon dose, a model trial was carried out using saline water as control. It reflects that, 2.5 g/100 mL is the optimum dose for the successful removal of 98.5% of TDS in saline water. The estimated contact time for the maximum color removal was 150 min.

Table 1 Characteristics of grey water

S. No.	Characteristics of grey water	Quality of grey water (mg/L)		Percentage (%)
		Before treatment	After treatment	
1.	pH	8.3	4.5	45.8
2.	EC (μmhos/cm)	1380	340	75.4
3.	Alkalinity	885	240	72.9
4.	Suspended solids	985	225	77.2
5.	Total dissolved solids	895	220	75.4
6.	Dissolved oxygen	8.5	5.5	35.3
7.	BOD	380	65	82.9
8.	COD	1450	280	80.7

Table 2 Growth characteristics of *S. glauca* and *S. melongena*

S. No.	Parameters	*S. glauca*	*S. melongena*
1.	Pot size	15 cm (diameter)	15 cm (diameter)
2.	Growth period	120 days	120 days
3.	Total biomass	3.0 kg/pot	3.5–4.5 kg/plant

5 Discussion

5.1 Characteristics of Biocarbon

The nature and the characteristics of the biocarbon are essential to understand the adsorptive behaviour on the surface of the biocarbon. The detailed characteristics of the biocarbon are reported in our previous work (Singanan 2013). It is observed that, the surface and particle size is relatively high and is responsible for potential removal of metal ions and organic molecules from grey water system. The result indicates that, the biocarbon has high capacity for metal ion adsorption and it is regenerative in nature.

5.2 Surface Morphology

To investigate surface condition for before and after adsorption of dissolved solids, SEM photograph was used. The SEM micrograph of biocarbon indicates the regular morphological structure of the particles and presents beautiful square like structures and large numbers of grains and cavities. This makes possible for the adsorption of inorganic and organic pollutants on different parts of the biocarbon. It is observed

Fig. 3 **a** SEM photograph of biocarbon—before adsorption. **b** SEM photograph of biocarbon—after adsorption

that, both SEM photographs have similar features and indicates the biocarbon has regenerative in nature.

5.3 *Removal of Total Dissolved Solids in Saline Water*

The removal of total dissolved solids present in the saline water was carried out in the pH of 4.5. The dissolved solids removal studies was conducted with the initial dissolved solids concentration of 3000 mg/L at the biocarbon dose of 2.5 g/100 mL. The percentage sorption increases with an increase in contact time and the maximum (98.5%) of dissolved solids are removed at the contact time of 180 min (Fig. 4). This experimental model clearly supports that, the biocarbon can be effectively used as a cheap adsorbent for the removal of pollutants in the real industrial water.

5.4 *Effect of Biocarbon Dose on TDS Removal*

The amount of adsorbent dose was a key parameter to control both availability and accessibility of adsorption sites in the sorption process of heavy metal ions as well as dissolved solids present in the synthetic wastewater (Rafeah et al. 2009). The effect of adsorbent dosage on the removal of dissolved solids was presented in Fig. 5. The percent removal of dissolved solids has reached up to 98.5% at the adsorbent dose of 2.5 g/100 mL. The results indicates that, the higher dosage of adsorbent will increase the adsorption due to more surfaces and functional groups are available on the biocarbon matrix.

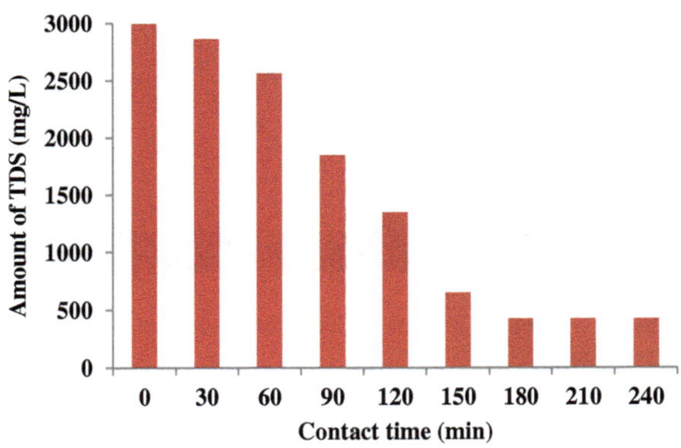

Fig. 4 Removal of total dissolved solids

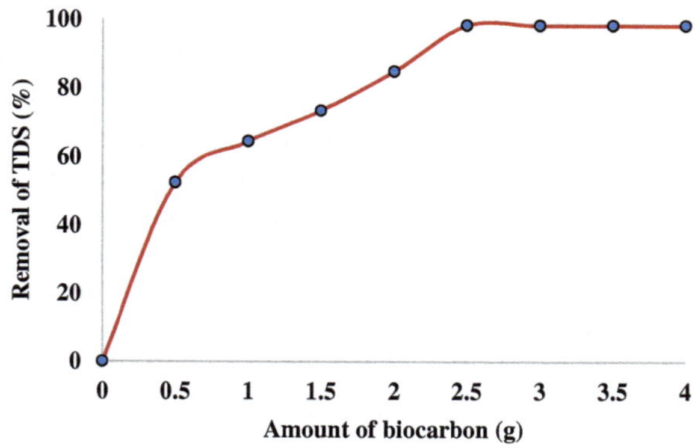

Fig. 5 Effect of biocarbon dose on TDS removal

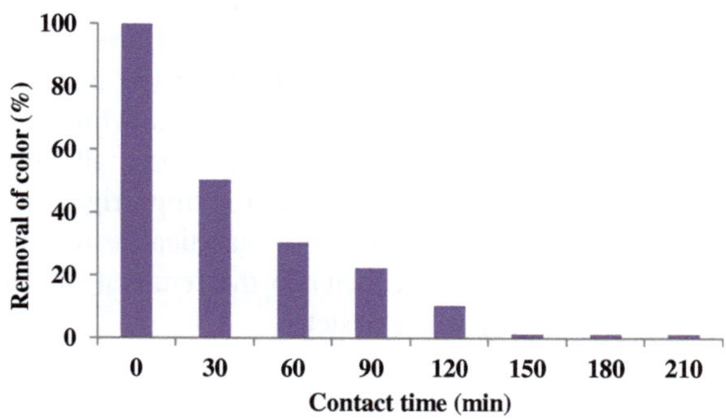

Fig. 6 Effect of contact time on color removal

5.5 *Effect of Contact Time on Color Removal*

The role of contact time is very important in color removal process in wastewater. It is evident from the experimental results that the contact time required to attain the equilibrium depends on the concentration of coloring species present in the wastewater. In the present investigation, the removal of color of the wastewater gradually increased when increasing in contact time. It attains a maximum removal of 98.5% at 150 min and then, no much changes observed (Fig. 6).

5.6 *Characteristics of Grey and Treated Water*

The analytical results indicates that, the grey water is highly contaminated by organic pollutants which accelerates the higher level of COD in the grey water system. Higher organic load is mainly contributing the rise in biological oxygen demand (BOD) of the wastewater.

The results indicate that, the biocarbon technology works well and all the parameters are significantly reduced. The color of wastewater is mainly due to the presence of different types of organic and inorganic pollutants present in the wastewater. It will definitely cause toxic effects in humans and animals when using this grey water for any irrigation purpose. The dark color of grey water was well reduced to almost colorless solution. The main pollution parameter such as COD and BOD is reduced to 80.7 and 82.9% respectively. Lakdawala and Patel (2012) reported similar findings in sugar mill effluent for the COD and BOD removal using bagasse fly ash as an adsorbent. It is observed in their studies that, removal of BOD more favorable then COD. However, Kulkarni et al. (2011) reported that, the bagasse flyash is found to be an effective adsorbent for the removal of COD in distillery effluent. The percent COD removal obtained was 78–85. The adsorbent dosage, pH, contact time and initial concentration of organic matter have significant effect on the COD removal. Hami et al. (2007) investigated effective use of powdered activated carbon (PAC) on the performance of a pilot-scale laboratory dissolved air flotation (DAF) unit. It is observed that, the removal efficiencies for BOD in wastewater is increased from 70 to 94% while those for COD increased from 64 to 92.5% respectively. In the present study, alkalinity of the grey water is 72.9% reduced making the water suitable for crop irrigation. Relatively, the level of TDS is good in treated wastewater. The analytical results supports that, the biocarbon has excellent biosorption capacity to the removal of dissolved solids and the organic pollutants in grey water system.

5.7 Reuse of Treated Wastewater

The treated industrial wastewater is applied in pilot scale for cultivation of *S. glauca* a fodder grass for livestock applications and *S. melongena* for small scale irrigation. The productive results demonstrated that, the treated grey water has good nutrient capacity and hence, the species is steadily grown well and produced good yield.

6 Conclusion

The biocarbon technology is an effective technology for the treatment of industrial wastewater. It does not produce any major secondary effluent and sludge. The treated wastewater can be reused in industries mainly for cooling and cleaning operations. Wastewater reuse management can help to minimize the use of available freshwater resources. The yield of selected fodder grass and the food crop is very good. Hence, it can be used well in agricultural practices for sustainable development.

References

APHA (1985) Standard methods for the examination of water and wastewater. American Public Health Association, Washington, D.C.

Bernard E, Jimoh A, Odigure JO (2013) Heavy metals removal from industrial wastewater by activated carbon prepared from coconut shell. Res J Chem Sci 3(8):3–9

Guixia Z, Xilin W, Xiaoli T, Xiangke W (2011) Sorption of heavy metal ions from aqueous solutions: a review. Open Coll Sci J 4:19–31

Hami ML, Al-Hashimi MA, Al-Doori MM (2007) Effect of activated carbon on BOD and COD removal in a dissolved air flotation unit treating refinery wastewater. Desalination 216(1–3):116–122

Kulkarni SJ, Patil SV, Bhalerao YP (2011) Flyash adsorption studies for organic matter removal accompanying increase in dissolved oxygen. Int J Chem Eng App 2(6):434–438

Kumar S, Meikap BC (2014) Removal of chromium (VI) from waste water by using adsorbent prepared from green coconut shell. Desal Water Treat 52(16–18):3122–3132

Lakdawala MM, Patel YS (2012) The effect of low cost material bagasse fly ash to the removal of COD contributing component of combined waste water of sugar industry. Arch App Sci Res 4 (2):852–857

Mahmoodi NM, Hayati B, Arami M, Lan C (2011) Adsorption of textile dyes on pine cone from colored wastewater: Kinetic, equilibrium and thermodynamic studies. Desalination 268(1–3):117–125

Naseem Z (2012) Lead removal from water by low cost adsorbents: a review. Pak J Anal Env Chem 13(1):01–08

Nomanbhay S, Palanisamy K (2005) Removal of heavy metal from industrial wastewater using chitosan coated oil palm shell charcoal. Ele J Biotech 8(1):43–53

Oghenerobor BA, Gladys OO, Tomilola DO (2014) Heavy metal pollutants in wastewater effluents: Sources, effects and remediation. Adv Biosci Bioeng 2(4):37–43

Rafeah W, Zainab N, Veronica UJ (2009) Removal of mercury, lead and copper from aqueous solution by activated carbon of palm oil empty fruit bunch. World App Sci J 5 (Special Issue for Environment):84–91

Singanan M (2013) Defluoridation of drinking water using metal embedded biocarbon technology. Int J Env Eng 5(2):150–160

Singanan M (2015) Biosorption of Hg(II) ions from synthetic wastewater using a novel biocarbon technology. Env Eng Res 20(1):33–39

Part VI
Agriculture and Irrigation Management

Adaptation and Performance of Six Eucalypt Species Irrigated with Qom Sewage

Hossein Sardabi and Hossein Tavakkoli-Neko

Abstract The aim of the study was to evaluate adaptation and performance of six eucalypt species and provenances under sewage irrigation in arid condition. For this reason, a trial was conducted in 2007 and continued to 2014 in a site located 5 km north east of Qom city in Iran, near sewage release station under statistical design of Randomized Complete Blocks with four replicates, six treatments and 5 × 5 m spacing. Sewage was used for irrigation at one week interval for four years, continued at rainfed condition. The treatments consisted of Six eucalypt species and provenances, including: *Eucalyptus camaldulensis* Qom, *E. camaldulensis* 41-ch, *E. camaldulensis* *41*-zh, *E. microtheca* 62-Qom, *E. microtheca* 62-Ahwaz and *E. rubida* 166-sh. Growth and morphological parameters consisted of: total height, diameter at breast height, survival, height increment, diameter increment at breast height and height/diameter ratio. Tree inventory was made at two times per year: spring (after cold period) and autumn (after warm and dry season). Data were analyzed, using variance analysis and Duncan test methods. *E. camaldulensis* species achieved a high survival and growth values, whereas *E. rubida* species achieved the worst values. *E. microtheca*, particularly the 62-Qom provenance obtained a greatest survival value (100%).

Keywords Survival · Diameter · Height/diameter ratio · Height and diameter increment · Sewage

1 Introduction

Wood production in natural forests is globally decreasing due to different reasons, so it's necessary to find out another substitute for wood production via afforestation and wood farming. Forest plantation with exotic fast growing species which is

H. Sardabi (✉)
Research Institute of Forests and Rangelands, P.O. Box 13185-116, Tehran, Iran
e-mail: hosseinsardabi@gmail.com; sardabi@rifr-ac.ir

H. Tavakkoli-Neko
Qom Province Agricultural and Natural Resources Research Center, Qom, Iran

O. Abdalla et al. (eds.), *Water Resources in Arid Areas: The Way Forward*, Springer Water, DOI 10.1007/978-3-319-51856-5_25

common at many countries of the temperate and tropical regions, is one of the solutions to increase wood production. Most eucalypts are capable to produce huge amount of wood and other bi-products in short periods due to their fast growth characteristics. Considering growth rate of native species of Iran and the exotic fast growing species which is 0.5 and 25–30 m^3/ha/year, respectively (Assareh and Sardabi 2007), forest plantation with exotic fast growing species can reduce rotation period and increase people's income as well. Furthermore, eucalypts can improve soil properties. Sardabi et al. (2010) reported that eucalypt plantation reduced soil compaction and increased its porosity at site located on lowland Caspian forest of Guilan province of Iran. Soil Acidity increased after 22 years, but was less than one unit.

Due to water resources limit in arid and semi-arid areas of Iran, wastewater irrigation may be the best solution for eucalypt plantation. According to the authors experiences, using raw sewage in such conditions may not cause environmental problems for some reasons, including: (1) organic sediments usually deplete under high radiation of sun and (2) heavy metals and pollutants often are absorbed by eucalypt roots and stored in their leaves and stems under phytoremediation process. Furthermore, row sewage is not harmful for people in eucalypt plantation, because wood is not food (Sardabi et al. 2013).

Literature review will be made only for *E. camaldulensis* due to its fine adaptation at different ecological conditions in Iran, its fast growth rate and its wood quality, particularly for paper and particle board industries.

Eucalyptus camaldulensis Dehnh (River Red Gum) is the most widely distributed species and it grows at different climate conditions, including warm to moderate and sub-humid to semi-arid. Average air temperature of warmest and coldest months of the year is 27–40 and 3–15 °C, respectively. Average number of frost days per year is 20. Average annual rainfall is 250–600 mm. When rainfall is not enough, it depends on seasonal flood water and presence of high level water table. Rainfall distribution varies at different locations of Australia. For instance, maximum rainfall at southern and northern parts of Australia is in winter and summer, respectively. Rainfall is usually between November and March. Rainfall variation is very high at inland section of Australia and sometimes is accompanied with long dry period. Soil type is sandy alluvial. River Red Gum is sensitive to calcareous soils. It is typically a riverine species and has a ribbon-like distribution across the landscape (Boland et al. 2002, 2007).

River Red Gum has been tested widely at many countries of the word, including Iran and due to its successful adaptability, fast growth and resistance to cold, warm and dry climate conditions, has been planted at wide extent for economical, ornamental and soil and water conservation purposes. For instance, this species was used successfully for sand dune stabilization in India (Rajasthan desert), using seed origins from Morocco and Australia (Sinha et al. 1999). Saleheh Shoushtari and Rouhipour (2005) announced that four eucalypt species and provenances, including; *E. camaldulensis* 9616, *E. microtheca*, *E. camaldulensis* and *E. sargentii* were planted under species trial on West Karkheh sandy dunes of Khuzestan province of Iran in order to stabilize and fix the dunes, biologically. Results showed that *E. camaldulensis* 9616 obtained the greatest survival, diameter and total height (73%, 10 cm and 9.7 m, respectively).

Survival and growth rate of River Red Gum differs according to variation in ecological conditions. For instance, in Madagascar height growth of one and four year old seedlings was 1.14 and 2.38 m, respectively and height increment was 71 cm (Rarivoson et al. 2008). Few eucalypt species were planted under pilot species trial and floodwater condition at Garebigan Fasa area of Fars province of Iran. Results showed that *E. camaldulensis* and *E. microtheca* achieved the greatest survival (>80%); whereas the greatest amount of diameter and total height belonged to *E. camaldulensis* (Mortazavi Jahromi and Kousar 2010). Hamzepour et al. (2012) announced that 15 eucalypt species and provenances were planted under species trial at Kazeroun Experimental Station (Fars province of Iran). Results showed that although the greatest survival belonged to five species and provenance, including *E. camaldulensis* 15195, *E. camaldulnesis* 15272, *E. camaldulensis* 15023, *E. camaldulensis* 20709 and *E. maidenii* 300-sh, but their diameter and total height growth rate was moderate. Eucalypt species trial was conducted at central littoral of Caspian Sea of Iran and *E. camaldulensis* achieved the greatest survival in comparison to *E. saligna* and *E. grandis*, but its height and diameter performance was less than *E. saligna* (Sadati et al. 2004).

Water consumption by *E. camaldulensis* differs at different locations and periods. A trial was conducted by Rad et al. (2010b) to study water use efficiency of three year old seedlings of *E. camaldulenis* under drainage lysimetric condition and different regimes of irrigation at Yazd city suburb of Iran (dry climate). Results showed that water use efficiency (dry mass/t) improved by increase in drought stress and the highest value belonged to 40% field capacity treatment by 2.5 g mass dry weight per one kilogram water consumption in transpiration process. Furthermore, He indicated that biomass production (shoot, root and leaf) and root/shoot ratio significantly decreased when soil water content reduced. Regarding wood production, water use efficiency increased significantly by 70% field capacity treatment, compared to other treatments (Rad et al. 2011). Root distribution of three year old *E. camaldulensis* seedlings was determined under different moisture regimes in Yazd City of Iran. Results showed that root dry weight and length decreased significantly when soil water content reduced ($\rho < 0.001$). Intensive soil drought stress inhibited root growth in high depth of soil (Rad et al. 2010a).

Few eucalypt species are able to tolerate water and soil salinity, including: *E. camaldulensis*, *E. microtheca*, *E. sargentii*, *E. occidentalis*, *E. gomphocephala*, *E. largicornis* and *E. platypus* (AbbasAlian and Zafarani 2010; El-Juhany et al. 2008). Rad et al. (2012) reported that among eight eucalypt species and provenances planted on saline soils close to Yazd city of Iran, under species trial, irrigated with industrial and urban sewages, *E. microtheca* and *E. camaldulensis* in comparison to other species achieved the greatest survival, diameter and total height at first and

Table 1 List of eucalypt species and provenances under Qom trial

No.	Eucalypt species and provenances	Seed origin
1	*E. camaldulensis* (41-Qom)	Iran, Qom
2	*E. microtheca* (62-Qom)	Iran, Qom
3	*E. camaldulensis* (41-zh)	Iran, Mazandaran, Zhagmarz
4	*E. camaldulensis* (41-ch)	Iran, Mazandaran, Noor, Chamestan
5	*E. microtheca* (62-Ahwaz)	Iran, Khuzestan, Ahwaz
6	*E. rubida* (166-sh)	Iran, Guilan, Shafaroud, Sheikhneshin

second year of the trial. *E. sargentii* was recommended for soil conservation purposes at desert areas due its resistance to saline soil and water, somewhat coldness and high shoot growth, particularly diameter using sewage irrigation.

2 Materials and Methods

Soil and wastewater physical and chemical test, climatical data analysis for 20 years and plant cover identification were applied prior to trial conduction. Soil morphology was studied up to 150 cm depth and sampling was made at each horizon or layer.

The trial was conducted in 2007 under experimental design of Randomized Complete Blocks with three replicates and six treatments, including six eucalypt species and provenances (Table 1). Refined wastewater was used for irrigation for four years (2017–2010), without applying control treatment due to lack of natural water resources in the trial site. Fifteen seedlings were planted at each plot in each replicate at 5 × 5 m spacing.

Eucalypt performance was studied, including survival, total height, diameter at breast height and health condition. Survival was recorded two times, after summer and winter seasons, whereas the growth parameters were measured only one time (after growth period) and health condition was estimated at the end of the trial. Data analysis was applied, using Variance analysis and Duncan test methods and SAS, SPSS and MSTAT computer programs.

3 Results

Table 2 displays wastewater physical and chemical characteristics. Table 3 shows the climate characteristics in detail. Dry period is eight months (Fig. 1). Climate type is dry. Average annual rainfall, absolute minimum and maximum temperature was 126 mm, −11 and 47 °C, respectively.

Table 2 Physical and chemical characteristics of Qom's wastewater

Test items	Characteristics
Temperature at time of testing	25°C
Color (Pt-Co)	Positive
EC	2930 µS/cm
pH	7.9
Turbidity (NTU)	30.5
Total dissolved solid at 180 °C	1770 mg/L
Total hardness ($CaCO_3$)	1095 mg/L
Chloride (Cl)	750 mg/L
Sulphate (SO_4)	790 mg/L
Ammonia (NH_3)	36 mg/L
Phosphate (PO_4)	18 mg/L
COD	120 mg/L
BOD_5	66 mg/L
TSS	70 mg/L
P-Coliform (MPN/100)	16,000

Table 4 shows soil physical and chemical properties of the trial location in Qom. Soil texture, EC and pH are sandy, 8.6 ds/m and 7.6, respectively, whereas pH and EC of irrigation water is 7.4 and 9.3 ds/m, respectively. Soil orders in this site are Entisols and Aridisols, in which suborders are Torrifluvent and Calcigypsids. Torrifluvent are layered soils, in which organic carbon percentage differs at different soil layers. Soil elements, including P, K, Fe, Zn, Cu and Mn differed at various layers. Furthermore, $CaCO_3$ and pH in these soils are relatively high.

Table 5 shows results of variance analysis. Effect of treatments, replicates, treatment × replicates interaction and year on total height and diameter was significant ($\rho < 0.01$) and their CV was 4.18 and 5.04%, respectively which is very satisfaction due to the trial type and condition. Effect of treatments on height increment was significant ($\rho < 0.05$), whereas effect of replicates, replicate × treatments interaction and year was not significant and its CV was 40.6 which is quite high due to the trial type and condition. Effect of treatments and years on diameter increment was significant ($\rho < 0.01$) and as same as for replicate x treatment interaction ($\rho < 0.05$) and its CV was 31.1% which is high due to the trial type and condition. Effect of treatments, replicates and treatment x replicates interaction on height/diameter ratio was significant ($\rho < 0.01$) and its CV was 6.02 which is satisfaction due to the trial type and condition. Figure 2 compares eucalypt species and provenances of Qom trial in respect to their mean growth parameters from 2011 to 2013. The greatest and the lowest values of total height and diameter belonged to *E. camaldulensis* 41-ch (5.6 m and 9.6 cm, respectively) and *E. rubida* 166-sh (3.9 m and 7 cm, respectively). The highest value of height increment belonged to *E. camaldulensis* with Qom and Chamestan origins (70 cm) and the lowest value belonged to *E. microtheca* 62-Ahwaz and *E. rubida* 166-sh (47 and 43 cm, respectively). The greatest and the lowest values of diameter

Table 3 Climatical data for Qom City (1993–2013)

Parameters	Months											
	JAN	FEB	MAR	APR	MAY	JUN	JUL	AUG	SEP	OCT	NOV	DEC
Avrg. daily temperature °C	3.5	5.5	10.8	18	23.4	29.4	32.1	30.5	24.9	17.8	11	4.9
Avrg. Minim. temperature °C	−2.1	−0.3	1.4	9.9	14.8	19.5	22.6	20.7	14.5	9.3	3.7	−0.8
Avrg max. temperature °C	10.1	12.2	17.5	25.8	31	37.5	40.2	39	34.4	26.4	19.2	11.9
Abs. Min. temperature °C	−12.6	−11	−11	0.4	5.4	8	15	13.5	6.5	0.6	−7	−10.5
Abs. Max. temperature °C	23.4	23.2	27.5	35.5	40.8	44.2	45.6	44.5	41	36.6	28.4	22
Avrag. air humidity %	65	60	52	42	34	23	23	24	26	38	50	63
No. of frost days	23.9	16.4	4.6	0	0	0	0	0	0	0	6.5	19
Annual rainfall (mm)	17.4	29.5	29.8	15.3	16.1	1.3	1.2	0	0.6	8.6	12.3	15.2
No. of rainfall days	6.8	7	7.3	5.8	4.6	0.4	0.7	0.1	0.2	3.2	3.6	4.7
No. of snowfall days	3.2	1.8	0.5	0	0	0	0	0	0	0	0.1	1.3

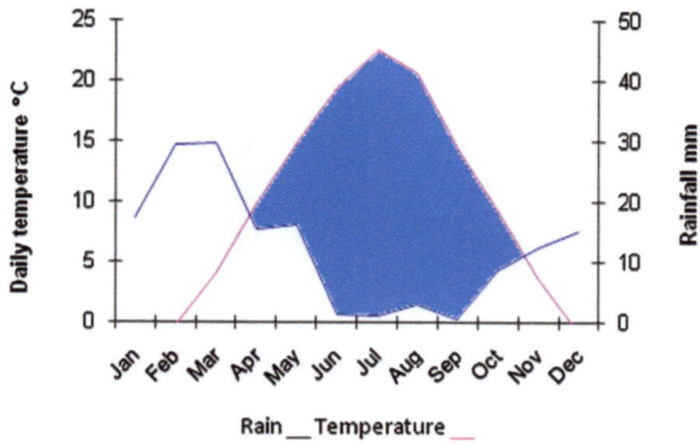

Fig. 1 Average monthly temperature and precipitation curves (1986–2013) for Qom City, showing the dry period

increment belonged to *E. camaldulensis* 41-Qom and *E. rubida* 166-sh (17 and 10 mm, respectively). Finally, the greatest and the lowest values of height/diameter ratio belonged to *E. camaldulensis* 41-zh and *E. rubida* 166-sh (6.1 and 5.5, respectively).

Figure 3 compares the significance of eucalypts characteristics of Qom trial at three years (2012–2014). Total height (483–547 cm) and diameter (72–98 mm) increased significantly from 2012 to 2014. Although there was no significant difference among the years in respect to height increment, but there was significant difference among the years in respect to their diameter increment and height/diameter ratio. Diameter increment in 2013 (10 mm) was lower than in 2011 and 2012 (6.1 and 5.6 mm, respectively). The highest and the lowest values of height/diameter ratio belonged to 2011 and 2013 (6.1 and 5.7, respectively). Furthermore, trends of total height, diameter, height increment, diameter increment and survival performances are displayed in Figs. 4, 5, 6, 7 and 8. Actually the total height and diameter of eucalypt species and provenances had an accelerate trend during the trial period in Qom. Survival percentage of all eucalypt species and provenances was excellent (100%) from 2007 to 2012, but in 2013 it decreased slightly for 41-Qom, 62-Ahwaz and 166-sh (94, 88 and 94%, respectively). In 2014 the highest values belonged to 41-Qom and 62-Qom (94 and 100%, respectively), whereas the lowest values belonged to 41-ch, 62-Ahwaz and 166-sh (44, 50 and 50%, respectively) (Fig. 8). Height and diameter increment in different eucalypt species and provenances varied dramatically at different years and there was not regular trend in that case (Figs. 6 and 7).

Table 4 Soil and water physical and chemical characteristics of the trial site in Qom City

Soil depth cm	Texture	K (absorbable) (Mg/Kg)	Na (absorbable) (Mg/Kg)	Ca + Mg (Meq/L)	Cl (Mg/Kg)	TNV (%)	N (%)	OC (%)	EC (ds/m)	pH
0–40	Sandy	176	80.5	12.0	84	15.7	0.02	0.18	1.6	7.7
41–50	Sandy	148	598.0	18.0	745	20.9	0.02	0.23	4.0	7.5
>50	Sandy	108	1270.0	35.5	3464	11.8	0.02	0.18	8.5	7.7
Water sample	–	19	667.0	12.0	1276	–	–	–	4.9	7.5

Table 5 Results of variance analysis of Qom eucalypt trial

Sources	df	Height/diameter	Diameter increment	Height increment	Diameter	Total height
Treatments (T)	5	**0.65	**78	*1616	**965	**41,041
Replicates (R)	3	**2.05	ns 7	ns 193	**1424	**24,482
Years (Y)	2	ns 1.4	**252	ns 802	**3915	**70,980
T × R	15	**1.01	*41	ns 470	**765	**16,179
T × Y	10	ns 0.05	ns 35	ns 176	ns 43	ns 800
Y × R	6	ns 0.05	ns 2	ns 950	ns 2	ns 393
Error	30	–	–	–	–	–
CV %	–	6.02	31.1	40.6	5.04	4.18

* and ** significant at 0.05 and 0.01% levels, respectively. ns = not significant

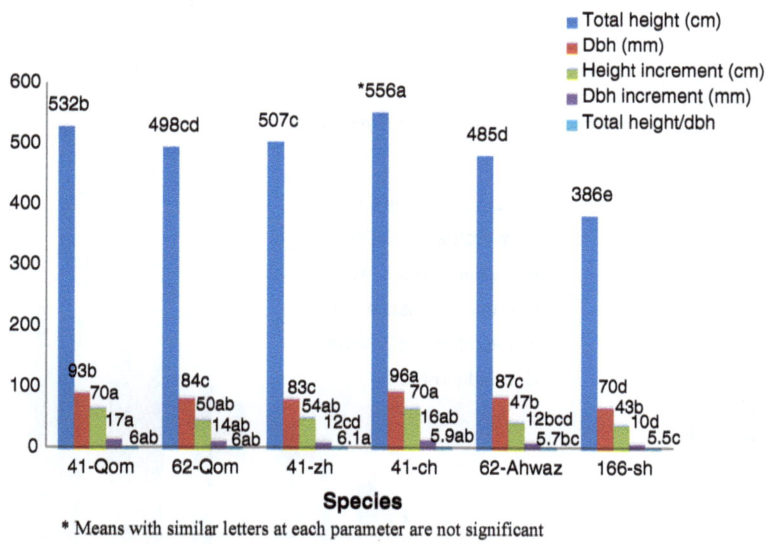

Fig. 2 Comparing eucalypt species and provenances of Qom trial in respect to their mean growth parameters

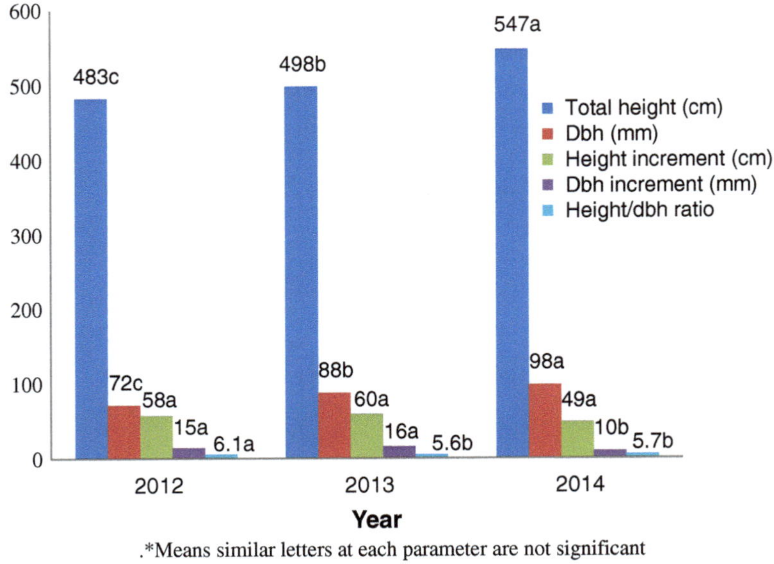

Fig. 3 Eucalypt characteristics of Qom trial at three years

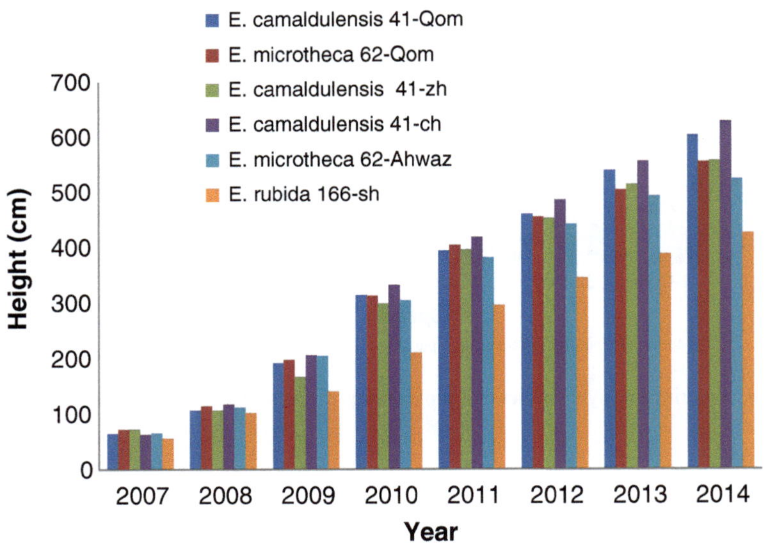

Fig. 4 Trend of total height growth rate at Qom trial period (cm)

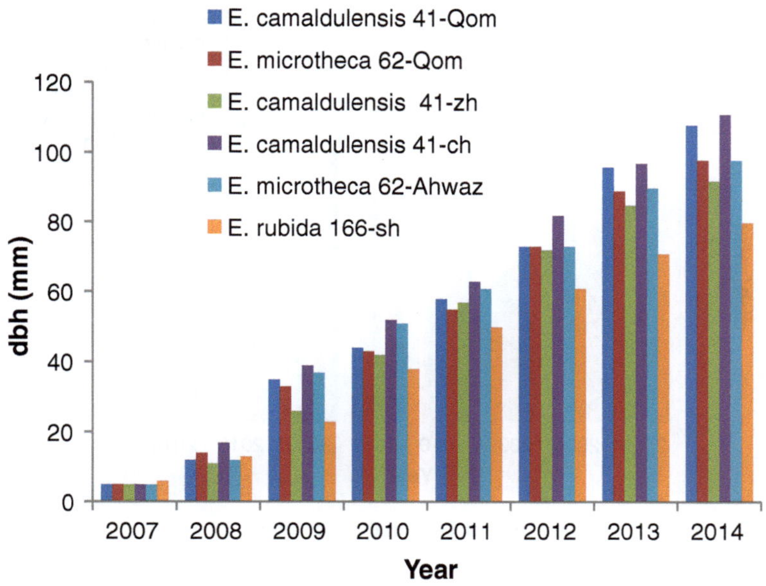

Fig. 5 Trend of diameter growth rate at Qom trial period (mm)

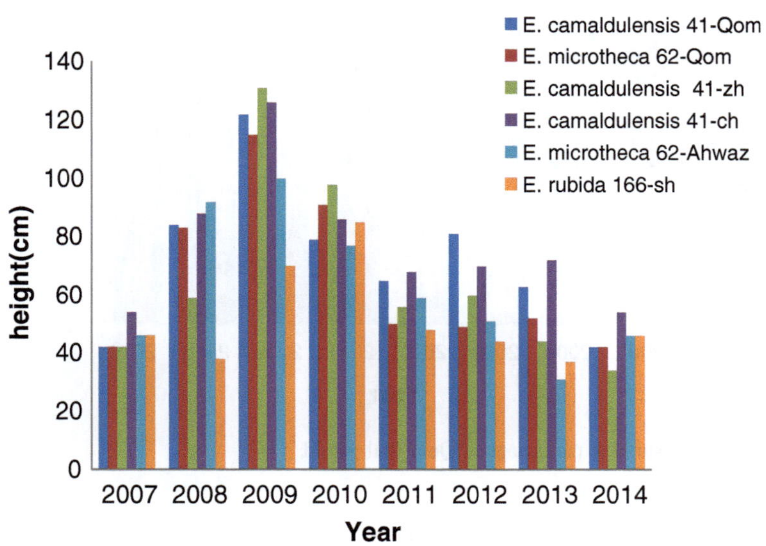

Fig. 6 Trend of height increment at Qom trial period (cm)

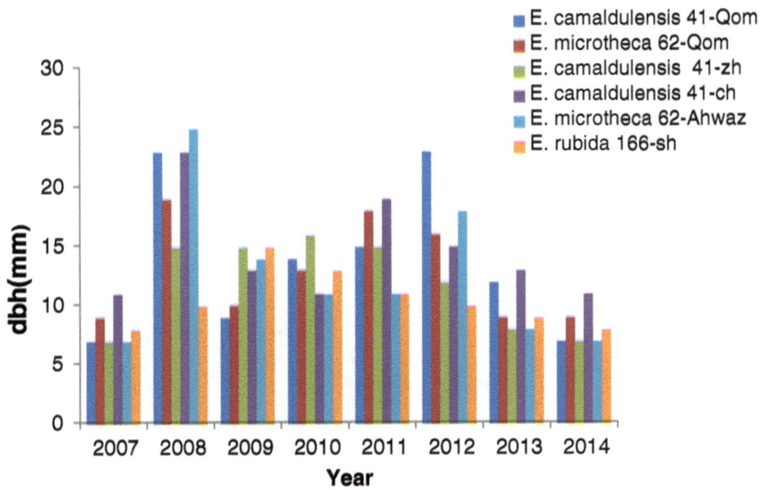

Fig. 7 Trend of diameter increment at Qom trial period (mm)

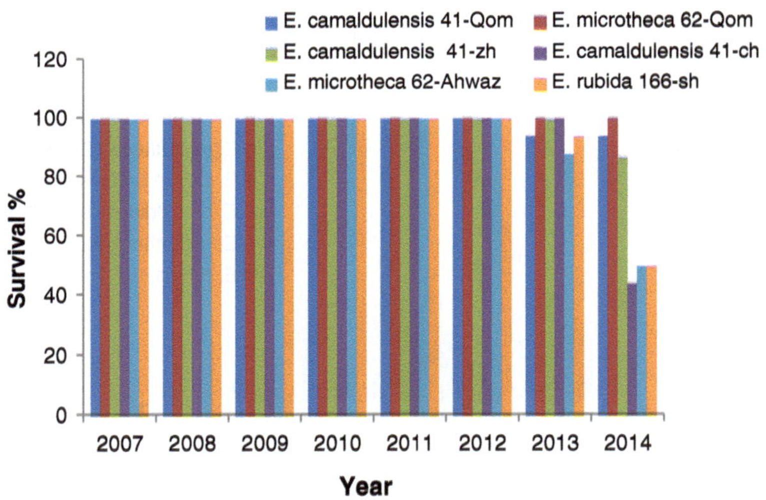

Fig. 8 Trend of survival percentage at Qom trial period

4 Discussion and Conclusion

Actually the greatest growth performance in 2014 at Qom environmental condition belonged to *E. camaldulensis* species, particularly Chamestan provenance (41-ch). The greatest values of total height and diameter of 41-ch provenance were 5.6 m and 9.6 cm, respectively. Furthermore, the greatest values of height increment (70 cm) achieved by *E. camaldulensis* species (41-ch and 41-Qom provenances). In addition, the highest values of diameter increment (17 mm) and height/diameter ratio (6.1) achieved again by *E. camaldulensis* species (41-Qom and 41-zh, respectively). Although irrigation was not applied from 2011 to 2014 due to lack of financial support and due to lack of labor service, most of *E. camaldulensis* provenances made desirable adaptation to dry condition under rain fed condition, except 41-ch provenance which was not able to completely tolerate the dry condition due to its humid seed origin in Iran (its survival was less than 50%), whereas the provenances 41-Qom and 41-zh were able to tolerate the harsh conditions very well and their survival was 94 and 87%, respectively. It might be concluded that *E. camaldulensis* is the most successful and adaptable eucalypt species at Qom city and its suburbs and for similar environmental conditions in Iran, particularly if the required irrigation water can be applied. Many references acknowledge this conclusion (Sinha et al. 1999; Sadati et al. 2004; Saleheh Shoushtari and Rouhipour 2005; Boland et al. 2007; Rarivoson et al. 2008; El-Juhany et al. 2008; Mortazavi Jahromi and Kousar 2010; Abbasalian and Zafarani 2010; Hamzepour et al. 2012; Zahid et al. 2010; Flores et al. 2013). McMahon et al. (2010) reported that *E. camaldulensis* is resistant to salinity, drought, coldness and soil water logged condition and is utilized for wood industry, forest plantation, soil conservation and honey production. Therefore, results of my colleagues in relation to eucalypt species trials at different provinces of Iran confirm the species high adaptability at different ecological conditions (Esfandiarpour et al. 2015; Sardabi et al. 2014; Rostamikia et al. 2015; Hosseinzadeh et al. 2015; Rashidi et al. 2015; Ramazani et al. 2015; Tavakkoli-Neko et al. 2015; Hamzepour et al. 2015; Sharafieh et al. 2015; Karamian et al. 2015; Goodarzi et al. 2015a, b; Farajipoul et al. 2015; Rad et al. 2015). This species has high reputation for its adaptability to arid and semi-arid regions and has been planted at many countries. Its wood is used for fuel consumption; carpentering; mining; building construction; particle board, veneer and paper industries (McMahon et al. 2010; Alavijeh et al. 2014; Anonymous 2014). The species resistance to coldness is relatively desirable and it is able to tolerate −9 to −10 °C and if it suffers sever frost, it can continue its life and growth by sprouting from collar part of the stem. Of course shoot growth is outstanding. Furthermore, *E. camaldulensis* was able to tolerate −10°C in the humid mountainous lands of China (Arnold et al. 2004). However, if the species is under stress of high salinity, drought and heat conditions, it is usually not able to tolerate sever cold and frost climate at winter (Rad et al. 2015).

In contrast, *E. rubida* made a worst growth and survival performance in Qom trial site (see Figs. 2 and 7). This finding is in accordance with my colleagues findings in arid and semi-arid sites of I ran where survival varied from 1 to 49.7% and the average value was 23.6% (Hosseinzadeh et al. 2015; Ramazani et al. 2015; Hamzepour et al. 2015; Goodarzi et al. 2015a, b), whereas its value in Mazandaran province of Iran where climate is humid, was 100% (Farajipoul et al. 2015) due to: (1) its resistance to high snow fall and freezing climate and (2) appropriate soil physical condition in terms of aeration and drainage which guaranteed its satisfactory growth and (3) enough soil and air moisture (Anonymous 2015). Natural habitat of this species on mountainous region of south eastern Australia, including south New South Wales, Victoria and Tasmania Island, where winter is snowy and cold, is in accordance with the previous reference (Boland et al. 2007). A botanist from the Australian Botanical Garden included *E. rubida* in list of species resistant to cold climate (Lyne 2003).

Although adaptability of *E. microtheca*, particularly at arid and semi-arid climates and saline soils was excellent (100%), but due to its shrub and coppicing form, its non-industrial wood production and its desirable shooting potential, is suggested for water and soil conservation purposes. Mortazavi Jahromi and Kousar (2010) reported that *E. camaldulensis* and *E. microtheca* achieved the greatest survival (>80%) under dry and subtropical conditions at Garebigan Fasa area of Fars province of Iran. Furthermore, Rad et al. (2012) reported that among eight eucalypt species and provenances planted on saline and dry soils close to Yazd city of Iran, *E. microtheca* and *E. camaldulensis* achieved the greatest survival, diameter and total height at first and second year of the trial.

References

Abbasalian H, Zafarani H (2010) Eucalyptus, a tree species suitable for saline and waterlogged lands. Sci Prof J Olive 206:62–66 (Persian)

Alavijeh ESH, Habibpour B, Moharramipour S, Rasekh A (2014) Bioactivity of *Eucalyptus camaldulensis* essential oil against *Microcerotermes diversus* (Isoptera: Termitidae). J Crop Prot 3(1):1–11

Anonymous (2014) *Eucalyptus camaldulensis* (red gum). CABI Web. http://www.cabi.org/isc/datasheet/22596. Accessed 17 Aug 2015

Anonymous (2015) *Eucalyptus rubida*, Candle Bark Gum. Yarra Ranges Shire Council. http://fe.yarraranges.vic.gov.au/files/B4E1B906-6AFF-45B2-89C5-9DA50108707C/Eucalyptus_rubida.pdf. Accessed 7 Apr 2015

Arnold RJ, Clarke B, Luo J (2004) Trials of cold tolerant eucalypt species in cooler regions of south central China. CSIRO, Australian Centre for International Agricultural Research, Canberra

Assareh MH, Sardabi H (2007) Eucalypts, volume 1: Description, illustration & propagation by advanced techniques. Research Institute of Forests and Rangelands, Tehran (Persian)

Boland DJ, Brooker MIH, Chippendale GM, Hall N, Hyland BPM, Johnston RD, Kleinig DA, Turner JD (2002) Forest trees of Australia. CSIRO Publishing, Victoria, Australia

Boland DJ, Brooker MIH, Chippendale GM, Hall N, Hyland BPM, Johnston RD, Kleinig DA, Turner JD (2007) Forest trees of Australia. CSIRO Publishing, Victoria, Australia

El-Juhany LI, Aref IM, Ahmed AIM (2008) Response of *Eucalyptus camaldulensis*, *Eucalyptus microtheca* and *Eucalyptus intertexta* seedlings to irrigation with saline water. World J Agric Sci 4(S):825–834

Esfandiarpour P, Sardabi H, Shakarchian M, Poursafari B (2015) Investigation on adaptability and performance of industrial eucalypt provenances at different ecological zones, second stage (Kerman province). Final Report of National Research Project, Research Institute of Forests and Rangelands of Iran, Tehran (English abstract)

Farajipoul RA, Sardabi H, Kiadaliri SH, Khazaei T, Souri GA (2015) Investigation on adaptability and performance of industrial eucalypt provenances at different ecological zones, second stage (Mazandaran province). Final Report of National Research Project, Research Institute of Forests and Rangelands of Iran, Tehran (English abstract)

Flores HJM, Avalos VMC, Rincon MN, Magana JG, Reyes JTS (2013) Development of three commercial plantations of *Eucalyptus camaldulensis* Dehnh., established in the municipality of Buenavista, Michoacan, Mexico. Foresta Veracruzana 15(2):23–30

Goodarzi GR, Sardabi H, Mirdavoodi HR, Ahmadlou F, Shirmard M, Haghshenas M (2015a) Investigation on adaptability and performance of industrial eucalypt provenances at different ecological zones, second stage (Markazi province, Saveh). Final Report of National Research Project, Research Institute of Forests and Rangelands of Iran, Tehran (English abstract)

Goodarzi GR, Sardabi H, Mirdavoodi HR, Azdoo Z, Ahmadlou F, Haghshenas M, Arabi A (2015b) Investigation on adaptability and performance of industrial eucalypt provenances at different ecological zones, second stage (Markazi province, Mahallat). Final Report of National Research Project, Research Institute of Forests and Rangelands of Iran, Tehran (English abstract)

Hamzepour M, Sardabi H, Bordbar K, Joukar L, Abbasi AR (2012) Investigation on establishment of some industrial *Eucalyptus* species and provenances in Kazeroon, Fars province. Iran J For Poplar Res 20(2): 217–232 (English abstract)

Hamzepour M, Sardabi H, Bordbar SK, Nejabat M, Abbasi AR, Joukar L, Ghaedi MS (2015) Investigation on adaptability and performance of industrial eucalypt provenances at different ecological zones, second stage (Fars province). Final Report of National Research Project, Research Institute of Forests and Rangelands of Iran, Tehran (English abstract)

Hosseinzadeh J, Sardabi H, Mohammadpour MA, Zarrin-Kaviani H (2015) Investigation on adaptability and performance of industrial eucalypt provenances at different ecological zones, second stage (Ilam province). Final Report of National Research Project, Research Institute of Forests and Rangelands of Iran, Tehran (English abstract)

Karamian R, Sardabi H, Mohammadian A, Mehdifar D, Jahanpour F, Hasani-Moghaddam A (2015) Investigation on adaptability and performance of industrial eucalypt provenances at different ecological zones, second stage (Lorestan province). Final Report of National Research Project, Research Institute of Forests and Rangelands of Iran, Tehran (English abstract)

Lyne A (2003) Eucalypts for cold climates: Northern hemisphere cultivation. Information about Australia's flora, growing native plants. Australian National Botanical Gardens, Australian National Herbarium, 1 page report.https://www.anbg.gov.au/gnp/cold-climate/eucalypts-cold-climates.html/17.08.2015

McMahon L, George B, Hean R (2010) *Eucalyptus camaldulensis*. Primefact 1054:1–6

Mortazavi Jahromi SM, Kousar SA (2010) A pilot trial on drought resistant tree species irrigated by flood water. Iran J For Poplar Res 18(1): 90–106 (English abstract)

Rad MH, Assareh MH, Soltani M (2010a) Response of the root of *Eucalyptus camaldulensis* Dehnh. to drought stress. Iran J For Poplar Res 18(2): 285–296 (English abstract)

Rad MH, Assareh MH, Meshkat MA, Dashtegian K, Soltani M (2010b) Water requirement and production function of *Eucalyptus* (*E. camaldulensis* Dehn) in arid environment. Iran J For 2 (1):61–71 (English abstract)

Rad MH, Assareh MH, Meshkat MA, Soltani M (2011) Effects of drought stress on biomass, several growth parameters and water use efficiency of Eucalyptus (*Eucalyptus camaldulensis* Dehn) in response to drought stress. Iran J Rangelands For Plant Breed Genet Res 19(1):13–27 (English abstract)

Rad MH, Sardabi H, Soltani M, Ghelmani SV (2012) Compatibility of different *Eucalyptus* species and provenances under sewage irrigation using Yazd city wastewater treatment plant effluent. Water Wastewater 25(1):85–94 (English abstract)

Rad MH; Sardabi H; Soltani M (2015) Investigation on adaptability and performance of industrial eucalypt provenances at different ecological zones, second stage (Yazd province). Final Report of National Research Project, Research Institute of Forests and Rangelands of Iran, Tehran (English abstract)

Ramazani M, Sardabi H, Rahimi H, Gheisari NA (2015) Investigation on adaptability and performance of industrial eucalypt provenances at different ecological zones, second stage (Khorasane-Razavi province). Final Report of National Research Project, Research Institute of Forests and Rangelands of Iran, Tehran (English abstract)

Rarivoson C, Vincelette M, Sitandy T, Mara R (2008) Growth results of five non-native fast growing species used to reforest sandy and nutrient poor soils, biodiversity, ecology and conservation of littoral ecosystems in southeastern Madagascar, Tolagnaro

Rashidi MJ, Sardabi H, Kazerouni H (2015) Investigation on adaptability and performance of industrial eucalypt provenances at different ecological zones, second stage (Boushehr province). Final Report of National Research Project, Research Institute of Forests and Rangelands of Iran, Tehran (English abstract)

Rostamikia Y, Sardabi H, Samadzadeh A, Sharifi-Ziveh P, Nikkhah R, Noori AW (2015) Investigation on adaptability and performance of industrial eucalypt provenances at different ecological zones, second stage (Ardabil province). Final Report of National Research Project, Research Institute of Forests and Rangelands of Iran, Tehran (English abstract)

Sadati SA, Dastmalchi M, Rezaei SAA, Mostafanezhad SR (2004) Investigation on adaptation and performance of three 20 year-old eucalypt species at Chamestan (Mazandaran province of Iran). Iran J For Poplars 12(1):61–78. (English abstract)

Saleheh Shoushtari MH, Rouhipour H (2005) Eucalypt species trial on sandy dunes of Khuzestan province (Iran). Iran J For Poplar Res 13(4):475–499 (English abstract)

Sardabi H, Rahmani A, Hamze B, Assareh MH, Ghorany M (2010) Impact of different Eucalypt species on forest soil properties in Guilan province. Iran J For Poplar Res 18(1):116–131 (English abstract)

Sardabi H, Saleheh S, Banj Shafiei SH, Jafari AA, Thoghraie N, Shariat A, Assareh MH (2013) Investigation on potential of few eucalypt species for absorbing pollutants and reserving them in their leaves. Iran J For Poplar Res 21(2):357–372 (English abstract)

Sardabi H, Jafari M, Farahani A, Beizaeinezhad AM, Ghamghami F (2014) Investigation on adaptability and performance of industrial eucalypt provenances at different ecological zones, second stage (Tehran province). Final Report of National Research Project, Research Institute of Forests and Rangelands of Iran, Tehran (English abstract)

Sharafieh H, Sardabi H, Moradi GR, Nejatian AF, Mirakhorli R (2015) Investigation on performance of few eucalypt species at different ecological sites of Semnan province. Final Report of Research Project, Research Institute of Forests and Rangelands of Iran, Tehran (English abstract)

Sinha RK, Bhatia S, Vishnoi R (1999) Desertification control and rangeland management in the Thar desert of India—desertification and rangeland management in India. RALA report No. 200, pp 115–123

Tavakkoli-Neko H, Sardabi H, Pourmeidani A, Adnani SM, Moradi MR (2015) Investigation on adaptability and performance of industrial eucalypt provenances at different ecological zones, second stage (Qom province). Final Report of National Research Project, Research Institute of Forests and Rangelands of Iran, Tehran (English abstract)

Zahid DM, Shah FR, Majeed A (2010) Planting *Eucalytpus* in arid environment-is it useful species under water deficit system? Pak J Bot 42(3):1733–1744

Applications of Magnetic-Water Technology, A Novel Tool for Improving Chick-Pea Crop and Water Productivity

M. Hozayn, A.A. Abd El Monem, M.A. Darwish and Ebtihal M. Abd Elhamid

Abstract Optimization of crop productivity from land and water unit is considered the main target for agriculture scientist. Our previous results under screen house and/or field conditions on other crops reported clear increasing in yield of tested crops. So let us to continuously test this technique on chick-pea. Two field trials using chick-pea (Var. Giza-3) were conducted at Research and Production Station, National Research Centre, Alemam Malek village, Al Nubaria district, Al Beheira Governorate, Egypt in 2009/10 and 2010/11 winter seasons to study and evaluate the effects of magnetizing irrigation water on water and chick-pea productivity. Results indicated that, irrigation chick-pea plants with water passed through static magnetic device (2″, produced by Magnetic-Technologies LCC, Russia, branch UAE) caused clear positive significant effect on the most of tested parameters. The percentage of increment ranged from 3.29 to 33.54% in growth parameters, photosynthesize property of plants (3.91–10.85%), yield components (9.36–36.65%), yield per fed (26.83–28.36%). As well as magnetic treatment increased nutrition value in chick-pea yielded seeds (0.75–56.25%) and amino acids (0.59–33.33%). It could be concluded that magnetic water treatment could be used to improve Chick-pea and water productivity under newly reclaimed sandy soil conditions.

M. Hozayn (✉)
Agricultural and Biological Research Division, Field Crops Research Department, National Research Centre, 33 El Behouth St., (Former El-Tahrir St.), Dokki, Giza 12622, Egypt
e-mail: m_hozien4@yahoo.com

A.A. Abd El Monem · E.M. Abd Elhamid
Agricultural and Biological Research Division, Botany Department, National Research Centre, 33 El Behouth St., (Former El-Tahrir St.), Dokki, Giza 12622, Egypt

A.A. Abd El Monem
Department of Biology, Faculty of Science, Tabuk University, Branch Tayma, Tayma, Saudi Arabia

M.A. Darwish
Fertilization Technology Department, Agricultural and Biological Research Division, National Research Centre, 33 El Behouth St., (Former El-Tahrir St.), Dokki, Giza 12622, Egypt

O. Abdalla et al. (eds.), *Water Resources in Arid Areas: The Way Forward*, Springer Water, DOI 10.1007/978-3-319-51856-5_26

Keywords Chickpea · Magnetized-water · Growth · Chemical constitute · Yield · Water productivity

1 Introduction

In Egypt, the most of people depend mainly on chick-pea and other pulses crops as a source of protein. Moreover, chick-pea crop is lower competitive than other winter crops like wheat and clover on planted area so it's annually decreasing. Researchers take an interest not only traditional agricultural practices i.e., cultivation methods, fertilizers, weed control, etc. but also un-common factors that improvement productivity of crops from unite either area and/or water like magnetic technology.

Several studies in many countries, like Russia, China, Poland and Bulgari, etc. indicating that magnetic treatment of irrigation water offered many benefits in agriculture such as improved germination, growth, yield, early maturity of crops, reduced plant diseases, reduced salinity stress, improved crop quality, increased fertilizers efficiency and reduced cost of farm operations (e.g., Maheshwari and Grewal 2009; Babu 2010; Suchitra and Babu 2011; Hozayn et al. 2011, 2013, 2014, 2015, 2016; Yousf and Ogunlela 2015). Moreover, in macro trials, many authors recorded that application of magnetic treatment caused an increase in economic yield reached up to 144.80% in potato, rice by 18%, pepper by 64.90%, soybean 15.00%, sunflower by 15.10%, cereal by 20.00%, wheat by 10.50%, broad bean by 10%, pea by 15%. (e.g. Vashisth and Nagalajan 2008; Aladjadjiyan 2010).

In Egypt, our clear and promising studies under field conditions, Hozayn et al. (2013, 2014, 2015, 2016a, b) reported that faba bean, wheat, sugar beat, lentil, canola, flax and potato irrigated with magnetized-water gave more value of economic yield compared to control treatment. These increases ranged between 8.04 and 42.23% depended on tested crop. Therefore, the present work aim to test application of magnetic-water treatment on chick-pea productivity under newly reclaimed sandy soil.

2 Materials and Methods

During two winter successive seasons (2009/10 and 2010/11), two field experiments wear conducted at Agricultural Research Station, NRC, Alemam-Malek Village, Al-Nubaria District, Al-Beheira Governorate, Egypt to comparison between irrigation with magnetized and normal water on productivity of chick-pea crop. The soil and water samples in the site of experiments were analyzed according to Chapman and Pratt (1978; Table 1)

Table 1 Analysis of soil and water in site experiments

Parameters	Soil depth (cm) 0–15	Soil depth (cm) 15–30	Irrigation water
Particle size distribution			
Coarse sand	48.21	54.75	–
Fine sand	49.12	41.43	–
Clay + silt	02.67	3.82	–
Texture	Sandy	Sandy	–
PH (1:2.5)	8.23	7.94	7.25
EC(dS m^{-1}) (1:5)	0.21	0.15	0.50
Organic matter (%)	0.66	0.43	–
Soluble cations (mg/l)			
Ca^{++}	0.61	0.52	2.15
Mg^{++}	0.52	0.32	0.50
Na^{++}	0.91	0.81	3.00
K^{+}	0.21	0.11	0.31
Soluble anions (mg/l)			
CO^{-3}	–	–	0.01–
HCO^{-3}	0.60	0.40	2.33
Cl^{-}	0.75	0.70	2.17
SO^{-4}	0.85	0.60	1.45

Cultivation method and layout of experiment: Recommended rates of chick-pea (var. Giza 3) seeds were sown in plots (120 m^2) at the first week of November in both seasons. Magnetic treatment was applied through passing the irrigation water in static magnetic device [2 in., UT3, produced by Magnetic-Technologies, LLC, PO-Box, 27559, Dubai, UAE] while control treatment was irrigated with normal-water. Each treatment was contained four replications. Recommended of NPK fertilizers and other agricultural practices for sowing chick-pea crop under Nubaria region conditions were applied through the period of experiment under sprinkler irrigation system. Experiment layout was shown in Fig. 1.

2.1 Data Recorded

Growth parameters: Plant height (cm), fresh and oven dry weight (g) of ten plants from each treatment were determined after 85 days from sowing. According to Henson et al. (1981), water content in shoot was calculated using the following formula: Water contents (WC) = 100 × (fresh mass − dry mass)/fresh mass.

Photosynthetic pigments: In fresh leaves, chlorophyll a, b and carotenoids were estimated according to Lichtenthaler and Buschmann (2001).

Chick-pea yield and its components: At harvest stage, twenty plants were selected randomly in each treatment to determine chick-pea yield components. Biological,

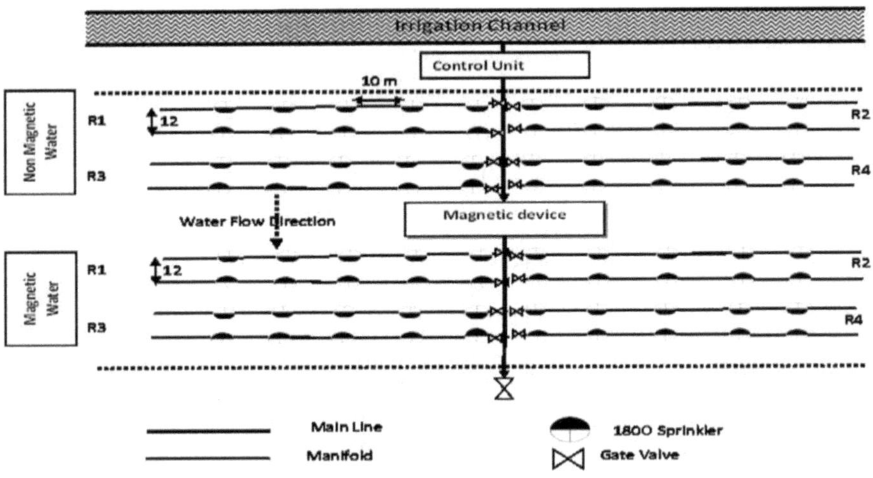

Fig. 1 Experiment layout under sprinkler irrigation system

seed and straw yields were determined in area (120 m^2) and calculated to feddan. Harvest and crop indexes were calculated by dividing seed yield/biological and straw yield, respectively

Water productivity (WP): It was calculated with the following Eqs.: WP = [(E_y/ E_t], (Howell et al. 1990).

Where WP is the water productivity ($kg_{chick\text{-}pea\ seeds}/m^3$ water); Ey = seed yield (kg/fed); E_t = total applied of irrigation water, (m^3/fed/season).

Nutritional value of yielded seeds: In dried seeds, NPK, Ca and Mg as Macro-elements and Fe, Mn, Zn and Cu as micro-elements were determined. Total N was determined using micro-Kjeldahl method (AOAC 1970). Phosphorus was determined calorimetrically using Spekol-spectrocolorimeter (VEB-Carl Zeiss, Jena, Germany). Potassium (K^+) was determined using flame photometer. Iron (Fe), Zink (Zn), Copper (Cu), Manganese (Mn) and Magnesium (Mg) were determined using the Atomic Absorption Spectrophotometer (Perkin Elemer 100 B).

Free amino acids: Amino acids in seeds were estimated according to Catalog of Amino Acid analyzer (1999).

Statistical analysis: Unpaired *t*-test was used to statistically comparison between magnetic and normal water treatments using SPSS statistics software package, Version16.

3 Results and Discussion

3.1 Chick-Pea Growth

Table 2 show that application of magnetic-water treatment significantly increased tested growth morphological criteria compared to ordinary water. The percentage of increasing reached to 33.54, 18.56 and 3.29% in fresh and dry wt. (g plant^{-1}) and water contents (%), respectively. Improving in photosynthetic pigments (Fig. 2) may be is resulted in increasing vegetative growth (Table 2). Previous studies (Hozayn and Abd El-Qodos 2010a, b; Abd El-Qodos and Hozayn 2010a, b) suggested that stimulation growth of lentil, flax, chick-pea and wheat under magnetic water treatment may be due to formation of new protein bands. In this regards, Fomicheva et al. (1992a, b), Belyavskaya (2001), Hozayn et al. (2015) showed that magnetic treatment induces mitosis meristematic cells and cell metabolism of pea, lentil, flax and onion. Generally, many scientists reported that magnetic field can be

Table 2 Fresh and dry weight (g plant^{-1}) and water content (%) at 85 days after sowing under magnetic and ordinary water treatments

Treatment	Mean ± SE		*p-value*	Increase (%) over control
Character	Normal water (control)	Magnetic water		
Fresh weight/plant (g)	07.96 ± 0.17	10.63 ± 0.23	0.0001	33.54
Dry weight/plant (g)	01.94 ± 0.05	02.30 ± 0.04	0.0001	18.56
Water contents (%)	75.63 ± 0.47	78.12 ± 0.67	0.0040	3.29

Average; 2009/10 and 2010/2011 winter seasons

n = 20, $P < 0.01$ and 0.001 levels are highly significant according unpaired *t*-test

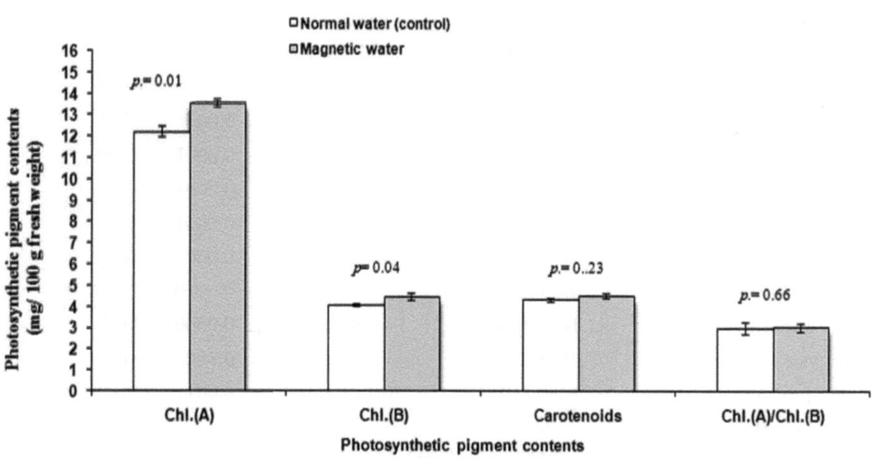

Fig. 2 Photosynthetic pigment contents in chick-pea shoot at 85 days after sowing under magnetic and normal water treatments (Average; 2009/10 and 2010/2011 winter seasons). n = 8; $P < 0.05$ and 0.01 levels are significant according unpaired *t*-test, $P > 0.05$ is non-significant

affecting on mitosis meristematic cells, cell division, protein biosynthesis which reflecting on change in plant growth and development (Reina et al. 2001; Celik et al. 2008; Shabrangi and Majd 2009). Nasher (2008) obtained taller chick-pea plant under magnetic-water treatment compared tap water. Recently, Hozayn et al. (2015, 2016a, b) obtained similar results on wheat, potato and canola plant.

3.2 Photosynthetic Pigment

Pigments contents are considered good indicator for healthy and productivity of plants and the most important issue for farmers. Significant increases in photosynthetic pigments parameters i.e., chl. a, chl. b, carotenoids and ratio (chl. a/chl. b) in shoot plants at 85 days after sowing were recorded under magnetized water compared with ordinary water treatment (Table 3). The increases in these parameters reached to 10.85, 9.25, 3.91, and 0.99%, respectively over control treatment. Improvement of photosynthetic pigments under magnetic water treatment were reported and explained due to increasing growth promoters indole acetic acid

Table 3 Chick-pea yield (ton fed^{-1}), yield components and water productivity ($kg_{chick\text{-}pea\ seeds}/m^3$ water) under magnetic and normal water treatments

Treatment	Mean ± S.E.		*p-value*	Increase (+) (%) over control
Character	Normal water (control)	Magnetic water		
Plant ht (cm)	40.30 ± 1.03	45.90 ± 1.59	0.0050	13.90
Branches (number plant^{-1})	3.88 ± 0.10	4.68 ± 0.10	0.0001	20.62
Pods (no. plant^{-1})	7.53 ± 0.16	10.29 ± 0.22	0.0001	36.65
Pods wt. (g plant^{-1})	2.43 ± 0.06	3.20 ± 0.07	0.0001	31.69
Pods wt. (g)	0.35 ± 0.08	0.42 ± 0.10	0.0001	20.00
Seeds wt. (g pod^{-1})	0.29 ± 0.01	0.33 ± 0.01	0.0030	13.79
Biological yield (g plant^{-1})	4.79 ± 0.07	6.08 ± 0.07	0.0001	26.93
Seed yield (g plant^{-1})	1.88 ± 0.03	2.21 ± 0.06	0.0001	17.55
Straw yield (g plant^{-1})	2.91 ± 0.04	3.87 ± 0.07	0.0001	32.99
100-seed weight (g)	8.55 ± 0.06	9.35 ± 0.05	0.0001	09.36
Biological yield (ton fed^{-1})	1.64 ± 0.04	2.08 ± 0.05	0.0001	26.83
Seed yield (ton fed^{-1})	0.67 ± 0.02	0.86 ± 0.03	0.0001	28.36
Straw yield (ton fed^{-1})	0.96 ± 0.03	1.23 ± 0.06	0.0020	28.13
Harvest index (%)	41.10 ± 0.96	41.38 ± 1.68	0.9500	00.68
Crop index (%)	72.48 ± 2.75	72.54 ± 4.95	0.8600	00.08
Water productivity (WP)	0.272 ± 0.009	0.345 ± 0.014	0.001	26.83

Means of 2009/10 and 2010/11 seasons
n = 20 for yield parameters, n = 8 for yield per feddan and WP, feddan = 4200 m^2, $P > 0.05$, 0.01 and 0.001 levels are significant according to *t*-test, $P > 0.05$ is non-significant

(IAA) (Hozayn and Abd El-Qodos 2010a, b), induces mitosis meristematic cells and cell metabolism (Hozayn et al. 2015), increase in cytokinin synthesis (Atak et al. 2003). The last author also added that "cytokinin play an important role on chloroplast development, shoot formation, axillary bud growth, and induction of number of genes involved in chloroplast development nutrient metabolism". In this connection, Zude et al. (1999) showed that Fe and Mg may be play a vital role in chlorophyll synthesis. Improvement of photosynthetic pigments were recorded in *Paulownia species* (Atak et al. 2000), sunflower (Oldocay 2002), *Glycine max* (Atak et al. 2003) when seeds or explants exposed to magnetic field (3.8–4.8 mt) for a short time and Nasher (2008) on chick-pea irrigated with magnetized water. Recently, Hozayn et al. (2015, 2016a) obtained similar results on wheat and canola irrigated with magnetic water.

3.3 Chickpea Yield and Its Components

Significant increases regarding application of magnetic-water treatment in tested parameters (i.e., plant height, branches number/plant, pods number and weight/plant, seed number/pod, biological, seed and straw yields/plant, and 100-seed weight) reached to 13.90, 20.62, 36.65, 31.79, 20.00, 13.79, 26.39, 17.55, 32.99 and 9.36% compared to irrigation with normal water, respectively (Table 3). Similar trends were observed in yield (ton fed^{-1}) and yield index where the percent of increments reached to 26.83, 28.36, 28.13, 0.68 and 0.08% in biological, seed and straw yield; and harvest and crop index, respectively (Table 3). Improvement chick-pea yield and its components under magnetic treatment are logical due to increasing value of vegetative growth criteria (Table 2) and pigments (Fig. 2). These results confirmed previous experiments under screenhouse condition on chick-pea, flax, wheat and lentil (Hozayn et al. 2011). As well as are in good harmony with our studies under field conditions on different crops i.e., wheat, sugar beat, lentil, flax, faba bean and canola (Hozayn et al. 2013, 2014, 2015, 2016a, b). In abroad, also several studies had clear and positive effects regarding application of magnetic treatments on different crops i.e., cereal, (Vakharia et al. 1991); *Helianthus annuus* (Pietruszowski 1993); *Linum usitatissimum* (Namba et al. 1995), *Triticum* aestivum, *Solanum lycopersicum*, *Glycine max*, *Solanum tuberosum* and *Beta vulgaris* (e.g., Atak et al. 1997; Özalpan et al. 1999; Pietruszewski 1999a, b; Reina et al. 2001; Oldaçay 2002; Takac et al. 2002; Crnobarac et al. 2002; Marinkovic et al. 2002).

3.4 Water Productivity (WP)

Water productivity increased significantly under magnetic treatment by 26.83% compared to control (Table 3). This result is logically regarding increasing seed yield (Table 3). Similar positive effects recorded on wheat, lentil, faba bean, canola and sugar beat under magnetic-treatment compared to control (Hozayn et al. 2013, 2014, 2016a, b). Alkhazan et al. (2011) also were recorded similar tends on jojoba under normal and drought stress.

3.5 Nutritional Value of Yielded Seeds

Application of magnetic-water treatment caused significant increases in contents of chick-pea yielded seed with macro (NPK, Ca and Mg) and Micro-nutrients (Fe-Mn, Zn and Cu) compared to control treatment (Table 4). These increases differed according types of elements, where ranged between 0.75 and 56.25% and from 3.33 to 32.43% for macro and micro-nutrients elements, respectively. In this regards, Esitken and Turan (2003) observed that magnetic-field caused a positive effect on immobile nutrient uptake which reflection in increasing the nutrition value in strawberry plant. It could be proposed that magnetized water increased the water absorption and hence increased Ca^{+2} and K^{+} uptake (Reina et al. 2001), resulting in improved germination and growth parameters (Table 2). The authors interpreted this effect to be due to increased nutrient uptake resulted from magnetic-field which induced changes in cell membrane parameters, enzyme activities, gene-expression, and protein biosynthesis (Omran et al. 2014).

3.6 Amino Acids Contents in Yielded Seeds

Table 5 show that magnetic-water treatment significantly increased amino acid contents compared to normal water. The percent of increments in essential, non-essential, total and the ratio (essential/non-essential amino acids) reached to

Table 4 Percentage of macro and macronutrients elements in seed yielded of chick-pea under magnetic and normal water treatments

Treatment		Means		*t sig.*	Increase % over control
Character		Ordinary water (control)	Magnetic water		
Macro-nutrients (%)	N	2.66	2.68	ns	0.75
	P	0.76	0.86	*	13.16
	K	1.15	1.28	*	11.30
	Ca	1.08	1.13	ns	4.63
	Mg	0.16	0.25	*	56.25
Micro-nutrients (ppm)	Fe	99.90	132.30	**	32.43
	Mn	30.00	31.00	Ns	3.33
	Zn	16.30	20.70	**	26.99
	Cu	10.50	12.00	*	14.29

*, **, *t* is significant at the $P > 0.05$ and 0.01 levels, respectively, ns: non significant

Table 5 Amino acids contents in seed yielded of chick-pea under magnetic and normal water treatments

Treatment Amino acids (g/100 g seed dry weight)	Ordinary water (control)	Magnetic water	Increases % over control
Aspartic	0.599	0.663	9.57
*Threonine	0.217	0.265	17.96
Serine	0.319	0.361	11.72
Glutamic Acid	1.020	1.185	13.86
Glycine	0.122	0.151	19.23
Alanine	0.452	0.492	8.18
*Valine	0.364	0.403	9.82
*Methionine	0.026	0.039	33.33
*Isoleucine	0.223	0.289	22.90
*Leucine	0.574	0.699	17.88
Tyrosine	0.239	0.241	0.59
*Phenylalanine	0.612	0.668	8.38
*Histidine	0.273	0.365	25.02
*Lysine	0.369	0.473	21.86
**Ammonia	0.490	0.505	3.00
Arginine	0.898	0.914	1.77
Proline	0.444	0.644	30.97
*Essential	2.66	3.20	16.93
Non-essential	4.09	4.65	11.96
Total amino acid	6.75	7.85	13.99
Ess./non-Ess.	0.65	0.69	5.65

16.93, 11.96, 13.98 and 5.65%, respectively. The improvement differed also according type of amino acids, where methionine came in the first order by 33.97%, followed by, proline (30.97%), isoleucine (22.90%), histidine (25.02%), lysine (21.86%), glycine (19.20%) compared others which the increasing ranged between 0.59 and 17.97%. Positive effects regarding to application of magnetic field on amino acids contents reported by Sharaf El-Deen (2003) on *Agaricus bispours*, Hozayn et al. (2015, 2016a) on canola and wheat. Proline is one of the most important amino acids; which accumulated under magnetic water in the present work (Table 5). Similarly, results recorded on seedlings of date palm when exposed to low dose of magnetic field (Dhawi and Al-Khayri (2008).

4 Conclusion

The present study concluded that, chick-pea irrigated with magnetized-water improvement all studied parameters. The percentage of increment ranged from 3.29 to 33.54% in growth parameters, photosynthesize property of plants (3.91–10.85%), yield components (9.36–36.65%), yield per fed (26.83–28.36%), nutrition value and amino acids in yielded seeds (0.75–56.25%) and (0.59–33.33%), respectively. It could be concluded that, irrigation with magnetic-water could be considered valuable modern and applicable technologies that can improve chick-pea and water productivity under newly reclaimed sandy soil.

Acknowledgements This work was funded by project no. (9050102), National Research Centre, Dokki, Giza, Egypt.

References

Abd El-Qodos AMS, Hozayn M (2010a) Magnetic water technology, a novel tool to increase growth, yield, and chemical constituents of lentil (*Lens esculenta*) under greenhouse condition. Am Eurasian J Agric Environ Sci 7(4):457–462

Abd El-Qodos AMS, Hozayn M (2010b) Response of growth, yield, yield components and some chemical constituent of flax for irrigation with magnetized and tap water. World Appl Sci J 8(5):630–634

Aladjadjiyan A (2010) Influence of stationary magnetic field on lentil seeds. Int Agrophys 24: 321–324

Al-Khazan M, Abdullatif BM, Al-Assaf N (2011) Effects of magnetically treated water on water status, chlorophyll pigments and some elements content of Jojoba (*Simmondsia chinensis L.*) at different growth stages. Afr J Environ Sci Tech 5(9):722–731

AOAC (1970). Official methods of analysis of association agriculture chemists, 11th edn. Association of Official Agricultural Chemists, Washington, p 777

Atak C, Danilov V, Yurttas B, Yalçn S, Mutlu D, Andrzakoulieva A (1997). Effects of magnetic field on soybean (*Glycine max* L.Merrill) seeds. Com JINR. Dubna, pp 1–13

Atak Ç, Danilov V, Yurttafl B, Yalçn S, Mutlu D, Rzakoulieva A (2000) Effect of magnetic field on Paulownia seeds. Com JINR. Dubna, pp 1–14

Atak C, Emiroglu Ö, Aklimanoglu S, Rzakoulieva A (2003) Stimulation of regeneration by magnetic field in soybean (Glycine max L. Merrill) tissue cultures. J Cell Mol Biol 2:113–119

Babu C (2010) Use of magnetic water and polymer in agriculture. Tropical Research ID 08-806-001

Belyavskaya NA (2001) Ultrastructure and calcium balance in meristem cells of pea roots exposed to extremely low magnetic field. Adv Space Res 28(4):645–650

Çelik Ö, Atak Ç, Rzakoulieva A (2008) Stimulation of rapid regeneration by a magnetic field in Paulownia node cultures. J Central Eur Agric 9(2):297–304

Chapman HO, Pratt PE (1978) Methods of analysis for soils, plants and water. University of California, Division of Agricultural Sciences Priced Publication, 4034, p 50

Crnobarc J, Marinkovic B, Tatic M, Malesevic M (2002) The effect of REIS on startup growth and seed yield of sunflower and soybean. Biophysics in agriculture production. University of Novi Sad, Tampograf

Dhawi F, Al-Khayri JM (2008) Proline accumulation in response to magnetic fields in date palm (*Phoenix dactylifera* L.). Open Agric J 2:80–83

Esitken A, Turan M (2003) Alternating magnetic field effects on yield and plant nutrient element composition of strawberry (*Fragaria xananassa* cv. Camarosa). Acta Agric Scand Sect B Soil Plant Sci 54:135–139

Formicheva VM, Govorun RD, Danilov VT (1992a) Proliferative activity and cell reproduction in the root meristem of pea, lentil and flax in the conditions of screening the geomagnetic field. Int Biophys 37:645–648

Formicheva VM, Zaslavskii VA, Govorun RD, Danilov VT (1992b) Dynamics of RNA and protein synthesis in the cell s of the root meristems of the pea, lentil and flax. Biophysics 37:649–656

Henson IE, Mahalakshmi V, Bidinger FR, Alagars-Wamy G (1981) Genotypic variation in pearl miller (*Pennisetum americanum* L.) Leeke in the ability to accumulate abscisic acid in response on water stress. J Exp Bot 32:899–910

Howell TA, Cuence RH, Solomon KH (1990) Crop yield response. In: Hoffman GJ, Howell TA, Solomon KH (eds) Management of farm irrigation systems, ASAE, St. Joseph, MI, USA, p 312

Hozayn M, Abd El-Qodos AMS (2010) Magnetic water application for improving wheat (*Triticum aestivum* L.) crop production. Agric Biol J N Am 1(4):677–682

Hozayn M, Abd El-Qodos AMS (2010) Irrigation with magnetized water enhances growth, chemical constituent and yield of chickpea (*Cicer arietinum* L.). Agric Biol J N Am 1(4): 671–676

Hozayn M, Abdel-Monem AA, Abdul Qados AMS, Abd El-Hameed HM (2011) Response of some food crops for irrigation with magnetized water under green house condition. Austr J Basic Appl Sci 5(12):29–36

Hozayn M, Abd El Monem AA, Abdelraouf RE, Abdallam MM (2013) Do magnetic water affect water use efficiency, quality and yield of sugar beet (*beta vulgaris* l.) plant under arid regions conditions? J Agron 34(1):1–10

Hozayn M, Abdel-Monem AA, Elwial TAE, EL-Shater MM (2014) Future of magnetic agriculture in arid and semi-arid regions. Series A. Agronomy, vol LVII, pp 197–204

Hozayn M, El-Bassiouny HMS, Abd El-Monem AA, Abdallah MM (2015) Applications of magnetic technology in Agriculture, a novel tool for improving crop productivity (2). Wheat Int J Chem Tech Res 8(12):759–771

Hozayn M, Abdallha MM, Abd El-Monem AA, El-Saady AA, Darwish MA (2016a) Applications of magnetic technology in agriculture, a novel tool for improving crop productivity (1): Canola. Afr J Agric Res 11(5):441–449

Hozayn M, Salama AM, Abd El-Monem AA, Hesham AF (2016b) The impact of magnetized water on the anatomical structure, yield and quality of potato (*solanum tuberosum* l.) grown under newly reclaimed sandy soil. Res J Pharm Biol Chem Sci 7(3):1059–1072

Lichtenthaler HK, Buschmann C (2001) Chlorophylls and carotenoids: measurement and characterization by UV-VIS spectroscopy. In: Wrolstad RE, Acree TE, An H, Decker EA, Penner MH, Reid DS, Schwartz SJ, Shoemaker CF, Sporns P (eds) Current protocols in food analytical chemistry (CPFA). Wiley, New York, pp F4.3.1–F4.3.8

Maheshwari BL, Grewal HS (2009) Magnetic treatment of irrigation water: its effects on vegetable crop yield and water productivity. Agric Manag 8:1229–1236

Marinkovic B, Ilin Z, Marinkovic J, Culibrk M, Jacimovic G (2002) Potato yield in function variable electromagnetic field. Biophysics in agriculture production. University of Novi Sad, Tomograf

Namba K, Sasao A, Shibusawa S (1995) Effect of magnetic field on germination and plant growth. Acta Hortic 399:143–147

Nasher SH (2008) The effect of magnetic water on growth of chick-pea seeds. Eng Tech 26(9)

Oldacay S, Erdem G (2002) Evaluation of chlorophyll contents and peroxides activities in (*Helianthus annuus* L.) genotypes exposed to radiation and magnetic field. Pak J Appl Sci 2(10):934–937

Omran WM, Mansour MMF, Fayez KA (2014) Magnetized water improved germination, growth and tolerance to salinity of cereal crops. Int J Adv Res 2(5):301–308

Özalpan A, Atak C, Yurttas B, Alikamanoglu S, Canbolat Y, Borucu H, Danilov V, Rzakoulieva A (1999) Effect of magnetic field on soybean yield (*Glycine max* L. Merrill). In: XI national biophysics congress, Turkish Association of Biophysics, Abstract Book, p 60
Pietruszewski S (1999a) Influence of pre-sowing magnetic biostimulation on germination and yield of wheat. Int Agrophysics 13:241–244
Pietruszewski ST (1999b) Effect of alternating magnetic field on germination, growth and yield of plant seeds. Int Agrophysics 5(11):209–215
Pietruszweski S (1993) Effects of magnetic seed treatment on yields of wheat. Seed Sci Techol 21:621–626
Renia FG, Pascual LA, Fundora IA (2001) Influence of a stationary magnetic field on water relations in lettuce seeds. Part II: experimental results. Bioelectromagnetics 22:596–602
Shabrangi A, Majd A (2009) Effect of magnetic fields on growth and antioxidant systems in agricultural plants. In: PIERS proceedings, Beijing, China, March, 23–27
Sharaf El-Deen S (2003) Improvement of some characters of edible mushroom with magnetic field. Bull NRC Egypt 28:709–717
Suchitra K, Babu EA (2011) A pilot study on silt magnetized and non-magnetized water in the on-farm water use efficiency management. Centre for Water Resources, Anna University, Chennai, India
Takac A, Gvozdenovic G, Marinkovic B (2002) Effect of resonant impulse electromagnetic stimulation on yield of tomato and pepper. Biophysics in agriculture production, University of Novi Sad, Tampograf
Vakharia DN, Davariya RL, Parameswaran M (1991) Influence of magnetic treatment on groundnut yield and attributes. Indian J Plant Physiol XXXIV:131–136
Vashisth A, Nagarajan S (2008) Exposure of seeds to static magnetic field enhances germination and early growth characteristics in chickpea (*Cicer arietinum* L.). Bioelectromagnetics 29(7):571–578
Yusuf KO, Ogunlela AO (2015) Impact of magnetic treatment of irrigation water on the growth and yield of tomato. Not Sci Biol 7(3):345–348
Zude M, Alexander A, Ludders P (1999) Influence of feddha and sugar derivatives on curing iron chlorosis in citrus. Gesunde pflanzen 51:125–129

Socio-hydrological Framework of Farmer-Drought Feedback: Darfur as a Case Study

Nadir Ahmed Elagib, Ammar Ahmed Musa and Hussein M. Sulieman

Abstract This paper attempts to conceptualize key socio-hydrological feedback loops between farmers and drought phenomena. To this end, two-fold aspects are considered for the Darfur region in the west of Sudan. The first one is the understanding of farmers' perception of the major drought conditions, i.e. changes in patterns of rainfall and increase in temperature, in the region in comparison to the observed trends and variability (meteorological records). The second aspect is the identification of the types of concordant adjustments in the farming practices those farmers have made in response to drought conditions. Scheduled interviews composed of 98 farmers formed the basis of establishing the socio-hydrological framework. The interview results are discussed under the reality of human conflict in the context of a prototype cause-effect approach to defining the framework of drivers, pressures, state, impacts and responses (DPSIR) in the farmer-drought system. Climate in the region shows different features of drought, including increasing rainfall variability, changes in wet-season structure to which crop growth is sensitive, and increasing temperatures. Farmers have stronger perception of changes in rainfall patterns than in temperature. They rank the former as the main factor behind the declining crop; however, the results also suggest poor soil fertility as additional important agent. To reduce the drought risk to farming in the region and enhance crop productivity, farmers use a variety of strategies, such as early planting, frequent weeding, diversification of crops, cultivation of early maturing

N.A. Elagib (✉)
Institute for Technology and Resources Management in the Tropics and Subtropics (ITT), Technische Hochschule Köln, Betzdorferstr. 2, 50679 Cologne, Germany
e-mail: elagib@hotmail.com

A.A. Musa
Environment, Health & Safety Department, King Fahd University of Petroleum and Minerals (KFUPM), Dhahran 31261, Saudi Arabia

H.M. Sulieman
Centre for Remote Sensing and GIS, University of Gadarif, P.O. Box 449, 32211 Gadarif, Sudan

O. Abdalla et al. (eds.), *Water Resources in Arid Areas: The Way Forward*, Springer Water, DOI 10.1007/978-3-319-51856-5_27

varieties and soil conservation. However, these steps are challenged mainly, in order of priority, by lack of finance, lack of machinery, invasion of insects and pests and shortage of labour. Urgent support from policy-makers is thus needed to enable an environment for sustainable agricultural practices in the region.

Keywords Socio-hydrology · Drought · Rainfall · Farmers' perception · DPSIR framework · Darfur

1 Introduction

Though socio-hydrology is considered by some researchers (Sivakumar 2012) to be a long-standing discipline of hydrosociology, several studies have recently emerged in the literature building on the definition of socio-hydrology as the science aiming at understanding the dynamics and co-evolution of coupled human-water systems (Sivapalan et al. 2012). Among these studies, which underpin the principles of socio-hydrological systems, one can note the assessment of the changing norms of duty or use of water in irrigation by Wescoat (2013), the historical analysis of co-evolution of humans and water performed by Liu et al. (2014), water sustainability problems (Sivapalan et al. 2014), linking the science and policy in a context of significant environmental and societal change (Gober and Wheater 2014) and the illustration of the balance of water dynamics between agricultural development and environmental degradation (Kandasamy et al. 2014).

Climate issues pertinent to the African regions have been recognized to give more visibility to African global change issues, including among others extreme events and climate variability (Desanker and Justice 2001). These issues are in turn relevant to African water resources, which are plagued by three critical problems, namely natural distribution and reliability of the resources, human population distribution and growth, and intimate link between water and human health (disease) due to the lack of access to safe drinking water. In this context, rainfall is a critical climatic element for traditional farmers who utilize marginal agricultural lands. The close and detailed observations and understanding of its various aspects pave the way for measuring the reliability of rains and the magnitude of risk facing those farmers (El-Tom 1986; Hendrix and Glaser 2007). The farming settings can, thus, be taken successfully within the concept of socio-hydrology.

During the past decade, the region of Darfur has come to the forefront of the worldwide concern and coverage by major media outlets (Parks 2009), owing to the devastation of livelihood of significant proportion of the population in the region resulting from the conflict that flared in 2003. Apart from the ongoing civil war, the climate and environment of the region are already unpleasantly harsh as it triggers humanitarian catastrophes due to natural hazards. Darfur is part of Sudan where water resources assessment is beset by a number of problems, including high population growth, land-locked location, vast area (over 510,000 km^2), remotely separated and inaccessible rural areas, long hours of sunshine which enhance evaporation rate, and large-scale farming and grazing (Omer 2002). Statistics show that the human population in the region increased during the period 1983–2008 by $\sim$2.66 fold, the growth in livestock population was $\sim$2.74 fold from 1973 to 2007 whilst the average cropland area for 2003–2007 increased by 3.44 times that for 1971–1976 (CBS 1993, 2008; Faki et al. 2008; PCC 2009). All of the above factors clearly reflect the high pressures exerted upon the regional water and land resources.

Extensive agricultural and pastoral activities shape the economy of the region. Agricultural production is predominantly rain-fed. These characteristics make the region particularly sensitive to small changes in rainfall in the absence of major permanent irrigation sources. Hence, climate stresses, such as drought and declining rainfall, have the potential to affect the livelihoods of rural residents, increase migration from rural/agricultural areas to towns and cities (rural-urban migration) and increase internally displaced people (IDP) camps, especially that agriculture is heavily relying on rainfall for the provision of water for crops. This is particularly true in view of the low capacity of soils to retain moisture that renders agricultural lands infertile (Hope 2009). Perpetual drought in Darfur has already aggravated the degradation and lessened the fertility of soil through over- and sedentarized-cultivation, especially millet cultivation beyond the agronomic dry boundary leading to diminution of crop yield, grazing as well as depletion of forest (Ibrahim 1978; Mangouri 2006). The number of at-risk people over the next two decades could thus increase (Funk et al. 2011) since the above factors effect land degradation, which in turn could compound the difficulties and threats posed normally by climate (Sivakumar 1992; Balogun 2011). Degradation of vegetation through overgrazing strongly affects the surroundings of settlements, the water supply sources and the lands along the main grazing routes of herds (Ibrahim 1978) and, more recently, the expansion of agriculture into rangeland (Selby and Hoffmann 2014).

Climatic stresses place further threats on the natural resources in areas which are already suffering from severe droughts and land degradation (Balogun 2011). However, Man in Darfur tends to ignore, or at least forgets, the dynamic nature of the physical environment that is capable of causing considerable agony, human suffering and socio-economic complications by extending the developmental activities up to the maximum limit of the allowable carrying capacity of the natural system (El-Tom 1986). Since rainfall is an essential resource for both agricultural and domestic purposes, knowledge of rainfall variation in time and space and the associated drought in the region of Darfur is of great concern for management of water resources and sustainable agriculture. As farming is practiced under arid and semi-arid rain-fed conditions, drought risk in the region is found to significantly reduce the levels of crop yield (Elagib 2014).

The objective of this paper is, therefore, to understand the socio-hydrological framework of traditional Darfurian farmers, who utilize marginal agricultural lands under critical dry conditions. Hence, an answer is given to the question as to "How do they perceive the major climate vagaries that feature drought and respond to them through adjustments?"

2 Conflict in the Region

A better understanding of the recent socio-hydrological framework of the farming system in Darfur can only be provided within the context of the current conflict in the region. To have a broader understanding of the current crises in Darfur, attention should be given to the cumulative political and environmental factors that shaped the events which produced the current situation. According to Manger (2006), the escalation of the contemporary conflict in the region is on one side related to the political development in the country and the central government, which is becoming more and more an independent player in the conflict, rather than playing a neutral and appeasing role. This situation leads to increasing problems between local people. Moreover, the ineffective response by successive governments to recurring drought has contributed to the environmental stress (De Juan 2015).

Barnett and Adger (2007) classify Sudan, among others, as a kind of countries where climate change seems to increasingly undermine human security and induce violent conflict due to the following reasons:

1. the weak capacity of the country to provide the people the opportunities and services that sustain their livelihood,
2. access to education and health care and income is poor and
3. much of the population is dependent for their livelihood on primary resources.

In the course of the current conflict, the resources could be a trigger of conflict. Undisputable competition over land and water resources between farmers and herders has escalated to become one of the main causes of the conflict taking place in the region. In the absence of original owners who fled to IDP camps, a violent dispute occurred within the pastoralist groups over pastures and resources (Satti 2009). It is true that societies in this region coped in the past with conditions of low rainfall and high variability through adaptive strategies, such as agricultural intensification (Fadul 2006), rural-rural migration (Ibrahim 1991) and movement across borders (Elasha and El Sanjak 2009), rather than large-scale violence (Kevane and Gray 2008). However, a gradual loss of mobility in favour of settlement that started since the beginning of the twentieth century has been reinforced by the establishment of infrastructures and permanent drinking water supply (Ibrahim 1988). Peaceful relations and coexistence between southward immigrants and their hosts have been considerably disturbed in the past few decades due in part to drought disasters, resulting in immigrants outnumbering the indigenous population, thus alarming the host populations (Ibrahim 1998).

Linkage between climate and conflict is identified by Young (2009) as a "local conflict livelihood cycle" that he explains as follows. When drought exerts pressure on the livelihood system, local groups tend to adapt by competing on land, pasture and water. In the absence of a strong local government which efficiently manages the access to resources, pressure on the livelihood mechanism increases resulting in conflicts between the competing groups. This undermines further the local governance. The situation complicates further due to degradation of the environment caused by population pressure. As a result, the livelihood of the various population groups would be strained more, and the cycle carries on (Young 2009).

Therefore, the relation between conflict and environmental change in Darfur is best understood as a risk factor, the actual consequences of which are strongly shaped by prevailing political and social contexts (Selby and Hoffmann 2014; De Juan 2015). To deal with risk and uncertainty that dominate livelihood strategies, flexible livelihood adaptation strategy remains one of the optimal solutions for peaceful coexistence of different groups of land users (Chavunduka and Bromley 2011).

3 Data and Methods

The data for this study consist of three sets. First, the paper used a structured interview, which was conducted in August 2013 with 98 randomly-selected farmers in Northern Darfur State and Southern Darfur State, precisely 50 and 48 farmers from each state, respectively. The structured interview was conducted to explore the smallholder's recognition of the drought conditions in the region. Background information collected from famers included gender, age, household size and level of education. The main questions asked to famers were under for major themes, namely agricultural practices, perception of changes in main climate patterns that feature the drought phenomenon, adaptation measures that they try to follow and barriers to their adaptation strategies. Second, the farmers' perceptions of the climate were contrasted to observations of meteorological data from Sudan Meteorological Authority while those of state of crop yield were compared to yield data from the Ministry of Agriculture and Irrigation (MAI 2013). The meteorological data comprise daily rainfall data over 1941–2013 and monthly maximum and minimum temperatures over the period 1941–2012. As lack of dense areal coverage of climatic data is an inherent problem in Sudan, the observed data for this study came from two main stations in the region (Fig. 1), namely El Fasher and Nyala.

Several rainfall behaviours were studied herein. Changing variability of monthly and annual rainfalls was investigated using the coefficient of variation for 30-year running periods. Shifts in the rainfall seasonal start and end were investigated between two periods (1971–1990 and 1991 onwards) by calculating the difference in probability of occurrence of a wet day (defined as the day receiving >1 mm of rain). To overcome the effect of inherent dry spells between falls, running 10-day averages of the probability were calculated prior to obtaining the difference.

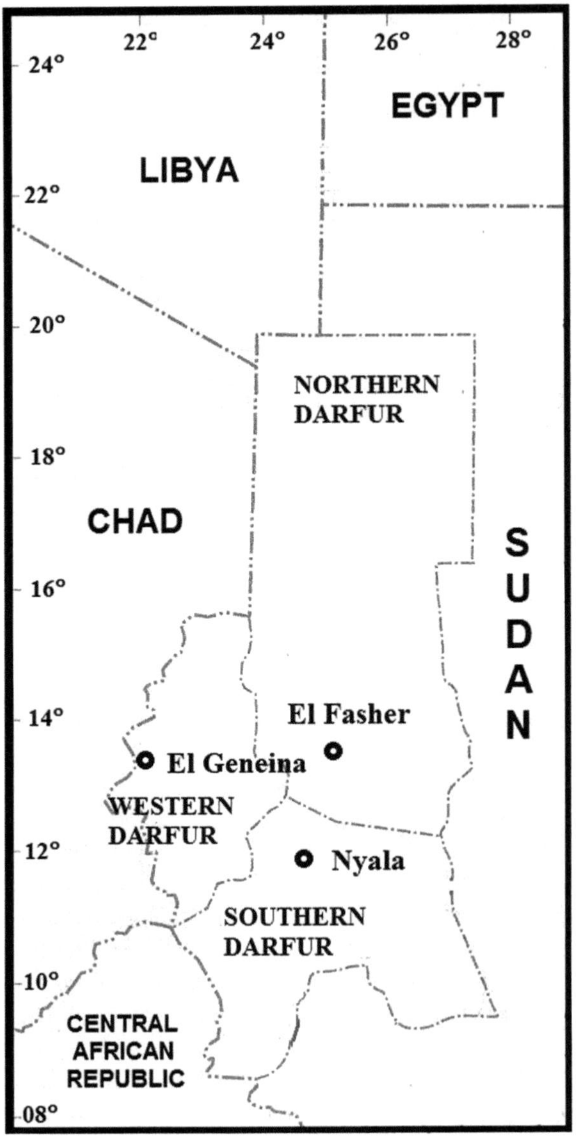

Fig. 1 Map of geographical location of Darfur States and meteorological stations

These two periods were selected on the basis of a wide acceptance that the former represents the drought period and the latter refers to the rainfall recovery period (see e.g. Lebel and Ali 2009). In this study, standardized precipitation index (SPI) was calculated based on e.g. McKee et al. (1993) on time scales of a month (maximum monthly), 3 months (running 3 months starting from May and ending in October) and 12-months (annual). Standardized anomaly indices of temperatures were also developed for the wet (summer: June to October) and dry (winter: January,

February, November and December) seasons. These standardized time series allow investigating the year-to-year anomalies of rainfall and temperatures from the long-term averages. However, the Regime Shift Detector (RSD) software (Rodionov 2004) was used to detect statistically significant shifts in the standardized series before a trend rate has been found. The trend is analysed to find out its direction and significance using the non-parameteric Kendall tau rank correlation test (Kanji 1997). In this paper, the period chosen for trend rate analysis was basically the last portion of the long-term series which showed significant regime shift. The same analysis was performed on the crop yield data.

To contribute to a better understanding of the interactions of farmer and drought in the farming system, the survey is guided by the driving force-pressure-state-impact-response (DPSIR) framework. Major driving forces (D) represent the household characteristics and their perception of the indicators of drought (changes in rainfall patterns and temperature levels). Under the conflict conditions upon land and water, the primary driving forces and the perception of farmers to drought indicators then manifest related barriers to capacity to cope with the drought stress (P). The development of pressures (e.g. finance, machinery, seeds, labour, information, etc.), which represent the processes affecting the resources, produces agents that describe the state (S) behind the consequence on productivity. This consequence is identified as impacts (I) in the form of decline in productivity. As a result, farmers adopt a range of adaptation strategies that are figured out as responses (R).

4 Results and Discussion

In this section, the results obtained from the analysis of the farmers' opinions will be described within the context of the DPSIR framework shown in Fig. 2.

4.1 Drivers

4.1.1 Characteristics of the Respondents

Table 1 shows the household characteristics of the sample. The number of male farmers represented 73% while female farmers constituted 24% of the sample. Sixty one percent of the farmers were in the middle age of 25 to less than 50 years. Eighty three percent of the interviewed farmers had family size of five or more members. In 81% of the cases, the father was the head of the family. The level of education of 60% of the farmers was secondary school or university graduate.

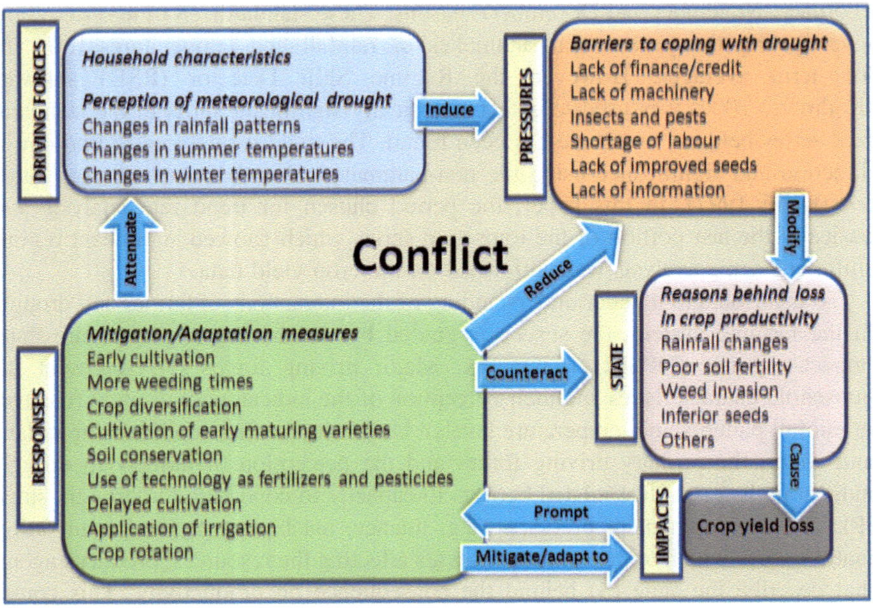

Fig. 2 Farmer-drought feedback in the DPSIR framework

Table 1 Household characteristics of the farmers

Item	Description	Percentage of respondents
Gender	Male	73
	Female	24
	No response	2
Age group	Young (<25 years)	13
	Middle aged (25–50 years)	61
	Old ($\geq$ 50 years)	23
	No response	2
Household size	Small (1–4 Persons)	9
	Medium (5–8 Persons)	43
	Large (>8 Persons)	40
	No response	8
Head of the family	Father	81
	Mother	9
	Brother	3
	No response	7
Education level	Illiterate	5
	Primary	20
	Intermediate	7
	Secondary	21
	Graduate	39
	No response	7

4.1.2 Perception of Recent Changes in Rainfall

As regards rainfall, changes in rainfall indicators of drought were recognized by most of the respondents. However, the recognition of details of these changes varied among the farmers. Out of the 98 farmers, 90% of the respondents across the study area did think that changes in rainfall patterns have occurred. The main perceptions of change were in the order of erratic pattern, later onset of the rainy season and decrease of rainfall amount. These changes were perceived respectively by 55, 52 and 44% of the farmers. Only 6% of the interviewed farmers believed that rainfall has increased, and 2% had the notion of early start of the rainy season. Citation of decreasing amount of rainfall by farmers were reported by farmers in many countries of Africa, such as Ethiopia, Kenya and South Africa (Ovuka and Lindqvist 2000; Bryan et al. 2009; Rao et al. 2011; Gebrehiwot and Veen 2013).

As explained below, it is unequivocal from the comparison of the above perceptions with scientific observations that farmers' knowledge of changes and variability of rainfall is advanced. Table 2 gives the results of the regime shift test for the rainfall time series. Although all time series exhibit significant regime shift, only one time series (May-July) shows significant trend following the shift year. Thus, it would be necessary to view the anomalies individually during the recent past. Figure 3 shows the SPIs for El Fasher and Nyala. First, the year-to-year variability is striking for both stations. El Fasher in Northern Darfur has seen many negative SPIs during the 21st century, especially during the main rainy season (July-September), May-July, June-August and August-October, which also reflect in the annual time series. These SPIs are larger in absolute magnitude than the counterpart positive SPIs. On the other hand, Nyala's time series show in general opposite behaviour despite one or two years with salient below-average rainfalls.

In Fig. 4, the 30-year running coefficient of variation indicates no change in the inter-annual variability of annual rainfall during the last 40 years. In contrary to this picture, increased variability is very clear during the beginning and end of the rainy season. This trend is exhibited in, for instance, May, September and October for El Fasher and in May, June and October for Nyala.

Analysis of the occurrence of a rainy day (Fig. 5) demonstrates an increase in the corresponding probability during June, early July and the whole of August for El Fasher. There is almost insignificant change in the probability otherwise during the

Table 2 Year of shift in the SPI time series

Series	El Fasher		Nyala	
	Year	p	Year	p
May-Jul	1994[a]	0.011	1976	0.012
Jun-Aug	1991	0.013	2003	0.032
Jul-Sep	1972	0.000	1997	0.007
Aug-Oct	1972	0.000	1999	0.003
Annual	1972	0.000	2003	0.008
Maximum	1971	0.000	2003	0.005

[a]1994–2013: Significant decreasing trend ($p = 0.032$)

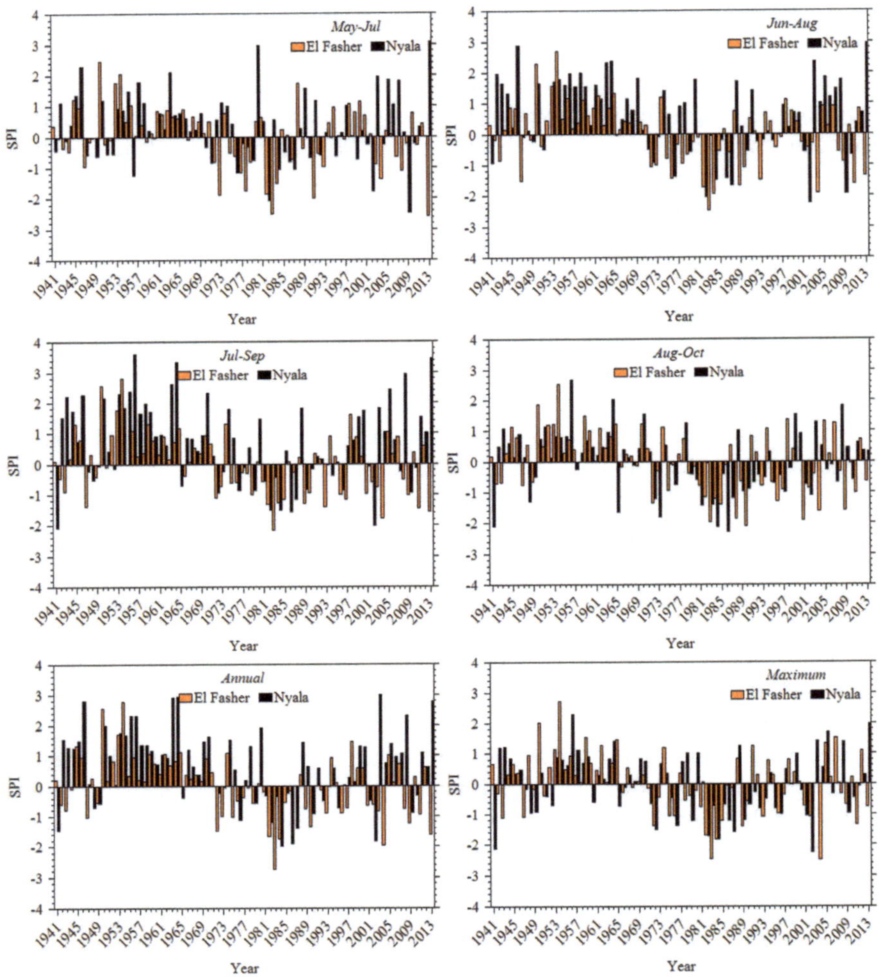

Fig. 3 Temporal evolution of the standardized precipitation index for the region

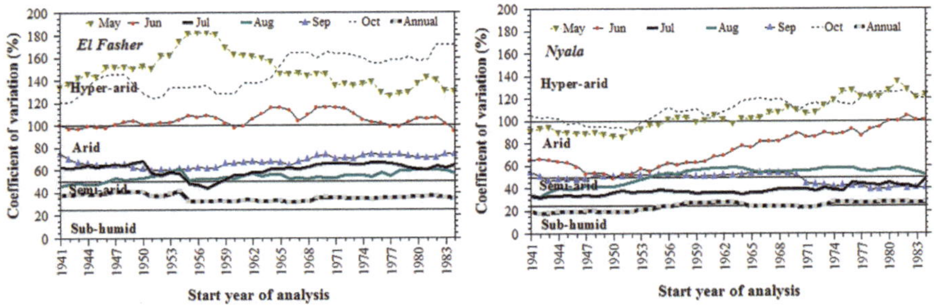

Fig. 4 30-year running coefficient of variation

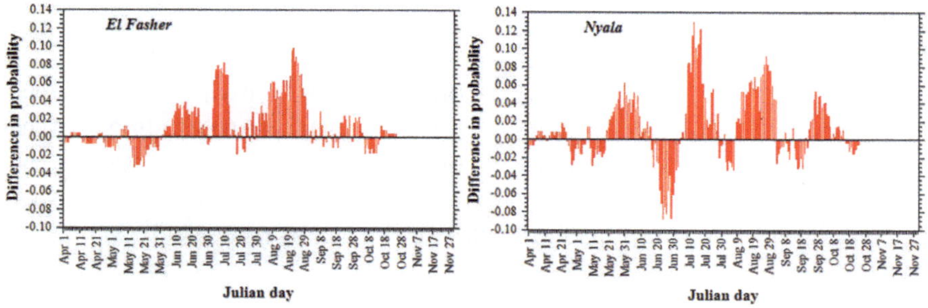

Fig. 5 Difference in 10-day running average probability of occurrence of a wet day (>1 mm) between dry period 1971–1990 (dry) and the rainfall recovery period 1991 onwards calculated as the probability for the recovery period minus that of the dry period. See text for more details

rest of the season. However, the rainfall in Nyala shows noticeable intra-annual variability of occurrence of a rainy day. This is revealed by two distinct behaviours. Firstly, there is an increase in the probability of occurrence: (1) of a false start in May and early June, (2) during the second and third dekads of both July and August and (3) in the first half of October. Secondly, other periods of the rainy season indicate less probable rainy day, particularly during the onset of the season (second half of June). This latter result points out late onset of the rainy season in Nyala, i.e. shrinking length of the season, and also more occurrence of dry spells during the season in the last two decades. Farmers in the African Sahel region attach great significance to the timing of start of the rainy season for deciding when to plant crops (Rao et al. 2011; Bussmann et al. 2016) since it is important for maximizing yield (Marteau et al. 2011; Waha et al. 2013). In the Sahel region of Sudan, including Darfur, Elagib (2015) reported significant negative impact of drought on the yield when it occurs during the early rainy season.

4.1.3 Perception of Recent Changes in Temperatures

There was a general perception among the farmers that levels of temperatures have changed recently. For summer temperatures, almost all farmers (93%) said they have observed changes while the other 7% denied any occurrence of change. Among the 98 farmers, 73% believed that the daytime temperatures have increased while 15% noticed a decrease in daytime temperatures. On the other hand, night-time temperatures were understood by 44% of the informant farmers to have decreased as opposed to 12% of the informants who observed increasing nighttime temperatures.

The results reveal that a large number of participants (82%) had the view of changing winter temperatures, 17% did not notice any change recently whereas 1% did not respond. Commonly observed trends were decreasing daytime and night-time temperatures during the summer by 37 and 43% of the farmers, respectively.

Table 3 Year of shift in the mean temperature time series

Series	Season	El Fasher		Nyala	
		Year	p	Year	p
Maximum	Dry	1998	0.001	1961	0.000
	Wet	1996	0.000	1972	0.000
Minimum	Dry	1993[a]	0.000	1998	0.000
	Wet	1992	0.000	1972	0.000

[a]1993–2013: Significant increasing trend (p = 0.008)

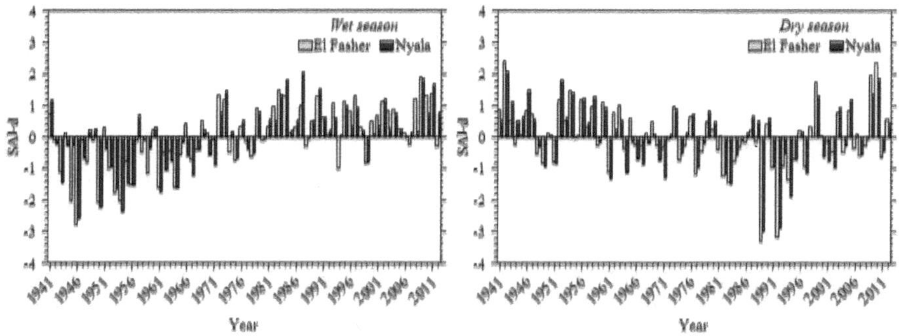

Fig. 6 Temporal evolution of the standardized daytime temperature anomaly index (SAI-d) for the wet (summer) and dry (winter) seasons

Moreover, increasing daytime and nighttime temperatures during the winter were perceived by 30% and only 9% of the respondents, respectively.

In contrasting these perception results with actual temperature observations to see how knowledgeable the farmers in this region are of the temperature trends, the mean temperature levels during the wet season (summer months: June to October) and the dry season (winter months: January, February, November and December) are of more importance. Table 3 shows evidence of shifts in the temperature regime in all time series. However, only the dry minimum temperature has exhibited significant positive trends from the date of shift onwards. Similar to rainfall, one has to examine the year-to-year anomalies in quest for any recent warming or cooling phenomenon. Figures 6 and 7 display the corresponding temperature anomalies for both seasons during the daytime and nighttime, respectively. The daytime temperature anomalies of the wet season reveal successive positive anomalies since the early 1980s in comparison to the early 1970s for the nighttime series. Few years in the second half of the 1980s and early 1990s were the exception of the nighttime temperatures only at the station of El Fasher in Northern Darfur. In the dry part of the year, on the other hand, the time series of the daytime show above- and below-average anomalies throughout the data period; nevertheless, the above-average anomalies tend to be of higher magnitude in absolute term since the mid-1990s. The temperatures during nights show continuous, higher-than-average levels since the early 1990s.

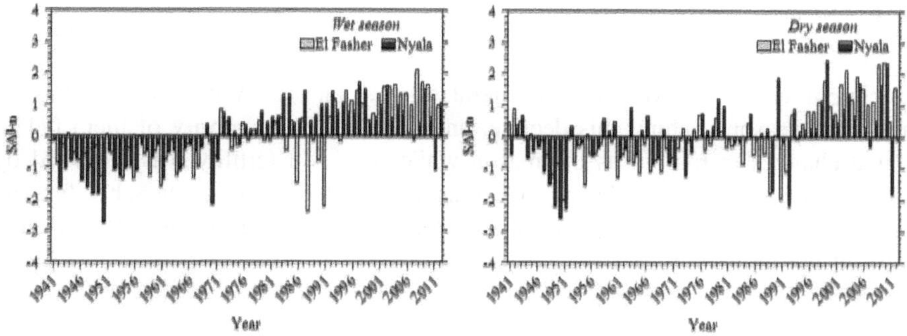

Fig. 7 Temporal evolution of the standardized nighttime temperature anomaly index (SAI-n) for the wet (summer) and dry (winter) seasons

The above results mean that the climate in the study areas has become constantly warmer during the last decades. Such results further suggest continued drying conditions as reported earlier by Elagib (2009) and Elagib and Elhag (2011). However, the majority of the farmers in this region seem to have poor realization of temperature levels. Elsewhere (e.g. Ethiopia), farmers' understanding of trends in temperature seem to correspond better with the climatic data (Gebrehiwot and Veen 2013).

4.2 *Pressures*

Smallholder farmers face multiple barriers to their adaptation strategies against climate vagaries (Fig. 2), including lack of finance/credit, lack of machinery, invasion of insects and pests and shortage of labour. Lack of finance/credit was seen by two-third (66%) of the farmers as the most critical barrier. Gebrehiwot and Veen (2013) reported lack of access to financial resources as the main constraint to adjusting to changing climate in the Ethiopian Highlands. Lack of machinery came second in the list. This was reported by 53% of the Darfurian farmers. Invasion of insects and pests came third (reported by 51% of the farmers) followed by shortage of labour (43%). The need for labour is never a major factor hindering adaptation to changing climate in a neighbouring country like Ethiopia (Bryan et al. 2009; Gebrehiwot and Veen 2013), likely due to the differential population number between the two countries. Despite the demand for additional labour, African farmers are unlikely to be attracted by highly labor-intensive practices (Thapa and Yila 2012), avoiding increases in the cost of farming activities. Additional less critical constrains in the present study area were lack of improved seeds and lack of information, reported respectively by only 20 and 14% of the respondents.

4.3 State

Decline in crop yield was markedly iterated by all farmers. When they were asked about the reasons behind this decline, an overwhelming majority of them (71%) stated changes in rainfall as the reason while poor soil fertility was indicated by 47% of the farmers. Similar observations were noted by western Sahel farmers (Zorom et al. 2013). Weed invasion and inferior seeds were mentioned by 18 and 12% of the farmers, respectively.

4.4 Impact

The observation made by farmer that yield had declined is strikingly supported by the yield data of the two main crops for the two states, as shown in Fig. 8. Since the beginning of this century, sorghum yield has fluctuated, with many years showing below long-term average. Marked decline is particularly seen for Northern Darfur during the last 15 years; however, Southern Darfur did record slight below-average yields in the last couple of years of the time series. Millet yield, on the other hand, has been declining dramatically since the early 1980s. It has never improved in Northern Darfur. In Southern Darfur, only in 9 years during the last 30 years had yields been higher than the long-term average, but again the yield levels were down in the last three years of record.

4.5 Responses

Based on the perception among farmers that climatic conditions have changed over the recent years, those farmers have made changes in their farming practices (Fig. 2) to reduce their vulnerability to drought conditions or to adjust to changes in

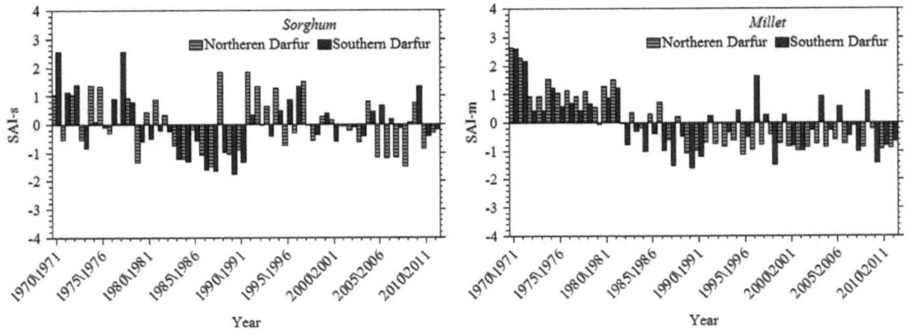

Fig. 8 Temporal evolution of the standardized anomaly index for sorghum (SAI-s) and millet (SAI-m) for the two states

climatic conditions. The farmers considered to cultivate earlier, as indicated by 39% of the respondents, to weed more frequently (30% of respondents), to diversify crops (29%), to cultivate early maturing varieties (28%) and to implement soil conservation practices (27%). Sowing date adapted to a shifted start of the rainy season is already implemented by farmers in some African countries (e.g. Waha et al. 2013). For eastern Sudan, Bussmann et al. (2016) recommended relatively earlier sowing date in order for farmers not to miss earlier rainy days or face more frequent dry spells later in the growing season. While adapting the sowing date is considered by farmers in many African countries as a strategic option to changes in climate (Hassan and Nhemachena 2008), shifting sowing date is seen ineffective for condition of short rainy season and frequent dry spells (Tingem and Rivington 2009). Several types of crops are grown in Darfur. Millet is the main staple crop in Darfur while sorghum comes next. However, cash crops, such as peanut and sesame are also grown in the region. Our results show that these crops are cultivated respectively by 77, 57, 50 and 17% of the sample of farmers interviewed in the present work. In the western part of the African Sahel, a vast majority of farmers adopt soil conservation techniques, and a much lesser number of them use shorter-cycle rainfed crop varieties in case of persistent drought (Zorom et al. 2013). Other options reported by the farmers of in our study area include the use of technologies as fertilizers and pesticides (17%), delayed cultivation (16%), application of irrigation (13%) and crop rotation (12%). It is worth mentioning that some farmers having lands along wadis use surface (runoff) water for pump irrigation following the rainy period of the year. Such practices, however, seem to be confined to vegetables readily marketable in the nearby sites. In the occasion of good rainy season, rainwater replenishes groundwater aquifers, thus allowing wealthy farmers to practice groundwater irrigation. However, this kind of farmers is of small number and such practices are only of small scale. Hence, the potential for large-scale irrigation apart from the wadi runoff irrigation following a good rainy season is too little in view of the fact that the majority of farmers of the region are poor.

5 Conclusions and Policy Implications

The region of Darfur represents one extreme example of the African drylands. Climatic factors contribute to the devastation of livelihood in Darfur. The whole population of Darfur usually suffers from food shortage when drought occurs (Ibrahim 1988) due to consequent reduction of water resources and crop yield.

The data presented in this study illustrate that the farming households are concerned about changes in climate elements that describe drought, especially rainfall, and take steps to respond to these changes in a socio-hydrological essence. This rainfed farming-drought system further interacts in a DPSIR framework. Somehow, there are pitfalls in the farmer understanding and knowledge of temperature trends. Farmers are, on the other hand, well aware of changes in rainfall patterns than in temperatures. Farmers seem to be better informant of rainfall variability and rainfall patterns during the critical time for crop than of long-term trends. The improvement of understanding of the changes and fluctuations of the critical climatic elements in this region is particularly a crucial prerequisite for enhancing vulnerability assessments, mitigation and adaptive capacities and ability to implement response options for the challenging water- and agricultural-management activities. However, these improvements require the enhancement of the data network, monitoring facilities and early warning systems in the region.

This work has shed light on a range of responses that rural societies have identified and implemented in the face of forms that drought phenomena take. Primary agricultural management practices include early cultivation, more weeding times, crop diversification of early maturing varieties and soil conservation. Adoption of technologies, such as fertilizer, pesticides and irrigation, are found to be among the least favoured practices.

Notwithstanding the wealth of knowledge that the farmers have developed from their experiences and the presented responses, farmers communities in the region have highlighted several constraints that limit their ability to adapt, among which changes in rainfall patterns and poor fertility of the soil were given high ranks. These observations on the farmers' perceptions offer support to the statement made by Rao et al. (2011) that "farmers' perceptions about climate are a combination of various factors that affect production and are not entirely based on climatic observations". Thus, overcoming bitter rivalry over limited land and water resources necessitates the involvement of the different stakeholders in formulating pertinent solutions, strategies and laws. This is particularly true in view of the fact that drought occurrences in a region with high rainfall variability, such as Darfur, will continue to be unpredictable (Tiffen and Mortimore 2002) given the wide range of drought forms demonstrated in the present work.

References

Balogun OL (2011) Sustainable agriculture and food crisis in sub-Sahara Africa. Chapter 20. In: Behnassi M, Draggan S, Yaya S (eds) Global food insecurity: rethinking agricultural and rural development paradigm and policy, pp 283–297

Barnett J, Adger WN (2007) Climate change, human security and violent conflict. Polit Geogr 26:639–655

Bryan E, Deressa TT, Gbetibouo GA, Ringler C (2009) Adaptation to climate change in Ethiopia and South Africa: options and constraints. Environ Sci Policy 12(4):413–426

Bussmann A, Elagib NA, Fayyad M, Ribbe L (2016) Sowing date determinants for Sahelian rainfed agriculture in the context of agricultural policies and water management. Land Use Policy 31(52):316–328

Central Bureau of Statistics (CBS) (1993) Guidebook of population and development. Dar Al-Azhar, Khartoum, Sudan (in Arabic)

Central Bureau of Statistics (CBS) (2008) Statistical year book for the year 2007. Currency Printing Press Company Ltd, Khartoum, Sudan

Chavunduka C, Bromley DW (2011) Climate, carbon, civil war and flexible boundaries: Sudan's contested landscape. Land Use Policy 28:907–916

De Juan A (2015) Long-term environmental change and geographical patterns of violence in Darfur, 2003–2005. Polit Geogr 45:2–33

Desanker PV, Justice CO (2001) Africa and global climate change: critical issues and suggestions for further research and integrated assessment modeling. Clim Res 17:93–103

Elagib NA (2009) Assessment of drought across central Sudan using UNEP dryness ratio. Hydrol Res 40(5):481–494

Elagib NA (2014) Development and application of a drought risk index for food crop yield in Eastern Sahel. Ecol Indic 43:114–125

Elagib NA (2015) Drought risk during the early growing season in Sahelian Sudan. Nat Hazards 79(3):1549–1566

Elagib NA, Elhag MM (2011) Major climate indicators of ongoing drought in Sudan. J Hydrol 409 (3):612–625

Elasha BO, El Sanjak A (2009) Global climate changes: impacts on water resources and human security in Africa. In: Leroy M (ed) Environment and Conflict in Africa: Reflections on Darfur. University for Peace, Africa Programme, Addis Ababa, Ethiopia, pp 27–40

El-Tom MA (1986) Climate, environment and development in the Sudan. GeoJournal 12(4): 399–402

Fadul AA (2006) Natural resources management for sustainable peace in Darfur. In Proceedings of conference on environmental degradation as a cause of conflict in Darfur, Khartoum, December 2004. University for Peace, Switzerland, pp 33–44

Faki H, El-Dukheri I, Mekki M, Peden D (2008) Opportunities for increasing livestock water productivity in Sudan. Proceedings of the CPWF 2nd international forum on water and food, Addis Ababa, Ethiopia, 9–14 November 2008. Retrieved 15 April 2011 from http://www.ifwf2.org/addons/download_presentation.php?fid=1059

Funk C, Eilerts G, Verdin J, Rowland J, Marshall M (2011) A climate trend analysis of Sudan. Geological Survey Fact Sheet 2011–3072, US. Retrieved 20 July 2012 from http://pubs.usgs.gov/fs/2011/3072

Gebrehiwot T, Veen VDA (2013) Farm level adaptation to climate change: the case of farmer's in the Ethiopian Highlands. Environ Manage 52(1):29–44

Gober P, Wheater HS (2014) Socio-hydrology and the science–policy interface: a case study of the Saskatchewan River basin. Hydrol Earth Syst Sci 18(4):1413–1422

Hassan R, Nhemachena C (2008) Determinants of African farmers' strategies for adapting to climate change: multinomial choice analysis. Afr J Agric Resour Econ 2(1):83–104

Hendrix CS, Glaser SM (2007) Trends and triggers: climate, climate change and civil conflict in Sub-Saharan Africa. Polit Geogr 26:695–715

Hope KR Sr (2009) Climate change and poverty in Africa. Int J Sustain Dev World Ecol 16(6): 451–461

Ibrahim F (1978) Anthropogenic causes of desertification in western Sudan. GeoJournal 2(3): 243–254

Ibrahim FN (1988) Causes of the famine among the rural population of the Sahelian zone of the Sudan. GeoJournal 17(1):133–141

Ibrahim FN (1991) The role of rural-rural migration as a survival strategy in the Sahelian zone of the Sudan—a case-study in Burush, N. Darfur. GeoJournal 25(1):31–38

Ibrahim F (1998) The Zaghawa and the Midob of North Darfur—a comparison of migration behaviour. GeoJournal 46:135–140

Kanji GK (1997) 100 Statistical tests. SAGE Publications, London, p 216

Kevane M, Gray L (2008) Darfur: rainfall and conflict. Environ Res Lett 3(3). doi:10.1088/1748-9326/3/3/034006

Lebel T, Ali A (2009) Recent trends in the central and western Sahel rainfall regime (1990–2007). J Hydrol 375:52–64

Liu Y, Tian F, Hu H, Sivapalan M (2014) Socio-hydrologic perspectives of the co-evolution of humans and water in the Tarim River basin, Western China: the Taiji-Tire model. Hydrol Earth Syst Sci 18(4):1289–1303

Kandasamy J, Sounthararajah D, Sivabalan P, Chanan A, Vigneswaran S, Sivapalan M (2014) Socio-hydrologic drivers of the pendulum swing between agricultural development and environmental health: a case study from Murrumbidgee River basin, Australia. Hydrol Earth Syst Sci 18(3):1027–1041

Manger L (2006) Resource Conflict as a Factor in the Darfur Crisis in Sudan. Colloqueinternational "Les frontières de la question foncière – At the frontier of land issues", Montpellier

Mangouri HA (2006) Combating desertification: experience from Umm Kaddada District in East Darfur. In: Proceedings of conference on environmental degradation as a cause of conflict in Darfur, Khartoum, December 2004. University for Peace, Switzerland, pp 46–55

Marteau R, Sultan B, Moron V, Alhassane A, Baron C, Traore′ SB (2011) The onset of the rainy season and farmers' sowing strategy for pearl millet cultivation in Southwest Niger. Agric Meteorol 151(10):1356–1369

McKee TB, Doesken NJ, Kleist J (1993) The relationship of drought frequency and duration to time scales. The proceedings of the 8th conference on applied climatology, 17–22 January 1993. Anaheim, California, pp 179–184

Ministry of Agriculture and Irrigation (MAI) (2013) Time series of area planted, harvested, production and yield of the main food and oil crops by production centre and type of irrigation 1970/1971-2011/2012, vol 2, 2nd edn. General Administration of Planning and Agricultural Economics, Department of Agricultural Statistics

Omer AM (2002) Water-Supply Introduction Management in Sudan: 2000 onwards (Abridged). J Chartered Instit Water Environ. Manage 16:25–27

Ovuka M, Lindqvist S (2000) Rainfall variability in Murang'a District, Kenya: meteorological data and farmers' perception. Geogr Ann A, Phys Geogr 82(1):107–119

Parks L (2009) Digging into google Earth: an analysis of "Crisis in Darfur". Geoforum 40:535–545

Population Census Council (PCC) (2009) 5th Sudan population and housing census—2008. Khartoum, Sudan

Rodionov SN (2004) A sequential algorithm for testing climate regime shifts. Geophys Res Lett 31(9)

Rao KPC, Ndegwa WG, Kizito K, Oyoo A (2011) Climate variability and change: farmer perceptions and understanding of intra-seasonal variability in rainfall and associated risk in semi-arid Kenya. Exp Agr 47(02):267–291

Satti YH (2009) Pastoralists, land rights and migration routes: the case of West Darfur State. In: Leroy M (ed) Environment and Conflict in Africa: reflections on Darfur. University for Peace, Africa Programme, Addis Ababa, Ethiopia, pp 259–270

Selby J, Hoffmann C (2014) Beyond scarcity: rethinking water, climate change and conflict in the Sudans. Glob Environ Change 29:360–370

Sivakumar B (2012) Socio-hydrology: not a new science, but a recycled and re-worded Hydrosociology. Hydrol Process 26(24):3788–3790

Sivakumar MVK (1992) Climatic change and implications for agriculture in Niger. Clim Change 20:297–312

Sivapalan M, Savenije HH, Blöschl G (2012) Socio-hydrology: a new science of people and water. Hydrol Process 26(8):1270–1276

Sivapalan M, Konar M, Srinivasan V, Chhatre A, Wutich A, Scott CA, Wescoat JL Rodríguez-Iturbe I (2014) Socio-hydrology: use-inspired water sustainability science for the Anthropocene. Earths Future 2(4):225–230

Thapa GB, Yila OM (2012) Farmers' land management practices and status of agricultural land in the Jos Plateau, Nigeria. Land Degrad Dev 23(3):263–277

Tiffen M, Mortimore M (2002) Questioning desertification in dryland sub-Saharan Africa. Nat Resour Forum 26(3):218–233

Tingem M, Rivington M (2009) Adaptation for crop agriculture to climate change in Cameroon: turning on the heat. Mitig Adapt Strat Glob Change 14(2):153–168

Waha K, Müller C, Bondeau A, Dietrich JP, Kurukulasuriya P, Heinke J, Lotze-Campen H (2013) Adaptation to climate change through the choice of cropping system and sowing date in sub-Saharan Africa. Glob Environ Change 23(1):130–143

Wescoat JL Jr (2013) Reconstructing the duty of water: a study of emergent norms in socio-hydrology. Hydrol Earth Syst Sci 17:4759–4768

Young H (2009) The conflict-livelihoods cycle: reducing vulnerability through understanding maladaptive livelihoods. In: Leroy M (ed) Environment and conflict in Africa: reflections on Darfur. University for Peace, Africa Programme, Addis Ababa, Ethiopia, pp 193–209

Zorom M, Barbier B, Mertz O, Servat E (2013) Diversification and adaptation strategies to climate variability: a farm typology for the Sahel. Agr Syst 116:7–15

Water and Energy Use Efficiency of Greenhouse and Net house Under Desert Conditions of UAE: Agronomic and Economic Analysis

Abdelaziz Hirich and Redouane Choukr-Allah

Abstract The GCC (Gulf Cooperation Council) countries are considered one of the most water scarce region in the world, and facing over the coming years the most severe intensification of water scarcity in history. Protected agriculture area in the GCC countries is close to 13,000 ha and most of it are using Pad-fan cooling system which lead to high energy and water consumption. This research aims to assess the water and energy use efficiency between a high technology greenhouse equipped by pad-fan and sun screen system and a low technology net house equipped by a mist system. Three crops were cultivated, cherry tomato and sweet pepper under greenhouse and cucumber under net house. Greenhouse presented the highest water consumption used for cooling process. In fact, cooling consumes 2.6 and 3.5 times more water than the required irrigation water for sweet pepper and cherry tomato respectively. However, the fogging system in the net house was consuming less water, about 75% of consumed irrigation water used for cucumber. Data related to energy use were tremendously high where greenhouse consumed 32 times the energy used under net house. This study showed also that cooling cost in the total production cost is much higher and heavier under greenhouse resulting in high production cost and loss of competitiveness of the local product in the market where imported products seems to be more competitive than local produced products. Therefore, there is a need to improve energy and water use efficiency in the protected agriculture in GCC region and to reduce the water and energy footprint under protected agriculture in GCC region.

Keywords Yield · Cucumber · Cherry tomato · GCC region · Net house

A. Hirich (✉) · R. Choukr-Allah
International Center for Biosaline Agriculture, Dubai, UAE
e-mail: h.aziz@biosaline.org.ae

O. Abdalla et al. (eds.), *Water Resources in Arid Areas: The Way Forward*,
Springer Water, DOI 10.1007/978-3-319-51856-5_28

1 Introduction

Middle East countries are most water-scarce countries in the world, such as Saudi Arabia and Jordan, have per capita annual water resources less than 200 m^3. Overall, it is also expected that by 2025, due to population increase, the regional average water availability is projected to be just over 500 m^3 per person per year (Abdel-Dayem and McDonnell 2012). The overall objective of protected cultivation is to modify the natural environment by practices or structures to achieve optimal productivity of crops by enhancing yield (Wittwer and Castilla 1995). The GCC (Gulf Cooperation Council) countries are considered one of the most water scarce region in the world, and facing over the coming years the most severe intensification of water scarcity in history. Agriculture is the sector using by far the majority of available fresh water resources (>85%) of whish 92% is used for dates and forages production (Kotilaine 2010). Agriculture is the prime target for water conservation efforts, as it plays an important socio-cultural role (heritage) and within food security considerations. So, improving water management, performances and productivity in major agricultural systems is major issue within the strategy of most gulf countries. The Arabian Peninsula is also high food-importing, with abundant solar energy conditions is highly conducive to the use of renewable-energy (solar) (Alnaser and Alnaser 2011; El Katiri and Husain 2014). Protected horticulture can offer problem-solving approaches to challenges involving water and food supplies. Cultivation on substrates under protected conditions is one of the spearheads in this process (Wittwer and Castilla 1995).

Greenhouse production is considered as the most water conservative solution in the agriculture sector. In fact, protected crop production is now a growing reality throughout the world with an estimated about 4 Millions ha of greenhouses spread over all the continents (Choukr-Allah 2015). The degree of sophistication and technology depends on local climatic conditions and the socio-economic environment. The greenhouse production development in northern Europe stimulated the expansion of protected agriculture in other areas, including the Mediterranean, North America, Oceania, Asia and Africa, with various rates and degrees of success. It has been shown that a mere transposition of north European solutions to other parts of the world is not a valid process. Each environment requires further research, development, extension, training and new norms of application to meet local requirements.

The protected agriculture in the GCC countries is close to 13,000 ha and due to hot climate conditions the greenhouses are cooled which lead to high energy and water consumption (Al-Nasser and Bhat 1998). Greenhouse energy consumption is the largest component of the system's environmental impact, and this is true in particular during the hot season (May to September). For example the energy consumption in the greenhouse cooling in the mediterranean region is about 100,000 kWh/ha/year which leads to high energy cost (Kittas et al. 2012). This number should be bigger under GCC conditions where temperatures during the summer region are much higher than Mediterranean regions. Goals for reductions in energy consumption and a larger contribution from sustainable energy sources are of great importance for the Greenhouses growers.

In UAE, the number of greenhouses increased over the period 2005–2011 by 48% and 14,777 had been installed by 2011. This was accompanied by a 78% increase in the area so that greenhouses covered 493 ha in 2011. There were some regional differences, however, Abu Dhabi registering an increase in number while all other regions showing a reduction. The total crop production was 2.1 million tons in 2010 and 74% of this was from forages and field crops. Fruit trees contributed 19% while vegetables accounted for only 7%. The situation of protected agriculture changed significantly in 2011 when total production fell to 1.2 million tons and this may be due to salinity and water scarcity problems which call for better use of saline water and fresh water saving (NBS 2013). UAE is very dependent to foreign markets for its needs in fruits and vegetable, it imports 62 and 47% of vegetables and fruits needs respectively which indicates that there is a great potential to increase the horticultural production in UAE as well as the GCC region (Woertz et al. 2008).

The aims of this study is to compare between two cropping system, high-tech greenhouse cultivated with high cash crop cherry tomato and low tech net house cultivated with low cash crop cucumber. The comparison consists of yield, water and energy use and economic benefice.

2 Materials and Methods

2.1 Structure Description

The greenhouse and net house area is 560 m^2 each with two spans, each span has a width of 8 m, length of 35 m and height of 5 m (Fig. 1). The net house is equipped with a mist system consisting of nozzles with an hourly discharge of 32 l/h. In order to reduce the temperature and evaporation a shade net was installed above the mist system. The used net is an insect proof net with stitch dimensions equal to 1 × 0.5 mm. while the greenhouse is cooled through fan-pad system and equipped by automatic sun screen system. Both greenhouse and net house contain close drainage system where drainage water is recycled.

2.2 Climate Control

Climate under greenhouse and net house is controlled by computer where the temperature and humidity are monitored through Synopta software. Both greenhouse and net house are equipped by temperature and humidity sensors. The start conditions of the mist and fan-pad system set in the Hortimax system are presented in Table 1.

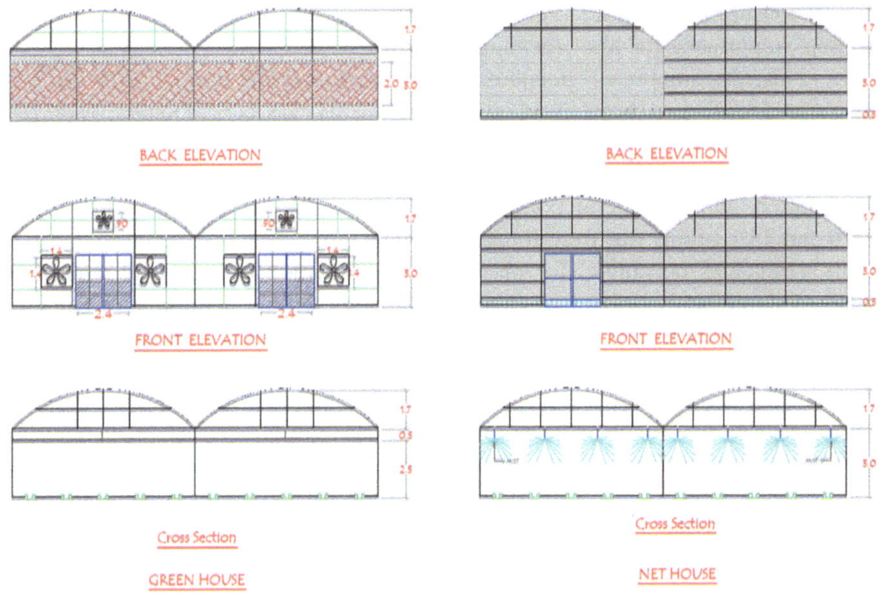

Fig. 1 Greenhouse and net house technical specifications

Table 1 Start conditions of cooling system in net house and greenhouse

Net house	Greenhouse
• Temperature set point for fogging: 29 °C • Relative humidity set point: 75% • Min air temperature for fogging: 15 °C • Min fogging duration: 20 s • Max fogging duration: 30 s • Min pause between 2 fogging cycles: 300 s • Start fogging at 10:00 a.m • Stop fogging at 17:00 p.m	• Cooling temperature set point: 26 for cherry tomato and 27 °C for sweet pepper • Relative humidity set point: 75% • Min temperature for top fan: 30 °C

2.3 *Fogging Operation*

Figure 2 shows the outside and inside net house temperature variation during the day. It is obvious that during the fogging period (10:00–17:00) the mist system with the shade net reduced temperature by about 6 °C. The shade net without mist system reduced temperature by only 3 °C. Under hot climate it is the high temperatures which cause more damage to cucumber especially when temperature exceeds 40 °C which leads to plant wilting and water loss. An appropriate irrigation management accompanied with fogging system can reduce significantly the heat damage on cucumber plants and thus avoid yield losses.

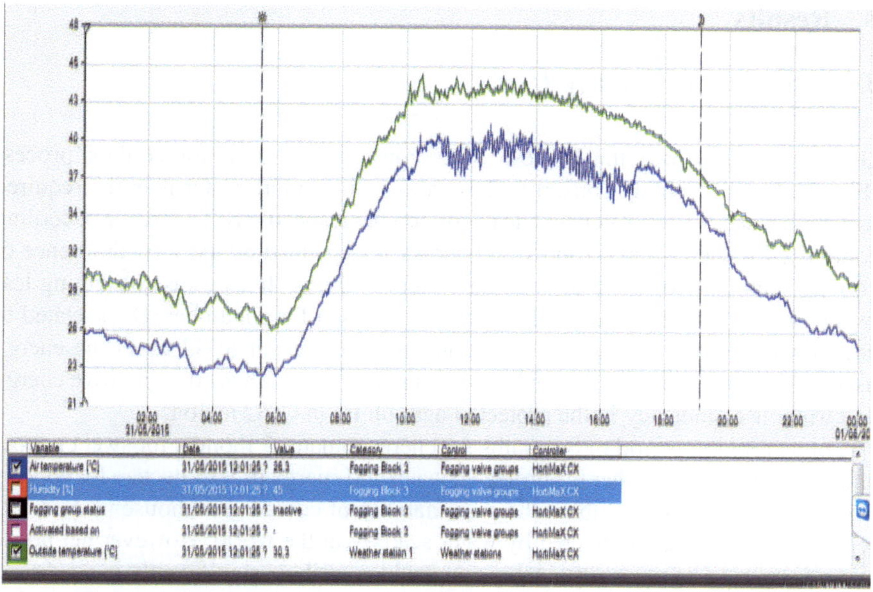

Fig. 2 Net house and outside temperature variation during the day

2.4 Growing Medium and Technical Practices

The whole system was controlled by computer through the Hortimax system. Irrigation and fertigation as well as climate control under greenhouse and net house were monitored automatically via Synopta software. Experimental design consisted of a completely randomized block design with 4 replications. Red dunes sand mixed with peat moss (2/3 sand + 1/3 peat) has been used as substrate in a polystyrene pot of 6 L. In the bottom of pots a layer of 3 cm has been filled with gravel to facilitate drainage. For plant fertilization the modified hoagland solution has been supplied for plants (Hoagland and Arnon 1950; Cooper 1988; Hochmuth and Hochmuth 2001). Drainage water was recycled and treated through a UV system (VITALITE) before reuse. Irrigation scheduling was carried out automatically based on accumulated radiations. Agronomic practices for both cucumber under net house and cherry tomato under greenhouse were carried out according to commercial farming practices.

3 Results

3.1 Water and Energy Use

Greenhouse presented the highest water consumption used for cooling process (Fig. 3). In fact, cooling consumes 2.6 and 3.5 times more water than the required irrigation water for sweet pepper and cherry tomato respectively. Deference of cooling water use for cherry tomato and sweet pepper is explained by the 1 °C difference of cooling setpoint. However the fogging system in the net house was consuming less water, about 75% of consumed irrigation water used for cucumber. Data related to energy were tremendously high where greenhouse consumed 32 times the energy used for fogging system in the net house. Therefore, there is a need to improve energy and water use efficiency in the protected agriculture in GCC region.

For a cropping period of 8 months for cherry tomato, 5 months for sweet pepper and 4 months for cucumber the projected water and energy use per hectare for cooling is raising more questions about the sustainability of cooled greenhouse in the GCC region, the most region affected by water scarcity in the world. However net house system seems to be more sustainable and could contribute significantly in water and energy saving. Further research is needed to explore more options for cooling using alternative water resources as saline and treated wastewater and to upgrade the existing greenhouse to be more efficient in terms of water and energy use.

3.2 Water and Energy Productivity

Figure 4 shows the obtained crop irrigation water productivity (CIWP) under both greenhouse and net house. Data indicate that 1 m^3 of irrigation water produced about 16 kg of cucumber (Zeco Variety), 10 kg of cherry tomato (Sarah variety) and 12 kg of sweet pepper (Red Mountain variety). This difference in terms of

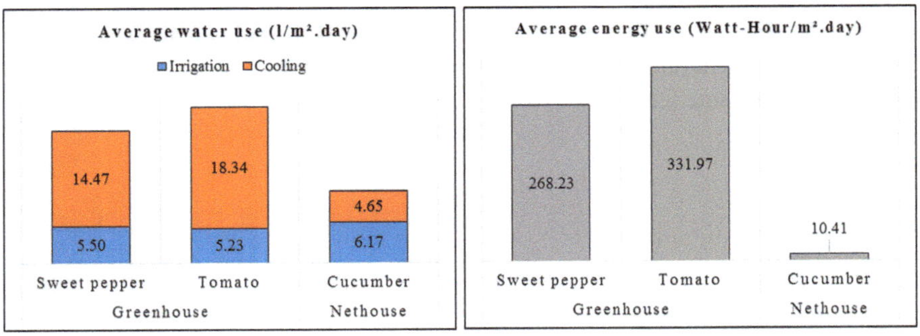

Fig. 3 Energy and water use under greenhouse and net house

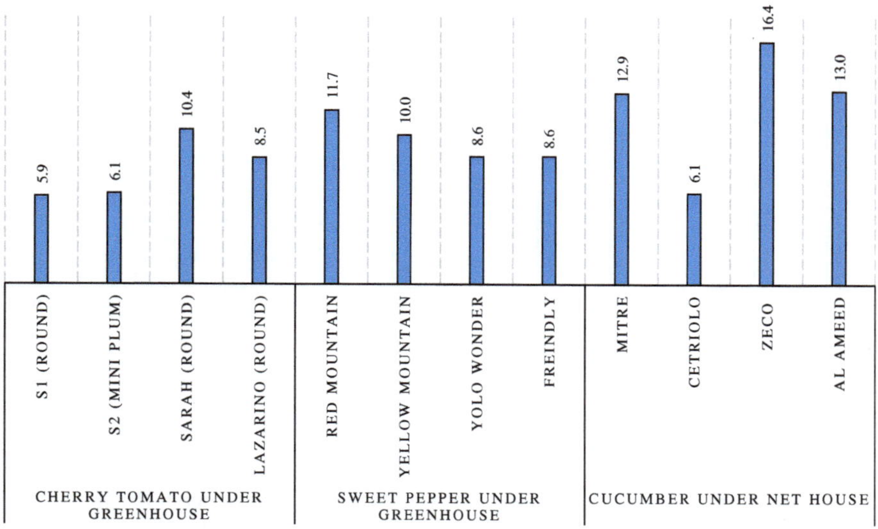

Fig. 4 Irrigation crop water productivity (kg/m^3) under greenhouse and net house conditions

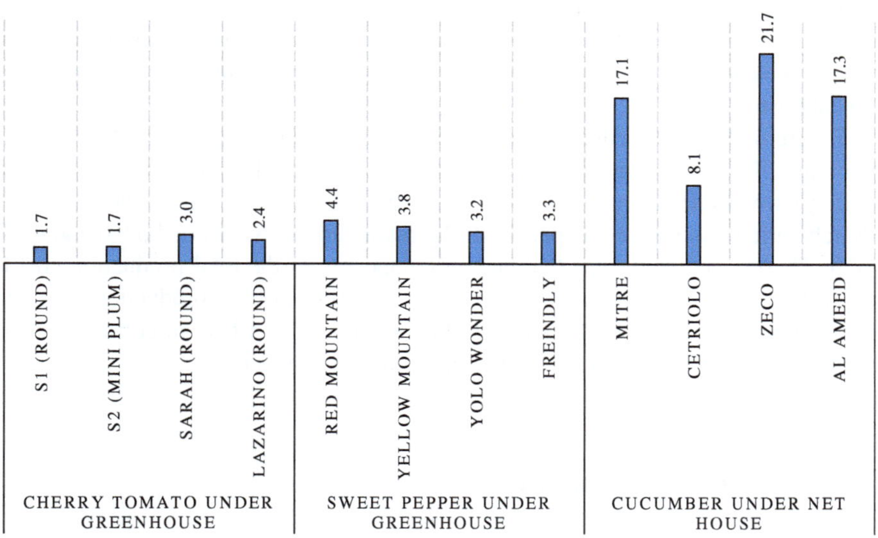

Fig. 5 Cooling crop water productivity (kg/m^3) under both greenhouse and net house

CIWP is explained by difference in plant species and cropping period (3 months for cucumber and 5 months for both cherry tomato and sweet pepper).

Crop cooling water productivity (CCWP) under net house and greenhouse is presented in Fig. 5. Data shows that the difference between net house and greenhouse was obvious in terms of CCWP where 1 m^3 of cooling water produced more

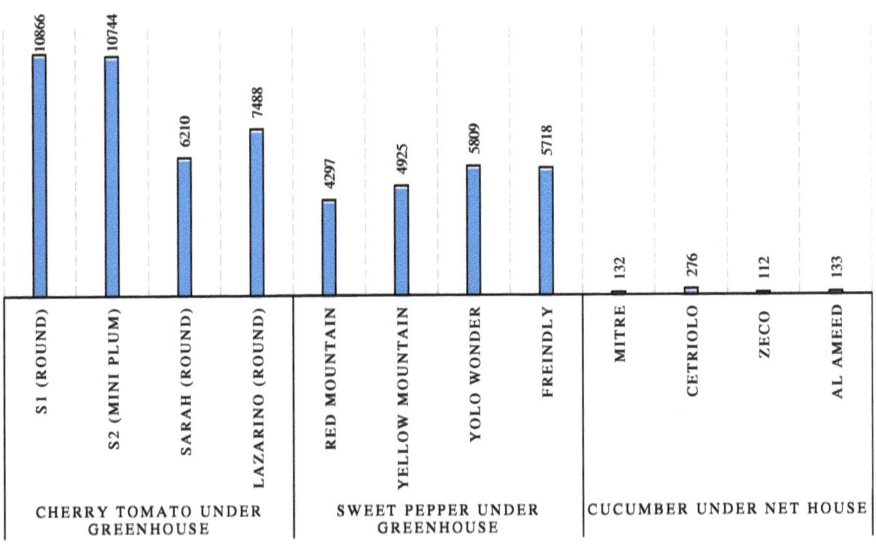

Fig. 6 Energy use efficiency under greenhouse and net house conditions

than 21 kg of cucumber (Zeco variety) and only 3 and 4 kg of cherry tomato (Sarah variety) and Sweet pepper (Red Mountain variety) respectively. Obtained result indicates clearly that net house is more efficient and more water saving compared to greenhouse.

Figure 6 shows the energy productivity under greenhouse and net house for 4 varieties of cucumber and cherry tomato. To produce 1 kg of cucumber (Zeco variety) we needed about 112 W-h, however to produce 1 kg of cherry tomato (Sarah variety) and sweet pepper (Red Mountain variety) we needed 6210 and 4297 W-h. These results indicates that energy component in the total production cost is much higher and heavier under greenhouse resulting in high production cost and loss of competitiveness of the local product in the market where imported products seems to be more competitive than local produced products.

4 Costs and Benefice Analysis

4.1 Investment

Table 2 presents the detailed costs of greenhouse (high technology) and net house (low technology). The data show that greenhouse cost per 1 m^2 is 1.5 times compared to net house and this difference is mainly due to greenhouse expensive components as the structure coverage. However the other components related to pumping and drainage recycling are both required for both structures. The table presents also the effective area can be served by greenhouse and net house

Table 2 Investments costs for greenhouse and net house

	Actual cost for 280 m^2	Effective served area (m^2)	Cost for 1 m^2
Greenhouse components			
Span (35 × 8) covered by 6 mm tick 2 layers PC sheet equipped with pad and fans cooling system	47,500	280	169.6
Drip irrigation system	4875	280	17.4
Drainage water system	28,000	10,000	2.8
Dosing unit	47,075	10,000	4.7
Technical room	43,200	50,000	0.9
Ground cover by 5 cm thick white gravel	6375	280	22.8
Total greenhouse investment cost	177,025	–	218
Net house components			
Span net house (35 × 8)	14,000	280	50.0
50 mesh white insect proof net	2000	280	7.14
Mist system	10,920	280	39.0
Drip irrigation system	4875	280	17.4
Drainage water system	28,000	10,000	2.8
Dosing unit	47,075	10,000	4.7
Technical room	43,200	50,000	0.9
Ground cover by 5 cm thick white gravel	6375	280	22.8
Total net house investment cost	156,445	–	145

components. This parameter is important when costs are projected at farm level because investment cost at experimental level per surface unit are very high compared to investment costs at farm level.

4.2 *Investment Depreciation*

Depreciation will be calculated by dividing the initial investment on total structure life period. All depreciation will be reported by month to be able to calculate investment depreciation equivalent for each growing period (3 months for cucumber and 5 months for cherry tomato and sweet pepper). Since the calculation will concern only the first research year, no discounted value will be updated. Table 3 shows the depreciation for all structures as well as the depreciation for each growing period. The net house structure depreciation is less than greenhouse structure and furthermore is the half when producing cucumber under net house compared to cherry tomato under greenhouse.

Table 3 Investment life period and depreciation

Components	Actual cost for 280 m^2	Life period (year)	Cost AED (m^2)	Depreciation AED/month	Depreciation AED/cropping period	
					Cherry tomato and sweet pepper	Cucumber
Greenhouse components						
Span (35 × 8) covered by 6 mm tick 2 layers PC sheet equipped with pad and fans cooling system	47,500	15	169.6	0.942	4.711	–
Drip irrigation system	4875	5	17.4	0.290	1.450	–
Drainage water system	28,000	15	2.8	0.016	0.078	–
Dosing unit	47,075	15	4.7	0.026	0.131	–
Technical room	43,200	15	0.9	0.005	0.025	–
Ground cover by 5 cm thick white gravel	6375	15	22.8	0.127	0.633	–
Total greenhouse investment cost	177,025	–	218	1.406	7.028	–
Net house components						
Span net house (35 × 8)	14,000	15	50	0.278	–	0.833
50 mesh white Insect proof net	2000	5	7.14	0.119	–	0.357
Mist system	10,920	10	39	0.325	–	0.975
Drip irrigation system	4875	5	17.4	0.290	–	0.870
Drainage water system	28,000	15	2.8	0.016	–	0.047
Dosing unit	47,075	15	4.7	0.026	–	0.078
Technical room	43,200	15	0.9	0.005	–	0.015
Ground cover by 5 cm thick white gravel	6375	15	22.8	0.127	–	0.380
Total net house investment cost	156,445	–	145	1.185	–	3.555

4.3 Running Costs

4.3.1 Labors

In general labor requirements vary according to the cropping system and crop species and technological level. In most of cases cherry tomato requires more labors compared to cucumber and sweet pepper and this due to its technical practices which should be carried out at daily basis.

The estimated time for the technical practices of cherry tomato, cucumber and sweet pepper is presented in Table 4. Presented data indicate that cucumber and sweet pepper require 3 permanent labors/ha while cherry tomato requires 6 labors/ha.

4.3.2 Fertilizers

For all crops we have adopted a Hoagland fertigation solution in its modified formula by Cooper (1988) and adjusted according to Hochmuth and Hochmuth (2001). The adopted fertigation solutions is presented in Table 5.

The used fertilizers quantity as expressed in g/m^2 is presented in Table 6.

The fertilizers price as well as the fertigation cost per each crops is presented in Table 7.

4.3.3 Pots and Substrates

Our calculation were based on results obtained under substrate containing 2/3 sand and 1/3 peat. Table 8 presents the substrates requirement as well as related costs.

Pot cost is related to pot market price, however one pot can be used for 3 years, it means 9 cropping periods of cucumber and 6 cropping periods of cherry tomato and sweet pepper. So the final cost will be divided either on 6 or 9. So the total cost related to substrate and pots is equal to 1.92 for cucumber and 2.19 for cherry tomato and sweet pepper.

4.3.4 Pesticides, Seeds and Other Costs

Table 9 presents the pesticides used quantity and equivalent costs for all tested crops. Presented results indicate that cherry tomato has consumed nearly 40% of used quantity and cucumber and sweet pepper have consumed about 25 and 35% respectively of the total used quantity of pesticides.

Table 10 presents the seeds price and costs and other costs related to the tested crops. Cherry tomato presents the highest costs in terms of seeds and other costs (rope, seedling trays, accessories, etc.).

4.3.5 Energy Costs

Energy costs concern the energy used for cooling, irrigation pumping and fresh water production. In our case fresh water is produced by Reverse Osmosis machine as most of groundwater in UAE is saline so many farmers have installed R.O machine for fresh water production. Farmers in UAE are paying 0.03 AED/kWh as energy price which is subsidized by the government. However the actual cost of

Table 4 Labors requirement for 1 ha per crop and cropping practice

Technical practice	Cucumber			Cherry tomato			Sweet pepper		
	Required labor	Nb of hours per labor	Working days/ha	Required labor	Nb of hours per labor	Working days/ha	Required labor	Nb of hours per labor	Working days/ha
Sowing and transplanting	2	8	2	2	8	2	2	8	2
Plant training	4	175	88	10	125	156	4	160	80
Plant pruning	2	116	29	8	210	210	5	80	50
Leaf removal	–	–	–	9	125	141	3	180	68
Harvesting	5	180	113	9	150	169	7	195	171
Plant inclination	–	–	–	8	60	60	–	–	–
Phytosanitary treatment	1	20	3	2	80	20	1	20	3
Total working days per cropping period			234			758			371
Total labors (ha_			3.11			6.06			2.96
Cost/day			112			112			112
Total cost (m^2)			2.52			8.4			4.2

Table 5 Fertigation solution, elements and fertilizers concentration

Fertilization unit	mg/l	Fertilizers (in mg/l)	mg/l
N	100	MAP	188
P_2O_5	115	Potassium nitrate	391
K_2O	180	Magnesium sulphate	415
CaO	210	Calcium nitrate	808
MgO	66	Fe	20
S	50	Cu	1
Fe	1	Mn	4
Cu	0.1	Zn	1
Zn	0.1	B	2
Mn	0.5	Mo	1
B	0.3		
Mo	0.1		

Table 6 Fertilizers consumption (g/m^2)

Fertilizers (g/growing period m^2)	Cucumber	Cherry tomato	Sweet pepper
Mono-ammonium phosphate	53.84	70.54	74.18
Potassium nitrate	112.25	147.05	154.64
Magnesium sulphate	119.05	155.95	164.00
Calcium nitrate	231.69	303.52	319.19
Iron chelate 6%	1.43	1.88	1.98
Copper	0.19	0.25	0.26
Manganese	1.10	1.45	1.52
Zinc	0.19	0.25	0.26
Boron	0.57	0.75	0.79
Molybdenum	0.36	0.47	0.49
Phosphoric acid	3.57	3.92	4.29

energy in UAE is much higher (0.25 AED/kWh) (Afshari and Friedrich 2016) and in order to encourage farmers government subsidized the energy used in agriculture. Table 11 shows the water and energy use as well as energy related costs. Obtained results indicates that to produce 1 m^3 of desalinated water we need 2.6 kWh of energy. It was shown also that cherry tomato was the highest consumer in terms of energy and especially the cooling part and this was due to cooling temperature which was equal to 26 °C less than the cooling temperature under sweet pepper greenhouse (27 °C). This result indicates that reducing cooling temperature increased the energy and related cost with about 24%.

Table 7 Fertilizers price and costs per each crop

Fertilizers	Package	Unit price	AED/kg or L	Cost AED/m^2		
				Cucumber	Cherry tomato	Sweet Pepper
Mono-ammonium phosphate (25 kg/bag)	Bag	145	5.8	0.312	0.409	0.430
Potassium nitrate (25 kg/bag)	Bag	125	5.0	0.561	0.735	0.773
Magnesium sulphate (25 kg/bag)	Bag	75	3.0	0.357	0.468	0.492
Calcium nitrate (25 kg/bag)	Bag	80	3.2	0.741	0.971	1.021
Iron chelate 6% (1 kg/bag)	Bag	45	45.0	0.065	0.085	0.089
Copper (1 kg/bag)	Bag	45	45.0	0.009	0.011	0.012
Manganese (1 kg/bag)	Bag	45	45.0	0.050	0.065	0.068
Zinc (1 kg/bag)	Bag	40	40.0	0.008	0.010	0.011
Boron (1 L/bottle)	Bottle	35	35.0	0.020	0.026	0.028
Molybdenum (1 L/bag)	Bag	35	35.0	0.013	0.016	0.017
Phosphoric acid (20 L/container)	Container	150	7.5	0.054	0.054	0.054
Total				2.189	2.851	2.995

Table 8 Substrates and pots costs

Substrates	Substrate volume in liter per pot of 6 L				Substrate cost (AED/pot)	Pot cost (AED/pot)	Total cost (m^2)
	Sand	Peat moss	Gravel	Total			
2/3 sand + 1/3 peat	3.33	1.67	0.75	5.75	0.68	2.5	6.36
Price of substrate AED/L	0.037	0.3	0.075				

4.4 *Summary of Running Costs and Investment Depreciation*

Table 12 shows that the lowest production costs are corresponding to cucumber grown under net house. Costs related to labor were the highest among running cost (under subsidized energy scenario) especially for cherry tomato which requires a high level of conduct and maintenance. However cucumber does not require such as number of labor as its cropping practices are not requiring much care.

Table 9 Pesticides used quantity and equivalent costs

Pesticides	Unit price (AED/l)	Used pesticides quantity (ml)			Pesticides costs (AED)		
		Cucumber	Cherry tomato	Sweet pepper	Cucumber	Cherry tomato	Sweet pepper
Ouardin (insecticide)	100	125	200	175	0.04	0.07	0.06
Lorsban (insecticide)	100	125	200	175	0.04	0.07	0.06
Prefex (fungicide)	180	250	400	350	0.16	0.26	0.23
Prisma (insecticide)	220	125	200	175	0.10	0.16	0.14
Tina (acaricide)	220	125	200	175	0.10	0.16	0.14
Chlorpyriphos (insecticide)	85	82	132	115	0.03	0.04	0.04
Total (AED/m^2)					0.47	0.75	0.66

Table 10 Seeds and other costs

Costs	Cucumber	Cherry tomato	Sweet pepper
Seeds price (AED/grain)	0.36	0.9	0.2
Seeds costs (AED/m^2)	0.72	1.8	0.4
Other cost	0.34	0.56	0.41

Table 11 Energy use and costs under both greenhouse (cherry tomato and sweet pepper) and net house (cucumber) for both scenario (energy subsidized and not subsidized)

Energy consumption		Unit	Cucumber	Cherry tomato	Sweet pepper
Irrigation		l/m^2 CP[a]	432.2	627.4	659.5
Cooling		l/m^2 day	4.7	18.3	14.5
Cooling energy		Wh/m^2 day	10.4	332.0	268.2
Fresh water production energy		Wh/m^2 CP	1124	1631	1715
Irrigation energy		Wh/m^2 day	12.5	12.0	12.1
Energy subsidized	Irrigation and cooling energy cost	AED/m^2 CP	0.062	1.548	1.261
	Fresh water production energy cost	AED/m^2 CP	0.034	0.049	0.051
	Total energy cost	AED/m^2 CP	0.096	1.597	1.313
Energy not subsidized	Irrigation and cooling energy cost	AED/m^2 CP	0.515	12.898	10.511
	Fresh water production energy cost	AED/m^2 CP	0.281	0.408	0.429
	Total energy cost	AED/m^2 CP	0.796	13.306	10.940

[a]Cropping period

Table 12 Summary of running costs and investment depreciation

Crops	Cucumber	Cherry tomato	Sweet pepper
Labors	2.52	8.40	4.20
Fertilizers	2.19	2.85	3.00
Pesticides	0.47	0.75	0.66
Substrates and pots	1.92	2.19	2.19
Others	0.34	0.56	0.41
Energy (subsidized)	0.096	1.597	1.313
Energy (not subsidized)	0.80	13.31	10.94
Seeds	0.72	1.80	0.40
Running costs (energy subsidized)	8.26	18.15	12.17
Running costs (energy not subsidized)	8.96	29.86	21.80
Investment depreciation	3.56	7.03	7.03
Total (energy subsidized)	11.81	25.18	19.20
Total (energy not subsidized)	12.51	36.88	28.82

4.5 *Benefice*

According to Table 13 the highest net benefice based on one cropping period is corresponding to cherry tomato followed by sweet pepper and this due to mainly to their high market price compared to cucumber and the subsidized energy. However the highest net benefice based on the whole year production (where cucumber can be produced 3 times during the year) is corresponding to cucumber. Those numbers indicate that producing cucumber under net house with 3 cropping periods allowed to maximize the farmer benefice compared to cherry tomato under greenhouse and

Table 13 Yield, market price, turnover and net benefice

<table>
<tr><th colspan="2">Crops</th><th>Cucumber</th><th>Cherry Tomato</th><th>Sweet pepper</th></tr>
<tr><td colspan="2">Yield (kg/m^2)</td><td>9.35</td><td>8.38</td><td>8.78</td></tr>
<tr><td colspan="2">Price (AED/kg)</td><td>2.50</td><td>4.75</td><td>3.51</td></tr>
<tr><td colspan="2">Total turnover (AED/m^2)</td><td>23.37</td><td>39.79</td><td>30.82</td></tr>
<tr><td rowspan="4">Energy subsidized</td><td>Total cost (AED/m^2)</td><td>11.81</td><td>25.18</td><td>19.20</td></tr>
<tr><td>Net benefit (AED/m^2)</td><td>11.56</td><td>14.61</td><td>11.62</td></tr>
<tr><td>Net benefit (AED/ha)</td><td>104,038</td><td>131,519</td><td>104,598</td></tr>
<tr><td>Net benefit for the whole year</td><td>312,114</td><td>263,037</td><td>209,196</td></tr>
<tr><td rowspan="4">Energy not subsidized</td><td>Total cost (AED/m^2)</td><td>12.51</td><td>36.88</td><td>28.82</td></tr>
<tr><td>Net benefit (AED/m^2)</td><td>10.86</td><td>2.90</td><td>1.99</td></tr>
<tr><td>Net benefit (AED/ha)</td><td>97,736</td><td>26,141</td><td>17,955</td></tr>
<tr><td>Net benefit for the whole year</td><td>293,208</td><td>52,283</td><td>35,910</td></tr>
</table>

further more save water and energy. In case of not subsidized energy the greenhouse is not rentable as the energy cost increase the total cost and reduce the net benefice. Cucumber production under net house remains very profitable even if energy is not subsidized and allows to earn more than 5 and 8 times compared to cherry tomato and sweet pepper grown under greenhouse.

5 Discussion

It has been shown that greenhouse farming system performed better than the open farming system in terms of crop yield, irrigation water productivity and fruit quality. The results revealed that the crop evapotranspiration inside the greenhouse matched the 75–80% of the crop evapotranspiration computed with the climatic parameters observed in the open environment (Harmanto et al. 2005). Under desert conditions as the case of UAE greenhouse cooling is a must in order to produce especially during the summer season (May-September). However cooling process consume a considerable amount of water and energy especially the pad-fan system widely used in UAE and GCC countries. Our results indicates that cooled greenhouse consume 4 times more cooling water than net house. Whereas there is no significant difference between irrigation water consumption. Obtained results for tomato irrigation water consumption (5.23 mm/day) confirm the finding reported by Harmanto et al. (2005) who showed that tomato actual water consumption under greenhouse conditions was about 5.6 mm/day.

According to Franco et al. (2014) the main drawback of the pad-fan system was the horizontal temperature gradients, with a maximum difference of 11.4 °C between the pads and the fans. According to López et al. (2012) the fog system required higher energy consumption (7.2–8.9 kWh) than the pad-fan system (5.1 kWh) for continuous operations over 1 h. If we assume, based on López et al. (2012) that the fogging system in the net house was operating 20 s per 5 min, it means 36 min during 10 h. Thus the energy consumed during the day for fogging system is equal to 4.32 kWh/day. While the cooled greenhouse during summer season is operating continuously with a total energy consumption of 122 kWh/day, which means that the cooled greenhouse consume about 28 time more energy compared to fogging system. Our results indicate that the greenhouse consumes 32 time energy compared to net house equipped with fogging system which confirm the results obtained by López et al. (2012). Canakci and Akinci (2006) analyzed energy use pattern in greenhouse for vegetable. They found that the operational energy and energy source requirements in greenhouse vegetable production were varying from 23,883.5 to 28,034.7 and 45,763.3 to 49,978.8 MJ/1000 m^2, respectively. The energy cost ratio from total production cost of four major greenhouse vegetable crops—such as tomato, pepper, cucumber, and eggplant- was found about 0.32, 0.19, 0.31, 0.23, respectively which confirm our finding where energy cost (not subsidized) represent 0.36 and 0.37 of total running cost for tomato and sweet pepper respectively. However energy cost ration in the case of cucumber under net house does not exceeds 0.06.

A study aimed to determine the cost and return of soilless greenhouse cucumber has been carried out by Engindeniz and Gül (2009) in Turkey. The study reported that total production cost was equal to 3.47 USD/m^2 which is in agreement with our finding, where cucumber production cost was equal to 11.81 AED/m^2, equivalent to 3.21 USD/m^2. However the net benefit reported by Engindeniz and Gül (2009) is equal to 1.84 USD/m^2 although the total yield was equal to 31 kg/m^2. However in our case net benefit was equal to 3.14 USD/m^2 with a total yield of 9.35. This difference in terms of benefit is explained by the lowest market price of 0.17 USD in Turkey compared to 0.68 USD/kg in UAE.

6 Conclusion

Giving the current situation of water resources and energy aggravated by climate change in UAE as well as GCC countries, cooled greenhouse is not a sustainable option for horticultural crops production. It was shown that this protected agriculture system consumes a considerable amount of water and energy in cooling process. However UAE climate allows to produce a large number of fruits and vegetables crops between October and May, that is a total of 8 months with a mild climate. During this period horticultural crops can be grown under net house even without cooling allowing, thus, increasing crop water productivity, energy saving and income improvement. Agricultural policies such as subsidized energy leads to use more cooled greenhouse rather than encouraging sustainable protected agriculture as net house for better water and energy saving. Therefore, there is a need to improve energy and water use efficiency in the protected agriculture in GCC region and to reduce the water and energy footprint under protected agriculture in GCC region.

References

Abdel-Dayem S, McDonnell R (2012) Water and food security in the Arab Region. In: Choukr-Allah R (ed) Integrated water resources management in the mediterranean region: dialogue towards new strategy. Springer, Berlin

Afshari A, Friedrich L (2016) A proposal to introduce tradable energy savings certificates in the emirate of Abu Dhabi. Renew Sustain Energy Rev 55:1342–1351. http://dx.doi.org/10.1016/j.rser.2015.05.086

Al-Nasser AY, Bhat N (1998) Protected agriculture in the State of Kuwait. In: Proceedings of the workshop on protected agriculture in the arabian peninsula, Doha (Qatar), 15–18 Feb 1998, pp 15–18

Alnaser W, Alnaser N (2011) The status of renewable energy in the GCC countries. Renew Sustain Energy Rev 15:3074–3098

Canakci M, Akinci I (2006) Energy use pattern analyses of greenhouse vegetable production. Energy 31:1243–1256

Choukr-Allah R (2015) Overview of protected agriculture worldwide. In: Choukr-Allah R (ed) Workshop on protected agriculture: "Unlocking the potential of Protected Agriculture in the GCC countries: cutting water consumption while supporting improved nutrition and food security". ICBA, Dubai, UAE

Cooper A (1988) 1. The system. 2. Operation of the system. The ABC of NFT. Nutrient Film Technique: 3–123

El Katiri L, Husain M (2014) Prospects for renewable energy in GCC States: opportunities and the need for reform. Oxford Institute of Energy Studies 10

Engindeniz S, Gül A (2009) Economic analysis of soilless and soil-based greenhouse cucumber production in Turkey. Scientia Agricola 66:606–614

Franco A, Valera DL, Peña A (2014) Energy efficiency in greenhouse evaporative cooling techniques: cooling boxes versus cellulose pads. Energies 7:1427–1447

Harmanto, Salokhe VM, Babel MS, Tantau HJ (2005) Water requirement of drip irrigated tomatoes grown in greenhouse in tropical environment. Agric Water Manag 71:225–242. doi:10.1016/j.agwat.2004.09.003

Hoagland DR, Arnon DI (1950) The water-culture method for growing plants without soil. Circular. California Agric Exp Stat 347

Hochmuth GJ, Hochmuth RC (2001) Nutrient solution formulation for hydroponic (perlite, rockwool, NFT) tomatoes in Florida. HS796. Univ. Fla. Coop. Ext. Serv., Gainesville

Kittas C, Katsoulas N, Bartzanas T (2012) Greenhouse climate control in mediterranean greenhouses. Cuadernos de estudios agroalimentarios: 89–114

Kotilaine JT (2010) GCC agriculture. Econ Res 27

López A, Valera DL, Molina-Aiz FD, Peña A (2012) Sonic anemometry to evaluate airflow characteristics and temperature distribution in empty Mediterranean greenhouses equipped with pad–fan and fog systems. Biosys Eng 113:334–350

NBS (2013) Agricultural statistics. National Bureau of Statistics, United Arab Emirates, UAE

Wittwer SH, Castilla N (1995) Protected cultivation of horticultural crops worldwide. HortTechnology 5:6–23

Woertz E, Pradhan S, Biberovic N, Koch C (2008) Food inflation in the GCC countries. Gulf Research Center, Dubai

Agricultural Water Use and Technologies for Adaptations to Water Deficits

Ashok Alva, Mohammed Albaho, Gautam Kumar, Mohamed Annabi, William Stevens and Kris Dodge

Abstract The water withdrawals for agriculture across the United States account for 25–50%, unlike in the dry regions of the world this proportion can be >75–90%. About 57% of the total irrigated acreage in the US is in six states (California, Nebraska, Texas, Arkansas, Idaho, and Colorado), with the remaining states accounting for only 43% of the irrigated acreage. Western states represent the high percent of farmland under irrigation, i.e. 25–40% with some areas >40%. Surface water accounts for a larger portion of total irrigation water withdrawals in the western states, unlike the mid-western and the southern states where groundwater is the major source of irrigation water. The north-eastern states show mixed trend, with both surface and groundwater contributing for irrigation water withdrawals. Real time, automated measurement of soil water, temperature, and electrical conductivity in the soil profile, within and below the root zone, is a critical component of best management of nutrients and irrigation aimed to optimize nutrient and water uptake efficiency while minimizing potential leaching of nutrients and water below the root zone. A case study described in this paper was conducted in a potato (cv. Umatilla Russet) and corn field on Quincy fine sand (mixed, mesic, Xeric Torripsamments) under center pivot irrigation in eastern Washington State. Annual precipitation in this region is about 160 mm with only 20% of this amount received during the growing season, i.e. May through August. Therefore, careful management of irrigation is critical to enhance water use efficiency and water productivity. Sensors were used for automated monitoring of soil water content, temperature, and electrical con-

A. Alva · W. Stevens
USDA-ARS, Sidney, MT, USA

A. Alva (✉) · M. Albaho
Kuwait Institute for Scientific Research, Kuwait city, Kuwait
e-mail: alva300@gmail.com

G. Kumar
Village Shop, Gurgaon, India

M. Annabi
Agronomy Lab, National Agricultural Research Institute of Tunisia, Ariana, Tunisia

K. Dodge
International Center for Agricultural Research in Dry Areas, Amman, Jordan

O. Abdalla et al. (eds.), *Water Resources in Arid Areas: The Way Forward*,
Springer Water, DOI 10.1007/978-3-319-51856-5_29

ductivity on a real-time basis at different depths in the soil profile to represent root zone (0–60 cm) and below the root zone (>60 cm). The utility of real time monitoring of soil water data for developing the irrigation set points to maintain the soil water content in the root zone within the management allowed deficit (MAD) to mitigate any impact of water stress while minimizing the water percolation below the root zone are discussed. Similar technologies using user friendly but low cost sensors have been tested in small farms in India (Punjab and Haryana) over the past year. The sensors use capacitance to determine volumetric soil water content in the root zone (0–30 cm). Farmers were advised on the soil water threshold values to be maintained for different crops (spinach, okra, cauliflower, bitter gourd, and rice). While these sensors resulted in initial high adoption of 80% of advisories, the efficacy was reduced after a short period. A new technology, where the sensors automatically control irrigation set points is now being tested. Initial feedback suggests that farmers have a greater level of confidence in the automation, leading to increased and longer adoption rates. This technology is now being tested in only a few sites; will be expanded across three more states in India.

Keywords Capacitance probe · Climate variability · Deficit irrigation · Full point · Irrigation set point · Low cost irrigation decision support · Management allowed water deficit · Real time soil profile water monitoring · Refill point · Soil temperature · Water productivity · Water use efficiency

1 Introduction

Agriculture is a major water user in different parts of the world. The proportion of water withdrawal for agriculture is about 25–50% in the United States, unlike 50–75% in most of South America, parts of Africa, and China (FAO 2016; World resources sim center 2010) (Fig. 1).

However, in Central America, most of Central and South Asia, Africa, and Australia, the water withdrawal for agriculture accounts for 75–90% and in some cases >90% of total withdrawal. Rapid increase in world population will result in further demand for increasing the food production within the available resources, land and water. Therefore, the overall fresh water use for agriculture in the world will attain 70% of the total fresh water withdrawal. To make the matters worse, predictions of water availability due to climate variability show greater water limitation in the Mediterranean region which is already a region of severe water scarcity and some degree of political unrest and instability. This region is subject to severe land degradation due to desertification and increased salinization of groundwater due to excess pumping with low recharge. The population growth predictions shown in Fig. 2 reveal that the world population will attain 9 billion within a few years from the current population of about 7 billion (Onesimpleidea.com).

Despite a number of factors influence population growth, the above prediction suggests that a big increase in world population is imminent within the very near future. This brings about the need to increase food production with very little

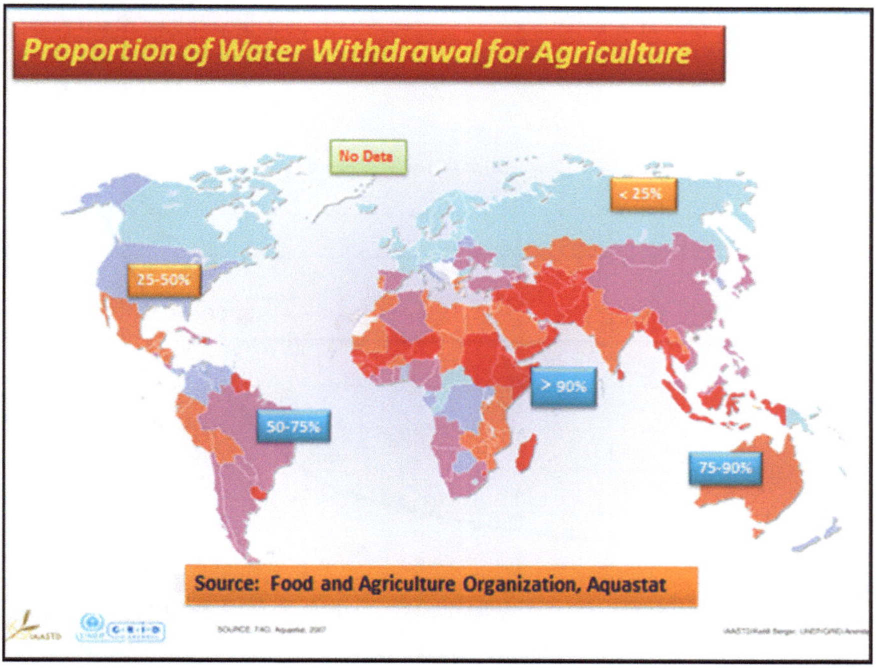

Fig. 1 Water use for agriculture as proportion of total water withdrawal in different parts of the world. Extracted from FAO (2016), Aquastat

prospects for significant increase in resources, primarily land and water. Therefore, the agricultural production systems have to be intensified to ensure adequate food production for increasing population and maintain sustainability. A recent study by the Rothamsted Research Station in England shows that if the current intensive production continues, by 2050 nearly 70% of the earth's surface area need to be used for agricultural production to meet the food needs of the increasing population as compared to 40% being used now. The soil degradation will result in 30% less soils productivity than their current level. The estimate by the United Nations Food and Agriculture Organization (FAO) reveals that 25 and 8% of the agricultural lands are highly and moderately degraded, respectively.

During the recent years, the disposable income of the emerging economy countries has increased rapidly as a result of the overall economic boom of these countries. This results in greater buying power of the consumers in those emerging economy countries of the world including greater demand for more nutritious and high quality food and also their food habits will gradually change to increased meat based diet unlike the traditional plant based diet. This in turn enhances the pressure on resource constraints. As a result, despite the fact that the economic engines in the emerging economy countries are contributing to increased world economic growth, this is exerting increased pressure on already resource strapped agricultural production systems.

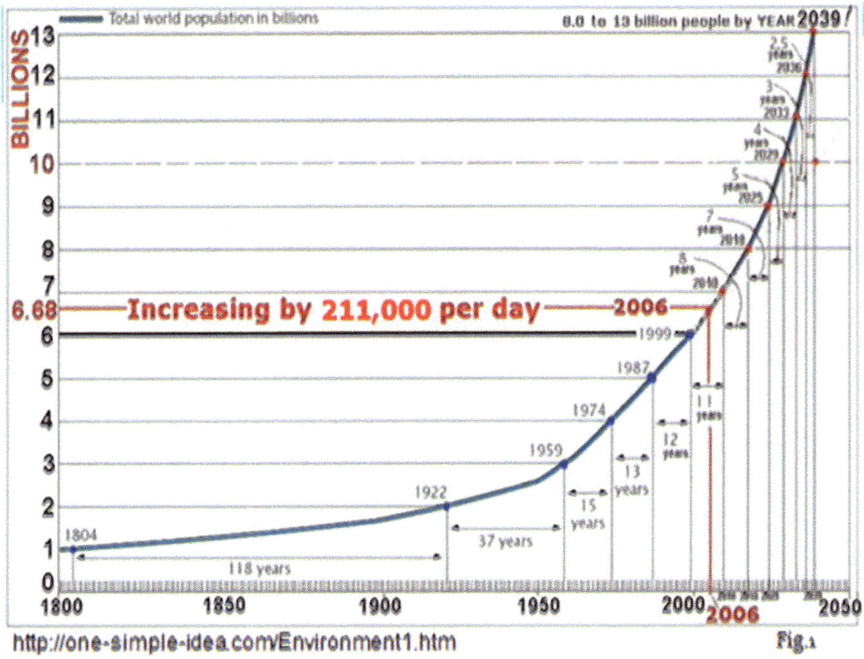

Fig. 2 Predicted population growth through 2050. The choice of this model prediction is only by convenience, and authors have no basis or preferential judgement to this prediction or exclusion of other predictions

The objectives of this paper are to: (i) summarize the status of water availability and use for agriculture with specific examples from the United State of America; (ii) draw attention to potential future crisis related to predicted water scarcity in some parts of the world impacted by climate variability and population growth; and (iii) describe the technologies to improve agricultural water use efficiency and increase water productivity.

1.1 Agricultural Water Use

Population in North and South America represent 6–8% of the global population, while fresh water resources in this continent represent 15–26% of the global available fresh water resources (The crop site 2015) (Fig. 3). The respective proportions for Asia are 60 and 36%. This shows the tremendous pressure on available fresh water in Asia on per capita basis.

As a result per capita water use is high in North America, and Europe, while low in Asia and Africa (University of Michigan 2000) (Fig. 4). The challenge of water

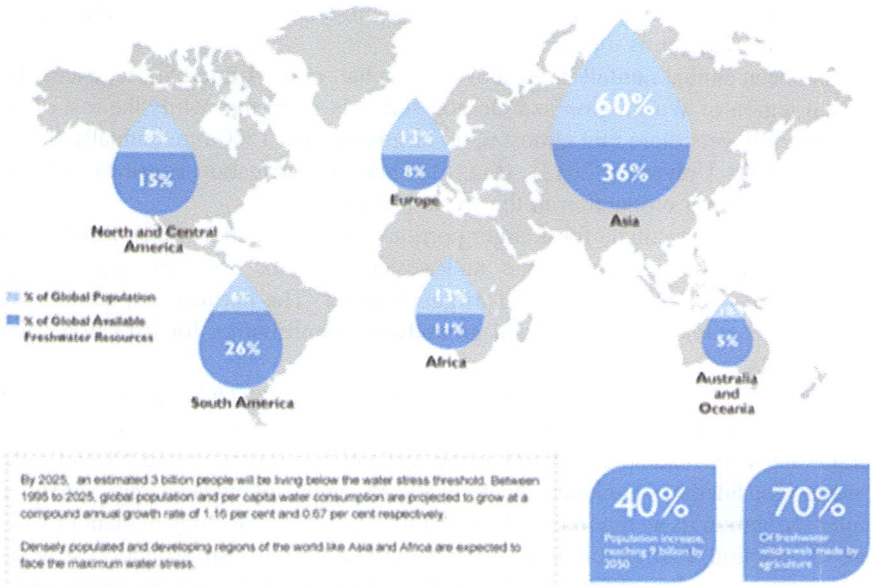

Fig. 3 Comparison of percent of global population versus percent of global available freshwater source in different parts of the world. *Source* The crop site (2015)

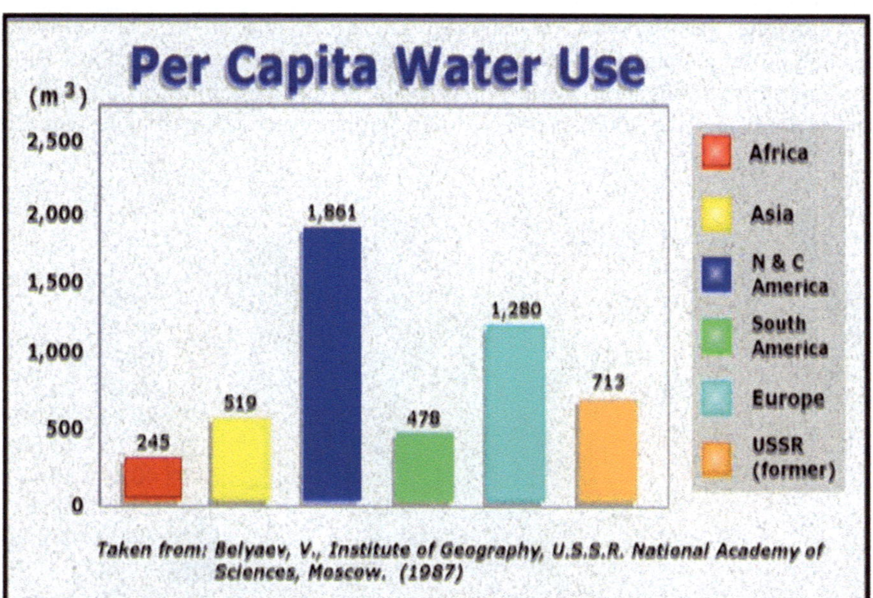

Fig. 4 Per capita water use in different parts of the world. *Source* http://www.globalchange.umich.edu/globalchange2/current/lectures/freshwater_supply/freshwater.html

management gets worst with a rapid increase in world population in the regions already stretched for available resources.

The mean annual rainfall in most of western half of the United States, except the coastal region of Washington and Oregon states, is <500 mm while the rest of the US generally gets more rain almost up to or greater than 1500 mm annually (North Carolina State University 2012) (Fig. 5). Therefore, irrigation is critical for sustained agricultural production in the western states.

Surface water accounts for a larger portion of total irrigation water withdrawals in the western states, unlike the mid-western and the southern states where groundwater is the major source of irrigation water. The northeastern states show mixed trend, with both surface and groundwater contributing for irrigation water withdrawals.

In the US, only 18% of total agricultural land is irrigated and only about 20% of the irrigated land is under fertigation. These proportions may be different in different states, for example in Florida; the respective proportions are 68 and 53%. Despite considerable evidence of increased efficiency of nutrient and water use under fertigation its adaption is still rather low. Several factors may explain this low rate of adaption. Fertigation works well under sprinkler or drip irrigation systems. A significant portion of irrigation in several states is by furrow irrigation which doesn't lend itself for fertigation.

About 57% of the US total irrigated land is in six states, i.e. Nebraska, California, Texas, Arkansas, Idaho, and Colorado (USDA—National Agricultural Statistics Service 2009) (Fig. 6).

Therefore, the proportion of irrigated acreage in relation to total cultivated acreage in the remaining states is often <5%. Surface water accounts for a larger portion of total irrigation water withdrawals in the western states, unlike the mid-western and the southern states where groundwater is the major source of

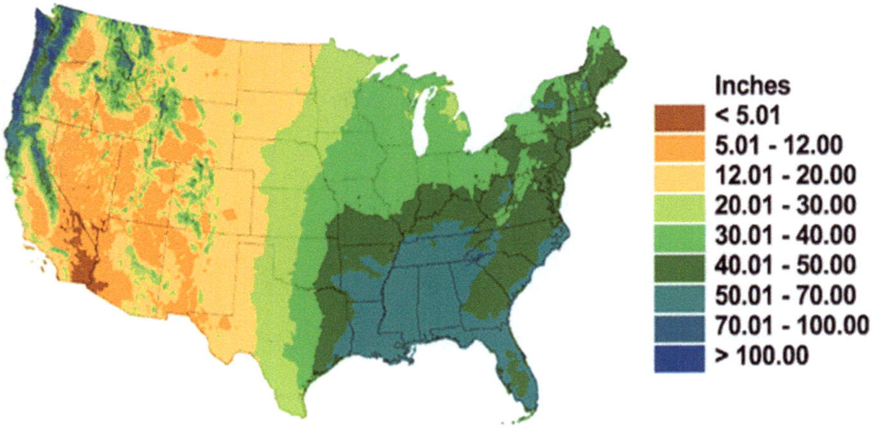

Fig. 5 Annual mean total precipitation in different parts of the United States. *Source* https://climate.ncsu.edu/edu/k12/.SEPrecip

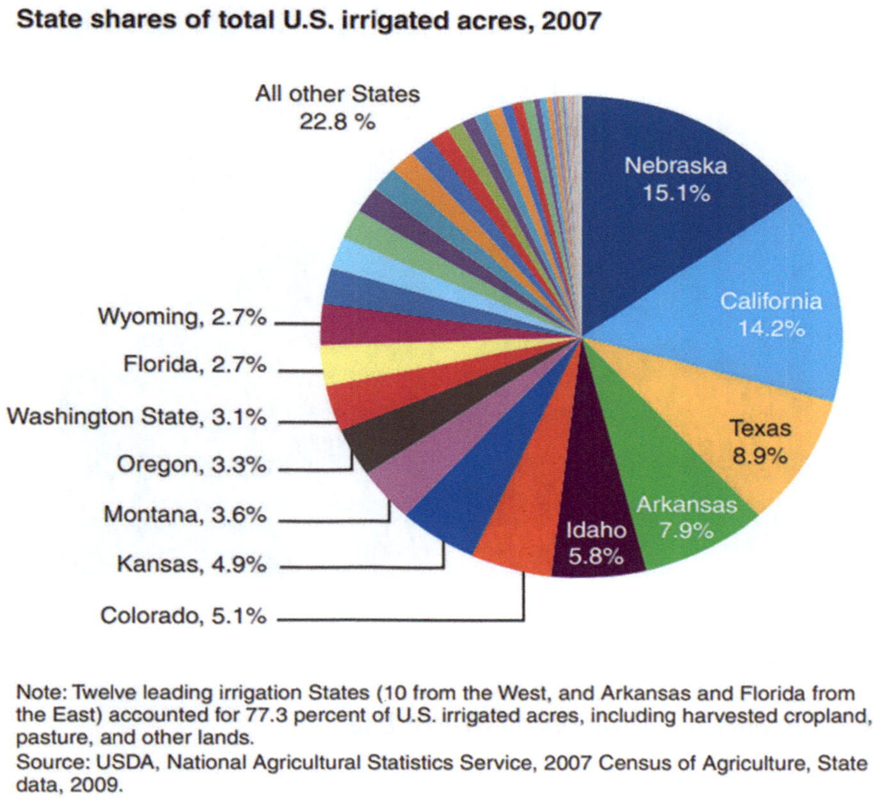

Fig. 6 Irrigated acreage in different states as percent of total US irrigated acreage

irrigation water. The northeastern states show mixed trend, with both surface and groundwater contributing for irrigation water withdrawals.

Agricultural production in California is important for the whole country because of its significance in production of fruits and vegetables, as a major supplier for the rest of the country particularly in the winter. Therefore, recent drought in California is of great concern for the whole country. The state authorities have taken aggressive measures to conserve water to mitigate the drastic negative effects of drought on the agricultural production. NASA studies indicate that droughts during 2011 and 2014 resulted in loss of 4 trillion gallons (15 km^3, or 12 million acre-feet) of water each year in combined river basins in California, an amount far greater than the annual water use by California's 38 million residents (Fig. 7).

Fig. 7 Impact of drought in California on water storage in river basins from 2002 to 2014. *Source* NASA GRACE (2014)

1.2 Impact on Agricultural Water Use

Water requirement to produce unit amount of crops or animal products vary considerably. For example, average water requirements to produce one kilogram each of cereals, pulses, nuts, chicken, and beef are 1644, 4055, 9063, 4325, and 15415 L, respectively (Mekonnen and Hoekstra 2012). Therefore, water requirement for meat based diet is several fold greater than that for crop based diet. This is of concern as the food system in the emerging economy countries gradually transition into increased meat based as compared to the traditional crop based food system as an outcome of gradual increase in disposable income of the population in those countries in the recent years.

Water use efficiency depends on number of factors including; irrigation scheduling, method of irrigation, deficit or supplemental irrigation, laser leveling, drip irrigation, improved furrows, and tail water return flow etc. Irrigation efficiency can be as low as 40–70% for furrow irrigation, while it can be as high as 90–95% for drip irrigation or pivot irrigation with low energy precision application (LEPA) system (Fig. 8).

Fig. 8 Differences in water distribution and evaporative loss in pivot systems with low energy precision application (LEPA) versus traditional pivot sprinkler package without LEPA

1.3 Sample Technologies to Improve Water Use Efficiency and Water Productivity

Three innovative techniques described in this paper include: (i) real time soil water monitoring; (ii) real time soil temperature monitoring; and (iii) low cost user friendly irrigation decision support aid. The ease of adaptation of a given technology depends on user friendliness of the technology, demonstrated benefits to increase water productivity, and high return on extra investment.

The first two techniques are evaluated in center pivot irrigated potato-corn rotation experiment in the Pacific Northwest (PNW) of US. The soil in the study area was Quincy fine sand (mixed, mesic, Xeric Torripsamments) in Benton County, Washington. Sand content in 0–120 cm depth varied from 917 to 948 g kg^{-1}, with bulk density range of 1.33–1.60 kg m^{-3}.

1.3.1 Soil Water Monitoring

Real-time automated continuous monitoring of water content in the soil profile, within and below the root zone, is important to manage irrigation to ensure adequate soil water available for plant uptake to avoid negative impacts of water stress while minimizing leaching losses below the rooting depth (Fig. 9) (Alva and Fares 1998, 1999; Buss 1993; Fares and Alva 1999, 2000a, b; Mead et al. 1994; Paltineanu and Starr 1997; Starr and Paltineanu 1998).

The aim is to improve irrigation scheduling to increase water use efficiency and water productivity. Capacitance probe was used in this study to measure the dielectric constant of the soil within the zone of influence. This parameter is influenced by the water content of soil-water-air mixture. The dielectric constant of

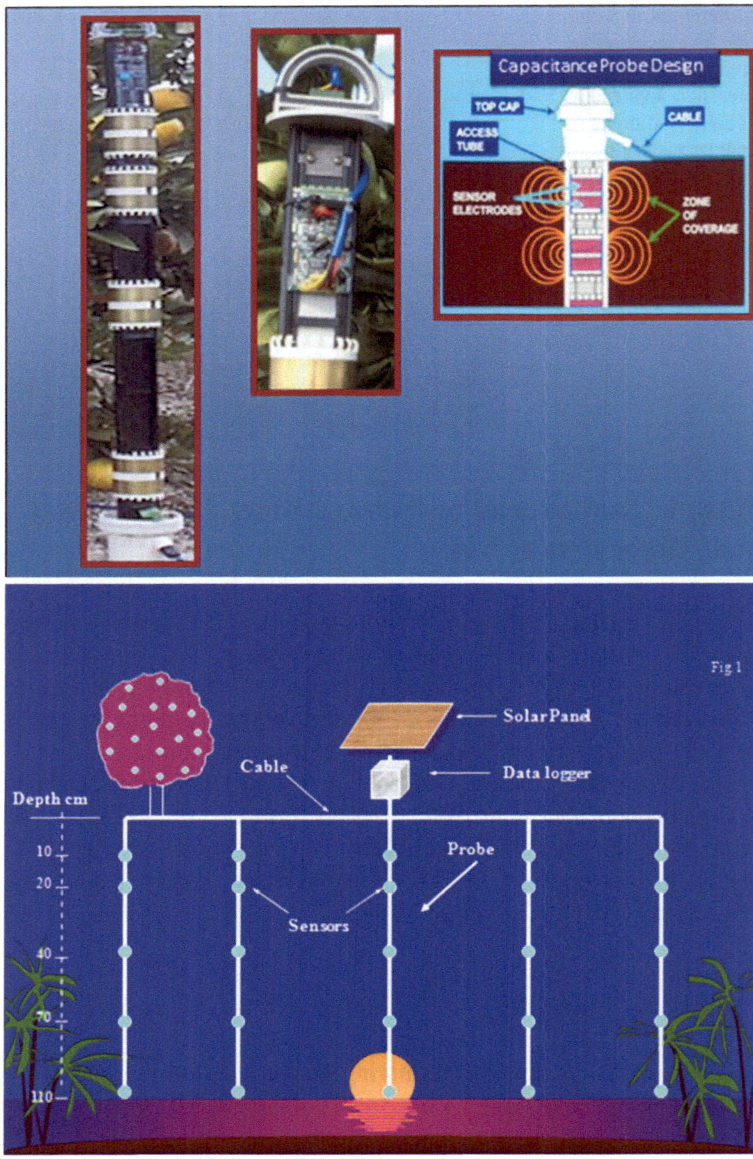

Fig. 9 Capacitance probes for real time, automated monitoring of water content in soil profile at various depths within and below the root zone

water is 80, which is greater than that of either soil matrix (<10) or air (1). Accordingly, dielectric constant of the soil-water-air mixture is greatly influenced by small changes in soil water content.

1.3.2 Irrigation Set-Points

Irrigation scheduling can be improved by availability of appropriate information to guide: (i) when and how much to irrigate, and (ii) management allowed depletion (MAD) of soil water in the root zone without causing any negative effects on plant growth and/or production. Irrigation based on these information provides adequate water for plant uptake to mitigate negative effects of water stress on the plant, while avoids application of excess water that causes water and nutrients leaching below the root zone which leads to low water and nutrient use efficiency. These guidelines are termed as 'Irrigation Set-points' (Alva et al. 2003; Alva 2008) (Figs. 10 and 11). Real time measurement of soil water content within and below the root zone is important to evaluate the soil water content in relation to the set-points to trigger irrigations to be scheduled as needed.

1.3.3 Full Point

Refers to the amount of water held by the soil against gravity, i.e. after excess water is drained following irrigation. This is somewhat equivalent to the field capacity and is the target water content to be attained in the rooting depth soil at each irrigation event. As an example, for Quincy fine sand used for potato production in the Northwest of US, the field capacity soil water content was 12% (v/v). The rooting depth of potato is 60 cm. Therefore, the full point for the rooting depth of 60 cm is $60 \times 0.12 = 7.2$ cm or 72 mm (Figs. 10 and 11). Thus, the maximum amount of water that can be retained in the top 60 cm depth soil after leaching of excess soil water for 24 h under the conditions of this experiment is 72 mm. Application of water in excess of this target point will result in leaching below the 60 cm rooting depth, which should be avoided or minimized.

1.3.4 Driest Point

This represents the soil water content below which plants cannot uptake water from the soil, thus resulting in wilting of the plants and other negative effects on plant growth and production associated with water stress. This is close to the 'wilting point' of the soil, which is 3% (v/v) for this soil. Thus, amount of water in the top 60 cm rooting depth for this soil corresponding to this setpoint is: $60 \times 0.03 = 1.8$ cm or 18 mm.

1.3.5 Refill Point

Refers to management allowed depletion (MAD) of soil water content in the root zone at which irrigation must be scheduled to avoid possible negative effects of mild water stress on plant growth and/or production. An irrigation event aimed to

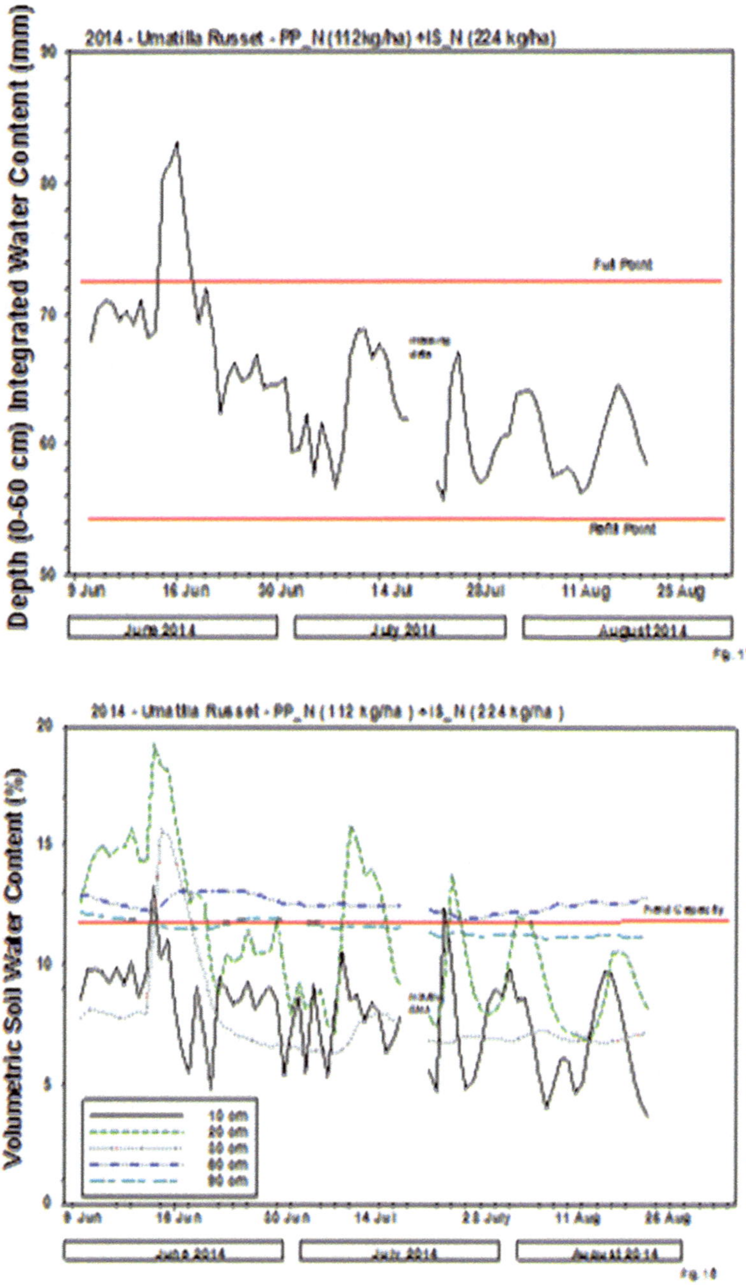

Fig. 10 Volumetric soil water content at different depths monitored by capacitance probes installed in a potato experiment and depth integrated soil water content within the root zone (0–60 cm) using Umatilla Russet cultivar with 336 kg/ha nitrogen (N), combination of pre plant and in season application in a sandy soil under center pivot irrigation in the Northwest United States. The field capacity water content and irrigation set points are also shown

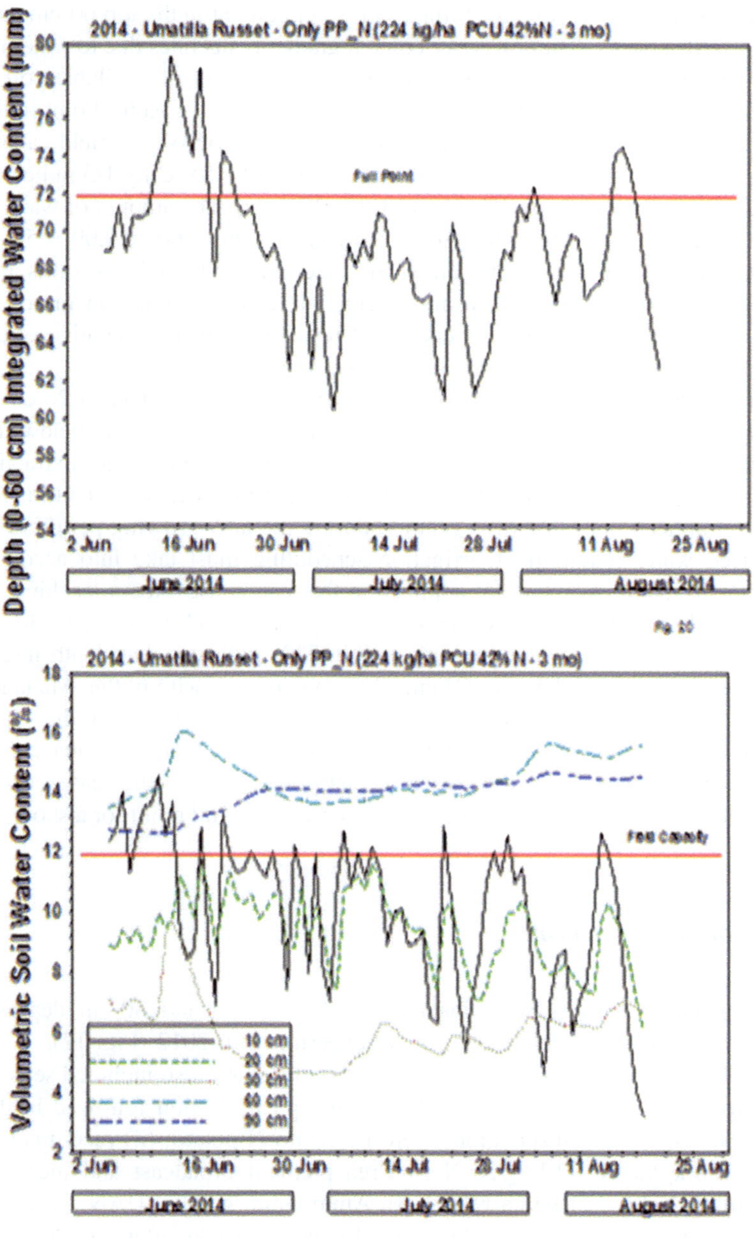

Fig. 11 Volumetric soil water content at different depths monitored by capacitance probes installed in a potato experiment and depth integrated soil water content within the root zone (0–60 cm) using Umatilla Russet cultivar with 224 kg/ha nitrogen (N), only pre plant application as controlled release N (polymer coated urea, PCU, 3 months release of N), in a sandy soil under center pivot irrigation in the Northwest United States. The field capacity water content and irrigation set points are also shown

attain Full Point in the root zone brings soil water content in the top 60 cm depth to close to field capacity. The soil water content gradually declines due to water uptake by the plants. Assuming that the MAD for potato is 1/3 of available soil water content, the following calculation shows the water content at Refill Point in the top 60 cm rooting depth. The available soil water content (ASWC) = field capacity—wilting point, i.e. 12 − 3 = 9% (v/v). The depletion of ASWC by 1/3 indicates that water content can decline by 3%, i.e. from 12–9% (v/v). The amount of water in the top 60 cm depth soil at 9% volumetric soil water content is: 60 × 0.09 = 5.4 cm or 54 mm (Fig. 10). Therefore, the soil water content in the top 60 cm soil is allowed to decline to 54 mm to schedule next irrigation. Further reduction in soil water content below 54 mm water content could result in potential negative effects of water stress on the plant growth and/or production.

The capacitance probe data showed peak water contents in 10 and 20 cm depth only indicating the irrigation was carefully done not to leach water below the top soil (Figs. 10 and 11). The 30 cm sensor showed evidence of water extraction by the plant roots most noticeably in June and July, but rather lesser extent in August. This is an indication of loss of vegetation during the later growing period resulting in reduced water requirement. Irrigation scheduling must take into account this difference to avoid excess water application that can cause excess leaching.

The depth integrated water content in the top 60 cm soil profile ranged from 55 to 70 cm during most of the growing period. This indicates that depth integrated soil water content in the top 60 cm was maintained within the Management Allowed Deficit (MAD) to ensure adequate water availability for the plants to mitigate any stress effects. The example described shows that soil water content in the root zone was maintained between the set points most of the season except in the middle of June when the water content exceeded the full point for a short period.

1.4 Soil Temperature

Soil temperature sensors were installed at 20, 40, 60 and 80 cm depths in a potato-corn rotation nutrient management experiment in 2012 (Fig. 12).

The photo for illustration shows only potato, the study also included sensor data from corn in rotation with potato, which is a typical rotation followed in the US Northwest. The sensor installations were in: (i) no N potato; (ii) recommended N potato (336 kg/ha N; 112 kg/ha N as Urea preplant broadcast and incorporated during tillage, plus 224 kg/ha N as Urea Ammonium Nitrate, UAN, 32%N, fertigated in 5 applications at 2 weeks interval began 4 weeks after emergence); and (iii) corn with 448 kg/ha N as fertigation. Both the crops were irrigated by center pivot irrigation to replenish the full evapotranspiration (ET) using the weather data in the field. Soil temperature measurements showed major diurnal changes only in 20 and slightly in 40 cm depth (Fig. 13).

As the season progressed the range of variation became narrow as a result of increased canopy cover shielding the soil from direct exposure to sun. In June the

Fig. 12 Installation of sensors for real-time monitoring of temperature and salinity at different depths in soil in a potato experiment in a sandy soil with center pivot irrigation in the northwest United States

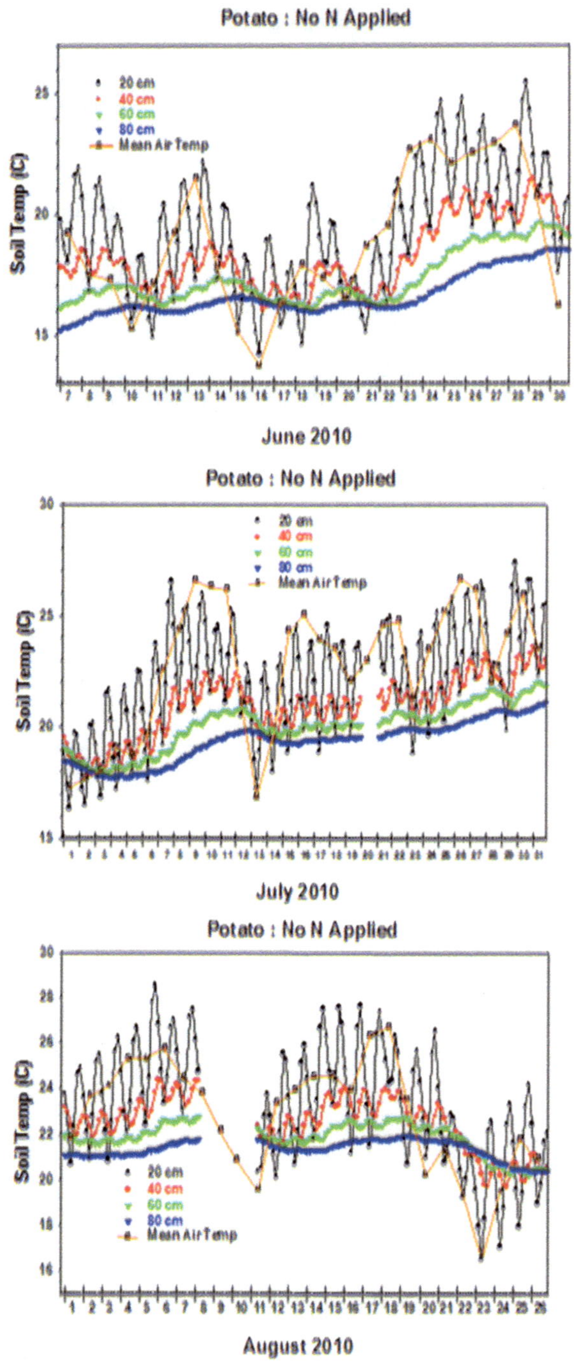

Fig. 13 Real-time measurement of soil temperature at different depths over potato growing season with no nitrogen (N) applied, i.e., poor canopy growth, in a sandy soil with center pivot irrigation in a sandy soil in northwest United States

surface soil temperature was very close to that of ambient air. As the season progressed, the surface soil temperature was much lower than that of ambient air, due to the canopy interception of heat.

In June, the surface 20 cm soil temperature was relatively similar in both corn (with 448 kg/ha N) and potato (both no N and 336 kg/ha N) fields, as well as ambient air temperature (Fig. 14). In July, however, 20 cm soil temperature was similar under corn and potato with 336 kg/ha N plots, which were much lower than that of ambient air or in potato plots with no N. In the latter, poor vegetation cover exposed the soil for direct heat unlike the other vegetation cover which shielded the soil from direct sunlight and heat.

1.5 Low Cost Soil Water Sensor

New technologies to aid the decisions are indeed very important in both high and low income farming systems, to support sustainable farming systems and for judicious use of natural resources. However, cost of technology is a determining factor for farmer's acceptance and adaptation of the new technologies. The case study described in this section is from Haryana, in northern India, where most farmer's annual income may range from $500 to $1000 on small holdings of about 2 ha. Most common crops are spinach and cauliflower. Use of pesticides and fertilizers to support high production are rare, in part, due to the high cost of these inputs and uncertainties of appropriate returns for the added cost.

The soil moisture sensor tested in this region was developed by the Northeast Center for Technology Application and Reach (NECTAR), an autonomous research group within the Indian Department of Science and Technology (Fig. 15).

This sensor when inserted into the soil reads instantly the volumetric water contents on an LCD screen, and provides irrigation decisions. In a pilot study this sensor was given to one farmer, who in turn was able to help a community of 200 farmers on irrigation decisions. However, this was rather overwhelming for that one farmer to serve all the farmers in his community. This is an ideal situation where either government or international aid agencies should play a role to finance the initial pilot study, which in turn can become attractive for private investors to take over the commercialization and distribution of this technology and provide needed training to the farmers to develop confidence in use and adaptation of the new technology.

However, the sensor did advise farmers on one thing: They were irrigating too infrequently and too much in a given event. After farmers used the sensor, they learned that their field was drying during the mid-week, especially during the summer months. With the aid of sensor information, farmers began irrigating twice a week at reduced amounts at each irrigation. Overall the use of sensors facilitated 25–30% reduction in water use as compared to the amount of irrigations without using the sensors. This is a substantial demonstration of benefits of using sensors for increased water productivity.

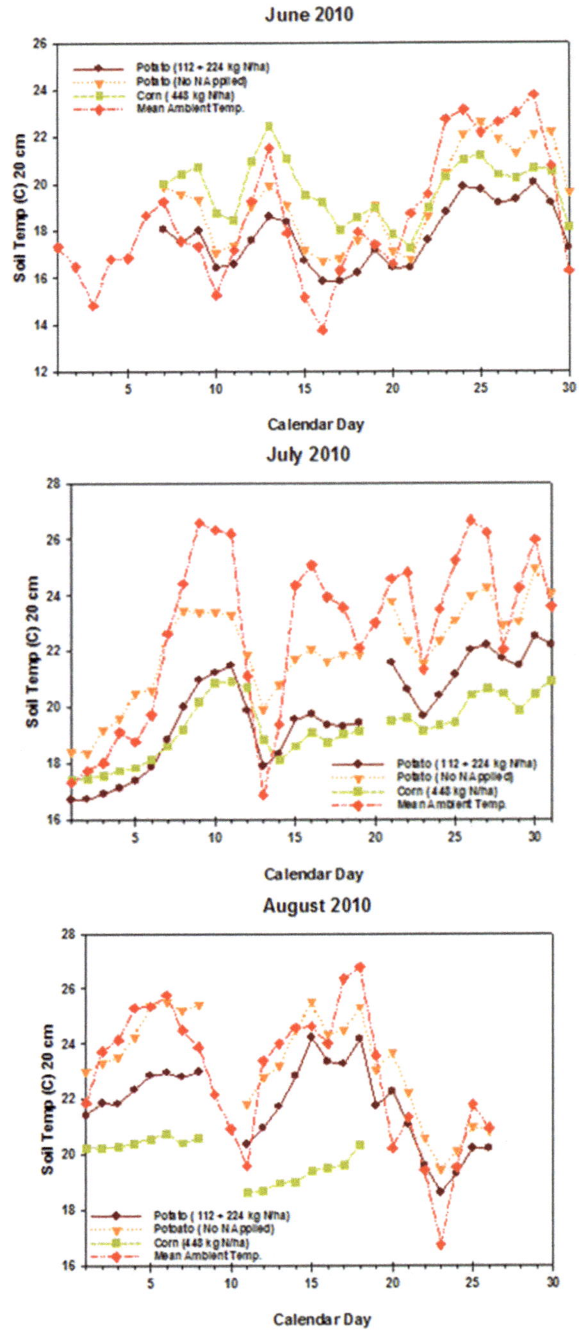

Fig. 14 Comparison of real-time measurement of soil temperature at 20 cm depth in a corn field with 448 kg/ha nitrogen (N) applied, vs. potato field with no N (poor canopy growth) and with recommended N application (336 kg/h N as combination of pre plant and in season applications) in a sandy soil with center pivot irrigation in a sandy soil in northwest United States

Fig. 15 Low cost soil water sensors being evaluated for irrigation decisions in India

The use of this sensor was also demonstrated in another farm with sprinkler irrigation system. The overall response was a greater utility of the sensor in sprinkler irrigated production system as compared to that in furrow irrigated system, because of greater control for precision water application in the former as compared to that in the latter system.

A new solution is now being tested, which makes irrigation valves 'smart', by having them open and close, when the moisture level decreases beyond a certain preset value. This technology solution is an add-on to current irrigation systems, which farmers can purchase with significant government subsidy. One element that discourages the farmers from buying automated irrigation system is the cost, but the larger reason is that they are concerned they will not know how to operate it. The initial feedback is positive: farmers not only enjoy the precision of the system, but they also enjoy the freedom it gives them. No longer do they have to be at the field, which gives the farmers the time and options to do something else to increase their revenue.

2 Conclusions

Agriculture is a major but important user of fresh water withdrawal, both surface and ground water sources. Water use for agriculture as percent of total water withdrawal varies in different parts of the world, depending on many factors. Predictions of climate models suggest severe water shortage particularly in the Middle East and North Africa region. Precision technologies are needed to minimize water losses therefore, enhance water use efficiency and water productivity. In this paper, we discussed state of the art technologies for real time monitoring of soil water content and temperature in the soil profile to develop irrigation set points to carefully maintain soil water content within the root zone at desirable range to support plant growth and production, while minimizing leaching losses below the root zone. Transfer of this technology to low income countries has been demonstrated by providing example of low cost sensors, which is still in experimental stage. The initial feedback from the farmers is highly positive for acceptance of this low cost technology to increase water use efficiency.

References

Alva AK (2008) Setpoints for potato irrigation in sandy soils using real-time, continuous monitoring of soil-water content in soil profile. J Crop Improv 21:117–137

Alva AK, Fares A (1998) A new technique for continuous monitoring of soil moisture content to improve citrus irrigation. Proc Fla State Hort Sci 111:113–117

Alva AK, Fares A (1999) Precision irrigation scheduling in sandy soils using capacitance probes. In: Proceedings of international Dahlia Greidinger symposium on nutrient management under salinity and water stress, Haifa, Israel, Feb 28–Mar 4, pp 267–269

Alva AK, Collins HP, Fraisse C, Boydston RA (2003) Evaluation of Enviroscan capacitance probes for monitoring soil moisture in center pivot irrigated potatoes. J Appl Irrig Sci 38:93–110

Buss P (1993) The use of capacitance based measurements of real-time soil water profile dynamics for irrigation scheduling. In: Proceedings of National Conference Irrigation Association, Australia and National Committee Irrigation Drain, Launceston, Tasmania. Irrigation Association of Australia, Homebush, NSW, May 17–19

Fares A, Alva AK (1999) Estimation of citrus evapotranspiration by soil water mass balance. Soil Sci 164:302–310

Fares A, Alva AK (2000a) Evaluation of capacitance probe for monitoring soil moisture content in a sandy Entisol profile with citrus trees. Irrig Sci 19:57–64

Fares A, Alva AK (2000b) Soil water components based on capacitance probes in a sandy soil. Soil Sci Soc Am J 64:311–318

Food and Agriculture Organization of the United Nations (2016) AQUASTAT. http://www.fao.org/nr/water/aquastat/water_use/index.stm

Mead RM, Paltineanu IC, Ayars JE, Liu J (1994) Capacitance probe use in soil moisture measurements. In: American society of agricultural engineers (ASAE) summer meeting, Chicago, IL

Mekonnen MM, Hoekstra AY (2012) A global assessment of the water footprint of farm animal products. Ecosystems 15:401–405

NASA, Jet Propulsion Laboratory, California Institute of Technology (2014) PIA18816: NASA's GRACE sees a drying California. http://photojournal.jpl.nasa.gov/catalog/pia18816

North Carolina State University (2012) Climate education for K-12. https://climate.ncsu.edu/edu/k12/.SEPrecip

One simple idea.com. Environment. http://one-simple-idea.com/Environment1.htm

Paltineanu IC, Starr JL (1997) Real-time soil water dynamics using multisensor capacitance probes: laboratory calibration. Soil Sci Soc Am J 61:1734–1742

Starr JL, Paltineanu IC (1998) Soil water dynamics using multisensor capacitance probes in nontraffic interrows of corn. Soil Sci Soc Am J 62:114–122

The Crop Site (2015) Growing more food with less water. http://www.thecropsite.com/focus/5m/50/growing-more-food-with-less-water-improving-water-usage-in-agriculture

University of Michigan (2000) Human appropriation of the world's fresh water supply. http://www.globalchange.umich.edu/globalchange2/current/lectures/freshwater_supply/freshwater.html

USDA—National Agricultural Statistics Service (NASS) (2009) Census of agriculture, 2007. State data

World resources sim center (2010) Agricultural water use, global, 2001. http://www.wrsc.org/attach_image/agricultural-water-use-global-2001

MIX
Papier aus verantwortungsvollen Quellen
Paper from responsible sources
FSC® C105338

If you have any concerns about our products,
you can contact us on
ProductSafety@springernature.com

In case Publisher is established outside the EU,
the EU authorized representative is:
Springer Nature Customer Service Center GmbH
Europaplatz 3, 69115 Heidelberg, Germany

Printed by Libri Plureos GmbH
in Hamburg, Germany